Open Channel Flow

Open Channel Flow

F. M. HENDERSON

Professor of Civil Engineering
University of Canterbury
Christchurch, New Zealand

MACMILLAN PUBLISHING CO., INC.
New York

COLLIER MACMILLAN PUBLISHERS
London

Library of Congress catalog card number: 66-10695

Macmillan Publishing Co., Inc.
866 Third Avenue, New York, New York 10022

Collier Macmillan Canada, Ltd.

Printed in the United States of America

PRINTING **25 24**   YEAR   89

ISBN   0-02-353510-5

To My Wife

Preface

Although this book was originally conceived as a text for use by the civil engineering student in advanced courses either in his senior year or at graduate level, it is also designed to have some appeal to the practicing engineer.

Open channel flow, like any topic of engineering interest, is defined and classified partly by its possession of certain characteristic applications and partly by the principles that are invoked to deal with them. This particular subject is so rich in the variety and interest of its practical problems that any textbook on the subject is in danger of becoming a mere catalogue of applications and routine techniques devised for dealing with them. But it has to be remembered that mastery of this subject, as of any other, demands a grasp of basic principles no less than a facility in routine operations. The practicing engineer is reminded of this fact whenever he turns from the familiar numerics of backwater curves and flood-routing procedures to some unusual transition problem whose solution requires a good grasp of fundamentals.

The importance of basic principles is recognized in this text in two ways: first, by devoting the opening chapters to a fairly leisurely discussion of introductory principles, including a recapitulation of the underlying arguments derived from the parent subject of fluid mechanics; and second, by taking every opportunity in the later chapters to refer back to this earlier material in order to clarify particular applications as they arise. It is hoped that the practicing engineer, as well as the student, will find this kind of treatment helpful, and a compensation for the fact that not every application is pursued through every possible variant that occurs in practice. Further compensation will, it is also hoped, be found in the fairly complete system of references and in the unusually large number of applied topics dealt with.

This insistence on the importance of principles does not imply that they should be given a status and significance independent of the applications they possess. The engineer invokes principles in order to deal with problems that arise in practice, and when dealing with these general principles he still remains in touch with the physical events which have prompted the need to generalize. This notion has dictated the structure of many chapters in this book, particularly Chapters 2 and 3. In each of these, a typical basic problem is discussed first; the theory is then developed to solve this problem, and is finally generalized to cover other problems as well.

It is generally agreed that in order to read a textbook effectively the student must to some extent help to write it. For this reason the reader is invited to complete certain aspects of the main line of argument in the form of set problems; the results of these problems are subsequently referred to as if they were an established part of the text. Although basic principles are of first importance in the treatment of the problems generally, prominence must also be given to methods of numerical computations, including trial solutions. For this reason some of the early chapters have Appendices dealing with the details of algebraic and numerical manipulation. Moreover, the growing importance of the high-speed computer is recognized by a special section in Chapter 5, and by the inclusion, after a brief introduction in Chapter 2, of computer programs in the problems at the end of most chapters.

Some particular features of the text material may be remarked on here. In the early chapters care has been taken to develop the formal proofs of the occurrence of critical flow in circumstances broadly characterized by the term "restraint-and-release"; such proofs are known but are often neglected. Novel methods of backwater computation are presented in Chapter 5, in particular an extension to the Ezra method which develops the full potentialities of the method. At the end of Chapters 4 and 5 effort is devoted to grouping the preceding material on controls and longitudinal profiles into a synthesis aimed at the better solution of practical steady-flow problems, in particular the determination of the discharge. In Chapter 8 an attempt is made to bring a more fundamental treatment of unsteady flow, including the method of characteristics, more completely into the realm of engineering practice, for these methods are relevant to a greater variety of real engineering problems than flood routing alone. The scope and significance of the various flood-routing methods in Chapter 9 are discussed in relation to the basic character of the motion, in particular to the number of slope terms that are significant in the dynamic equation of motion. In Chapter 10, on sediment transport, the vexed question of channel stability and formation is touched on in sufficient detail to show what issues are at present in doubt; under model theory in Chapter 11 the verification of models, seldom discussed explicitly in the literature, is given a brief airing. Wave theory is treated in enough detail in Chapter 8 to enable harbor models and their possible distortion to be discussed competently in Chapter 11.

The book has grown out of courses given since 1956 to fourth-year students of civil engineering at the University of Canterbury, and an occasional refresher course given to practicing engineers. From the experience so gained it appears that a three-hour, one-semester undergraduate course could conveniently be formed from Chapters 1 through 5 (excluding some of the middle sections of Chapter 5), and from selected topics in Chapters 6 and 7, with possibly some of the elementary material from Chapters 8 or 9 included, e.g., Sec. 9.2. A graduate course of similar length could be formed from the remaining material in Chapters 5 through 8, together with a substantial amount of the remaining four chapters. The total content of the book would occupy between two and three one-semester courses. While an effort has been made to avoid a too encyclopedic treatment of all applications, some topics have been pursued at such length that they may be beyond the scope of even a graduate course. It is hoped, however, that the treatment of these topics will have some reference value for the practicing engineer.

New Zealand practice in this field is substantially influenced by American trends, in particular by the American development over the last thirty years of a more scientific approach to problems in engineering hydraulics. So the writing of this book has happily been uncomplicated by any need to resolve conflicts between British and American usage, except in one small respect. The British abbreviation "cusec," meaning one cubic foot per second, is used throughout the book, in the hope of persuading American readers that this simple and expressive term deserves to be better known outside the British Commonwealth.

My interest in many of the topics treated in this book has been inspired and made more effective by contacts with friends and colleagues in engineering practice, and by discussions with them about their problems and observations. It is a pleasure to acknowledge the helpfulness of these contacts, and to express the belief that they have given the book a stronger flavor of physical reality that it would otherwise have had. It is hoped that the effect has been to make the result more digestible. Chapters 1 and 8 have been materially improved as the result of some searching criticism by Dr. T. Brooke Benjamin; he is of course in no way responsible for faults that remain.

No one can write on the subject matter of this book without becoming indebted to Ven Te Chow's recent authoritative treatise on the subject. While I have attempted to make my acknowledgments as specifically as possible in the body of the text, the more general acknowledgment made here recognizes the general illumination which his treatise has cast on so many areas of the subject, and from which every subsequent writer benefits. Frequent reference was also made to the authoritative text *Engineering Hydraulics*, edited by Hunter Rouse.

Finally, thanks are due to the University of Canterbury Civil Engineering Department secretaries, Mrs. D. E. Ball and Mrs. A. T. Perfect, who typed the manuscript and dealt ably with its many difficulties.

F. M. H.

Christchurch, New Zealand

Contents

List of Symbols

The following definitions of symbols used in this book conform in all essential respects to current usage, in particular to "American Standard Letter Symbols for Hydraulics" (ASA Z10.2—1942), prepared by the American Standards Association.

The wide range of topics that comes under the heading of open channel flow requires the definition of more parameters than the combined total of characters in the Roman and Greek alphabets. For this reason the repeated use of some symbols cannot be avoided. There seems little point, therefore, in seeking to avoid the conflicts in usage arising out of differing conventions in different branches of the subject—for example the use of the letter c for wave velocity and also for concentration of suspended material. Any attempt to avoid conflicts of this kind would only cause confusion by introducing unfamiliar associations of symbol and concept, without producing any compensating advantage.

As far as possible the letters used to indicate the more important parameters such as A, P, Q, R, S, v, etc., have not been used for other purposes. Repeated use of the same letter for trivial manipulative purposes such as the condensation of equations has been confined to other letters such as G, K, T, a, b, k, and r. In general, repeated use of the same letter need not cause any confusion, for the use of any limited particular definition is usually confined to a limited characteristic part of the text. Nevertheless, the following list is provided as a guide in order to resolve any doubts that may arise. The list gives every definition used and the source (usually an equation number) in which it first appears.

The use of superscripts and subscripts follows normal practice. The "prime" superscript is normally used for dimensionless numbers, as in the case of $y' = y/y_c$ or $y' = my/b$, although Eq. (8–72) provides an exception; the zero subscript indicates conditions which are initial, uniform, undisturbed, or in some way standardized. A notable example is the symbol y_0 [subscript zero], which usually (but not exclusively) indicates uniform depth; it is occasionally used in a different sense, as in Fig. 10-20, Eq. (10–58).

Letter subscripts are formed in the conventional way from the initial letter of a descriptive or qualifying word. For example, s for "standard" is used instead of the zero subscript in Eq. (2–37), where the use of the zero subscript itself might cause confusion. Two letter subscripts, c for critical and r for ratio, are of sufficiently general importance to be listed separately below; others are used only attached to particular letters and are listed accordingly.

Subscripted symbols are listed only when the subscript gives a distinct new meaning to the parameter, and not in those cases where the subscript merely attaches the parameter to a particular place or time, except for the important zero subscript. When the parameter occurs in the text only with a time- or place-subscript attached—for example, G_1 in Prob. 6.15—it is listed here without the subscript.

A	cross-sectional area of flow, Eq. (1–1).
A_1, A_2	particle shape factors in Sec. 10.5, before Eq. (10–43).
a	acceleration, Eq. (1–4).
	a constant, Appendices to Chaps. 2 and 3.
	$\tau_0/\rho v^2$, Eq. (4–5).
	Tainter gate dimension, Fig. 6-22.
	half amplitude of oscillatory waves, Eq. (8–69).
	$2\pi/T_0$, Eq. (9–8).
	height of standard section above bed, Eq. (10–23).
	a constant, Eq. (10–52).
a_1, a_2	constants in Eqs. (9–31) and (9–32).
B	channel surface width, Eq. (1–3).
	conduit width, Fig. 6-2b.
	width, transverse to flow, of the broad-crested weir, Eq. (6–52); of the Parshall flume throat, Eq. (6–53); and of the box culvert, Eq. (7–31).
b	width of rectangular channel, Eq. (2–16).
	base width of trapezoidal channel, Eq. (2–25).
	a constant, Appendices to Chaps. 2 and 3.
	orifice width, Fig. 6-2b.
	wave-crest length, Example 8.2.
	$gS_0(1 - \tfrac{1}{2}\,\mathrm{Fr_0})/v_0$, Eq. (9–85).
	average channel width, Eqs. (10–81) and (10–82).
	bridge-pier width, Eq. (10–96).
C	Chézy coefficient $v/\sqrt{RS}$, Eq. (4–6).
	circulation constant, Eqs. (6–26) and (7–20).
	broad-crested weir coefficient, Eq. (6–51).
	side-weir coefficient, Eq. (7–53).
C_1, C_2	characteristic curve labels, Fig. 8-2.

C_B	width-contraction coefficient, Eqs. (6–52) and (7–31).
C_c	general contraction coefficient, Eq. (6–1).
C_d	discharge coefficient, Eq. (6–2).
C_D	drag coefficient, Eq. (1–22).
C_f	overall surface-drag coefficient, Sec. 1.8.
C_h	vertical-contraction coefficient at box culvert inlet, Eq. (7–32).
C_L	energy-loss coefficient for eddy losses, Example 5.4; for channel bends, Eq. (7–27); and for bridge piers, Eq. (7–37).
C_s	DuBoys coefficient, Eq. (10–40).
c	(subscript): critical conditions, Eq. (2–7); an exception is r_c, Sec. 7.3 and Eq. (10–88).
	wave velocity relative to water, Eq. (2–11), also in Chap. 8 and Sec. 9.7.
	wave velocity relative to bank, rest of Chap. 9.
	concentration of air, Eq. (6–16), and sediment, Eq. (10–20).
c_d	dynamic (as distinct from kinematic) wave velocity, Eq. (9–24).
c_f	local surface drag coefficient, Eq. (1–16).
c_g	group (wave) velocity, Eq. (8–83).
D	diameter of culvert, Fig. 2-15, and of pipe, Eq. (4–7).
	vertical transverse culvert dimension, Sec. 7.4.
D_1, D_2	characteristic differential operators, Eqs. (8–13b) and (8–14b).
d	depth measured normal to bed, Fig. 2-1c, and (with various subscripts) in connection with aerated flow, Eqs. (6-19), (6-21), (6-22), and (6-25).
	grain size of sediment, Sec. 10.3 before Eq. (10–7).
E	specific energy, Eqs. (2–4) and (2–19).
	complete elliptic integral of second kind, Eq. (10–61).
E'	E/y_c, Eq. (2–17); mE/b, Eqs. (2–31) and (2–36).
E_c	critical specific energy.
E_{co}	bulk modulus of elasticity, Sec. 11.2 after Eq. (11–1).
E_L	energy loss at base of overfall, Fig. 6-17.
E_J	energy loss in hydraulic jump, Fig. 6-17.
e	the exponential number, 2.71828...
F	force, Eq. (1–4).
$F_A(h), F_B(h)$	stage functions in Ezra method, Eqs. (5–30) and (5–31).
F_s	Shields entrainment function, Fig. 10-3.
Fr	Froude number $v/\sqrt{gy}$, Sec. 2.4 before Eq. (2–13). $Q\sqrt{B/gA^3}$, Sec. 2.7 before Eq. (2–23).
f	Darcy resistance coefficient, Eqs. (4–7) and (7–50).
	Lacey silt factor, Eqs. (10–70) and (10–71).
f_b, f_s	Blench silt factors, Eqs. (10–76) and (10–77).
G	a hydraulic jump function, Prob. 6.15.
	S_0L/y_L in lateral-inflow problem, Eq. (7–51).
	variable wave-amplitude, Eq. (8–82).

G	exponent in Eq. (9–31).
	particle fall velocity function, Eq. (10–46).
	a function in Bogardi's method, Eq. (11–22b).
g	the acceleration of gravity, Eq. (1–12).
H	"total energy," Eq. (2–3).
	head over weir, Eq. (6–1), spillway, Fig. 6-8, and culvert entry, Eq. (7–31).
	total head over broad-crested weir, Eq. (6–49).
	wave height, crest to trough, Fig. 8-23.
H_E	error in estimate of total energy, Eq. (5–24).
h	stage, or height of water surface above datum, Eq. (1–3).
	height of sills or of teeth, Fig. 6-33.
	depth below total energy line, Probs. 6.6 and 6.7.
	depth of rainfall excess, Eq. (9–103).
h_f	head loss, Eq. (4–7).
I	rate of inflow to river reach, Eq. (9–1).
i	exponent in Eq. (2–37).
i_0	lateral inflow velocity, Eq. (9–96).
J	$N/(N-M+1)$, Ven Te Chow's analysis, Eq. (5–13).
j	exponent in Eq. (9–59).
K	conveyance $Q/S_f^{1/2}$, Eq. (4–25).
	a shape factor for the hydraulic jump profile, Prob. 6.15, and for bridge piers, Eq. (7–34).
	a coefficient in Muskingum V-O relation, Eq. (9–11).
	diffusion coefficient in flood routing, Eq. (9–82).
	constant in Exner's bed-wave equation, Eq. (10–2).
K'	dimensionless form of conveyance, Eq. (4–27).
$K_{1, 2, 3, 4}$	regime theory coefficients, Table 10–3.
K_α	bridge-pier scour coefficient for angle of attack α, Fig. 10-29.
K_s	bridge-pier scour coefficient for nose forms, Table 10-4.
k	number of basic quantities in Buckingham-π theorem, Sec. 1.7.
	a general purpose symbol for condensing algebraic arguments; equal to q_2/q_1, Eq. (3–16); $F(h)/(Q/Q_0)^2$, Eq. (5–32); $C_D(1-\sigma)/2$, Prob. 7.10; O/V, Eq. (9–5).
k_1	$1/K(1-X)$ ⎫
k_2	$1/KX$ ⎬ Eq. (9–13).
k_s	height of surface roughness projections, Eq. (4–11).
L	length of weir crest, Eq. (6–7) and Fig. 7-24.
	length of channel, Sec. 7.6 before Eq. (7–51).
	distance to surge front, Eq. (8–63).
	wavelength, Eq. (8–69).
	particle travel distance, Eq. (10–43).
	lift force on particle, Fig. 10-18.
	meander wavelength, Eq. (10–87).

$L_{II, III, IV}$	lengths of USBR stilling basins, Fig. 6-33.
L_B	length of SAF stilling basin, Table 6–2.
L_d	drop length, Fig. 6-16 and Eq. (6–39).
L_j	hydraulic jump length, Figs. 6-16 and 6-31.
l	pipe length, Eq. (4–7).
	Prandtl mixing length, Eq. (10–24).
M	momentum function, Eqs. (3–3) and (3–8).
	label for mild-slope profiles, Chap. 4.
	exponent in Eq. (5–6).
	arbitrary function in Prob. 7.14.
	$gS_0(\text{Fr}_0 - 2)/2c_0\text{Fr}_0$, Eq. (8–34).
M_1	$X_1/Y_1^{5/2}$, Eq. (2–27).
M_2	$Q/mE_c^2\sqrt{gE_c}$, Eq. (2–32), or $Q/E_c^2\sqrt{gE_c}$, Eq. (2–33).
Ma	Mach number, Eq. (7–5).
m	mass, Eq. (1–4).
	$\cot\theta$, Eq. (2–25).
	sidewall divergence angle, Fig. 6-35.
	$2\pi/L$, Eq. (8–81).
	exponent in Eq. (9–95).
	exponent in Eq. (10–85).
N	exponent in Eq. (5–3).
	arbitrary function in Prob. 7.14.
	storage routing function, Eq. (9–2).
n	coordinate direction, Sec. 1.5 before Eq. (1–7).
	number of parameters in Buckingham-π theorem, Sec. 1.7.
	exponent in Eq. (1–21).
	Manning's resistance coefficient, Eqs. (4–17) and (4–18).
	direction normal to flow, Eq. (7–15).
O	rate of outflow from river reach, Eq. (9–1).
P	wetted perimeter of flow cross-section, Eq. (4–1).
P_f	drag force, Eq. (1–22).
P_g	force on grain, Eq. (10–9).
P_w	wave resistance, Eq. (8–86).
p	pressure, Eq. (1–7).
	$i + j + 1$, Eq. (9–61).
	proportion by weight of grain size fraction, Eq. (10–50).
p'	pressure coefficient $(p - p_0)/\frac{1}{2}\rho v_0^2$, Eq. (1–23).
p_s	probability of grain dislodgement, Eq. (10–43).
Q	volumetric rate of discharge Eq. (1–1).
Q_0	discharge under original conditions, Eq. (5–36), or at uniform flow, Eq. (9–87).
Q_r	overrun discharge, Eq. (9–26).
Q_x	$dQ/dx = Q/x$, Eq. (7–45).
q	discharge per unit width, Sec. 1.7.

q_s	sediment discharge per unit width, Eq. (10–1).
q_x	$dq/dx = q/x$, Eq. (7–48).
R	hydraulic mean radius A/P, Eq. (4–2).
R_1	boundary radii, Eqs. (6–26) and (6–27).
Re	Reynolds number vL/v, Sec. 1.7; $4vR/v$, Eq. (4–9), $v_1 H/v$, Eq. (6–51).
Re*	particle Reynolds' number $v*d/v$, Fig. 10-3.
r (subscript)	ratio of prototype to model parameter, Sec. 1.7 and Eq. (11–4); an exception in Eq. (9–26).
	radius of streamlines and channel boundaries, Eqs. (6–27) and (7–15); of Tainter gate, Fig. 6-22.
	a general-purpose symbol, usually denoting a dimensionless ratio; equal to y_2/y_1, Eq. (3–12) and Appendix to Chap. 3; $Q^2/Q_0{}^2$, Eq. (5–33); b_2/b_1, Prob. 7.1; y_1/y_3, Prob. 7.10; open area : total area of bed, Probs. 7.20 and 7.21; c_g/c, Example 8.2; S_w/S_0, Eq. (9–50); bed-packing factor, Eq. (10–9); bed-coverage factor, Eq. (10–48); $(\tau_0 - \tau_c)/\tau_c$, Eq. (10–52).
r_c	radius at channel centerline, Sec. 7.3 and Eq. (10–88).
S	longitudinal slope; label for steep-slope profiles, Chap. 4.
S_0	bed slope, $-dz/dx$, Eq. (4–2).
S_a	acceleration slope, $\partial v/g\partial t$, Eq. (8–4).
S_e	energy slope, dH/dx, Eq. (8–4).
S_f	friction slope, $v^2/C^2 R$, Eq. (4–3).
S_w	water surface slope, Eq. (5–35), or wave slope, Eq. (9–49).
s	tooth spacing, Fig. 6-33.
	distance in the direction of flow, Eq. (1–5); y_3/y_1, Prob. 7.1.
s_s	solid : fluid density ratio, Sec. 10.3 before Eq. (10–10).
T	function of $y' = my/b$, Eq. (6–32).
	wave period, Eq. (8–70).
	duration of inflow, Eq. (9–8).
	function substituted from Eq. (8–42) to Eq. (9–85).
TW	tailwater depth, Eq. (6–57).
t	time, Eq. (1–3).
U	$E - \frac{1}{2}S_f\Delta x$, Eq. (5–21).
u	y/y_0, Eq. (5–9).
	horizontal velocity, Probs. 6.6 and 6.7.
V	$E + \frac{1}{2}S_f\Delta x$, Eq. (5–23).
	volume of fluid, Eqs. (9–1) and (10–20).
v	velocity of fluid, Eq. (1–5) and throughout text except in Sec. 3.5, where it denotes surge velocity.
	$u^{N/J}$, Eq. (5–13).
$v*$	shear velocity $\sqrt{\tau_0/\rho}$, Eq. (4–14).
W	height of weir crest above bed, Eq. (6–4).
	weight of bed grain, Figs. 10-6 and 10-18.

w	underflow gate opening, Eq. (6–42).
	tooth width, Fig. 6-33.
	speed of transverse plate, Fig. 8-6.
	particle-fall velocity, Eq. (10–22).
X	a coefficient in Muskingum V-O relation, Eq. (9–13).
	$x/2\sqrt{Kt}$, Eq. (9–82).
X_1	Z or $Q/D^2\sqrt{gD}$, Eq. (2–27).
x	distance, usually in the direction of flow.
x_p	particle penetration distance, Eq. (10–37).
Y	$y_1 y_2/(\sqrt{y_1}+\sqrt{y_2})^2$, Eq. (9–32).
Y_1	y_c/D or my_c/b, Eq. (2–27).
y	vertical depth of flow, or distance in the vertical direction, except in Eqs. (6–17) through (6–22), where it denotes *normal* elevation above the channel bed.
$\bar{y}$	depth from surface to centroid of section. Eq. (3–8).
y'	y/y_c, Eq. (2–17); my/b, Eq. (2–26), mean fluid particle height above bed, Eq. (8–72).
y_0	uniform depth, Sec. 4.3 and in most of text; exceptions are Prob. 6.18, Figs. 6-12, 7-16, 8-6, 10-20 and 10-27c.
y_c	critical depth, Eq. (2–7).
y_{cr}	critical depth relative to Q_r, Eq. (9–32).
y_s	scour depth, Fig. 10-27c.
	station depth, Eq. (6–34).
Z	$Q\sqrt{m^3/gb^5}$, Eq. (2–26).
	total head over spillway, Fig. 6-8.
	variable of integration, Eqs. (9–82) and (9–83).
Z_M	$Q/my^2\sqrt{gy}$, Eq. (3–12).
z	height above datum of fluid element, Eq. (1–7); of bed level, Eq. (2–3) and in most of text.
α	velocity (energy) coefficient, Eq. (1–24).
	longitudinal slope of steep channels, Eq. (2–2) and after Eq. (6–56).
	weir-notch angle, Eq. (6–8).
	bridge pier thickness ratio, Eq. (7–34).
	coefficient in Eq. (9–95).
	angle of attack to bridge piers, Fig. 10-29.
	$s_s - 1$, Eq. (11–17).
β	velocity (momentum) coefficient, Eq. (1–25).
	angle in culvert section, Prob. 2.8.
	Bakhmeteff's parameter $Fr^2 S_0/S_f$, Eq. (5–11).
	shock deflection angle, Eqs. (7–5) and (7–6).
	ratio of grain to total volume in granular bed, Eq. (10–1).
	ratio $\varepsilon_s/\varepsilon_m$, Eq. (10–31).
	Bogardi's parameter, $1/d^{0.88}F_s^2$, Eq. (11–21).

γ	specific weight of fluid, Eq. (1–7).
Δ	small increment of, before Eq. (1–7).
	departures of experiment from theory, Eq. (6–45).
δ	boundary layer thickness, Eqs. (1–15) and (6–44).
δ_1	B.L. displacement thickness, after Eq. (1–21).
δ^*	B.L. maximum displacement thickness, Eq. (6–52).
ε	kinematic eddy viscosity, or transfer coefficient, Eq. (10–21).
ε_b	air-bubble transfer coefficient, Eq. (6–18).
ε_m	momentum transfer coefficient $\Big\}$ Eq. (10–31).
ε_s	sediment transfer coefficient
η	height of water surface above mean level, Eq. (8–69).
	eddy viscosity, Eq. (10–19).
θ	bank side-slope angle, Fig. 2-14*b* and Eq. (10–15).
	$\tan^{-1}(Q/Q_0)^2$ in modified Ezra method, Fig. 5-8.
	Tainter gate lip angle, Fig. 6-22.
	angle of deflection: of spillway toe, Fig. 6-12*a*; of jet at base of overfall, Fig. 6-16; of channel bend, Eqs. (7–6) and (7–12), and near Eq. (7–27).
θ_0	angular distance to first wave crest at channel bend, Eq. (7–18).
κ	von Kármán's constant, Eq. (10–26).
λ	coefficient L/d in Einstein's bed-load argument, Eq. (10–43).
μ	dynamic viscosity, Sec. 1.7.
v	kinematic viscosity μ/ρ, Eq. (1–15).
π	circular circumference-diameter ratio; also in name of Buckingham-π theorem, Sec. 1.7.
ρ	fluid density, Eq. (1–7) and in most of text.
ρ_s, ρ_f	solid and fluid densities, Sec. 10.3 before Eq. (10–7).
Σ	summation symbol, Eqs. (1–26) and (5–25).
σ	submergence factor for broad-crested weir, Fig. 6-27*a*.
	width contraction ratio b_2/b_1, Fig. 7-22.
	fluid surface tension, Eq. (8–97) and after Eq. (11–1).
τ	shear stress anywhere in fluid, Eq. (10–19).
	$t - x/(v_0 + c_0)$, Prob. 8.7 and before Eq. (9–85).
τ_0	shear stress at solid boundary, Eq. (1–17).
τ_c	critical shear stress on the threshold of motion of a granular bed, Sec. 10.3 after Eq. (10–10).
Φ	Bresse function, Eq. (5–8).
	Einstein bed-load function, Eq. (10–44).
ϕ	angle of repose of granular material, Eq. (10–10).
Ψ	reciprocal of Shields' entrainment function, Eq. (10–44).
ω	fluid vorticity, Sec. 1.8.
	Escoffier's stage variable, Eq. (8–52).

Chapter **1**

Basic Concepts of Fluid Flow

1.1 Introduction

In the text of this book it is generally assumed that the reader is familiar with the basic laws of fluid flow. However, a brief discussion of some of the fundamental laws of fluid motion is given in this chapter in order both to recapitulate material with which the reader is assumed to be already familiar, and to emphasize certain points that are of particular interest in later applications to open channel flow.

1.2 Definitions

A *streamline* is a line, drawn at any instant, across which there is no flow component, so that at every point on it the resultant fluid velocity is in a direction tangential to the streamline. A *streamtube* may be thought of as a bundle of streamlines; it is a tube whose walls are made up of contiguous streamlines, and across which there can therefore be no flow. Neither streamlines nor streamtubes have any physical substance: they are geometrical figures which the observer imagines to be drawn within the flowing fluid. They are illustrated in Fig. 1-1.

Unsteady flow changes with time: *steady* flow does not. The difference is not an absolute one, but may be dependent on the viewpoint of the observer.

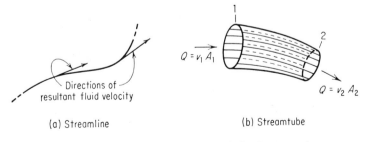

(a) Streamline

(b) Streamtube

Figure 1-1. *The Streamline and the Streamtube.*

1

Suppose for example that a landslide falls into a river and partially blocks it, sending a surge wave upstream as shown in Fig. 1-2. A *surge wave*, often simply called a surge, is a moving wave front which brings about an abrupt change in depth; another example of this phenomenon is the tidal "bore" by which the tide invades certain rivers, e.g., the River Severn in England.

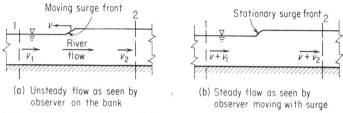

Figure 1-2. *Alternative Views of the Moving Surge, Seen as Unsteady Flow and as Steady Flow*

Now an observer on the bank would see this as an unsteady-flow phenomenon, since the flow changes its velocity and depth as the surge passes him. However, an observer who is moving with the surge sees the situation as one of steady flow, at least in the first stages of the movement before the surge begins to decay. He is level with a stationary wave front, and there is flow of unchanging velocity and depth upstream of him (assuming the river has a uniform slope and cross section) and downstream of him.

The distinction being made here is not an academic one, for the equations of motion are very much easier to write down and manipulate for steady flow than they are for unsteady flow. It is one of the most interesting features of fluid mechanics that one may greatly simplify the analysis of a problem by changing one's viewpoint from, say, that of a stationary to that of a moving observer, and so changing the flow situation from an unsteady to a steady one.

There are, of course, many cases in practice where there is no such dependence on the viewpoint of the observer, and the flow would be classified as steady (or unsteady as the case may be) by any observer. Such a case is the progress of a flood wave down a river: a man standing on the bank would clearly see the phenomenon as unsteady and so would another man moving downstream and keeping pace with the peak of the flood, since the magnitude of the peak discharge itself tends to reduce as the flood moves downstream. In a problem such as this one cannot take the easy way out by transposing to a steady-flow case, and the problem must be treated as one of unsteady flow.

When the flow is steady, a streamline is also the path followed by an individual fluid particle, but when the flow is unsteady this coincidence no longer holds good, and streamlines are distinct from *pathlines*.

In *uniform* flow, as strictly defined, the velocity stays the same, in magnitude and direction, throughout the whole of the fluid; in *nonuniform* flow it may change from point to point. Usually, however, a somewhat less restrictive definition of uniform flow is adopted, according to which flow is said to be uniform if it does not change *in the direction of the flow*. For example, if water flows in an open channel of uniform section at a mean velocity and depth which remain the same at all sections along the channel (Fig. 1-3), the flow is said to be uniform although the velocity may vary across the flow as shown in this figure.

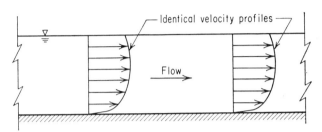

Figure 1-3.   *Uniform Flow as Usually Defined, i.e., Uniform in the Direction of Motion*

1.3   Continuity

Consider a case of steady flow through the streamtube shown in Fig. 1-1. Since there is no flow across the side walls of the tube, and since matter cannot be created or destroyed, fluid must be entering one end of the tube at the same rate at which it leaves the other. The term " rate " must imply a rate of mass transfer, which in English units would be measured in slugs per second. If the fluid is incompressible, so that its density remains constant, " rate " can be interpreted as a rate of volumetric transfer, measured in cubic feet per second. Now the rate at which volume is transferred across a section equals the product vA, where A is the area of the section and v is the mean velocity component at right angles to the section. If the subscripts 1 and 2 are applied to the two ends of the streamtube then we can write

$$v_1 A_1 = v_2 A_2 = Q, \text{ the discharge} \qquad \textbf{(1–1)}$$

which is the *equation of continuity* for steady flow of an incompressible fluid.

The most common use of the streamtube concept in practice is to apply it to the whole region of flow, so that the boundaries of the streamtube are also the physical boundaries of the pipe, the river, or whatever it may be. When the waterway (as the flow cross section is normally called in open channel flow) divides into branches carrying discharges $Q_2, Q_3, \ldots Q_n$, then the equation of continuity must clearly take the form

$$Q_1 = Q_2 + Q_3 + \cdots + Q_n \qquad \textbf{(1–2)}$$

where Q_1 is the flow in the main waterway upstream of the branches.

Equations (1–1) and (1–2) relate to incompressible fluids, i.e., liquids, with which the material in this book will be exclusively concerned except for some remarks on the gas-flow analogy in Chap. 2.

Application of the continuity principle to unsteady flow is rather more difficult. Unsteadiness implies change of many kinds: the velocity may be increasing, and the streamlines themselves may be shifting. However in open channel flow only one complication arises, namely that of a changing water-surface level. We consider the streamtube to embrace the whole cross section of the channel, and to be of very short length Δx, as shown in Fig. 1-4. The

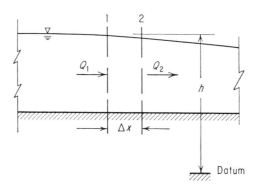

Figure 1-4. *Definition Sketch for the Equation of Continuity*

discharges at the two ends are not necessarily the same, but will differ by the amount

$$Q_2 - Q_1 = \frac{\partial Q}{\partial x} \Delta x$$

and this term gives the rate at which the volume within this region is decreasing. (The partial derivative is necessary because Q may be changing with time as well as with the distance x along the channel.)

Now if h is the height of the water surface above datum, then the volume of water between sections 1 and 2 is increasing at the rate

$$B \frac{\partial h}{\partial t} \Delta x$$

where B is the water-surface width. The two terms derived must therefore be equal in magnitude but opposite in sign, i.e.,

$$\frac{\partial Q}{\partial x} + B \frac{\partial h}{\partial t} = 0 \qquad \qquad (1\text{–}3)$$

which is the equation of continuity for unsteady open channel flow, such as occurs in the movement of a flood wave down a river.

As mentioned previously, the unit of discharge is one cubic foot per second, usually abbreviated as cfs in the United States. In British countries this unit is commonly termed the "cusec" (pl. cusecs)—a simple and expressive abbreviation, which will be used throughout this book. The terms "cfs" and "cusec" are both used in the literature and may be used interchangeably.

1.4 Equations of Motion—General

The discussion so far has dealt with the geometrical concepts of fluid motion, without considering the forces that may give rise to the motion. The study of these forces is essentially one of dynamics, and must therefore be founded on Newton's laws of motion; of these, the second law

$$F = ma \qquad\qquad (1\text{-}4)$$

which defines the force F required to accelerate a certain mass m at a certain rate a as being equal to the product ma, is the major source of working equations applicable to real physical problems.

If both sides of Eq. (1-4) are multiplied by the component of length s parallel to the direction of the force and the acceleration (or in general integrated with respect to that length), we have

$$\int_{s_1}^{s_2} F \, ds = m \int_{s_1}^{s_2} a \, ds = \tfrac{1}{2} m (v_2{}^2 - v_1{}^2) \qquad\qquad (1\text{-}5)$$

which is the energy equation, stating that the work done on a body as it moves from $s = s_1$ to $s = s_2$ is equal to the kinetic energy acquired by that body. An important difference between Eqs. (1-4) and (1-5) is that Eq. (1-4) deals with vector quantities, whereas the terms of Eq. (1-5) are obtained by multiplying vector quantities in such a way as to form a scalar product. Energy is a scalar quantity.

Now if both sides of Eq. (1-4) are multiplied by, or integrated with respect to, time elapsed, we have

$$\int_{t_1}^{t_2} F \, dt = m \int_{t_1}^{t_2} a \, dt$$
$$= m(v_2 - v_1) \qquad\qquad (1\text{-}6)$$

which is the momentum equation, stating that the impulse (force × time) applied to a body is equal to the momentum (mass × velocity) acquired by it. Multiplication of the vector quantity force by the scalar quantity time produces the vector quantity, momentum.

The concepts of energy and momentum embodied in Eqs. (1-5) and (1-6) are basic to all dynamics, whether of solids or fluids. While each equation is removed one step from Eq. (1-4), Eq. (1-6) is more closely similar in form to Eq. (1-4), sharing with it the property of dealing with vector quantities.

Indeed it is hardly necessary to formulate Eq. (1–6) in order to invoke the momentum principle; it is necessary only to refer to Eq. (1–4), interpreting the product *ma* as a rate of change of momentum. This approach is particularly convenient in applications to fluid flow.

1.5   Equations of Motion—Fluid Flow

The term F in Eq. (1–4) is to be interpreted as a *net* impressed force, i.e., the difference between the total impressed force and any force such as friction or viscous drag which resists the motion of the body. In the following treatment these resistance forces will be neglected, to be taken into account at a later stage.

The difficulty in applying Eqs. (1–4) through (1–6) to fluid flow is that we are dealing not with a single body but with a continuous mass of moving fluid. The way out of the difficulty is to concentrate our attention on a single fluid element, as shown in Fig. 1-5, having unit thickness at right angles to

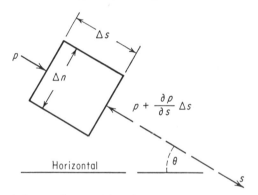

Figure 1-5.   *Definition Sketch, Forces on a Fluid Element*

the plane of the paper. On this element there can be only two kinds of impressed force in any chosen direction s; that due to the pressure gradient—i.e., $-(\partial p/\partial s)\Delta s\Delta n$—and that due to the weight of the element—i.e., $(\gamma\Delta s\Delta n)$ sin θ, or $-(\gamma\Delta s\Delta n)\,\partial z/\partial s$, where z is the vertical height above datum and γ is the specific weight of the fluid. If ρ is the mass density of the fluid, and a_s its acceleration in the s direction, then the *ma* term is equal to $(\rho\Delta s\Delta n)a_s$. Hence Eq. (1–4) applied to this situation becomes

$$\frac{\partial}{\partial s}(p + \gamma z) + \rho a_s = 0 \qquad (1\text{–}7)$$

which is the Euler equation. It is not so rich in direct applications as its integrated form, the Bernoulli equation, and furthermore it tends to repel engineers by the alarming presence of partial derivative signs; nevertheless it does help one's understanding of certain basic phenomena, and in particular it clarifies the scope and significance of the Bernoulli equation, which will now be dealt with.

We may remark first that the term $(p + \gamma z)$ in Eq. (1–7) is known as the *piezometric pressure*, and according to the principles of hydrostatics it remains constant throughout a body of still water, so that $\partial(p + \gamma z)/\partial s = 0$ whatever the direction of s may be. The presence of the acceleration term ρa_s indicates that if the water begins to move the hydrostatic pressure distribution is disturbed and $(p + \gamma z)$ no longer remains constant throughout the body of water.

To evaluate the term a_s we first note that the velocity varies both with time and position. From the theory of partial differentiation, we can write the equation

$$\frac{dv}{dt} = \frac{ds}{dt}\frac{\partial v}{\partial s} + \frac{\partial v}{\partial t}$$

which indicates the rate at which the resultant velocity v will appear to change in the eyes of an observer moving in the s direction with velocity ds/dt. If we are to interpret the derivative dv/dt as a fluid acceleration, the observer must be moving with the fluid itself, i.e., with the velocity v. We can write

$$\frac{dv}{dt} = a_s = v\frac{\partial v}{\partial s} + \frac{\partial v}{\partial t} \tag{1–8}$$

provided that s is the direction of fluid motion. It can also be shown that provided the flow is irrotational (which usually means that there is no energy dissipation), then

$$a_s = v\frac{\partial v}{\partial s} + \frac{\partial v_s}{\partial t} \tag{1–9}$$

whether or not s is in the direction of flow. In Eq. (1–9), v is the resultant velocity and v_s is its component in the s-direction. The two terms of this equation are named *convective* and *local* acceleration respectively.

The complete form of Eq. (1–7) obtained by substituting from Eq. (1–9) is that required for unsteady-flow problems such as that of a flood wave moving down a river; meanwhile we confine our attention to steady flow, in which case Eq. (1–7) becomes

$$\frac{\partial}{\partial s}(p + \gamma z) + \rho v\frac{\partial v}{\partial s} = 0 \tag{1–10}$$

which can be integrated directly to

$$p + \gamma z + \tfrac{1}{2}\rho v^2 = \text{constant} \tag{1–11}$$

or
$$\frac{p}{\gamma} + z + \frac{v^2}{2g} = \text{constant}, \; H \tag{1–12}$$

which are alternative forms of the well-known Bernoulli equation. Since it was obtained by integrating a force equation with respect to distance, it is an energy equation; in fact it may be deduced directly from Eq. (1–5). The derivation given here, however, is of some interest as it brings out clearly the dependence of Eqs. (1–11) and (1–12) on steady-flow conditions; if the flow were unsteady there would be an extra term obtained by integrating $\rho \, \partial v_s / \partial t$ with respect to s. This operation would involve some difficulties—enough to make their avoidance desirable if it is at all possible. And as pointed out in Sec. 1.2, it often is possible to transpose an apparently unsteady-flow case to one of steady flow by changing the observer's viewpoint.

The dimensions of Eq. (1–12) simply amount to length, e.g., feet, or they may be thought of as energy per unit weight, ft-lb/lb. In fact, the summation H in Eq. (1–12) is commonly termed the "total energy," but this usage is misleading, for the pressure term p/γ does not represent energy. It is more properly called the "flow-work" term, as the reader can verify for himself by working through the direct derivation of the Bernoulli equation (Prob. 1.19). In a particular steady-flow problem—for example, the pipe-flow case shown in Fig. 1-6, the size of the terms of Eq. (1–12) can be conveniently shown

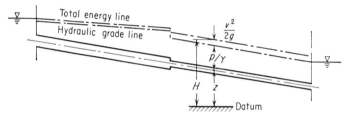

Figure 1-6. *Total Energy and Hydraulic Grade Lines in Pipe Flow*

graphically as indicated in the figure, in which the "total energy line" is drawn at a height H above datum and the "hydraulic grade line" is drawn at a height above datum equal to $(p/\gamma + z)$, the piezometric head.

If there is energy dissipation due to flow resistance, as there invariably is in practice, then either the fluid loses energy or more flow work has to be done on the fluid to maintain its energy. In either case the quantity H decreases by the amount of energy dissipated, and the total energy line dips steadily in the direction of flow. Particular features in the path of the flow may give rise to energy losses concentrated at certain localities, and hence to sharp local drops in the total energy line, as shown in Fig. 1-6.

To adapt the momentum principle to fluid flow, we again face the difficulty of dealing with a continuum of fluid rather than with a single body. Consider the flow through any streamtube, as in Fig. 1-7. In a time Δt the block of fluid originally contained between sections 1 and 2 moves to a new position bounded by sections 1' and 2'. It has therefore lost momentum equal to that of the fluid contained between 1 and 1', i.e.,

$$(\rho A_1 \Delta x_1)v_1, = \rho A_1 v_1{}^2 \Delta t = (Q\rho v)_1 \Delta t$$

and similarly has gained momentum equal to that of the fluid contained between 2 and 2', i.e., $(Q\rho v)_2 \Delta t$. The rate of change of fluid momentum is therefore equal to

$$(Q\rho v)_2 - (Q\rho v)_1$$

and this change can be accomplished only by the action of a forward force on the fluid equal to this rate of momentum increase:

$$F = (Q\rho v)_2 - (Q\rho v)_1 \tag{1-13}$$

One can think of the process shown in Fig. 1-7 as the transfer of momentum across sections 1 and 2 at the rates $(Q\rho v)_1$ and $(Q\rho v)_2$ respectively, due to

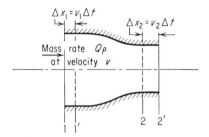

Figure 1-7.   *Definition Sketch for the Momentum Equation*

mass flow at the rate $Q\rho$ with momentum v per unit mass, as indicated in the figure. This view of the process is convenient in other contexts, for example in Sec. 1.8. Equation (1–13) relates to steady-flow conditions, and considers only changes from point to point, not changes in time. It can readily be extended to cover unsteady flow by the addition of a term giving the instantaneous rate of change of momentum integrated through the whole volume V contained between 1 and 2. It is

$$\frac{\partial}{\partial t} \int (\rho v) \, dV$$

but in many applications this term may not be easy to calculate.

From now on attention will be confined to steady-flow cases unless specifically stated otherwise.

1.6   Use of the Energy and Momentum Concepts

The energy equation always holds true, provided proper allowance is made for energy "losses" (more properly, the dissipation of kinetic energy into heat energy), in writing it down. Similarly the momentum equation always holds true, provided proper allowance is made for all forces acting. The simplicity of these two statements is rather misleading, for in some cases a certain dexterity is needed for the proper application of the equations. Moreover, there are limitations on what each equation is able to describe, so that their relative usefulness depends on just what information is being sought in each particular case.

Consider, for instance, the two flow situations shown in Fig. 1-8. In the smooth pipe contraction of Fig. 1-8a, it can be taken that there would be no

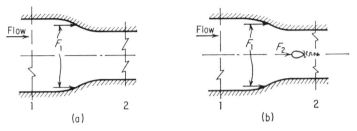

Figure 1-8.   *Force on a Pipe Contraction—a Problem in the Use of the Energy and Momentum Equations*

energy losses, so that the Bernoulli expression could be equated at sections 1 and 2, enabling us to calculate the pressure difference between these two sections for a given rate of flow. Supposing, however, that we wished also to calculate the net thrust F_1 of the flowing fluid on the contracting pipe walls; the energy equation would have nothing to tell us, first because it does not deal directly with forces as such, and second because it is concerned only with conditions at sections 1 and 2, not with points in between.

To discover anything about the size of F_1 we must have recourse to the momentum equation, having first established the pressure difference $(p_2 - p_1)$ by the use of the energy equation. This raises the question: is knowledge of the pressure difference enough, or do we also need to know the absolute values of p_2 and p_1? The question is left as an exercise for the reader (Prob. 1.4).

Consider now the pipe contraction of Fig. 1-8b, in which there is energy loss due to the presence of an obstruction in the middle of the pipe. In applying the momentum equation to this case, we have two kinds of force to consider; F_1, which causes no energy loss, and the drag force F_2 on the obstruction, which causes a definite energy loss. It is most important to realize that the momentum equation is concerned with all forces whatever their origin, and therefore makes no distinction between F_1 and F_2. The complete list of forces includes F_1, F_2, and the pressure forces at sections 1 and 2.

Here another difficulty arises: even if F_2 is specified in advance, the energy loss occasioned by the obstacle is still unknown, so we cannot apply the Bernoulli equation to obtain $(p_2 - p_1)$. The difficulty can be met by an approximate assumption, viz., that F_1 has the same magnitude whether the obstacle is present or not, and this assumption is reasonable provided that the obstacle is placed well into the contracted section, as in Fig. 1-8*b*. The details are left as an exercise for the reader (Prob. 1.5).

The general conclusion is that the energy and momentum equations play complementary parts in the analysis of a flow situation: whatever information is not supplied by one is usually supplied by the other. One of the most common uses of the momentum equation is in situations where the energy equation breaks down because of the presence of an unknown energy loss; the momentum equation can then supply results which can be fed back into the energy equation, enabling the energy loss to be calculated. Other applications are described in the problems at the end of this chapter.

1.7   Dimensional Analysis and Similarity

The theory of dimensional analysis has two major fields of application: the tidying up of arguments involving a large number of physical parameters, and the development of criteria governing dynamical similarity between two flow situations which are geometrically similar but different in size.

The basis of the theory is the Buckingham π theorem, which will be given here without proof. It states that if there is a functional relationship between n physical quantities, all of which can be expressed in terms of k fundamental quantities (e.g., mass length and time), then $(n - k)$ dimensionless numbers can be formed from the original n quantities, such that there is a functional relationship between them.

The term *functional relationship* implies that if values can be assigned to all but one of the variables concerned, the value of the remaining one can thereby be calculated, at least in theory. In practice it usually implies some determinate physical problem such that if all but one of the parameters have values given to them, Nature will fix the remaining one, even if our knowledge of the problem does not enable us to calculate it.

In problems of fluid flow there are always at least four parameters involved, Δp, v, L, and ρ, where L is a characteristic length dimension and Δp is a pressure difference, which may be replaced in some cases by a head difference Δh or an energy loss ΔH. Whatever its exact form may be, some such term is always present. Since three fundamental quantities are involved, these four terms give rise to $4 - 3 = 1$ dimensionless number, namely $\Delta p/\rho v^2$. The addition of more parameters introduces another dimensionless number for each new parameter added; the most important from the viewpoint of this text is the acceleration of gravity, g, which comes into play if the flow has a free surface. This important result can be deduced by considering first a completely enclosed fluid system, such as a closed pipe circuit round which a liquid is being driven by a pump. Energy is imparted to the liquid by the pump impeller and dissipated by the resistance of the circuit; provided the liquid remains a liquid, gravity does nothing to assist or retard this basic process,

which would proceed just as effectively if the circuit were disposed horizontally, vertically, were turned upside down or even removed to the surface
of the moon. The reason for this behavior is made clear by Eq. (1–10); the
flow is determined by the distribution of the " piezometric pressure "$(p + \gamma z)$,
of which gravity influences only the term γz. If the pressure is free to take any
value at all (which is true in an enclosed system), then gravity merely affects
the relative size of p and γz without affecting the total $(p + \gamma z)$. The distribution of this latter quantity would remain the same whether our hypothetical
circuit were placed on the earth or on the moon; only the pressures would
differ in the two cases, and the only effect this might have on the flow would
be to promote evaporation of the liquid at low pressures. If this possibility
is neglected there is no material difference between the two situations.

How, then, can gravity influence the flow pattern? Through the existence
of a free surface, on which the pressure p must take a prescribed value (usually
atmospheric). Once p is fixed, gravity determines $(p + \gamma z)$, and hence the flow
pattern, through its influence on the term γz. Many physical examples amplify
this algebraic argument; for example, in the flow of water over a weir there
is no pressure difference between the surface water upstream and downstream,
for both are exposed to the atmosphere. Any difference between the velocities
at the two sections can arise only through a difference in surface elevation and
the action of gravity.

If gravity is in fact significant, the appropriate dimensionless number is the
well-known Froude number

$$\text{Fr} = \frac{v^2}{gL}$$

If the viscosity μ plays an effective part, this fact introduces the Reynolds
number

$$\text{Re} = \frac{vL\rho}{\mu}$$

If more than one characteristic length dimension is involved, then further
dimensionless numbers can be formed from their ratios.

Since open channel flow is above all a free-surface phenomenon, we shall
find that the Froude number plays a very important part in the appropriate
flow equations. The Reynolds number plays a somewhat limited part in the
theory of open channel flow, particularly when the flow is on a large scale,
for since the Reynolds number is an inverse measure of the effect of viscosity,
it has least influence when it is largest. Its significance will be further discussed
in the next section.

Dimensional analysis is commonly thought of as a means of planning
experimental programs and model studies for dealing with problems that are
too complex or difficult for theoretical solution. However, it is also useful in
making tidy and convenient arrangements of any kind of equation, including

those derived from considerations of pure theory; it will commonly be used for this purpose in the succeeding chapters of this book.

Its usefulness can be further extended by taking a broader view of the concept of hydraulic similitude; it cannot be too strongly emphasized that the use of the similitude concept need not be confined to the operation and interpretation of hydraulic models, as in Chap. 11. To take an elementary example: anyone learning for the first time that the volume of a sphere varies as the cube of its diameter will find it helpful to visualize two spheres, one of which has twice the diameter and eight times the volume of the other. It will be shown in the following chapters that such mental exercises in model theory are helpful in visualizing the interplay and interdependence of the parameters governing a given flow situation.

For example, the theory of similitude states that for dynamical similarity to obtain between model and prototype, the Froude number must be the same in each case. It follows that, if the subscript r indicates a ratio of prototype quantity to model quantity, then

$$v_r = L_r^{1/2}$$

i.e.,

$$Q_r = L_r^{5/2}$$

or

$$q_r = L_r^{3/2}$$

where q is the discharge per unit width of channel. It will be found that expressions suggestive of these relationships are constantly occurring in the equations of open channel flow; even when the Froude number does not occur explicitly we shall find terms of the form Q^2/gL^5 and q^2/gL^3, which have essentially the same function. It should be kept in mind that the occurrence of such a term as one of the determinants of a flow situation has two interrelated meanings: (1) setting the value of this term fixes (or helps to fix) the detailed nature of the situation; (2) two flow situations having the same value of this parameter (and any other significant ones) will have the same detailed nature, i.e., they will be dynamically similar. Both these concepts are helpful in obtaining a grasp of the problem.

1.8 Flow Resistance

In the flow of any real fluid, energy is being continually dissipated, as indicated in Fig. 1-6. This occurs because the fluid has to do work against resistance originating in fluid viscosity. Whether the flow is laminar, as at low values of the Reynolds number Re, or turbulent, as at high values of Re, the basic resistance mechanism is the shear stress by which a slow-moving layer of fluid exerts a retarding force on an adjacent layer of faster moving fluid.

In this section a brief discussion will be given of the theoretical basis of this action. We note first an important feature of Eq. (1–13). The mass flow rate $Q\rho$ in that equation does not have to take place in the same direction as the velocities v_1 and v_2; indeed from the dynamical viewpoint the direction of $Q\rho$ is unimportant, for it is essentially a scalar quantity representing a rate of mass transfer from one region to another. Force and velocity are the vector quantities in the equation, which may accordingly be written as

$$\vec{F} = Q\rho\overrightarrow{(v_2 - v_1)} \tag{1-14}$$

indicating vectors by the usual notation.

Equation (1–14) will hold true even if $Q\rho$ is transverse to the flow, and if for some reason continuous exchange of material does take place between adjacent layers moving with different velocities there will, by Eq. (1–14), be a retarding force exerted on the faster-moving layer because it is continually receiving low-velocity fluid and losing high-velocity fluid. This retardation will be in the form of a shear force since it acts parallel to the interface between the two layers. In laminar flow this exchange of material and momentum occurs on a microscopic scale through the random movement of molecules, and in gases forms almost the sole basis of viscous shear. (In liquids the action of momentum exchange is reinforced by molecular cohesion.) In turbulent flow, momentum exchange alone is the basis of shear resistance; it occurs on a macroscopic scale, being caused by random fluctuations in velocity which are continually sending fluid particles back and forth between adjacent layers. Equation (1–14) makes it clear that in both laminar and turbulent flow the action is absolutely dependent on the existence of a velocity difference; such a difference usually arises from a continuous transverse velocity gradient, as in Fig. 1-3.

The Boundary Layer

In a large number of real flow situations, neither viscous nor turbulent shear can be said to *originate* a force resisting motion; they merely transmit such a force from its origin, which is often on a solid surface. It is well established by observation that fluid actually in contact with a solid surface has no motion along the surface; the transverse velocity gradient so created (Fig. 1-3) enables the solid surface to exert on the moving fluid a drag force which is transmitted outwards through successive faster-moving fluid layers. Flow resistance therefore depends as much on the presence of solid surfaces as on the strength of the viscosity or turbulence. Accordingly, if one solid surface is isolated in a large expanse of moving fluid (Fig. 1-9a), the effects of resistance are confined to a fluid layer of limited thickness—the "boundary layer"—adjacent to the surface. When the fluid is of small viscosity (e.g., air or water) this layer is quite thin and the pressure difference across it is negligible. The pressure distribution on the solid surface is therefore the same as it is outside the boundary layer, where the fluid behaves as if it had no viscosity.

Within the boundary layer the fluid is transmitting shear stress across the streamlines; each fluid element is therefore subject to opposing shear stresses on opposite faces, producing a torque which makes the element rotate. It is convenient to measure this rotation by the "vorticity" ω, equal to twice the angular velocity of the element. The strength of the rotation diminishes with increasing distance from the solid boundary, and in fact the vorticity ω can be thought of as a fluid property (like color, salinity, or temperature) which is diffused outwards from the solid boundary, and will be of appreciable magnitude only within·the boundary layer, just as in flow past a hot body there will be a layer of heated fluid close to the body, any hot fluid outside that layer being swept downstream. Examination of the equations of viscous flow and of heat conduction shows that the analogy described here is an exact one.

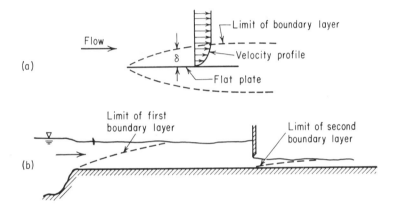

Figure 1-9.   *Boundary Layer Formation, (a) on a Flat Plate, (b) in a Channel, a Second Boundary Layer Growing Within the First*

It is quite possible for one boundary layer to form within another. As water flows from a lake into a channel (Fig. 1-9b) the boundary layer grows and eventually fills the whole channel. If now the velocity is suddenly raised by a channel contraction like the sluice gate in the figure, a fresh boundary layer must begin to grow. Reverting to the heat-diffusion analogy, we may compare the situation to one in which water flowing over a heated bed has reached a state of thermal equilibrium; if the flow now enters a region where the bed is distinctly hotter than before, the water must adapt to the new situation by the growth of a new heated band which will in its turn eventually fill the whole channel.

The leading parameters of the boundary layer must clearly be related to the Reynolds number, which in this context is defined in relation to distance x along the solid boundary. When this boundary is simply a flat plate held parallel to the flow the thickness δ and the local drag coefficient c_f are given by the following equations. When the boundary layer is laminar

$$\frac{\delta}{x} = 5\left(\frac{v_0 x}{\nu}\right)^{-1/2} \tag{1-15}$$

and

$$c_f = 0.66\left(\frac{v_0 x}{\nu}\right)^{-1/2} \tag{1-16}$$

where the local shear stress τ_0 at any point on the surface is given by the equation

$$\tau_0 = c_f \cdot \tfrac{1}{2}\rho v_0^2 \tag{1-17}$$

and v_0 is the free-stream velocity outside the boundary layer. For a turbulent boundary layer the corresponding equations are

$$\frac{\delta}{x} = 0.38\left(\frac{v_0 x}{\nu}\right)^{-1/5} \tag{1-18}$$

and

$$c_f = 0.059\left(\frac{v_0 x}{\nu}\right)^{-1/5} \tag{1-19}$$

for $\mathrm{Re} = v_0 x/\nu < 2 \times 10^7$ ($\nu = \mu/\rho$, the kinematic viscosity). Above this value of Re a logarithmic law holds. The boundary layer becomes turbulent on a smooth surface when Re reaches 5×10^5 approximately, i.e., when x becomes large enough; or it may be turbulent right from the leading edge if the edge or the surface is roughened. The total shear drag over the surface is termed *surface drag*. It is obtained by integrating the local shear stress τ_0 over the surface; more generally, an overall drag coefficient C_f may be obtained by integration from Eqs. (1–16) and (1–19) (Prob. 1.14).

The thickness δ in the above equations is defined as the distance y from the solid boundary at which v/v_0 approaches unity; this definition is not as vague as it may at first appear, for it is usually easy to fit equations such as

$$\frac{v}{v_0} = \sin\frac{\pi y}{2\delta} \tag{1-20}$$

(the sine curve approximating to the laminar flow profile), or

$$\frac{v}{v_0} = \left(\frac{y}{\delta}\right)^n \tag{1-21}$$

characteristic of turbulent flow, to observed velocity profiles. The fitting process itself determines the value of δ with very little error or uncertainty. The value of n in Eq. (1–21) is about $\tfrac{1}{7}$ for moderately high values of Re, although it may be less for very high values of Re.

An alternative definition, convenient for some purposes (e.g., Prob. 1.13) is the "displacement thickness" δ_1, which is the distance by which the boundary layer displaces the whole flow pattern laterally from the solid boundary. This definition implies that the discharge through any transverse section of the boundary layer is equal to $v_0(\delta - \delta_1)$, for the effect of the velocity reduction near the solid surface has been to increase by δ_1 the width required to accommodate this discharge. Given the above laws of velocity distribution for laminar and turbulent flow, it is easily shown (Prob. 1.12) that $\delta_1 = (\pi - 2)\delta/\pi$ or approximately $\delta/3$ for the laminar, and approximately $\delta/8$ for the turbulent, boundary layer.

Separation

Flow around the upstream, or leading, surface of a solid body, Fig. 1-10, is similar to flow along a flat plate except that outside the boundary layer the velocity is increasing and the pressure is decreasing. However there is still no pressure change *across* the boundary layer. Fluid within the boundary layer is steadily losing kinetic energy, with the result that when the downstream, or trailing, surface is reached this fluid may not have enough energy left to make its way into the region of increasing pressure which it finds there. In this event the flow will separate from the solid surface, as shown in Fig. 1-10a, and this separation is most likely to occur when the downstream surface curves sharply away from the flow as in the case of a thick imperfectly streamlined body—usually termed a "blunt" body. On a well-streamlined body, Fig. 1-10b, separation occurs well downstream, if at all.

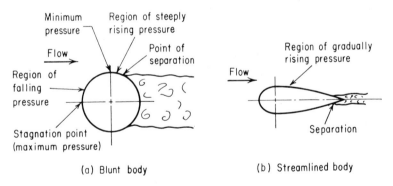

(a) Blunt body   (b) Streamlined body

Figure 1-10.   *Examples of Flow Separation*

If separation occurs, the pressure in the resultant downstream "wake" approximates to the low pressure at the comparatively high velocity point of separation. The result is a net thrust on the body due to the pressure difference between upstream and downstream faces—the "form drag," whose magnitude is substantially independent of Re. This is the major component of the drag force on blunt bodies; on well-streamlined bodies surface drag predominates. The distinction between these two extremes must clearly be based on the shape of the trailing rather than the leading surface. Thus a bridge pier may be long and slender, but if its downstream end is square it will undoubtedly behave as a blunt body.

The total drag force on any solid body is a combination of surface drag and form drag. The drag coefficient C_D is defined by the equation

$$P_f = C_D \cdot A \cdot \tfrac{1}{2}\rho v_0{}^2 \cdot \qquad (1\text{-}22)$$

where P_f is the drag force and A is the frontal area of the body— i.e., its area projected on a plane normal to the flow. At sufficiently high values of Re form drag predominates and C_D becomes virtually independent of Re. The drag is now due almost entirely to the pressure distribution round the body and it is readily shown (Prob. 1.16) that the drag coefficient C_D may be obtained by integrating the dimensionless pressure coefficient

$$p' = \frac{p - p_0}{\tfrac{1}{2}\rho v_0{}^2} \qquad (1\text{-}23)$$

over the frontal area of the body. In Eq. (1–23), p is the normal pressure on the body surface and the subscript 0 indicates free-stream conditions away from the influence of the body. If the free-stream velocity v_0 is very high the pressure decrease in the wake may be limited by the approach of absolute zero pressure (Prob. 1.17).

Typical values of C_D range from 0.03 for a well-streamlined body through about 1.0 for blunt axisymmetric bodies (e.g., 1.12 for a circular disk held normal to the flow) up to 1.5–2.0 for blunt "two-dimensional" bodies (those in which one dimension normal to the flow is very great, so that flow around the body is in two dimensions only.) An example is the long flat plate held normal to the flow, of which the C_D is 1.9.

The most common occurrence of flow resistance in open channel flow is in long channels where the boundary layer fills the whole channel section (Chap. 4). In this case the Reynolds number Re is usually so large that it has virtually no influence on the resistance coefficient. However, it will be seen in Chaps. 6 and 7 that there is occasion to consider also the resistance of bodies (e.g., bridge piers) placed in the stream, and the growth of the boundary layer on structures such as weirs and spillways. In this latter case Re is of some significance, as Eqs. (1–15), (1–16), (1–18), and (1–19) indicate.

1.9 Velocity Coefficients

In the preceding discussion of the energy and momentum equations it has been assumed that the velocity is constant across the whole section of the flow. This is never true in practice because viscous drag makes the velocity lower near the solid boundaries than at a distance from them. If the velocity does vary across the section, the true mean velocity head across the section, $(v^2/2g)_m$, will not necessarily be equal to $v_m^2/2g$ (the subscript m indicating the mean value). The true mean velocity head is obtained by the following argument:

The weight flow through an element of area dA is equal to $\gamma v dA$; the kinetic energy per unit weight of this flow is $v^2/2g$; hence the rate of transfer of kinetic energy through this element is equal to

$$\frac{\gamma v^3}{2g} dA$$

Hence the kinetic energy transfer rate of the whole flow is equal to

$$\int \frac{\gamma v^3}{2g} dA$$

and the total weight rate of flow is equal to

$$\gamma Q = \gamma v_m A$$

since the mean velocity v_m is by definition equal to Q/A. Hence the mean velocity head, or kinetic energy per unit weight of fluid, is equal to

$$\left(\frac{v^2}{2g}\right)_m = \frac{\displaystyle\int \frac{v^3}{2g} dA}{v_m A} = \alpha \left(\frac{v_m^2}{2g}\right)$$

where α is a correction coefficient to be applied to the velocity head as calculated from the mean velocity. It is sometimes known as the *Coriolis coefficient*. Hence finally

$$\alpha = \frac{\displaystyle\int v^3 dA}{v_m^3 A} \tag{1-24}$$

The same considerations apply to the calculation of the momentum term $(Q\rho v)_m$. The rate of transfer of momentum through an element of area dA is equal to $\rho v^2 dA$; thence by an argument similar to the preceding one we can deduce that the momentum correction coefficient is equal to:

$$\beta = \frac{(Q\rho v)_m}{Q\rho v_m} = \frac{\displaystyle\int v^2 dA}{v_m^2 A} \tag{1-25}$$

The coefficients α and β are never less than unity; they are both equal to unity when the flow is uniform across the section, and the further the flow departs from uniform, the greater the coefficients become. The form of Eqs. (1–24) and (1–25) makes it clear that α is more sensitive to velocity variation than β, so that for a given channel section, $\alpha > \beta$. Values of α and β are easily calculated for idealized two-dimensional velocity distributions such as those of Eqs. (1–20) and (1–21), replacing δ by the full depth of flow y_0. The results show (Prob. 1.18) that α and β individually are strongly dependent on the kind of velocity distribution, but that the ratio $(\alpha - 1)/(\beta - 1)$ varies only slightly, in the region 2.7–2.8, as the flow changes from laminar to turbulent.

The high value of α appropriate to laminar flow is of limited interest, since laminar flow is so rare in practical problems. For turbulent flow in regular channels α seldom exceeds 1.15, and it is doubtful whether the precision attainable with channel calculations warrants its inclusion, particularly as the experimental data on values of α are rather sparse and not always consistent.

However the above remarks are true only when we consider the smooth and gradual velocity variation within a single well-defined channel, as in Fig. 1-11a. Natural rivers, on the other hand, frequently flow in channels that can be subdivided into distinct regions, each with a different mean velocity. This is particularly true in time of flood, when the river overflows on to its flood plains, or "berms," producing the kind of channel section shown in Fig. 1-11b. In this case there are in effect three separate channels; over the berms the mean velocity will be materially less than in the main channel, because of higher resistance to the flow, partly due to the smaller depth over the berms, and partly to the rougher surface that they usually possess.

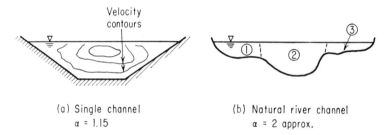

(a) Single channel
α = 1.15

(b) Natural river channel
α = 2 approx.

Figure 1-11. *Typical Cross-Sections with Values of the Velocity Coefficient* α

This stepwise variation in mean velocity among the different flow sections is in fact responsible for values of α much higher than those produced by gradual variation within a given section, so much higher as virtually to swamp any contribution to the value of α produced by gradual velocity variation. Accordingly, it is usually accurate enough to compute α by assuming the velocity to be constant within each subsection of the waterway; Eq. (1–24) then becomes

$$\alpha = \frac{v_1^3 A_1 + v_2^3 A_2 + v_3^3 A_3}{v_m^3(A_1 + A_2 + A_3)}$$

where

$$v_m = \frac{v_1 A_1 + v_2 A_2 + v_3 A_3}{A_1 + A_2 + A_3}$$

or we may write in a more compact form

$$\alpha = \frac{\Sigma v_1{}^3 A_1 (\Sigma A_1)^2}{(\Sigma v_1 A_1)^3} \qquad (1\text{--}26)$$

It will be shown in Chap. 5 how the above equation is combined with flow-resistance formulas so as to produce numerical values of α, whose magnitude may in some cases be well over 2.

Only in the above case of a divided waterway will the velocity coefficients be taken into account in this book. In other applications they will be neglected, since they represent only second-order corrections, and present knowledge of experimental values is imperfect. Problem 1.6 shows, as a matter of interest, a special experimental situation in which precise knowledge of the momentum coefficient is required; Prob. 1.7 shows, on the other hand, the way in which this requirement is avoided in a typical open channel problem.

Problems

(Take ν for water $= 1.2 \times 10^{-5}$ ft²/sec)

1.1. Classify the following flow cases as steady or unsteady, from the viewpoint of the appropriate observers:

Case	*Observer*
1. Ship steaming across a lake.	(a) Standing on ship.
	(b) Standing on shore.
2. Flow of river around bridge piers.	(a) Standing on bridge.
	(b) In boat, drifting with current.
3. Ship steaming upstream in a river.	(a) Standing on ship.
	(b) Standing on bank.
	(c) In boat, drifting with current.
4. Movement of flood down a valley after the collapse of a dam.	(a) Standing on bank at dam site.
	(b) Running downstream on bank, keeping pace with surge that forms at the front of the flood wave.

1.2. A small navigation canal of rectangular section contains still water 6 ft deep. The sudden opening of a lock gate sends a surge 1 ft high moving down the canal with a velocity of 15.6 ft/sec; find the mean velocity of the water behind the surge.

1.3. You are observing the rising stage of a major flood in a large river; at your observation point it is seen that at a certain time the discharge is 75,000 cusecs, and that the water level is rising at the rate of 1 ft per hr. The surface width of the river at this point, and for some miles upstream and downstream, is $\frac{1}{2}$ mile. You are asked to make a quick estimate of the present magnitude of the discharge at a point 5 miles upstream. What is your estimate?

1.4.   In the horizontal pipe contraction shown in Fig. 1-8a, water is flowing at the rate of 1 cusec, and the diameters at sections 1 and 2 are 6 in. and 3 in. respectively. Find the thrust F_1 on the walls of the contraction if the gage pressure at section 1 is (a) 10 lb/sq in.; (b) 20 lb/sq in. Neglect the small change in pressure across the diameter of the pipe.

1.5.   The situation is as in Prob. 1.4, except that there is a instrument mounted in the center of the contracted section, as in Fig, 1-8b. The drag force on this instrument is 10 lb; assuming that the thrust F_1 is the same as in Prob. 1.4, find the gage pressure at section 2 when the gage pressure at section 1 is (a) 10 lb/sq in.; (b) 20 lb/sq in.

1.6.   A water tunnel has a 1-ft square section, and runs full. A test is made on a 3-in. diameter cylinder (see figure) which runs the full width of the tunnel. It is found that there is a difference in pressure head of 8.57 in. of water between sections 1 and 2, and that the piezometric head is uniform across each section. At section 1 there is a uniform velocity of 10 ft/sec, and at section 2 the velocity distribution is as shown graphically in the figure. Find by graphical methods the momentum coefficient β at section 2, and the drag force on the cylinder.

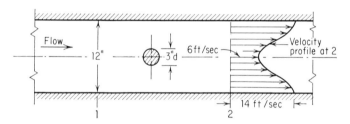

Problem 1-6

[The reader is assumed to be familiar with the differential-head and total-head Pitot tubes, one of which would have been used to obtain the velocity distribution shown in the figure. In order to determine the drag force on the cylinder, could the necessary information at section 2 have been obtained with a total-head tube alone, or were the separate pressure and velocity measurements (as given in the figure) necessary?]

1.7.   A bridge has its piers at a center distance of 20 ft. A short distance upstream of the bridge the water depth is 10 ft and the velocity 10 ft/sec; when the flow has gone far enough downstream to even out again after the disturbance caused by the piers, the depth is 9.5 ft. Neglecting the bed slope and bed resistance, and assuming that the pressure varies with depth just as it does in a tank of still water, find the thrust on each pier.

1.8.   As shown in the figure, water flows at a velocity v_0 in a pipe with a blind end, and flows outwards through lateral jets at right angles to the centerline of the pipe. The combined area of the jets is the same as that of the cross section of the pipe. The jet entries are smoothly rounded and there is no energy loss. Examine the following arguments:

(a) Applying Bernoulli's equation between B and C, we find that the pressure at B is atmospheric. Applying the same equation between C and the stagnation region D, we find pressure at $D = \frac{1}{2}\rho v_0^2$. Hence thrust on end plate $= \frac{1}{2}A\rho v_0^2$, where $A =$ cross sectional area of pipe;

(b) From the momentum theory, thrust on end plate = thrust required to deflect flow at right angles to original direction = force needed to destroy original forward component completely $= Q\rho v = A\rho v_0^2$;

and resolve the apparent contradiction between the two conclusions.

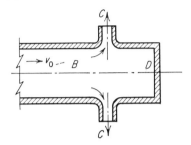

Problem 1-8

1.9. Reformulate and solve Prob. 1.8 for the case where the lateral branch pipes of Prob. 1.8 are replaced by sharp-edged circular holes in the wall of the main pipe. It may still be assumed that the combined cross-sectional area of the outflowing jets is equal to that of the pipe cross section.

1.10. A scoop made of 6-in. pipe is being pushed through a trough of still water in the direction shown in the figure, with a velocity v equal to 15 mph.

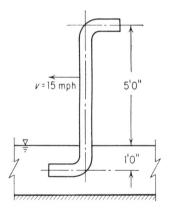

Problem 1-10

Assuming that enough water is picked up to make the pipe run full, and that no energy is dissipated, determine the horsepower necessary to push the scoop through the water. The answer is to be obtained by two separate methods: (a) using energy considerations alone, (b) using energy and momentum. Verify that the result is the same by either method, and produce an algebraic argument confirming that the two results are identical.

1.11. Water flows downward at the rate of 0.5 cusec through the 3-in. diameter
pipe shown in the figure, then radially outward through the space between
the two circular plates of 12-in. outer diameter at the end of the pipe, finally
discharging freely to the atmosphere. Find the magnitude and direction of
the thrust on the lower horizontal plate, and what the spacing between the

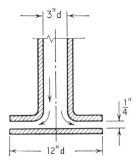

Problem 1-11

plates would have to be in order for this thrust to be zero. Assume that the
energy losses are negligible.

1.12. Derive from Eq. (1–21) a relationship between δ_1/δ and the exponent n.
Hence verify that $\delta_1/\delta = 1/8$ when n has the value $\frac{1}{7}$. Also prove from Eq.
(1–20) that $\delta_1/\delta = (\pi - 2)/\pi$ for laminar flow.

1.13. The figure shows an idealized version of flow over a wide " broad-crested
weir," controlling the outflow from a lake. The depth and velocity head will
be as shown if no boundary layer develops. The depth H_0 is 12 in. and the
length of the weir in the direction of flow is 4 ft. Assuming now that the
boundary layer does develop along the crest of the weir, calculate its displacement thickness δ_1 if it is (a) laminar, (b) turbulent. In each case calculate
the percentage difference between the actual discharge per unit width and that
obtained by neglecting the boundary layer. (*Note*: assume that the effective
depth of flow is unaltered by the existence of the boundary layer, which
simply lifts the water surface and reduces the velocity head by an amount δ_1.
To calculate δ_1, it is sufficiently accurate to assume that the velocity is constant and equal to the value obtained by neglecting the boundary layer.)

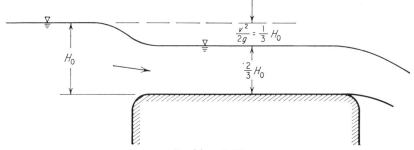

Problem 1-13

1.14. By integrating Eqs. (1–16) and (1–19) respectively, show that the total surface drag T on each side of a flat plate of length x_0 and unit width held parallel to fluid flow is given by the equation

$$T = C_f x_0 \cdot \tfrac{1}{2} \rho v_0{}^2$$

where C_f is equal to $1.32/\mathrm{Re}^{1/2}$ for a laminar, and $0.074/\mathrm{Re}^{1/5}$ for a turbulent, boundary layer.

1.15. If the piers in Prob. 1.7 are each 10 ft long, find the surface drag on each pier and express it as a fraction of the total drag. Assume a depth of 9.2 ft and a velocity of 11.5 ft/sec in the space between the piers.

1.16. The accompanying figure shows the variation of the pressure coefficient $p' = (p - p_0)/\tfrac{1}{2}\rho v_0{}^2$ round a sharp-edged plate held normal to the stream; the plate is effectively of infinite length, i.e., flow round the ends is suppressed. If surface drag is negligible, show that the drag coefficient C_D is equal to the

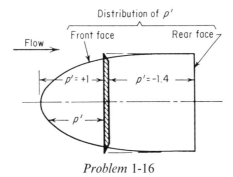

Distribution of p'

Problem 1-16

algebraic sum of the two mean values of p' taken over each face of the plate. Hence deduce the value of C_D from the figure, and compare it with the value given in Sec. 1.8.

1.17. Water is discharging to atmosphere under a head of 200 ft through a short horizontal undersluice of rectangular section, as shown in the figure. A "stop log," i.e., a fabricated steel section used for emergency closing of the tunnel, is

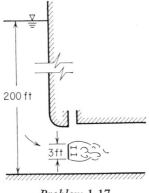

200 ft

3ft

Problem 1-17

lowered into the stream. The width of the tunnel is 15 ft; estimate the maximum drag force on the stop log and determine whether it is greater or less than the thrust under static conditions after the stop log, with others, has closed and sealed the opening.

1.18. Assuming a wide channel with the two-dimensional velocity distributions given by Eqs. (1–20) and (1–21), with δ replaced by the total depth y_0, determine the velocity coefficients α and β, in the latter case as functions of the exponent n. Hence show that the ratio $(\alpha - 1)/(\beta - 1)$ is equal to 2.76 for laminar flow, and equal to

$$\frac{(n+3)(2n+1)}{(3n+1)}$$

for turbulent flow.

1.19. Consider the flow of an incompressible fluid through the streamtube of Fig. 1-7, and prove the Bernoulli equation by equating the net gain in fluid energy to the "flow-work" done on the fluid by pressure forces as the body of fluid originally occupying the region 1-2 moves into the region 1′-2′. The streamtube need not be assumed to be horizontally disposed, and it must be assumed small enough in section for average values of p, v, and z to be applicable over the whole of each cross section.

1.20. As an approximation to the laminar boundary layer velocity profile, the parabolic distribution

$$\frac{v}{v_0} = \frac{2y}{\delta} - \left(\frac{y}{\delta}\right)^2$$

is sometimes suggested. Show that for this profile the displacement thickness δ_1 would be equal to $\delta/3$, and explain why the sine curve of Eq. (1–20) must provide a better approximation than the above equation in the close neighborhood of the solid surface. (*Note*: It is necessary to recall the most important respect in which boundary-layer flow differs from fully developed laminar flow in a closed pipe.)

Chapter 2

The Energy Principle in Open Channel Flow

2.1 The Basic Equation

The first problem is to determine how the Bernoulli expression

$$\frac{p}{\gamma} + z + \frac{v^2}{2g}$$

is applied to open channel flow. There is no difficulty about the velocity head term $v^2/2g$ provided we assume that the velocity is constant over the whole cross section, as implied in the conclusion of Sec. 1.9.

Each of the remaining two terms varies from point to point over the cross section; if we consider a case where the longitudinal slopes of both bed and water surface are small, as in Fig. 2-1a, then at any point A the pressure head p/γ simply equals the depth of A below the water surface, as in normal hydrostatic theory. The sum $(p/\gamma + z)$ must therefore be equal to the height of the water surface above datum, whatever may be the position of A; it follows that the free surface is the hydraulic grade line for all points in the cross section, and therefore for the flow in general.

Hydrostatic Pressure Distribution

This last statement is however subject to a most important qualification. In arriving at this conclusion it has been assumed that the pressure head at A is equal to the depth below the free surface, just as it would be in a tank of still water, i.e., that the pressure distribution is hydrostatic. This statement is true in very many cases of open channel flow, but not in all cases, so it should be recognized as a special assumption and not mistakenly supposed to be a universal truth.

In order to recognize the circumstances in which it is true we must refer to the Euler equation

$$\frac{\partial}{\partial s}(p + \gamma z) + \rho a_s = 0 \tag{2-1}$$

where s is measured along any straight line in the field of flow. In a tank of still water $(p + \gamma z)$ is constant; we are concerned here with the question of whether $(p + \gamma z)$ remains constant as we move along the vertical line OAB in Fig. 2-1a—i.e., with the question of whether $\partial(p + \gamma z)/\partial s = 0$, where s is the vertical direction. This will be true if $a_s = 0$, i.e., if there is negligible acceleration in the vertical direction.

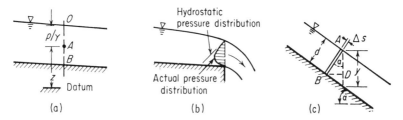

Figure 2-1. Typical Cases in the Application of the Bernoulli Expression to Open Channel Flow

Clearly this condition will be met in the very many situations where the flow is substantially horizontal with only small vertical flow components as in Fig. 2-1a. On the other hand, where there is marked vertical acceleration, as in the drawdown to a free overfall (Fig. 2-1b) the pressure distribution is clearly not hydrostatic. In such cases there is invariably pronounced curvature of the streamlines in the vertical plane.

The pressure distribution also departs from hydrostatic when the bed slope is very steep, as in Fig. 2-1c. The pressure at B in this case balances the component normal to the bed of the weight of an element AB,

$$p_B \cdot \Delta s = \gamma \cdot AB \cdot \Delta s \cdot \cos \alpha$$

or

$$p_B/\gamma = AB \cos \alpha$$

$$= y \cos^2 \alpha$$

$$= AD \text{ in Fig. 2-1}c \qquad (2\text{--}2)$$

Rivers and canals are regarded as very steep if $\sin \alpha$ is as high as 0.01, i.e., $\cos^2 \alpha = 0.9999$. It follows that in the great majority of cases the slope can be assumed small enough for $\cos^2 \alpha$ to approach unity; further, the vertical distance y may then be taken as equal to the water depth d at right angles to the bed. There will however be exceptional cases, such as spillways, where $\cos^2 \alpha$ departs materially from unity and the pressure distribution is, on this account, no longer hydrostatic.

This discussion can be summed up in the statement that in open channel flow the free surface coincides with the hydraulic grade line, provided that the pressure distribution is hydrostatic; that is, if vertical curvatures and accelerations are negligible, and if the slope is small.

When these conditions are satisfied, the expression for the total energy H may be written as

$$H = y + z + \frac{v^2}{2g} \qquad (2\text{--}3)$$

where y is the vertical distance from the bed to the water surface and z is defined now (and for all subsequent argument) as the height of the bed above datum. From this point onwards we shall be mainly concerned with the cases where this form of the energy equation holds true, but always bearing in mind that there are situations—for example, at the edge of a free overfall—where it breaks down.

2.2   The Transition Problem

We can now consider how Eq. (2–3) is to be used in practice; its use is quite simple in cases where the depth is specified in advance, as in the calculation of the flow under a sluice gate. The process is illustrated in the following example:

Example 2.1

As shown in the sketch, the depths a short distance upstream and downstream of a sluice gate in a horizontal channel are 8 ft and 2 ft respectively. The channel is of rectangular section and 10 ft wide; find the discharge under the gate.

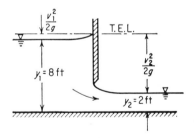

Example 2-1

From Eq. (2–3) we can write

$$8 + \frac{v_1{}^2}{2g} = 2 + \frac{v_2{}^2}{2g}$$

and by continuity

$$80 v_1 = 20 v_2$$

whence

$$\frac{v_2{}^2}{2g} = 16 \frac{v_1{}^2}{2g}$$

and
$$\frac{v_2{}^2}{2g} - \frac{v_1{}^2}{2g} = 15\frac{v_1{}^2}{2g} = 8 - 2 = 6$$

whence
$$v_1 = 5.07 \text{ ft/sec, and } Q = 406 \text{ cusecs.}\quad Ans.$$

The problem is essentially similar to the elementary one of calculating the discharge in a pipe from the upstream and throat pressures in a Venturi meter. However, in those problems where the depth at some section is not specified in advance but is to be calculated from our knowledge of some change in the channel cross section, we encounter the feature of open channel flow that lends it its special difficulty and interest. It is the fact that the depth y plays a dual role: it influences the energy equation (as we have seen), and also the continuity equation, since it helps to determine the cross-sectional area of flow. The problems involved are best appreciated by considering the two situations shown in Fig. 2-2, each of them amounting to a simple constriction in the flow passage, smooth enough to make energy losses negligible. Suppose that in each case the problem is to determine conditions within the constriction, if the upstream conditions are given.

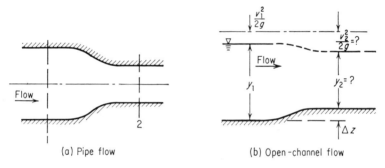

(a) Pipe flow   (b) Open-channel flow

Figure 2-2.   The Transition Problem

In the pipe-flow case we can, from the known reduction in area, readily calculate the increase in velocity and in velocity head, and hence the reduction in pressure. The open channel flow case, however, is not quite so straightforward. We have a smooth upward step in the otherwise horizontal floor of a channel having a rectangular cross section; since there is no energy loss, we can write

$$y_1 + \frac{v_1{}^2}{2g} = \Delta z + y_2 + \frac{v_2{}^2}{2g}$$

where y_2 and v_2 are both unknown, and all other terms are known. Introducing the discharge per unit width, $q = v_2 y_2$, which will be the same at sections 1 and 2, we have

$$y_2 + \frac{q^2}{2gy_2{}^2} = y_1 + \frac{v_1{}^2}{2g} - \Delta z$$

in which y_2 is the only unknown. The equation is, however, a cubic; setting aside the algebraic difficulties of solving it, there remain the questions of how many physically real solutions there will be, and whether all these real solutions are available or accessible to the upstream flow. In order to deal completely with this type of problem it is necessary to devise a special approach to the energy equation; this will now be dealt with.

Specific Energy and Alternate Depths

We define the specific energy E as the energy referred to the channel bed as datum, i.e.,

$$E = y + \frac{v^2}{2g} \tag{2-4}$$

and in this simple concept, introduced by B. A. Bakhmeteff in 1912, lies the key to even the most complex of open channel flow phenomena. We develop this concept first by considering, as in the previous paragraph, the simplest type of channel cross section—the rectangular shape, of width b. As before, we define $q = Q/b = vy$ as the discharge per unit width of channel, and we rewrite Eq. (2-4) in the form

$$E = y + \frac{q^2}{2gy^2} \tag{2-5}$$

We now consider how E will vary with y for a given constant value of q, i.e., we construct the graph of Eq. (2-5) on the E-y plane. We have

$$(E - y)y^2 = \frac{q^2}{2g} = \text{a constant}$$

so that the curve has asymptotes $(E - y) = 0$ and $y = 0$; in fact one section of the curve falls within the 45° angle between these two asymptotes in the first quadrant, as in Fig. 2-3b. There is another section of the curve shown as a broken line, but this is of no practical interest as it yields negative values for y.

If we regard this curve as a means of solving Eq. (2-5) for y, given E and q, the three solutions of the cubic are clearly shown by drawing a vertical line corresponding to the given value of E. Only two of them are physically real, so for given values of E and y there are two possible depths of flow, unless the vertical line referred to misses the upper curve altogether, a case which will be discussed later. When two depths of flow are possible for a given E and q, they are referred to as *alternate* depths.

Alternatively we may say that the curve represents two possible regimes of flow—slow and deep on the upper limb, fast and shallow on the lower limb—meeting at the crest of the curve, C.

Other curves might be drawn for other values of q; since, for a given value of y, E increases with q, curves having higher values of q will occur inside and to the right of those having lower values of q.

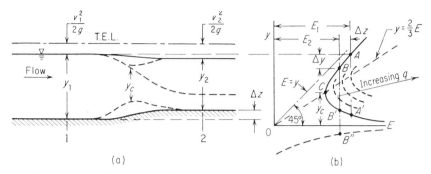

Figure 2-3. The Specific Energy Curve and Its Use in the Transition Problem

Specific Energy and the Transition Problem

With the aid of Fig. 2-3 it is now possible to see clearly what is involved in the problem shown in Fig. 2-2 and redrawn in Fig. 2-3a. Suppose that the flow upstream of the step (section 1) in the channel bed has such a velocity and depth as to be represented by the point A on the upper (slow and deep) limb of the E-y curve in Fig. 2-3b. We assume also that there is no change in channel width over the step, so that the value of q remains unaltered. Since the curve in Fig. 2-3b is drawn for a given constant value of q, it follows that the point representing the flow over the step (section 2) will also lie on the curve. Its position on the curve is easily located by recalling that the total energy must be the same at sections 1 and 2, and therefore that the specific energies differ by the height of the step, i.e., $E_1 - E_2 = \Delta z$.

Having obtained the value of E_2, we can now obtain solutions describing the flow at section 2; they are represented by the points where the line $E = E_2$ cuts the E-y curve. There are two physically real solutions, represented by the points B and B'; the third, represented by the point B'', is unreal and need not concern us further.

The two real solutions would also apply to the case where the upstream flow was represented by the point A', having the same specific energy E as the point A. The question now arises as to which of the two solutions represented by B and B' is more likely to occur in reality, and this question really reduces to one of accessibility; given a certain upstream flow (represented by A) are the two possibilities for the downstream flow (represented by B and B') equally accessible? The energy equation in itself appears to have little to tell us about this problem, but the form of the E-y curve is, as we shall see, a most valuable guide.

The Accessibility of Flow Regimes

We consider now the problem set in the last paragraph. As the flow moves over the step, the point representing the flow moves from A to either B or B': If we are concerned with the relative accessibility of B and B', it is reasonable to ask what kind of path this flow-point must follow. The answer is quite simple; since the width of the channel remains unchanged, so does the discharge per unit width q; the flow-point must therefore remain on the E-y curve which, it will be remembered, was drawn for a certain constant value of q. The point cannot jump across the curve straight from B to B', for if it did the value of q would have to change temporarily from its original magnitude and then return to it again, i.e., the width would have to undergo a similar temporary change. It is easily seen that this change in width would take the form of a constriction followed by an expansion as shown in Fig. 2-4. (For the sake of clarity this figure is drawn without including the step, as if the constriction were in fact a short way downstream of the step).

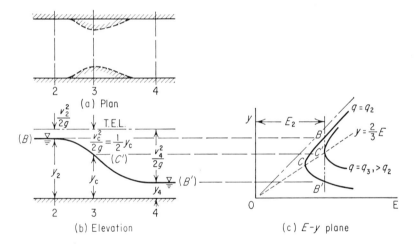

Figure 2-4. *The Change From One State of Flow to Another Via a Contraction in Width*

If the width remains constant the only available path from B to B' is round the curve itself; but if this path were followed the specific energy would have to drop below E_2 and return to it again. This could only happen if the bed level temporarily rose above the level of the step (just far enough to bring the point from B to C, the crest of the curve) and then dropped back to the level of the step, as shown in Fig. 2-3a.

The conclusion is that if the width remains unchanged, and if the bed rises steadily to the level of the step without rising above it at any point, the point B' is inaccessible when the upstream flow is represented by the point A.

Similarly the point B will be inaccessible if the upstream flow is represented by the point A'. The jump from one side of the curve to the other can be accomplished only by inserting a local constriction—either in width or in bed level—at or near the step. This statement of course relates only to this particular type of problem; there may be other conditions únder which a change from one regime to the other is possible.

One last remark may be made about this example. As we go from A to B the depth decreases and the velocity and velocity head must therefore increase. Since the total energy line remains at the same level, the water surface must therefore drop, as in Fig. 2-3a, as the flow moves over the step—a slightly surprising conclusion in view of one's natural expectation that the water surface would have to rise in order to get over the obstacle presented by the step. On the other hand, movement from A' to B' is accompanied by an increase in depth, so that the water surface rises.

The Gas-Flow Analogy

Readers who are acquainted with thermodynamics and gas flow will have noticed something familiar about the preceding arguments. First there is the notion of accessibility—that one of the points B and B' may be more accessible than the other, although from the energy viewpoint they are equivalent. This idea has its thermodynamic counterpart in the fact that the accessibility of a certain state in a gas depends not only on the energy available, but also in whether entropy would be gained or lost in reaching that state.

Second, there is the curious feature of the example discussed above—that a transition from slow deep flow to fast shallow flow could be accomplished only by some form of constriction in the channel, followed by an expansion. This principle is exactly analogous to a similar one in gas flow, according to which the transition from subsonic to supersonic flow requires a nozzle which first converges to a narrow throat section and then diverges again.

In fact there is a complete analogy between open channel flow and gas flow, the details of which are beyond the scope of this text. However, it may be remarked that according to the previous argument, the slow deep flow indicated by the upper limb of the E-y curve appears to have something in common with subsonic gas flow, and that of the lower limb with supersonic gas flow.

The Existence of a Solution to the Transition Problem

All of the above argument rests on the important assumption that a solution of some sort will be possible; it is, however, quite easy to visualize a situation in which no solution appears possible. It is that in which the height of the step, Δz, is great enough to make E_2 less than the minimum possible value of E, i.e., that which obtains at C, the crest of the E-y curve. In other words, in attempting to move from A to some point B, we miss the end of the curve.

Therefore the problem as originally set has no possible solution; that is, the three prescribed values of q, E_1, and Δz cannot exist simultaneously in the channel. There is nothing mysterious about this: we may think of the specific energy E_1 as maintaining a certain flow q against an obstruction of size Δz, and if the obstruction becomes too large the flow can no longer be maintained with the available specific energy. We may obtain a qualitative picture of events by imagining that in the first place there is no step in the channel, that q and E_1 have their initially given values, and that the step is then suddenly introduced into the channel. Clearly the water will back up and some kind of surge wave will move upstream until a new steady state is established; either q will be reduced or E_1 will be increased, but it is not possible to tell which of these will happen without more complete knowledge of the upstream conditions. The problem is no longer concerned with a localized transition in the channel but now must deal with the channel as a whole.

A complete treatment of this problem must therefore be postponed until a later chapter, after the question of flow resistance in long channels has been dealt with. Meanwhile we proceed in the next section with consideration of some questions about the E-y curve which have been suggested by the preceding discussion.

2.3 Critical Flow

Discussion of the transition problem has shown that for any given pair of values of specific energy and discharge per unit width q, there are two possible depths of flow, and that transition from one depth to the other can only be accomplished under certain special conditions. (Other conditions under which the transition is possible will be discussed at a later stage.) These two depths which are represented on the two different limbs of the E-y curve separated by the crest C, are characteristic of two different kinds of flow; a logical way to explore the nature of the difference between them would be to consider first the flow which is represented by the point C. Here the flow is in a critical condition, poised between two alternative regimes, and indeed the word "critical" is used to describe this state of flow; it can be defined as the state at which the specific energy E is a minimum for a given q.

Analytical Properties of Critical Flow

We set out first to derive equations defining critical flow, and to do this we start from Eq. (2–5)

$$E = y + \frac{q^2}{2gy^2}$$

(2–5)

E is a minimum at critical flow and we obtain this minimum by differentiation.

$$\frac{dE}{dy} = 1 - \frac{q^2}{gy^3}$$

$$= 0$$

when

$$q^2 = gy^3 \tag{2-6}$$

i.e.,

$$v_c^2 = gy_c \tag{2-7}$$

where the subscript c has been inserted, in accordance with the common practice for indicating critical flow conditions. Equation (2–7) is the most important defining equation for critical flow; another one can also be obtained from Eq. (2–6):

$$y_c = \sqrt[3]{\frac{q^2}{g}} \tag{2-8}$$

which conveniently expresses the critical depth y_c as a function of discharge alone. Further, since from Eq. (2–7)

$$\frac{v_c^2}{2g} = \tfrac{1}{2}y_c,$$

we have

$$E_c = y_c + \frac{v_c^2}{2g} = \tfrac{3}{2}y_c$$

whence

$$y_c = \tfrac{2}{3}E_c \tag{2-9}$$

The above equations are established by considering the variation of E with y for a given q; it is also of practical interest to study how q varies with y for a given $E = E_0$. Clearly the curve will be of the general form shown in Fig. 2-5; when $y \to E_0$, $v \to 0$ and hence $q \to 0$. Similarly when $y \to 0$, $q \to 0$,

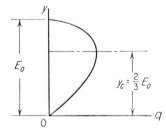

Figure 2-5.   The Discharge-Depth Curve for Given Specific Energy

and there will clearly be a maximum value of q for some value of y between 0 and E_0. (Manifestly y cannot be greater than E_0). We find this maximum by recasting Eq. (2–5) into the form

$$q^2 = 2gy^2(E_0 - y) \tag{2-10}$$

and differentiating with respect to y:

$$2q \frac{dq}{dy} = 4gyE_0 - 6gy^2$$

$$= 0$$

when $$6gy^2 = 4gyE_0$$

i.e., $$y = \tfrac{2}{3}E_0$$

which is essentially Eq. (2–9) representing critical flow; Eqs. (2–7) and (2–8) can be deduced from it. We have therefore established another important property of critical flow: it connotes not only minimum specific energy for a given discharge per unit width, but also maximum discharge per unit width for a given specific energy.

Any one of the three Eqs. (2–7) through (2–9) might be used to define critical flow, and all have their uses in the calculations required in problems concerning critical flow. Equation (2–9), for instance, shows that the crests of E-y curves drawn for all values of q can be joined by a straight line having the equation $y = 2E/3$, as in Fig. 2-3b; Eq. (2–8), showing that y_c increases with q, verifies our previous conclusion that curves having a high value of q are to the right of those having a low value of q (Fig. 2-3b). However our immediate interest centers on Eq. (2–7), for it has an important physical meaning.

Critical Velocity and Wave Velocity

Equation (2–7) states that at critical flow the velocity is equal to $\sqrt{gy}$. Now this term is equal to the velocity with which a long wave of low amplitude propagates itself in water of depth y, and herein lies the most important physical feature of critical flow.

Proof of this statement should be given before we proceed to discuss its significance. There are two kinds of waves to be considered—oscillatory waves, such as form in the sea at some distances from the shore (Sec. 8.6), and "surges," which were discussed in Sec. 1.2. The former of these move with substantially no energy loss; the latter may have broken turbulent fronts with substantial energy losses.

Considering first the oscillatory wave, it is shown in texts on hydrodynamics that the velocity c of waves of length L in water of depth y is given by the equation

$$c^2 = \frac{gL}{2\pi} \tanh \frac{2\pi y}{L} \tag{2–11}$$

now if L is large compared with y, $2\pi y/L$ is small, so that

$$\tanh \frac{2\pi y}{L} \approx \frac{2\pi y}{L}$$

whence
$$c^2 = \frac{gL}{2\pi}\frac{2\pi y}{L}$$

$$= gy \tag{2-12}$$

which is the required result.

For the surge wave, it is convenient to consider the wave as being generated by the horizontal movement of a vertical plate into a channel of unit width containing still water, as shown in Fig. 2-6. The argument is restricted to surges of small amplitude; the velocity of the plate is Δv, as is the velocity of the water immediately to the left of it and throughout all the water behind the surge which forms and moves to the left as soon as the plate begins to move. The height of the surge is Δy, and its velocity is c, which in general is much greater than Δv.

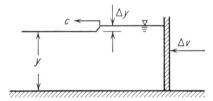

Figure 2-6.   The Elementary Surge of Small Amplitude

The normal method of treating this problem is that suggested by the discussion in Secs. 1.2 and 1.6; i.e., the situation is changed from unsteady to steady flow by changing the viewpoint of the observer, and the momentum equation is applied since energy may be dissipated at the face of the surge. This method will be developed fully in Chap. 3; however, since the surge in this case is of small amplitude the energy dissipation will be small and the energy equation may be used as an alternative to the momentum equation. It is left to the reader to show (Prob. 2.1) that this approach leads to the result

$$c^2 = gy \tag{2-12}$$

as for the oscillatory wave.

Both arguments are limited to long waves of low amplitude, and it is this type of wave which is most often generated in channels by the operation of controls and the existence of obstructions. We can conclude therefore that the wave velocity $c = \sqrt{gy}$ is the velocity with which a disturbance in open channel flow tends to move over the water surface, and this velocity is of course measured relative to the water, not to the banks.

2.4   Subcritical and Supercritical Flow

We have now established that at critical flow the water is moving just as fast, relative to the banks, as the wave resulting from a small disturbance would move relative to the water. One front of such a wave would therefore appear stationary to an observer standing on the bank; the opposite front (assuming the wave were spreading both upstream and downstream) would appear to the observer to be moving downstream twice as fast as the water. Indeed, "standing waves" are characteristic of flow which is at or near the critical condition, but a fact of more far-reaching interest is that when the stream velocity is less than critical, the wave from a disturbance can make its way upstream (from the viewpoint of the stationary observer) but when the velocity is greater than critical such a wave must be swept downstream, and no disturbance can propagate its influence in the upstream direction.

The former type of flow ($v < \sqrt{gy}$) is called *subcritical*, and the latter ($v > \sqrt{gy}$) *supercritical*. The two types of flow are clearly represented by the upper and lower limbs of the E-y curve respectively; the same statement can be made of the q-y curve drawn for a given E (Fig. 2-5). It should be just as clear, even to the reader unfamiliar with thermodynamics and gas flow, that since subcritical and supercritical flow are defined in relation to a natural wave velocity, they are closely analogous to subsonic and supersonic flow, which are defined in relation to the natural velocity of a compression wave in a gas, i.e., the velocity of sound. This conclusion has already been suggested in the remarks made in Sec. 2.2.

The Froude Number

We define the Froude number as the quantity $v/\sqrt{gy}$, i.e., the ratio of the stream velocity to the wave velocity. It will be designated by the letters Fr in all subsequent discussion.

The Froude number is analogous to the Mach number in gas flow, i.e., the ratio of gas velocity to sonic velocity. In the open channel case, the flow is supercritical or subcritical according as Fr is greater or less than unity; for this and other reasons Fr is a very convenient parameter in the manipulation of the equations of open channel flow. For example, we note that the derivative of Eq. (2–5) may be simply expressed as

$$\frac{dE}{dy} = 1 - \mathrm{Fr}^2 \qquad\qquad (2\text{–}13)$$

The Froude number as defined above is easily recognized as the square root of the Froude number as derived from dimensional analysis, i.e., v^2/gL, where L is some characteristic length dimension. This difference in definitions need not cause any confusion.

The arguments developed in this and subsequent chapters will demonstrate convincingly the truth of the conclusion arrived at in the theory of dynamic similarity—that the Froude number is a kind of universal indicator of the state of affairs in free-surface flow. Many equations can be expressed in terms of the Froude number in such a way as to make it clear that once the value of Fr is known, we know all that is required to describe a particular flow situation.

Upstream and Downstream Control

The preceding discussion about the ability of a disturbance to move upstream has a simple, practical, and most important significance. In subcritical flow, a disturbance can move upstream; what this means in practice is that a control mechanism such as a sluice gate can make its influence felt on the flow upstream, so that *subcritical flow is subject to downstream control.* Conversely, supercritical flow cannot be influenced by any feature downstream, and can therefore only be controlled from upstream.

These general principles are perfectly illustrated by the simple sluice gate, whose setting determines the velocity-depth relationship both upstream and downstream of the gate itself (Fig. 2-7). The upstream flow is subcritical and

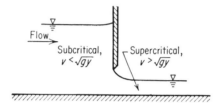

Figure 2-7.   The Sluice Gate as a Generator of Subcritical and Supercritical Flow

is controlled by a feature downstream of itself; the downstream flow is supercritical and is controlled by a feature upstream of itself. It is left as an exercise for the reader (Prob. 2.2) to prove that the two flow regimes are in fact subcritical and supercritical, and not, for instance, both subcritical.

A memory device often used by practicing engineers is the statement that in supercritical flow " the water doesn't know what's happening downstream," and this statement does indeed convey the truth of the matter in a way that is accurate as well as expressive.

2.5   The Occurrence of Critical Flow; Controls

The argument so far has dealt with the type of problem in which both q and E are prescribed initially; however, there is a further underlying problem which is of great practical interest. It is this: given a value of q, what factors determine the specific energy E, and hence the depth y? conversely, if E is given, what factors determine q?

The answer to these questions is, as might be expected, that there are many different kinds of control mechanism which can dictate what the depth must be for a given q, and vice versa. An obvious example is the sluice gate; for a given setting of the gate there is a certain relationship between q and the upstream depth, and similarly for the downstream depth. Weirs and spillways are further examples of this kind of mechanism. We might also expect that flow resistance due to the roughness of the channel bed will have some effect, and the detailed treatment of this subject in Chap. 4 will show that this is indeed the case.

In fact the flow situation in any particular channel is substantially determined by the control mechanisms operating within it. The notion of a "control"—any feature which determines a depth-discharge relationship—is therefore of paramount importance in the study of open channel flow, but it cannot be fully developed until flow resistance is treated in Chap. 4. Controls are mentioned here by way of introducing the fact that there are certain features in a channel which tend to produce critical flow, and are therefore controls of a rather special kind.

To determine the nature of these features, we first consider the general problem of flow without losses in a channel of rectangular section and constant width, whose bed level may vary. This is in a sense a general reconsideration of the transition problem introduced in Sec. 2.2, but with a somewhat different emphasis.

Since the total energy H and q are constant, we can write:

$$H = y + z + \frac{q^2}{2gy^2} = E + z = \text{constant} \qquad (2\text{--}14)$$

whence, differentiating with respect to x, the distance along the channel:

$$\frac{dE}{dx} + \frac{dz}{dx} = 0$$

a result which may be written as

$$\frac{dy}{dx}\frac{dE}{dy} + \frac{dz}{dx} = 0$$

whence from Eq. (2–13) we have:

$$\frac{dy}{dx}(1 - \text{Fr}^2) + \frac{dz}{dx} = 0 \qquad (2\text{--}15)$$

It is noteworthy that the Froude number Fr plays a prominent, indeed a critically important, part in this equation. The equation demonstrates in compact form a result deduced previously from the form of the E-y curve: if there is an upward step in the channel bed, i.e., if dz/dx is positive, then the product $(1 - \text{Fr}^2)dy/dx$ must be negative. Hence if the flow is subcritical (Fr < 1), dy/dx will be negative and the depth will decrease over the step. Similarly, if the flow is supercritical the depth will increase. Corresponding conclusions can easily be arrived at for a downward step, with dz/dx negative.

However, our main concern here is with the case $dz/dx = 0$, i.e., that of a horizontal channel bed. Since the product $(1 - \mathrm{Fr}^2)dy/dx$ is then equal to zero, it follows that either

$$\frac{dy}{dx} = 0$$

or $\mathrm{Fr} = 1$ (critical flow)

The first possibility clearly has a real physical meaning; in the step-transition problem $dz/dx = 0$ and $dy/dx = 0$ both upstream of the step and over the step, and in both cases $\mathrm{Fr} \neq 1$.

It is not so obvious that the second possibility is a physically real one. The question is: can we visualize a situation in which $dz/dx = 0$ and $dy/dx \neq 0$? The answer is found in the case shown in Fig. 2-8, in which water is released

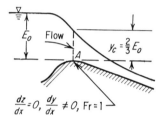

Figure 2-8.   *Free Outflow from a Lake—an Example of Critical Flow*

from a lake over a short (but not sharp-edged) crest so that it can fall freely downstream. This latter condition implies either a free overfall a short distance downstream or a slope so steep that the bed resistance imposes no effective restraint on the flow.

At the crest A, $dz/dx = 0$, but since the water is accelerating at this point, $dy/dx \neq 0$. The conclusion is that Fr must be equal to unity, and the flow will be critical. The cases of a sharp-edged (e.g., weir) crest, and a completely free overfall, are excluded because the pressure distribution would not then be hydrostatic; for the same reason the curvature of the crest should be fairly gentle so that vertical accelerations due to flow curvature will not be large. However even if the vertical accelerations are large, as at the edge of a completely free overfall, the flow is still quite close to the critical condition. Experimental evidence shows that the flow depth right at the brink of an overfall is approximately $5y_c/7$, and that $y = y_c$ at a distance back from the overfall equal to two to three times y_c, depending on the discharge. For the sharp-edged weir of infinite height, the discharge is remarkably close to that obtained by assuming critical flow at the crest, despite the pronounced vertical curvature of the flow. Details of this case are left as an exercise for the reader (Prob. 2.3). Both the free overfall and the sharp-edged weir will be discussed in more detail in Chap. 6.

Granted that the pressure distribution is hydrostatic, we conclude that when water is released from a lake without any downstream restraint critical flow occurs at the section of maximum vertical constriction: such a section is therefore a control. It is easily shown that critical flow will occur at a corresponding horizontal constriction; if we assume a horizontal channel bed and a variable width b, we can write, again differentiating Eq. (2–14) with respect to x, but this time taking z as a constant and q as a variable:

$$\frac{dy}{dx} - \frac{q^2}{gy^3}\frac{dy}{dx} + \frac{q}{gy^2}\frac{dq}{dx} = 0$$

and by continuity $\qquad qb = $ a constant, Q

whence $\qquad\qquad b\frac{dq}{dx} + q\frac{db}{dx} = 0$

Eliminating dq/dx, we have

$$\frac{dy}{dx}(1 - \mathrm{Fr}^2) - \frac{q}{gy^2}\frac{q}{b}\frac{db}{dx} = 0$$

i.e., $\qquad\qquad \frac{dy}{dx}(1 - \mathrm{Fr}^2) - \mathrm{Fr}^2\frac{y}{b}\frac{db}{dx} = 0 \qquad\qquad$ (2–16)

whence, as from Eq. (2–15), we can conclude that at the outflow from a lake critical flow will occur when $db/dx = 0$, i.e., at a section of maximum horizontal constriction. It is left to the reader to show that critical flow could not occur at a section of maximum width, but only at one of minimum width (Prob. 2.4).

It will be shown in Sec. 2.7 that Eqs. (2–15) and (2–16) can be generalized to fit any shape of channel cross section and that the same inference can be drawn—that critical flow tends to occur at a section of maximum constriction before the water is released into a region where there is no effective restraint on the flow. It will also be shown in Chap. 4 that this general concept of the occurrence of critical flow can be applied to cases where flow resistance is taken into account. In such cases, however, it will be found that less importance attaches to the section of maximum constriction (if any), and that critical flow occurs at the section where the release becomes effective—i.e., at the start of the region of no effective restraint, whether or not there is a constriction at this section.

It follows from the definition of a control that at any feature which acts as a control the discharge can be calculated once the depth is known. This fact makes control sections very convenient for flow measurement, and a critical section is particularly convenient. It will be seen in Chap. 6 that a number of flow measurement devices operate by producing critical flow at a certain section, usually by the mechanism described above of constriction followed by release.

We have now established the principles necessary for dealing with the transition problem: it remains to deal with particular problems in more detail, including the numerical and algebraic techniques necessary for their solution.

2.6   Applications in Rectangular Channels

The simplest application is the transition problem constituted by a smooth change in bed level or width. It has been shown in Sec. 2.2 how the solution can be traced on the E-y curve (Fig. 2-3). In a particular problem the numerical solution can be obtained either by scaling the solution from such a curve, or by algebraic means, as in the following examples.

Example 2.2

Water flows at a velocity of 3 ft/sec and a depth of 5 ft in a rectangular channel. There is a smooth upward step of 6 in. in the channel bed; find the depth of water over the step and the change in the absolute level of the water surface.

Describing as section 1 and 2 the sections upstream of the step and over the step respectively, we first calculate the upstream specific energy.

$$E_1 = 5 + \frac{3^2}{2g} = 5.14 \text{ ft}$$

Then the downstream specific energy is equal to

$$E_2 = E_1 - 0.5 = 4.64 \text{ ft}$$

and since $q = 15$,

$$E_2 = y_2 + \frac{q^2}{2gy_2{}^2} = y_2 + \frac{3.49}{y_2{}^2} = 4.64 \text{ ft}$$

This is a cubic equation, which can nevertheless be readily solved in a slide-rule operation†, giving the result

$$y_2 = 4.465 \text{ or } 0.977 \text{ ft}$$

and since, as is easily shown, the upstream flow is subcritical, then only the subcritical value of y_2 (i.e., 4.465) is possible. The water level is now $4.465 + 0.5 = 4.965$ ft above the upstream channel bed, so that the water level has dropped by only 0.035 ft.

In this example the upstream Froude number was quite low, so that the point representing section 1 would be well up the upper limb of the E-y curve, where the slope of the curve approaches 45° very closely. This means that Δy will only be slightly greater numerically than Δz, (Fig. 2-3b) and the water surface will drop by only a very small amount, as the calculation has in fact shown. If the upstream flow had been more nearly critical, the slope of the

† See Appendix on Mathematical Aids.

Figure 2-9.   *Standing Waves in Near-Critical Flow. The Waves Are*
Set Up by the Small Bed Feature at the Right of the Picture

E-y curve would have been steeper, Δy would have been much greater than Δz, and the change in the water surface level would have been much more pronounced. This argument points up an important property of flow which is at or near the critical condition—the large changes in water-surface level that can be brought about by small changes in bed level. In this fact lies a supporting explanation for the large "standing waves" already referred to as a feature of critical flow; such waves can be triggered off by very small irregularities in the channel bed, as shown in Fig. 2-9. These waves exemplify a general property of all mechanical systems, viz. that large-amplitude oscillations are likely wherever large displacements can occur with small changes in energy. It should be noted, however, that the E-y curve does not provide a complete description of the behavior of the waves, for they show vertical accelerations large enough to invalidate the assumption of hydrostatic pressure distribution made in the theory underlying the E-y curve.

The solution to the problem corresponding to Example 2.2 but involving width change can also be traced on the E-y curve, redrawn as Fig. 2-10. In

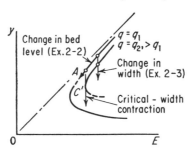

Figure 2-10.   *Changes in Width and Bed Level Traced on the*
E-y Curves

this case E does not change, since the bed level does not change, but q assumes a new value at section 2 because of the change in width. Thus an equation for y_2 can be obtained, as in the following example:

Example 2.3

Upstream conditions are as in Example 2.2. The width contracts smoothly from 6 ft to 5 ft; find the depth of water within the contraction.

$E_1 = 5.14$ as before, and on this occasion E_2 has the same value. But $q_2 = 15 \times \frac{6}{5} = 18$, whence:

$$E_2 = y_2 + \frac{q_2^2}{2gy_2^2} = y_2 + \frac{5.03}{y_2^2} = 5.14 \text{ ft}$$

yielding the solution

$$y_2 = 4.93 \text{ or } 1.12 \text{ ft}$$

Again, only the subcritical value of y_2 (i.e., 4.93) is possible. Also, the change in the absolute level of the water surface is again small because of the low upstream Froude number.

The E-y Relationship in Dimensionless Form

The solutions to both the above exercises can be traced on *E-y* curves, as in Fig. 2-10; numerical values may of course be obtained graphically from such curves, provided a stock is kept covering a large enough range of values of q. The inconvenience of keeping such a large stock can be avoided by E. Crausse's [1] dimensionless treatment of the specific energy equation, leading to a single curve covering all values of q. We have

$$E = y + \frac{v^2}{2g} = y + \frac{q^2}{2gy^2}$$

Dividing throughout by y_c, we obtain

$$\frac{E}{y_c} = \frac{y}{y_c} + \frac{q^2}{2gy^2 y_c}$$

Setting $E' = E/y_c$, $y' = y/y_c$, and making the substitution $q^2 = gy_c^3$, we have

$$E' = y' + \frac{1}{2y'^2} \tag{2-17}$$

which is a dimensionless form of the specific energy equation. If now we plot E' vs. y' we have, in effect, a general form of the *E-y* curve with a variable length scale (Fig. 2-11). We can apply it to any particular problem once we know the critical depth y_c, and this is readily obtained from q by using Eq. (2-8).

The foregoing mention of the term *length scale* raises the question of similitude, and Eq. (2-17) provides an instructive example of the importance of the Froude number in dealing with this question. We note first the relation

$$\text{Fr}^2 = \frac{v^2}{gy} = \frac{q^2}{gy^3} = \frac{gy_c^3}{gy^3} = \left(\frac{y_c}{y}\right)^3 \tag{2-18}$$

and consider now the problem of the simple upward step in the channel floor. The points A and B on the E'-y' curve (Fig. 2-11) can clearly relate to the solution of many different particular problems of this sort, each with its own particular value of q, and hence of y_c. It is also clear that any two of these

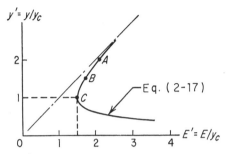

Figure 2-11.   *Dimensionless Representation of the Specific Energy*
—Depth Relation

flow situations will be geometrically similar in every respect, in that the ratios

$$\frac{y_1}{y_c}, \quad \frac{y_2}{y_c}, \quad \frac{\Delta z}{y_c}$$

are the same in each case. We now consider what dynamical conditions must exist in order to satisfy the above geometrical requirements, and Eq. (2–18) gives the answer immediately. If at corresponding points in the two flow situations the values of the ratio y/y_c are equal, then the values of the Froude number must also be equal, confirming the general conclusions of the theory of hydraulic similitude. We may note also from Eq. (2–18) a confirmation of another result of that theory, viz.,

$$q_r = y_r^{3/2}$$

where the subscript r indicates the ratio between corresponding quantities in two dynamically similar flow situations.

The Maximum Degree of Contraction; The Concept of a "Choke"

It has already been pointed out that in the type of problem treated above there is a certain degree of channel contraction beyond which no solution is possible. The amount of this contraction is readily calculated in the case of the upward step; the maximum height of step permissible is simply equal to the difference between the upstream specific energy and the minimum possible specific energy (that of critical flow), which is easily found from Eqs. (2–8) and (2–9).

It is important to notice that when the step is of this critical height, and is of limited length as in Fig. 2-12, the downstream flow may be either super-critical or subcritical, depending on the downstream conditions. Referring back to the discussion in Sec. 2.2, centered on Fig. 2-3, it can be seen that if the point representing the flow moves from A to C, it is free thereafter either to move back up the subcritical limb of the curve, or to carry on round into the supercritical limb. If there is a downstream control such as a sluice gate,

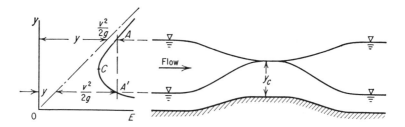

Figure 2-12.   *The Effect of a Short Upward Step of "Critical" Height*

the flow will tend to become subcritical; if not, supercritical. If there is any doubt about downstream conditions, the tendency will be towards super-critical flow; the convergence of flow towards the step will tend to carry on into a further convergence downstream of the step. This phenomenon is illustrated in Fig. 2-12; it is seen that the same principle operates when the upstream flow is supercritical. Either of the upstream flow regimes in this figure can pass to either of the downstream regimes.

Similar considerations apply in the case of a constriction in width. Here the critical condition is traced on the E-y plane as shown in Fig. 2-10; a vertical line drawn through the upstream point A just touches an inner curve, having a higher q, at its crest C'. The value of q on this inner curve, calculated from Eqs. (2–8) and (2–9), gives the amount of width contraction. Again, the downstream flow may be supercritical or subcritical.

The problem dealt with in this section is of considerable practical interest. It often happens that a local constriction must be introduced into a channel, e.g., to reduce bridging costs when the channel is being led under a highway. In such cases it is essential to know just how much constriction can be tolerated without influencing the flow upstream of the constricted section. The problem is simply that of sustaining a certain discharge for a given specific energy; the smallest section that can do this job is the one that will operate at critical flow, which gives maximum discharge for a given specific energy. Further, it has already been shown (for rectangular channels anyway) that if this smallest section is provided, the flow will automatically take up the critical condition.

The problem is not, of course, confined to rectangular channels; a common case is that in which flow passes from a trapezoidal channel (Sec. 2.7) into a circular culvert under a roadway. The principle, however, is still the same:

the limiting section is that which can take the required Q at the available E under critical flow conditions.

A constriction which is severe enough to influence the upstream flow becomes a special kind of control. It may be termed a *choke*; the same word may be used as a verb to describe the action of the constriction. As already mentioned in Sec. 2.2, the actual behavior of a choke cannot be completely determined without considering the channel as a whole, including the effects of resistance. Accordingly the complete discussion of its behavior is postponed until Chap. 5.

Limitations on the Use of the Energy Equation

In the whole of this chapter it has been assumed that we are dealing with channel transitions that are short and smoothly rounded, so that energy losses are negligible. This is in fact true of many transitions met with in practice, but there is a further limitation on the use of the energy equation which may in some cases be as important as an energy loss. This limitation is that in short transitions such as are discussed in this chapter, we are not strictly entitled to assume that the velocity and depth will be uniform across the section immediately the flow enters the contracted (or expanded) section. Uniform conditions may not become established until a point is reached some distance downstream. Consider, for example, the case of a contraction in width. The flow is essentially the same as flow around a number of similar and equally spaced bodies at a center distance equal to the upstream width of the channel (Fig. 2-13). Now it is a feature of flow around submerged bodies

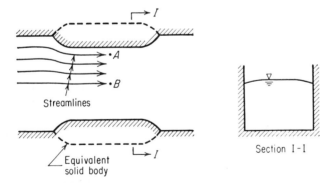

Figure 2-13.   *Variation of Velocity Across a Contracted Section*

(even without energy loss) that the velocity is much higher in the close neighborhood of the body (point A in Fig. 2-13) than at some distance out from it (point B). The resulting depth variation across the section is shown in the figure. This effect exists whether the body is alone or is grouped with others, but is much less pronounced when there is a group of bodies which produce a substantial reduction in the net width of the flow passage. In this case the effect referred to is swamped by the increase in average velocity produced by the reduction in net width.

Even when the effect is pronounced, it need not be of great concern in those cases where interest centers on the average velocity and depth across the section. In such cases the correction involved would not normally be large enough to matter. However, the effect is of considerable concern in cases where our interest is in the flow very close to the body, as when the resistance to flow of bridge piers is being considered. In such cases it would be quite wrong to assume that the velocity and depth near the body are the same as the average across the contracted section, and the elementary methods used in this chapter would be of little benefit as they deal with average velocities and depths. This particular question will be dealt with further in Chap. 7.

2.7   Nonrectangular Channel Sections

So far the theory presented has dealt only with channels of rectangular section, but such channels are not common in practice. In natural rivers the waterway may have a most irregular section, and even in artificial canals the rectangular section is relatively uncommon, the trapezoidal section (Fig. 2-14b) being often preferred in the interests of economy and bank stability.

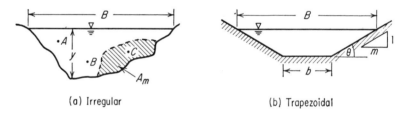

(a) Irregular   (b) Trapezoidal

Figure 2-14.   *Typical Nonrectangular Channel Sections*

The treatment must therefore be extended to cover channels of any possible section, such as the irregular one shown in Fig. 2-14a. The first question to resolve is the form of the specific energy equation; if we take our datum level at the lowest point on the section, and measure the depth upward from this level, we find that for every point in the cross section (e.g., points A, B, and C in Fig. 2-14a) the sum of the pressure head and potential head is still equal to the depth y, just as for the rectangular section. In other words, the irregularity of the section does not affect the hydrostatic pressure distribution.

We may therefore write the specific energy equation

$$E = y + \frac{v^2}{2g} \tag{2-4}$$

just as for a rectangular section. However to explore the dependence of E on y we can no longer use the discharge per unit width q, since it has lost its specific meaning. We must write

$$E = y + \frac{Q^2}{2gA^2} \tag{2-19}$$

where Q is the total discharge and A is the whole area of the cross section. Again we find the condition of minimum specific energy by differentiation:

$$\frac{dE}{dy} = 1 - \frac{Q^2}{gA^3}\frac{dA}{dy} \tag{2-20}$$

To assign a meaning to dA/dy, we consider the effect on the area A of small increases of depth. We have

$$dA = B\,dy$$

where B is the surface width of the waterway. Then Eq. (2-20) becomes

$$\frac{dE}{dy} = 1 - \frac{Q^2 B}{gA^3} \tag{2-21}$$

and the condition for minimum specific energy, or critical flow, is that

$$Q^2 B = gA^3$$

i.e.,
$$v_c{}^2 = g\frac{A}{B} \tag{2-22}$$

The length A/B is therefore the generalization of the depth y in the case of a rectangular section. The Froude number is defined as before in this way:

$$\text{Fr} = \frac{v}{v_c} = \frac{v}{\sqrt{gA/B}}$$

i.e.,
$$\text{Fr}^2 = \frac{v^2 B}{gA} = \frac{Q^2 B}{gA^3}$$

Equation (2-21) may therefore be rewritten

$$\frac{dE}{dy} = 1 - \text{Fr}^2 \tag{2-23}$$

as for a rectangular section. It follows from Eq. (2-23) that Eq. (2-15) is true for the general section as well as for the rectangular section. Equation (2-16) cannot be similarly generalized, since it hinges on the definition of $q = Q/b$, valid only for the rectangular section; however, a corresponding equation can be obtained from Eq. (2-19) having the same significance as Eq. (2-16). Details are left as an exercise for the reader (Prob. 2.5). In the end result, the conclusions of Sec. 2.5 about the occurrence of critical flow are equally true for all shapes of channel cross section.

It has not yet been proved explicitly that in a channel of unrestricted shape, Q is a maximum for given E at critical flow. This is left as an exercise for the reader (Prob. 2.6).

Divided Flow

It may often happen, as when a culvert or sluice gate discharges into a channel of larger section, that only part of the cross section (A_m in Fig. 2.14a) is occupied by moving water. However, the hydrostatic pressure distribution within that region will be fixed by the depth y just as if the whole section were moving, and the specific energy E of the flowing water will accordingly be given by the equation

$$E = y + \frac{v^2}{2g} = y + \frac{Q^2}{2gA_m{}^2} \qquad (2\text{–}24)$$

differing from Eq. (2–19) only in the term A_m. This form of equation is applicable to such problems as the "drowned" outflow from a sluice or culvert—a problem that is treated fully in Chap. 6.

Numerical Problems

There will naturally be more computational difficulties in problems where the channel is nonrectangular than when it is rectangular. In the Appendix on Mathematical Aids some methods are suggested which may ease the difficulties of dealing with trapezoidal sections, but for completely irregular sections there is no way out but to construct the necessary E-y relationship by numerical trial.

The very common use of the trapezoidal section in artificial canals and of the complete circular section (running part-full) in drainage work raises the question of whether the necessary computations could be simplified by the development of some general (preferably dimensionless) arrangement such as that plotted in Fig. 2-11 for rectangular channels. The problem is, of course, more complex: we must allow not only for variation in the side slope angle θ (Fig. 2-14b), but also for a characteristic width dimension such as the base width b of a trapezoidal section or the diameter of a circular culvert. This was not necessary in the case of the rectangular section, in which the act of defining q eliminated the width as a significant variable.

The most elementary problem is that of determining the critical depth y_c; in a trapezoidal channel section y_c will be a function of the discharge, the base width b and the side slope angle θ. To make a dimensional analysis of the problem we would write

$$y_c = f(Q, b, g, m)$$

where $m = \cot \theta$. The convenience of choosing this particular measure of θ will become apparent in the subsequent argument. There are only two basic dimensions, length and time; therefore there are three dimensionless parameters, which by inspection are the ones given in the following equation:

$$\frac{y_c}{b} = f\left(\frac{Q^2}{gb^5}, m\right) \tag{2-25}$$

The existence of three independent parameters suggests that the relationship among y_c, Q, b, g, and m could not be represented by a single curve, but only by a family of curves. However it has been shown [2, 3] that a two-parameter representation is possible by suitably combining the three parameters in Eq. (2–25). Application of Eq. (2–22) to the trapezoidal section leads (Prob. 2.7) to the result

$$Z^2 = \frac{Q^2 m^3}{gb^5} = \frac{y_c'^3(y_c' + 1)^3}{2y_c' + 1} \tag{2-26}$$

where $y_c' = my_c/b$. It follows that a single curve of Z vs. y', as on Fig. 2-15,

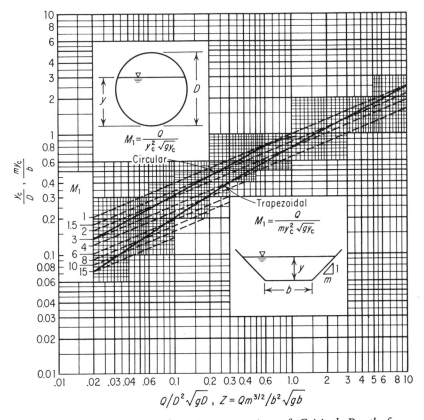

Figure 2-15. Dimensionless Representation of Critical Depth for Circular and Trapezoidal Channel Sections

conveys all the necessary information. On the same plane is drawn a curve of $Q/D^2\sqrt{gD}$ vs. y_c/D for the circular culvert, where D is the diameter. In this case there is no shape factor such as m, so that a two-parameter representation is directly available (Prob. 2.8).

These two curves, drawn as full lines in Fig. 2-15, enable us immediately to solve the following problems for a circular channel or a trapezoidal channel of known m:

1. Given y_c, and b or D, to obtain Q
2. Given Q, and b or D, to obtain y_c.

There is, however, a third problem:

3. Given y_c and Q, to obtain b or D; which could not be solved directly from the graph, but only by trial. This problem is relevant to that of determining the "choke" condition, which was discussed in Sec. 2.6; to deal with it let Y_1 be the ordinate (either y_c/D or my_c/b) in Fig. 2-15, and let X_1 be the abscissa (either Z or $Q/D^2\sqrt{gD}$). A family of lines can then be drawn having the equation

$$X_1 = M_1 Y_1^{5/2} \tag{2-27}$$

where M_1 takes a number of values. They are drawn as broken lines in Fig. 2-15.

Now if X_1 and Y_1 are given the values appropriate to the trapezoidal channel, it is easily shown that

$$M_1 = \frac{Q}{my_c^2\sqrt{gy_c}} \tag{2-28}$$

Similarly it can be shown that for the circular pipe

$$M_1 = \frac{Q}{y_c^2\sqrt{gy_c}} \tag{2-29}$$

From these results the method of solving the third problem follows immediately: from the given values of Q and y_c calculate M_1 and locate the appropriate broken line (or an interpolated line); the solution is given by the point where this line intersects the full line for the desired type of channel section.

In a number of practical problems the initial data relate not to the critical depth but to the critical specific energy—e.g., in the problem of outflow from a lake into a steep channel (Fig. 2-8) where the lake level and the channel-bed level at the outlet are given. The critical specific energy E_c is the difference between these two levels; it would be useful therefore to have a further set of curves in which E_c replaces y_c.

Such curves can be prepared by first obtaining the result (Prob. 2.9) for a trapezoidal channel

$$\frac{m}{b}\frac{v_c^2}{2g} = \frac{y_c'(y_c' + 1)}{2(2y_c' + 1)} \tag{2-30}$$

where $y'_c = my_c/b$ as before. From this it follows that

$$E'_c = \frac{mE_c}{b} = \frac{y'_c(5y'_c + 3)}{2(2y'_c + 1)} \qquad (2\text{-}31)$$

and although an explicit $Z - E'_c$ relation cannot readily be obtained, a $Z - E'_c$ curve can be plotted by calculating simultaneous values of Z and E'_c from Eqs. (2–26) and (2–31). Such a curve is drawn as a full line in Fig. 2-16,

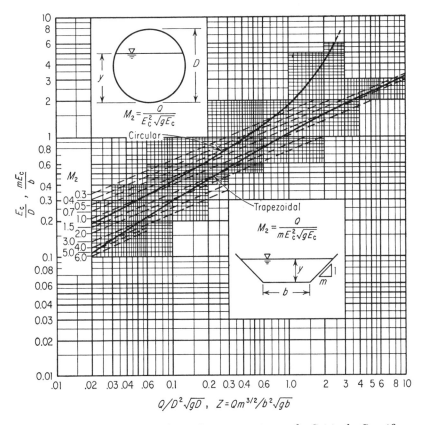

Figure 2-16. *Dimensionless Representation of Critical Specific Energy for Circular and Trapezoidal Channel Sections*

together with a curve of $Q/D^2\sqrt{(gD)}$ vs. E_c/D for a circular culvert of diameter D. The set of broken lines serves the same function as the corresponding set in Fig. 2-15; the constant M_2 is given by the equations

$$M_2 = \frac{Q}{mE_c{}^2\sqrt{gE_c}} \qquad (2\text{-}32)$$

for a trapezoidal channel, and

$$M_2 = \frac{Q}{E_c^2 \sqrt{gE_c}} \tag{2-33}$$

for a circular culvert. Figure 2-16 therefore yields the solution of the three problems corresponding to those solved by Fig. 2-15; in particular it becomes a simple matter to determine the size of a (steep) culvert or channel required to pass a certain discharge under a certain upstream head. Allowance, however, should also be made for contraction of the flow entering the channel or culvert; this matter is discussed more fully in Chap. 7.

The same considerations of similitude apply to Figs. 2-15 and 2-16 as applied to the $E/y_c - y/y_c$ curve in Fig. 2-11. Any one point on either of these two diagrams describes a large number of different flow situations, of which those having the same value of m are dynamically similar. The form of the numbers Z, M_1, and M_2 verifies the well known relationship

$$Q_r = L_r^{5/2}$$

While the curves for trapezoidal channels have much the same shape on Fig. 2-16 as on Fig. 2-15, there is a noteworthy difference between the two curves for circular culverts. On Fig. 2-15 the curve levels off towards its right-hand end and becomes asymptotic towards the line $y_c/D = 1$; this follows from the fact that y_c cannot exceed D if there is to be free surface flow within the culvert. On Fig. 2-16, however, the curve makes a sharp upturn at the right-hand end; this occurs because when $y_c \to D$, the surface width $B \to 0$ and therefore $v_c \to \infty$. Hence both E_c and Q tend to infinity.

In practice, critical flow could never occur when the water surface nears the top of the culvert because the standing waves which are a feature of critical flow would then touch the top of the culvert and cause it to run full, at least intermittently. This effect would occur when $y_c/D \to 0.9$ approx. ($E_c/D \to 1.5$ approx.), so beyond this point the curves for the circular culvert are of little practical interest.

Although the circular-culvert line on Fig. 2-16 is curved, parts of it are approximated very closely by straight lines. In the range $0 < E_c/D < 0.8$ such a straight line has the equation

$$\frac{Q}{D^2\sqrt{gD}} = 0.48\left(\frac{E_c}{D}\right)^{1.9} \tag{2-34}$$

and in the range $0.8 < E_c/D < 1.2$, the equation

$$\frac{Q}{D^2\sqrt{gD}} = 0.44\left(\frac{E_c}{D}\right)^{1.5} \tag{2-35}$$

which give results within 2 percent of those given by the curve.

All of the preceding argument, and Figs. 2-15 and 2-16, deal only with the critical-flow condition. There remains the question of whether E-y curves may be constructed in convenient dimensionless form. Investigation of the trapezoidal channel case shows that a similar reduction in the number of parameters may be made by using the composite parameters $y' = my/b$, $Z = Qm^{3/2}/b^2\sqrt{(gb)}$. It can be shown (Prob. 2.10) that

$$E' = y' + \frac{Z^2}{2y'^2(y' + 1)^2} \tag{2-36}$$

where $E' = mE/b$. (Note that in this case E and y are defined for *any* state of flow, not just for the critical condition). From Eq. (2–36) a family of E'-y' curves can be plotted, one for each value of Z. However, if Z covers a large range of values, the curves cannot all be fitted on to one diagram without a great sacrifice in precision through compression of the length scale.

A similar family of curves, having similar limitations, can be derived for the case of the circular culvert. For both the circular and the trapezoidal shapes it is clearly more profitable to relate E and y to a known length such as b, as in Eq. (2–36), than to the critical depth, as in Eq. (2–17); for in nonrectangular channels the critical depth is not usually known initially. In the case of the circular culvert it is logical therefore to plot E/D vs. y/D for a number of values of $Q/D^2\sqrt{(gD)}$ (Prob. 2.31). A complete set of curves has been plotted by Diskin [4].

Sections such as the triangular and parabolic, in which the surface width B is related to the depth y by an exponential form of equation

$$\frac{B}{B_s} = \left(\frac{y}{y_s}\right)^i \tag{2-37}$$

present no difficulty, for in this case Eq. (2–22) leads immediately to the explicit relations (Prob. 2.33)

$$\left(\frac{y_c}{y_s}\right)^{2i+3} = \frac{(i + 1)^3}{B_s^2 y_s^3} \frac{Q^2}{g} \tag{2-38}$$

and

$$\frac{E_c}{y_c} = \frac{2i + 3}{2i + 2} \tag{2-39}$$

whence if any two of E_c, y_c, and Q are given, the third may be calculated. It is clear that Eqs. (2–8) and (2–9) are the special cases of Eqs. (2–38) and (2–39) for which $i = 0$ and the section is rectangular.

Appendix on Mathematical Aids

The Specific Energy Equation—Rectangular Channels

It has been seen that the calculation of the depth y for given values of E and q involves the solution of a cubic equation of the form

$$y + \frac{a}{y^2} = b$$

where a and b are known. This form of equation may be solved in one slide-rule operation: the method will be illustrated for the particular example

$$y + \frac{40}{y^2} = 10$$

Following the usual practice, the four basic slide-rule scales are designated A, B, C, and D, starting from the top. Set the cursor hair line at 40 on the A scale; then for any position of the slide, if the hair line cuts the C scale at some number which we call y, the quotient $40/y^2$ appears on the A scale opposite the index mark of the B scale, as shown in the illustration. Thus the method simply consists of moving the slide until we find by trial a position in which the sum of y and $40/y^2$ is equal to 10.

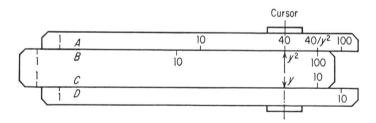

We have two solutions to find: one greater than $2/3 \times 10 = 6\frac{2}{3}$ and the other one less than this number. For the higher (subcritical) solution we could first try $y = 9$; this gives $40/y^2 = 0.493$ (note the position of the decimal point) so the sum $(y + 40/y^2)$ is too low. Clearly the slide must be moved to the left, since this operation will increase y faster than it will reduce $40/y^2$; finally we find

$$y = 9.56$$

$$\frac{40}{y^2} = 0.438$$

which is as close as we shall get to the correct solution. It is noteworthy that although we cannot read y directly off the C scale to an accuracy of three places of decimals, $40/y^2$ can be read to three places; we are therefore entitled to deduce a value of y correct to three places, and to write finally $y = 9.562$. This precision would of course be spurious unless the specific energy (10 in this case) were known to a similar degree of accuracy.

Since the subcritical depth y is substantially greater than the critical depth $y_c = 6\frac{2}{3}$, the supercritical depth will be substantially lower than y_c and we should look for a solution in the neighborhood $y = 2 - 3$. We find $y = 2.28$ when $40/y^2 = 7.72$, hence the second solution is $y = 2.28$.

The method works best with those slide rules in which the A, B, C, and D scales are all aligned with each other on one face of the rule; it will also work with those rules in which the C and D scales are aligned with A and B scales on the reverse face of the rule, but with the disadvantage that the rule must be continually turned over during the trial process.

With practice, the method will yield solutions in a few seconds.

The Specific Energy Equation—Trapezoidal Channels

The equation
$$E = y + \frac{Q^2}{2gA^2}$$

$$= y + \frac{Q^2}{2gy^2(b + my)^2}$$

must often be solved for y, given E, Q, b, and m. Since it is a quintic equation, it cannot be solved in any single operation such as described above. Satisfactory results can of course be obtained by the tabulation of trial values of y and the expressions deduced from it; however the trial process can be speeded up by observing that a certain percentage error in y will induce a smaller percentage error in $(b + my)$. We can therefore insert an approximate trial value of y in this expression, which will then be regarded as a constant. The remaining equation is of the form treated in the previous paragraph, which can be readily solved with the slide rule.

The process is best illustrated by an example. Take $Q = 1,000$, $b = 20$, $m = 2$, and $E = 7$. We therefore have

$$y + \frac{3880}{y^2(10 + y)^2} = 7$$

the subcritical solution will be a little less than 7; take $y = 6$ and insert in the term $(10 + y)^2$, making it equal to 256. We then have:

$$y + \frac{15.16}{y^2} = 7$$

whence by the previous method we find

$$y = 6.66$$

This becomes our second trial value, which we insert in $(10 + y)^2$, making it 277.5. The equation then becomes

$$y + \frac{13.96}{y^2} = 7$$

whence $y = 6.69$

which can be taken as the final answer: a third trial would make only a minute change in y. The process is at least as fast as tabulation, and usually somewhat faster. The procedure is of course equally applicable to the dimensionless form of the equation, i.e., Eq. (2–36).

The High-Speed Computer

This appendix has so far dealt with methods useful for solving individual problems as they arise, and requiring no special assistance from computing machinery or from information stored in the form of tables and charts. The understanding of these " manual " methods, as they may fairly be called, is a necessary part of every engineer's personal equipment, for they keep him in close touch with the underlying theory. He cannot, however, neglect the powerful alternative methods made possible by the high-speed computer, applied either to difficult individual problems or to the compilation of data for the solution of simpler problems in mass-production style.

Examples of both types of application are suggested by the material of this and later chapters and form useful exercises in the art of computer programming, to which most engineering students are now introduced when at the undergraduate level. Accordingly, some exercises of this sort will be listed at the end of this and some subsequent chapters, immediately after the list of set problems. In most cases the programs called for are relatively simple, particularly in the early chapters.

References

1. E. Crausse. "Sur une propriété des veines liquides horizontal en canal uniforme," *Compt. rend. de l'Acad. Française*, vol. 234 (May 26, 1952), p. 2152.
2. N. Rajaratnam and A. Thiruvengadam. "Critical Depth for Open Channels," *J. Inst. Eng. (India)* vol. 41, no. 8, part 1 (April 1961), p. 287.
3. R. M. Advani. "Critical Depth in Trapezoidal Channels," *Proc. A.S.C.E.*, vol. 88, no. HY3 (May 1962), p. 69.
4. M. H. Diskin. "Specific Energy in Circular Channels," *Water Power*, vol. 14 (July 1962), p. 270.

Problems

(*Note*: The notation 2H : 1V indicates a side slope angle of 2 units horizontally to 1 unit vertically—i.e., $m = \cot \theta = 2$.)

2.1.   For the situation shown in Fig. 2-6, adopt the viewpoint of an observer who is moving with the surge. Neglecting all squares and products of Δy and Δv, show that the continuity equation leads to the result

$$c = y \frac{\Delta v}{\Delta y}$$

and the energy equation to the result

$$g = c \frac{\Delta v}{\Delta y}$$

hence that $\qquad\qquad\qquad c^2 = gy$

2.2.   For the case of flow under a sluice gate, as in Fig. 2-7, prove that the upstream flow is always subcritical, and the downstream flow always supercritical, provided there is no energy loss.

2.3.   Obtain a discharge formula for the sharp edged weir of infinite height assuming that there is critical flow at the crest. Compare this with the standard weir discharge formula [Eq. (6–2) in Sec. 6.2] taking a discharge coefficient of 0.611, and determine the percentage difference between the two results.

2.4.   Show that when, in a rectangular channel with a horizontal bed, Fr $= 1$ and $db/dx = 0$, as indicated by Eq. (2–16), the width must be a minimum and not a maximum. (*Hint*: consider the variation of v and y with b for Fr > 1 and Fr < 1).

2.5.   Consider a channel of irregular and varying cross section, but with the low point of the section remaining at the same level along the channel. There is no resistance to the flow. Prove from Eq. (2–19) that

$$\frac{dy}{dx}(1 - \text{Fr}^2) - \text{Fr}^2 \frac{y}{B} \frac{db}{dx} = 0$$

where b is the average width of the section. Hence show that critical flow can occur at a section of minimum average width.

2.6.   Prove that for a given specific energy in a channel of unrestricted cross-sectional shape, the discharge is a maximum when the flow is critical.

2.7.   By expressing the area A and surface width B of a trapezoidal section as functions of b, y, and m, derive Eq. (2–26) from Eq. (2–22).

2.8.   If the angle β is defined as in the accompanying sketch of flow in a circular culvert, prove that if the flow is critical the following equation is true:

$$\frac{Q^2}{gD^5} = \frac{(\beta - \sin \beta \cos \beta)^3}{64 \sin \beta}$$

Choose three points on the appropriate curve of Fig. 2-15 and verify that they are in accordance with this equation.

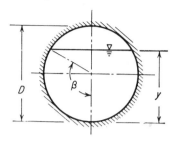

Problem 2-8

2.9.   Derive Eq. (2–30) from Eq. (2–22).

2.10.   Prove that Eq. (2–36) is true for flow in a trapezoidal channel. By differentiation from this equation, verify that Eq. (2–26) is satisfied at critical flow.

2.11.   Water is flowing at a velocity of 10 ft/sec and a depth of 10 ft in a channel of rectangular section. Find the change in depth and in absolute water level produced by (a) a smooth upward step of 1 ft; (b) a smooth downward step of 1 ft, in the channel bed. Also, (c) find the maximum allowable size of upward step for the upstream flow to be possible as specified.

2.12.   The same as Prob. 2.11 except that the upstream velocity is 15 ft/sec, the depth 2 ft, and the height of step 6 in.

2.13.   The upstream conditions are as in Prob. 2.11, and the width is 10 ft. Find the change in depth and in absolute water level produced by (a) a smooth contraction to a width of 9 ft; (b) a smooth expansion in width to 11 ft. Also (c) find the greatest allowable contraction in width for the upstream flow to be possible as specified.

2.14.   The same as Prob. 2.13 except that the upstream velocity is 15 ft/sec, the depth 2 ft, and the amount of contraction or expansion in width is 2 ft.

2.15.   The upstream conditions are as in Prob. 2.13 and there is then a smooth upward step of 2 ft in the channel bed. What expansion in width must simultaneously take place for the upstream flow to be possible as specified?

2.16.   Water flows from a lake into a steep rectangular channel 10 ft wide, and the lake level is 10 ft above the channel bed at the outfall. Find the discharge.

2.17.   Water flows from a lake into a steep rectangular channel over a broad crest which is 8 ft below the lake level; the bed then dips down 2 ft below the crest. Assuming that the flow is supercritical at this section just below the crest, find the velocity, depth, and Froude number there.

2.18.   A channel of rectangular section is to take a certain discharge Q at critical flow; prove that for the " wetted perimeter," $P = b + 2y$, to be a minimum, the water depth must be three-quarters of the width.

   A river diversion canal discharges into a concrete flume of rectangular section running down a steep slope. The walls and floor of the flume are to be 1 ft thick, and the allowance for freeboard is to be 1 ft. For $Q = 1,200$

cusecs, find the width of flume that will minimize the cross-sectional area of the concrete at the head of the steep slope, where the flow is critical. Calculate also the cross-sectional area of the concrete for this channel width.

2.19. A trapezoidal channel has a bottom width of 20 ft, side slopes of 2H : 1V, and carries a flow of 750 cusecs. (a) Find the flow depth at the head of a steep slope; (b) if there is a short but smooth transition to a rectangular section 20 ft wide just before the head of the steep slope, find the depth at the upstream and downstream ends of the transition, assuming that the specific energy remains unchanged through the transition.

2.20. The situation is as in Prob. 2.19, except that the width of the rectangular section is not yet fixed. What width must this section be in order to produce a depth of 5.20 ft at the trapezoidal end of the transition?

2.21. At the outfall from a lake into a steep channel, the channel is of trapezoidal section with a base width of 20 ft and side slopes of 2H : 1V, and the bed of the channel is 10 ft below the lake level. Find the discharge in the channel.

2.22. A lake is to discharge 2,000 cusecs into a steep channel, which at the lake outlet is to be trapezoidal in section, with side slopes of 2H : 1V. It is required that the channel bed at the outlet shall be 8 ft below the lake level: find the necessary base width of the channel and the cross sectional area of the channel from the bed up to lake level.

2.23. Water discharges from an irrigation pond through a short horizontal circular pipe 3 ft in diameter, and the entrance to the culvert is smoothly rounded. Find the discharge through the culvert for the following heights of pond level above the pipe "invert," i.e., the lowest point on the inner circumference of the pipe: 0.5 ft, 1.0 ft, 1.5 ft, 2.0 ft.

2.24. A circular pipe, discharging as in the previous problem, is to take a flow of 75 cusecs without running more than half full. Find the minimum diameter of pipe required.

2.25. A trapezoidal channel of base width 10 ft and side slopes of 2H : 1V carries a discharge of 250 cusecs at a depth of 4.8 ft. The channel is to pass under a highway, and a circular culvert is proposed with a smooth transition from the channel to the culvert; it is required that the culvert shall not act as a choke. (a) If the invert of the culvert is to be level with the channel bed, find the required culvert diameter. (b) If it is further required that the culvert "soffit" (i.e., the high point of its inner circumference) shall not be more than 6 ft above the channel bed, determine whether this condition can be met by dropping the culvert invert below the channel-bed level—also meeting the further condition that the water depth within the culvert shall not be greater than 0.8 of its diameter. Find (c) the minimum width required for a rectangular culvert.

2.26. A trapezoidal channel with a base width of 20 ft and side slopes of 2H : 1V carries a flow of 2,000 cusecs at a depth of 8 ft. There is a smooth transition to a rectangular section 20 ft wide accompanied by a gradual lowering of the channel bed by 2 ft. Find the depth of water within the rectangular section, and the change in the water surface level. What is the minimum amount by which the bed must be lowered for the upstream flow to be possible as specified?

2.27. With the situation as in the previous problem, find the amount by which the bed should be lowered to produce a drop in the water surface level of (a) zero, (b) 1.0 ft.

2.28. The upstream conditions are as in Probs. 2.26 and 2.27, and there is then a smooth contraction in width without any change in the bed level or the side slopes. Find the maximum allowable reduction in the base width if the upstream flow is to be possible as specified. If the width is actually reduced by only half this amount, find the downstream depth.

2.29. The upstream conditions are as in the previous problem, and there is then a smooth upward step in the bed with the sidewalls running on in their original alignment. Find the maximum allowable magnitude of the upward step if the upstream flow is to be possible as specified. If the actual size of the step is only half this amount, find the downstream depth and the change in the water level.

2.30. Water discharges from a lake, of surface elevation 97.0 ft above datum, into a steep channel of irregular section. At the lake outlet the channel section is as follows:

Height above datum, ft	91.0	91.5	92.0	92.5	93.0	93.5	94.0	94.5	95.0	95.5	96.0	96.5
Width, ft	0	9	15	18	19.5	22	27	29	31	34	37	40

Find the discharge in the channel.

2.31. For the circular culvert shown in Prob. 2.8, prove that

$$\frac{E}{D} = \frac{y}{D} + \frac{8Q^2}{gD^5(\beta - \sin \beta \cos \beta)^2}$$

Hence, using the result $y/D = (1 - \cos \beta)/2$, plot a curve of E/D vs. y/D for the case $Q/D^2\sqrt{(gD)} = 0.2$.

2.32. For what rate of flow in a 3-ft diameter culvert would the curve plotted in the previous problem be applicable? Given that water is flowing at this rate in the culvert, determine the critical depth and the depth alternate to $y = 2$ ft.

2.33. Derive Eqs. (2–38) and (2–39) from Eqs. (2–22) and (2–37).

2.34. A steep flume of triangular section and apex angle of 90° runs from a large tank in which the water surface level is 2 ft above the flume invert where it joins the tank. Find the discharge in the flume.

Computer Programs

C2.1. Write and operate a program to compute simultaneous values of E/D and y/D for a circular culvert, for values of $Q/D^2\sqrt{gD}$ ranging from 0.1 to 1.0 in steps of 0.1. Plot the appropriate curves of E/D vs. y/D and compare them with those of Diskin [4].

C2.2. Write and operate a program to compute simultaneous values of mE/b and my/b for a trapezoidal channel (Fig. 2-14b), for values of $Z = Qm^{3/2}/b^2\sqrt{gb}$ ranging from 0.1 to 1.0 in steps of 0.1. Plot the appropriate curves of mE/b vs. my/b for $Z = 0.3, 0.4, 0.5, 0.6$.

C2.3. Use the two sets of curves so obtained to rework Probs. 2.25 through 2.29.

Chapter **3**

The Momentum Principle in Open-Channel Flow

3.1 The Hydraulic Jump

We can begin by summarizing certain conclusions that arise out of the discussion in Secs. 2.4 and 2.5 on controls and the response to control of subcritical and supercritical flow. They are:

1. Subcritical flow is produced by downstream control, and supercritical flow by upstream control.

2. A control fixes a certain depth-discharge relationship in its own vicinity; it follows from the previous conclusion that it may also fix the nature of the flow for some distance upstream or downstream. If it does, it will therefore produce subcritical flow upstream and supercritical flow downstream.

3. A control will normally extend its influence both upstream and downstream, unless of course it stands at the end of a channel. This dual influence of a control is well illustrated by the sluice gate, discussed in Sec. 2.4.

Now the existence of both upstream and downstream influence from controls sets an interesting special problem. Consider the length of channel (Fig. 3-1) having a control at each end, typified by the sluice gates shown in

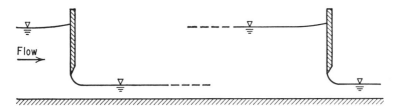

Figure 3-1. *The Conflict Between Upstream and Downstream Control*

the figure. The upstream gate produces supercritical flow downstream of itself, and the downstream gate produces subcritical flow upstream of itself.

The result is a conflict between the influence of the two controls, one of which is seeking to impose supercritical flow, the other subcritical flow, on the length of channel between them. The conflict can be resolved only if by some means the flow can pass from the one regime to the other; we have seen in Sec. 2.2 that this can happen if a specially proportioned constriction is provided in the channel, but we have no right to assume that such a feature will exist in the present situation, which is created solely by the two controls at the ends of the channel.

To solve this problem we must in fact appeal to experimental evidence, which shows quite clearly that flow can transfer abruptly from the super-critical to the subcritical condition through a feature known as the *hydraulic jump*, the change being accompanied by considerable turbulence and energy loss, as in the case of an abrupt expansion in a pipe. Figure 3-2 shows a

Figure 3-2.   *The Hydraulic Jump*

photograph of a hydraulic jump in a laboratory channel, in which the violent turbulence occasioned by the jump can be clearly seen.

The next problem is that of producing equations describing the jump. Since there is an unknown energy loss, this is clearly a case where the breakdown of the energy equation forces us to have recourse to the momentum equation.

3.2   The Momentum Function—Rectangular Channels

Although the following investigation of the momentum equation has been prompted by the particular problem of the hydraulic jump, the analysis can be made as general as possible, so as to cover other cases besides the hydraulic jump. We consider therefore the general situation shown in Fig. 3-3, in which there may or may not be an energy loss between sections 1 and 2, and there may or may not be some obstacle on which there is a drag force P_f. In Fig. 3-3 the direction of P_f is that of the force exerted by the obstacle on the flow; it is this force (not the drag on the obstacle) which is to be written in the momentum equation, dealing as it does with forces on the block of water between sections 1 and 2.

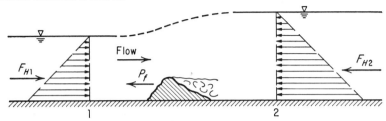

Figure 3-3. *Definition Sketch for Application of the Momentum Equation*

The following are some of the particular cases that will be covered by the general argument:

1. Energy loss $\Delta E = 0$, $P_f \neq 0$ (the sluice gate)
2. $\Delta E \neq 0$, $P_f = 0$ (the simple hydraulic jump)
3. $\Delta E \neq 0$, $P_f \neq 0$ (the hydraulic jump with its formation assisted by some obstacle in the flow such as a dentated sill).

Assigning the usual meanings to the symbols used, we have

$$\Delta(Q\rho v) = (Q\rho v)_2 - (Q\rho v)_1$$
$$= \text{Forward force on the block of water 1-2}$$
$$= F_{H_1} - F_{H_2} - P_f \tag{3-1}$$

where F_H is the hydrostatic thrust on one end of the block.

For the particular case of a rectangular channel, we need consider only a unit width of the channel, so that Eq. (3–1) becomes:

$$q\rho v_2 - q\rho v_1 = \frac{\gamma y_1^2}{2} - \frac{\gamma y_2^2}{2} - P_f$$

Where P_f is now defined as the force per unit width of the channel. This equation could be used in its present form to solve any particular problem, but it is more conveniently rearranged thus:

$$P_f = \left(\frac{q^2\rho}{y_1} + \frac{\gamma y_1^2}{2}\right) - \left(\frac{q^2\rho}{y_2} + \frac{\gamma y_2^2}{2}\right)$$

making the substitution $v = q/y$. Dividing throughout by γ, we now have:

$$\frac{P_f}{\gamma} = \left(\frac{q^2}{gy_1} + \frac{y_1^2}{2}\right) - \left(\frac{q^2}{gy_2} + \frac{y_2^2}{2}\right)$$
$$= M_1 - M_2 \tag{3-2}$$

where
$$M = \frac{q^2}{gy} + \frac{y^2}{2} \tag{3-3}$$

and is termed the *momentum function*.

Strictly speaking, the force P_f should include the resistance of the channel bed and the weight of the block of water resolved down the slope (if any). In many cases these forces are minor corrections only and may be neglected; they will not be taken account of in the present argument, but will be considered in connection with some special cases in a later chapter.

For the simple hydraulic jump, $P_f = 0$ and Eq. (3–2) can be written

$$\frac{q^2}{g}\left(\frac{1}{y_1} - \frac{1}{y_2}\right) = \tfrac{1}{2}(y_2{}^2 - y_1{}^2)$$

i.e.,
$$\frac{q^2}{gy_1y_2} = \tfrac{1}{2}(y_2 + y_1)$$

The substitution $q = v_1 y_1$ leads to

$$\frac{v_1{}^2}{g} = \frac{1}{2}\frac{y_2}{y_1}(y_2 + y_1)$$

or
$$\frac{v_1{}^2}{gy_1} = \mathrm{Fr}_1{}^2 = \frac{1}{2}\frac{y_2}{y_1}\left(\frac{y_2}{y_1} + 1\right) \tag{3-4}$$

which is the well-known equation of the hydraulic jump. The purpose of dividing the preceding equation by y_1 was to achieve a result expressed in terms of dimensionless numbers; again the Froude number Fr plays a key role. Equation (3–4) is a quadratic in y_2/y_1, whose solution is given by

$$\frac{y_2}{y_1} = \tfrac{1}{2}(\sqrt{1 + 8\mathrm{Fr}_1{}^2} - 1) \tag{3-5}$$

These two equations between them cover adequately the usual forms in which the problem occurs in practice; a third form can easily be deduced (Prob. 3.1) to deal with those cases in which only the downstream conditions are known. It is:

$$\frac{y_1}{y_2} = \tfrac{1}{2}(\sqrt{1 + 8\mathrm{Fr}_2{}^2} - 1) \tag{3-6}$$

In general, since each of Eqs. (3–4) through (3–6) contains three independent quantities, it is necessary to know two of them in advance; from these the third may be calculated. For example, if the upstream conditions (v_1 and y_1) are known, the downstream depth may be calculated. But it is most important to realize that this downstream depth is *caused* not by the upstream conditions but by some control acting further downstream. If this control produces the required depth y_2, a jump will form; otherwise it will not. The corresponding depths y_1 and y_2 are said to be *conjugate*, or *sequent*, to each other (both terms being in common use).

Since a jump will form only if there is a certain special relationship among v_1, y_1, and y_2, we are not really entitled to assume that a jump will always form in the situation shown in Fig. 3-1. We can visualize the event by imagining that in the first instance the correct relationship obtains, so that there is a clearly defined jump in the reach between the two controls. If y_2 now increases, the jump will be forced upstream until the outlet of the upstream gate is " drowned "; if y_2 decreases, the jump will be forced downstream until the upstream supercritical flow strikes the downstream gate directly.

Even if y_2 has the correct value, conjugate to y_1, it would seem that the jump could form anywhere in the channel between the gates and therefore would not have any one stable position, but would tend to drift up and down the channel. This unstable behavior can in fact be observed in the laboratory; we shall see in a later chapter that flow resistance tends to make the jump more stable, but over short distances, where the effect of flow resistance is slight, special devices may still be required to hold the jump stable in one position, in those cases where it is desirable to do so.

There is little that need be said about further applications of Eq. (3–2). Clearly it is applicable whether or not there is energy loss; it may, for instance, be used to calculate the force on a sluice gate, or the force on a step or sidewall contraction such as occurs in the problems dealt with in the last chapter. It can readily be applied to cases which involve energy loss and a force on some intermediate obstacle, such as flow around bridge piers, and the dentated sill as an aid to hydraulic-jump formation. However there are complications in these latter cases which make it desirable to postpone their full treatment until a later chapter.

Just as in pipe flow, there are certain cases (e.g., Prob. 3.10) where there may appear to be a conflict between the energy and momentum equations. The conflict can always be resolved by careful thought about the scope and applicability of each equation, as discussed in Sec. 1.6.

3.3 The M-y Relationship

There is a good reason for casting the momentum equation into the particular form of Eq. (3–2) and introducing the momentum function M. It is that for a given value of q, an M-y curve may be plotted (Fig. 3-4) which is similar in many ways to the E-y curve, except that it has only one asymptote, $y = 0$. It has upper and lower limbs representing subcritical and supercritical flow respectively, for it is easily shown (Prob. 3.2) that at the crest C, where M is a minimum, the flow is critical.

On this curve the solution of the hydraulic-jump problem can readily be displayed, for a vertical line will clearly cut the curve at points representing conjugate depths. By setting the E-y curve for the same q alongside the M-y curve (Fig. 3-4), the energy loss in the hydraulic jump can be shown as a lateral displacement on the E-y curve. The inverse case of the sluice gate can be traced on the same pair of curves, indicating the thrust on the gate as a

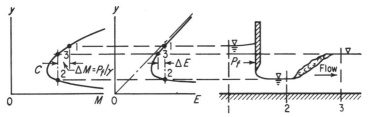

Figure 3-4. *Behavior of the Specific Energy and Momentum Function in Flow under a Sluice Gate and through a Hydraulic Jump*

displacement on the M-y curve. The situation which is actually traced on Fig. 3-4 is that of a sluice gate followed by a jump, so that:

$$E_1 = E_2, \text{ but } M_1 \neq M_2$$

$$M_2 = M_3, \text{ but } E_2 \neq E_3$$

Other conclusions brought out by the form of the M-y curve are:

1. Of two conjugate depths, y_1 and y_2, one must be subcritical and the other supercritical—a fact that can also be deduced from the algebraic argument (Prob. 3.3).

2. If the upstream depth y_1 is increased (q remaining the same) then its conjugate depth y_2 will decrease, and vice versa. This fact is of some importance in the location of the hydraulic jump when flow resistance is taken into account. The subject will be treated in Chaps. 4 and 5.

Dimensionless Treatment of the M Function

As with the specific energy equation, the momentum function

$$M = \frac{q^2}{gy} + \frac{y^2}{2} \tag{3-3}$$

can be reduced to dimensionless form. We divide throughout by y_c^2, obtaining

$$\frac{M}{y_c^2} = \frac{q^2}{gyy_c^2} + \frac{y^2}{2y_c^2}$$

whence, setting $M' = M/y_c^2$, $y' = y/y_c$, and making the substitution $q^2 = gy_c^3$, we obtain

$$M' = \frac{1}{y'} + \frac{y'^2}{2} \tag{3-7}$$

which may be plotted, as in Fig. 3-5, and used in the same general way as the E'-y' curve. An interesting fact emerges from the comparison of this equation with its specific energy counterpart:

$$E' = y' + \frac{1}{2y'^2} \qquad (2\text{--}17)$$

It is that M' is the same function of y' as E' is of $1/y'$, and vice versa. The effect is that Fig. 2-11, the E'-y' curve, might be made to do double duty by serving as an M'-$1/y'$ curve. The above development of M' as a function of y', including the observation that Eqs. (3–7) and (2–17) are related, is also due to Crausse (Ref. 1, Chap. 2).

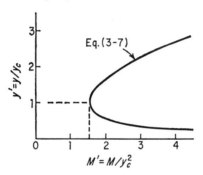

Figure 3-5.   Dimensionless Form of the Momentum Function

This reciprocal relation between specific energy and the momentum function is not confined to their dimensionless forms. It may be observed also in the normal forms of these functions; for example the equation for flow under a sluice gate may be put into a form very similar to that of the hydraulic jump (Prob. 3.4).

3.4   Nonrectangular Channel Sections

The momentum function may be adapted without difficulty to channels of nonrectangular section. As in the case of specific energy the channel must be treated as a whole, since a "unit width" is no longer representative of the whole area. The first term of the function is simply

$$\frac{Q\rho v}{\gamma} = \frac{Qv}{g} = \frac{Q^2}{gA}$$

and the second term—the hydrostatic thrust divided by γ—becomes

$$A\bar{y}$$

where $\bar{y}$ (Fig. 3-6) is the depth from the surface to the centroid of the section.

The whole function therefore becomes

$$M = \frac{Q^2}{gA} + A\bar{y} \qquad (3\text{--}8)$$

The second term is the moment of area of the section about the surface, and for irregular shapes must be obtained by numerical or graphical integration.

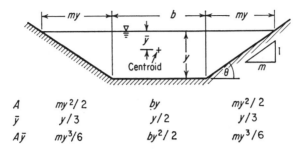

A	$my^2/2$	by	$my^2/2$
$\bar{y}$	$y/3$	$y/2$	$y/3$
$A\bar{y}$	$my^3/6$	$by^2/2$	$my^3/6$

Figure 3-6. The Term $A\bar{y}$ for the Trapezoidal Section

In the case of the trapezoidal section (Fig. 3-6) this term is obtained simply by adding the components representing the two triangular elements and the center rectangular element. It is

$$A\bar{y} = my^2\,\frac{y}{3} + by\,\frac{y}{2}$$

$$= \frac{y^2}{6}\,(2my + 3b) \qquad (3\text{--}9)$$

It is unnecessary to evaluate $\bar{y}$ separately, since we are concerned only with the whole term $A\bar{y}$.

It is readily shown (Prob. 3.2) that the condition for minimum M in Eq. (3–8) is given by the equation for critical flow

$$v^2 = g\,\frac{A}{B} \qquad (2\text{--}22)$$

which was derived from energy considerations in Sec. 2.7. Critical flow is therefore described by the same equations whether it is defined as the condition of minimum specific energy or of minimum value of the momentum function.

Divided Flow

When only part of the cross section is occupied by moving water, the hydrostatic thrust term $A\bar{y}$ will be unchanged [cf. the term y in Eq. (2–24)] but the momentum term Q^2/gA must now relate to the area A_m of moving water, not the total area A_t. Equation (3–8) therefore becomes

$$M = \frac{Q^2}{gA_m} + A_t\bar{y} \tag{3–10}$$

corresponding to Eq. (2–24) for specific energy.

Numerical Problems

Hydraulic-jump problems in rectangular channels are readily solved by Eqs. (3–4) and (3–5); however there is no simple counterpart of these equations for nonrectangular channels. As in the case of specific energy problems, the circular channel involves one more variable (D) and the trapezoidal channel two more (b and m) than the rectangular channel, and in the latter case the number of independent parameters can be reduced by one through a suitable combination of m with the dimensioned variables. Setting $y' = my/b$ as before, Eqs. (3–8) and (3–9) lead (Prob. 3.5) to the result

$$M' = \frac{Mm^2}{b^3} = \frac{Z^2}{y'(y'+1)} + \frac{y'^2(2y'+3)}{6} \tag{3–11}$$

where $Z^2 = Q^2m^3/gb^5$ as before. As in Eq. (2–36), there are only three independent dimensionless parameters.

Equation (3–11) suggests that a hydraulic-jump equation might be deducible as a function of three dimensionless numbers $-y_2/y_1, y_1'$, and a generalization of Fr_1. For this last it is convenient to use $Z_{M_1} = Z/y_1'^{5/2} = Q/my_1^2\sqrt{gy_1}$. Such an equation has been derived by Massey [1]; setting $r = y_2/y_1$ it can be written:

$$r^4 + r^3\left(\frac{5}{2y_1'}+1\right) + r^2\left(\frac{3}{2y_1'}+1\right)\left(\frac{1}{y_1'}+1\right) + r\left[\left(\frac{3}{2y_1'}+1\right)\frac{1}{y_1'} - \frac{3Z_{M_1}^2y_1'}{y_1'+1}\right]$$
$$- 3Z_{M_1}^2 = 0 \tag{3–12}$$

Massey gives the solution of this equation as a family of curves—y_2/y_1 vs. Z_{M_1} for various values of y_1'. Without these curves, direct trial solution of Eq. (3–12) would be lengthy, and it would probably be better to use the methods suggested in the Appendix to this chapter.

Equation (3–12), like Eq. (3–5), is applicable when the upstream conditions are known initially. Alternative solutions can readily be derived which are applicable when the downstream conditions are known (Probs. 3.33 and 3.34).

For the circular culvert, uncomplicated by a shape factor such as m, a hydraulic-jump equation should clearly be expressible in terms of three dimensionless numbers, y_2/y_1, $Q/y_1^2\sqrt{gy_1}$, and y_1/D. The equation does not have a simple explicit form, but a graphical solution has been presented by Thiruvengadam [2]. Dimensionless M-y curves can also be prepared for the circular culvert; details are left as an exercise for the reader (Probs. 3.26 and 3.27).

Sections described by the exponential equation

$$\frac{B}{B_s} = \left(\frac{y}{y_s}\right)^i \tag{2-37}$$

give rise to quite a simple form of the momentum function

$$M = \frac{Q^2(i+1)}{gB_sy}\left(\frac{y_s}{y}\right)^i + \frac{B_sy^2}{(i+1)(i+2)}\left(\frac{y}{y_s}\right)^i \tag{3-13}$$

(Prob. 3.30), but an explicit form of hydraulic-jump equation does not emerge unless i is a whole number (Prob. 3.31).

3.5 Unsteady Flow: Surges and Bores

There is one application of the momentum principle to unsteady flow that is simple enough to warrant its inclusion in this early chapter. It is that of the abrupt wave front, or surge, discussed in Secs. 1.2 and 2.3. When the surge is of tidal origin it is usually termed a *bore*; however, no difference in principle is indicated by this difference in name.

The analysis is easily accomplished by reducing the situation to one of steady flow, as in Sec. 2.3. However the present more general case is to cover surges of finite size and finite energy loss, so the momentum equation must be used instead of the energy equation. Figure 3-7a shows the surge as seen

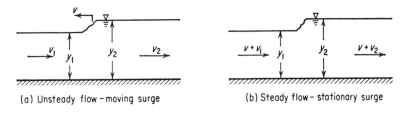

(a) Unsteady flow – moving surge   (b) Steady flow – stationary surge

Figure 3-7.   *Reduction of the Moving surge to a Stationary Surge, or Hydraulic Jump*

by an observer on the bank, as unsteady flow; Fig. 3-7b shows it as seen by an observer riding along with the surge, as a steady-flow case, since the surge is now stationary. The second picture is produced from the first by superimposing on the whole system a velocity equal and opposite to that of the surge v; the stream velocities are therefore increased to $(v + v_1)$ and $(v + v_2)$ as shown.

The state of affairs shown in Fig. 3-7b is simply the hydraulic jump, and it may be analyzed as such. From Eq. (3–4) we can write

$$\frac{(v + v_1)^2}{g y_1} = \frac{1}{2} \frac{y_2}{y_1} \left(\frac{y_2}{y_1} + 1 \right) \tag{3–14}$$

The continuity equation can also be applied directly to this situation, yielding the result

$$(v + v_1) y_1 = (v + v_2) y_2 \tag{3–15}$$

Alternatively this equation could be derived from the viewpoint of Fig. 3-7a by equating the difference in discharges, $v_1 y_1 - v_2 y_2$, to the rate of increase of water volume between the sections 1 and 2. With rather more difficulty, Eq. (3–14) could also be derived from the viewpoint of Fig. 3-7a (Prob. 3.35) but the effort involved is enough to make its avoidance by the device of Fig. 3-7b worthwhile.

It should be noted that in these equations the positive direction of v has been defined as opposite to that of v_1 and v_2; this convention is the easiest one to use in the case of a surge running upstream, as in Fig. 3-7, but when the surge is running downstream it may be convenient to define v, v_1, and v_2 as all having the same positive direction, in which case the sign of v would be reversed in Eqs. (3–14) and (3–15).

We now have two equations and a total of five physical quantities, v, v_1, v_2, y_1, y_2. In the case of a surge traveling upstream it is likely that only the upstream conditions would be known, i.e., v_1 and y_1. This leaves three unknowns, so that one further equation is required. A surge of this sort would be caused by a sudden blockage of the channel further downstream, such as a landslide, and if one were asked to predict the surge characteristics after such an event it would be natural to ask how big the landslide was, or to what extent the channel had been blocked. The information required therefore relates essentially to the relative size of the "absolute" discharges $v_1 y_1$ and $v_2 y_2$, the latter of these being anything from zero (if the channel is completely blocked) up to nearly $v_1 y_1$, if the partial blockage of the channel is small.

This information may be comparatively difficult to extract from the known data; this problem will be dealt with in detail in a later chapter. In the meantime it need only be pointed out that one further equation is required, and that one of the simplest forms this equation may take is a relationship between the discharges relative to the bank:

$$v_1 y_1 = k v_2 y_2 \tag{3–16}$$

where k is a known constant.

In general Eqs. (3–14) through (3–16) lead to a quartic in v, which must be solved by numerical trial. When the third equation happens to be of the form of Eq. (3–16), v_2 may be eliminated between Eqs. (3–15) and (3–16), leading to the results:

$$v = \frac{v_1 y_1 - v_2 y_2}{y_2 - y_1} = \frac{q_1 - q_2}{y_2 - y_1} \tag{3–17}$$

an equation of some general interest, which will be encountered in the later treatment of unsteady flow in Chaps. 8 and 9.

An examination of Eq. (3–14) leads to some interesting general conclusions. They are:

1. The term $(v + v_1)$ is the velocity of the surge relative to the water upstream. The form of the equation clearly shows that if $y_2/y_1 > 1$, then $(v + v_1) > \sqrt{gy_1}$. If the surge is small, i.e., if $y_2/y_1 \rightarrow 1$, then $(v + v_1) \rightarrow \sqrt{gy_1}$, confirming the result of Sec. 2.3 that $\sqrt{gy}$ is the velocity of a small surge relative to the body of water across which it moves.

2. Since $v + v_1 > \sqrt{gy_1}$ when $y_2 > y_1$, the surge could move upstream even when the upstream flow is supercritical ($v_1 > \sqrt{gy_1}$). This fact emphasises an important qualification to the previous statement (Sec. 2.3) that a disturbance cannot make its way upstream in supercritical flow. It has already been pointed out that this statement is true only of small disturbances: Eq. (3–14) makes it abundantly clear that a disturbance *can* make its way upstream against supercritical flow, *provided it is large enough*. However in so doing it will transform the flow to subcritical.

3. A further conclusion from the result $(v + v_1) > \sqrt{gy_1}$ is that the surge will overtake and absorb any small disturbances that may exist on the surface of the upstream water. On the other hand, since $(v + v_2) < \sqrt{gy_2}$ (Prob. 3.3) the surge is traveling more slowly, relative to the downstream water, than a small disturbance on the surface of that water. Hence any such disturbance will overtake the surge and be absorbed into it. It is this capacity for absorbing random disturbances on both sides of the surge that lends the surge its stable and self-perpetuating quality.

4. The fact that $(v + v_1) > \sqrt{gy_1}$ sets a lower limit for the value of v suggests a first approximation in the process of solving Eqs. (3–14) through (3–16) by trial. Theoretically an upper limit is set by the fact that $(v + v_2) < \sqrt{gy_2}$, but this limit can hardly be used as a guide in computation since v_2 and y_2 are both unknown in the first instance.

Appendix on Mathematical Aids

The Momentum Function in Rectangular Channels

The need for solving a cubic equation does not arise so often with the momentum function as it does with the specific energy equation. The reason is that in the derivation of the hydraulic-jump equation the cancellation of the term $(y_2 - y_1)$ has reduced the equation $M_1 = M_2$ (a cubic) to the essentially quadratic form of Eq. (3–4). However there are some cases (as in Probs. 3.28 and 3.29) where it is necessary to determine the depth y when M is given—that is, to solve a cubic equation of the form

$$\frac{a}{y} + y^2 = b$$

In the Appendix to Chap. 2 it was explained how to solve the specific energy equation by one slide-rule operation; with this as a guide it is left as an exercise for the reader (Prob. 3.7) to devise a similar technique for solving the momentum equation as given above.

The Momentum Function in Nonrectangular Channels

Problems corresponding to Probs. 3.28 and 3.29 for nonrectangular sections must normally be solved by trial; in the case of the trapezoidal section the appropriate equation is a quintic. The trial process may often be made a littler simpler by using the technique explained in the Appendix to Chap. 2— that of simplifying the whole equation by inserting the first trial values of y in those parts of the equation where the influence of y is slight, and retaining y as an unknown in the other parts. In the momentum equation there is no fixed rule for distinguishing these parts from each other, although it is usually satisfactory to follow the procedure laid down in the Appendix to Chap. 2, of inserting trial values of y in the second term of such expressions as

$$A\bar{y} = \frac{y^2}{6}(2my + 3b)$$

and leaving the other term, $y^2/6$, in its original form. This procedure leads to an equation of the form

$$\frac{a}{y} + y^2 = b$$

which can be solved by the slide-rule technique of Prob. 3.7.

In some cases (e.g., Prob. 3.25) the work can be made much simpler because one whole term of Eq. (3–8)—in this case the second term, $A\bar{y}$—is very much smaller than the other. Hence trial values of y can be inserted in the whole of this term, so that the remaining equation is merely a quadratic. In this case it will be found that the trial process converges very quickly and simply.

Small Differences between Large Numbers

To obtain the energy loss across a hydraulic jump in a rectangular channel, we have to compute $E_1 - E_2$ when $M_1 = M_2$. It will be found that unless $\text{Fr}_1{}^2$ and y_2/y_1 are quite large, the difference $E_1 - E_2$ is substantially smaller than either E_1 or E_2, making it difficult to calculate the difference with reasonable accuracy.

This is a common problem in computing work; the obvious but clumsy remedy is to calculate E_1 and E_2 to a higher degree of precision (e.g., with tables) than is required for the difference $E_1 - E_2$. A much better way out of the difficulty is to rearrange the algebra so that $E_1 - E_2$ is expressed in a form which involves products rather than differences. We have

$$E_1 - E_2 = \left(y_1 + \frac{v_1{}^2}{2g}\right) - \left(y_2 + \frac{v_2{}^2}{2g}\right)$$

$$= (y_1 - y_2) + \frac{q^2}{2g}\left(\frac{1}{y_1{}^2} - \frac{1}{y_2{}^2}\right)$$

The aim is to express $(E_1 - E_2)$ as a product: it is in line with this aim to factorize the expression as far as possible, e.g., by taking out the factor $(y_1 - y_2)$, leading to:

$$E_1 - E_2 = (y_1 - y_2)\left[1 - \frac{\text{Fr}_1{}^2}{2}\frac{y_1(y_1 + y_2)}{y_2{}^2}\right]$$

making the substitution $\text{Fr}_1{}^2 = q^2/gy_1{}^3$. We now insert the condition $M_1 = M_2$ by using the hydraulic-jump equation:

$$\text{Fr}_1{}^2 = \frac{1}{2}\frac{y_2}{y_1}\left(\frac{y_2}{y_1} + 1\right)$$

Setting $r = y_2/y_1$ and simplifying the expression within the second bracket, we have

$$E_1 - E_2 = (y_1 - y_2)\left[1 - \frac{r(r+1)}{4}\frac{(r+1)}{r^2}\right]$$

or
$$\frac{E_1 - E_2}{y_1} = (1 - r)\left[1 - \frac{(r+1)^2}{4r}\right]$$

$$= \frac{(r-1)^3}{4r} \tag{3–18}$$

Now this has not been an exercise in algebraic manipulation for its own sake: the aim has been to express a certain small quantity, not as the difference of two larger quantities, but as the product of other small quantities, and this aim is realized in the above equation. Numerical values can be inserted in this equation and a result obtained no less precise than the original data of the problem.

Although the preceding analysis was set up to deal with a particular difficulty, it is worthy of remark that computing work can often be made quicker and simpler by some rearrangement of the algebra before the computing is actually started. It is good practice, therefore, before doing any computation on an algebraic expression to inspect it with a view to possible rearrangement.

The problem corresponding to the one treated above is that of calculating $M_1 - M_2$ when $E_1 = E_2$, and here a similar difficulty arises. It is left as an exercise for the reader (Probs. 3.8 and 3.9) to develop expressions for this case which lend themselves to more precise computation. The development of such expressions may appear to require a considerable taste for algebraic manipulation, but anyone engaged in computing work must to some extent cultivate such a taste.

It will be found useful in working through subsequent problems to compare the results obtained by straight subtraction with those obtained by the results of this Appendix and Probs. 3.8 and 3.9.

The same difficulty can occur in analyzing a hydraulic jump in which the downstream conditions are given initially, and Eq. (3–6) is therefore applicable.

$$\frac{y_1}{y_2} = \tfrac{1}{2}(\sqrt{1 + 8\mathrm{Fr}_2{}^2} - 1) \tag{3–6}$$

Since $\mathrm{Fr}_2{}^2$ is always substantially less than one, the term $\sqrt{(1 + 8\mathrm{Fr}_2{}^2)}$ may be close to unity, and again we have the case of a small difference between two large numbers, with a consequent loss in precision. The remedy in this case lies in series expansion, which is often of great assistance in computing work. Using the binomial series, we can write

$$\sqrt{1 + 8\mathrm{Fr}_2{}^2} = (1 + 8\mathrm{Fr}_2{}^2)^{1/2} = 1 + \tfrac{1}{2} \cdot 8\mathrm{Fr}_2{}^2 + \frac{\tfrac{1}{2}(-\tfrac{1}{2})}{2!} 64\mathrm{Fr}_2{}^4$$

$$+ \frac{\tfrac{1}{2}(-\tfrac{1}{2})(-1\tfrac{1}{2})}{3!} 512\mathrm{Fr}_2{}^6 + \cdots$$

$$= 1 + 4\mathrm{Fr}_2{}^2 - 8\mathrm{Fr}_2{}^4 + 32\mathrm{Fr}_2{}^6 - \cdots$$

whence $$\frac{y_1}{y_2} = 2\mathrm{Fr}_2{}^2 - 4\mathrm{Fr}_2{}^4 + 16\mathrm{Fr}_2{}^6 - \cdots \tag{3–19}$$

so that the result is expressed as sums and differences of small numbers. The method need only be used when $\mathrm{Fr}_2{}^2$ is very small, less than, say, 0.05, and in this case only two, or at most three, terms of the series would be required.

A similar approximation can be used to determine the upstream Froude number of a hydraulic jump whose height, $\Delta y = y_2 - y_1$, is small compared with y_1. Application of the binomial series to Eq. (3–4) then leads to the result

$$\mathrm{Fr}_1 = 1 + \frac{3\Delta y}{4y_1} \tag{3–20}$$

which can also be used, by adding the appropriate terms, to give the velocity of the surge wave of small height.

Computer Programs

The formation of the hydraulic jump in nonrectangular channels lends itself very well to solution by means of data prepared on the high-speed computer, and a number of exercises of this sort are given at the end of this chapter.

Up until now it has been assumed that in hydraulic-jump problems the discharge, the channel dimensions, and either upstream or downstream depth are known initially. In practice, however, the problem may occur in many other forms. A typical one is that in which Q and the head loss ΔE are known initially—i.e., the problem is to determine the channel size so that the jump will dissipate a certain amount of energy. In this case the downstream depth y_2 may be known also. A solution to this problem has been given by Advani [3] but the power and flexibility of the high-speed computer make it quite easy to solve this or any other form of the problem without recourse to the algebraic solutions given by investigators such as Advani. The reader will see in the exercises given that quite simple extensions to a computer program will enable it to handle unusual forms of the problem such as the one discussed here.

Another type of problem well suited to computer solution is the development of a surge by a prescribed amount of sluice gate movement, as in Probs. 3.40 and C3.9.

References

1. B. S. Massey. "Hydraulic Jump in Trapezoidal Channels—an Improved Method," *Water Power*, vol. 13 (June 1961), p. 232.
2. A. Thiruvengadam. "Hydraulic Jump in Circular Channels," *Water Power*, vol. 13 (December 1961), p. 496.
3. R. M. Advani. "A New Method for Hydraulic Jump in Circular Channels," *Water Power*, vol. 14 (September 1962), p. 349.

Problems

3.1.   From either of the hydraulic jump Eqs. (3–4) and (3–5), derive the equation

$$\frac{y_1}{y_2} = \frac{1}{2}(\sqrt{1 + 8F_2{}^2} - 1) \tag{3-6}$$

3.2.   Prove that for a given discharge, the momentum function has its minimum value at critical flow, whether the channel is rectangular or nonrectangular in section. (*Hint*: for the latter case, prove first that $d(A\bar{y})/dy = A$.)

3.3.   Prove from the hydraulic-jump equation for a rectangular channel that the upstream flow is always supercritical and the downstream flow always subcritical.

3.4.   For the case of flow through a sluice gate, prove that

$$\mathrm{Fr}_1{}^2 = \frac{2}{\dfrac{y_1}{y_2}\left(\dfrac{y_1}{y_2} + 1\right)}$$

where the subscripts 1 and 2 apply to the upstream and downstream sections respectively. Two alternative proofs are to be obtained,
    (a) by working directly from the equation $E_1 = E_2$;
    (b) by modifying Eq. (3–4) as indicated by the reciprocal relationship between Eqs. (2–17) and (3–7)

3.5.   Derive Eq. (3–11) from Eqs. (3–8) and (3–9).

3.6.   Verify that Eq. (3–4) is the special case of Eq. (3–12) for which $m = 0$.

3.7.   Develop a method for solving equations of the form

$$\frac{a}{y} + y^2 = b$$

in one slide-rule operation.

3.8.   Water flows in a rectangular channel in which there is a smooth step (up or down) in the channel bed, without change in width. Prove that the force on the step, per unit width of channel, in the direction of the flow is equal to

$$\tfrac{1}{2}\gamma y_1{}^2 \frac{(1 - r)^3 + 4r\Delta z'}{1 + r}$$

(subscripts $1 =$ upstream; $2 =$ downstream) where $r = y_2/y_1$, $\Delta z' = \Delta z/y_1$, and $\Delta z =$ the change in the channel bed level, measured positive upwards.

3.9.   Water flows in a rectangular channel in which the width changes smoothly from b_1 to $b_2 = kb_1$, with no change in bed level. Prove that the force on the sidewall transition in the direction of flow is equal to

$$\tfrac{1}{2}\gamma b_1 y_1{}^2 \frac{1 - k^2 r^3 - 3kr(1 - r)}{1 + kr}$$

where $r = y_2/y_1$.

3.10.   As shown in the figure, water flows at a depth y and velocity v in a rectangular channel which is closed by a sluice gate, the water flowing smoothly down into a vertical shaft. Calculate the rise, Δy, in the water surface just upstream of the sluice gate, (a) by setting $E_1 = E_2$; (b) by setting $M_1 = M_2$ and assuming $\Delta y/y$ is small. Which answer is the correct one, and why?

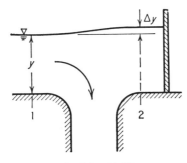

Problem 3-10

3.11. A sluice gate is mounted in a rectangular horizontal channel 10 ft wide. A short distance upstream the depth is 10 ft; the gate opening is 3 ft, and the contraction coefficient is 0.61. Calculate the thrust on the gate (a) by the correct method, (b) by assuming that the pressure distribution on the gate is hydrostatic. Verify that the difference between the two results is of the correct sign.

3.12. For a situation as in parts (a) and (b) of Prob. 2.11, find the magnitude and sense of the force per unit width (in the direction of flow) on the face of the step.

3.13. With the situation as in Prob. 2.11 the step has the maximum allowable height as in part (c) of that problem, and after a short distance the bed drops gradually to its original level. The flow passes to supercritical downstream; find the net force per unit width on the step, in the direction of flow.

3.14. The same as Prob. 3.12 except that the data are taken from Prob. 2.12.

3.15. The same as Prob. 3.13, except that the data are taken from Prob. 2.12, and the flow passes to subcritical downstream of the step.

3.16. For a situation as in parts (a) and (b) of Prob. 2.13, find the magnitude and sense of the force (in the direction of flow) on the sidewall transitions.

3.17. With the situation as in Prob. 2.13, the contraction has the minimum allowable width as in part (c) of the problem, and after a short distance the width expands gradually to its original value. The flow passes to supercritical downstream; find the net force in the direction of flow on the sidewall transitions.

3.18. The same as Prob. 3.16 except that the data are taken from Prob. 2.14.

3.19. The same as Prob. 3.17, except that the data are taken from Prob. 2.14, and the flow passes to subcritical downstream.

3.20. Water flows at a velocity of 20 ft/sec and a depth of 3 ft in a rectangular channel 20 ft wide. Find the downstream depth necessary to form a jump, the head loss, and the horsepower dissipated in the jump.

3.21.   Water discharges at the rate of 10,000 cusecs over a spillway 40 ft wide into a stilling basin of the same width. The lake level behind the spillway is 200 ft above datum; the river level downstream is 100 ft above datum. Assuming no energy is dissipated in the flow down the spillway, find the basin invert level required for a hydraulic jump to form within the basin.

3.22.   In the situation of Prob. 3.21, it is found impossible to set the basin invert level lower than 75 ft above datum. What width must the stilling basin be so that a jump can form within the basin?

3.23.   The situation is as in Prob. 3.22, but the basin is to be left at its original width of 40 ft. Jump formation is to be accomplished by placing a raised sill at the downstream end of the basin. Assuming critical flow over the sill, find the required level of the sill so that a jump can form within the basin.

3.24.   Water discharges from a lake under a sluice gate into a stilling basin 20 ft wide with an invert level 50 ft below the lake level. At the end of the basin is a sill 9 ft high, over which the water discharges into a steep river bed. For $y_1 = 2, 4, 6, 8$ and 10 ft, where y_1 is the depth just downstream of the gate, calculate the actual depth just upstream of the sill, and the depth required to form a hydraulic jump within the basin. Determine the range of discharges for which there is insufficient downstream depth to form a jump, and find the height of sill required if a jump is to form at all discharges.

3.25.   A hydraulic jump is to be formed in a trapezoidal channel with a base width of 20 ft and side slopes of 2H : 1V. The downstream depth is 8 ft and the discharge is 1,000 cusecs: find the upstream depth, the head loss, and the horsepower dissipated in the jump. [Do the first part of the problem in two ways: solve Eq. (3–12) by trial, and alternatively use the procedure given in the Appendix.]

3.26.   Water flows at a depth y in a circular culvert of diameter D, as shown in the figure. Prove that the product $A\bar{y}$ is equal to

$$\frac{D^3}{24}(3 \sin \beta - \sin^3 \beta - 3\beta \cos \beta)$$

and hence construct an M-y curve for a flow of 300 cusecs in a culvert 10 ft in diameter. A hydraulic jump is to be formed within the culvert; find the limiting value of upstream depth if the downstream depth is not to exceed 75 percent of the diameter. Is this limiting value a maximum or minimum?

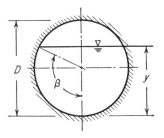

Problem 3-26

3.27. Choose three dimensionless numbers which are suitable generalizations of M, y, and Q in a circular culvert. Relabel accordingly the scales of the M-y graph drawn for the previous problem, and determine the appropriate value of the dimensionless Q number, so that this curve now shows the relationship among the three dimensionless numbers you have chosen. For what discharge could this curve be used in a 20-ft-diameter culvert, and what are the ratios between corresponding values of M and y in the 20-ft and 10-ft-diameter culverts?

3.28. A bridge has piers 2 ft wide at 20-ft centers. A short distance upstream the depth of the river is 10 ft and its velocity 10 ft/sec. Assuming that the piers have a drag coefficient, as defined by Eq. (1–22), of (a) 1.5, (b) 2.0 (with A and v based on the upstream flow) find the depth when the flow has gone far enough downstream for the local disturbances caused by the piers to be evened out. Neglect the bed slope and bed resistance.

3.29. Two rows of "sluice blocks" are to be installed in a stilling basin, as shown in the figure, in order to assist the formation of the hydraulic jump within the basin. It is found that such an arrangement of blocks has an effective

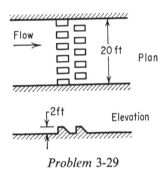

Problem 3-29

drag coefficient of 0.3, based on the upstream velocity and on the combined frontal area of the blocks—provided that the upstream depth is no less than the height of the blocks. If the discharge is 1,000 cusecs, and the upstream depth 2.0 ft, find the downstream depth required to form a jump (a) if the sluice blocks are installed, and (b) if they are not. In each case, find the head loss in the jump.

3.30. Prove that the momentum function, as in Eq. (3–8), takes the particular form given by Eq. (3–13) when the cross-sectional shape is given by Eq. (2–37).

3.31. For a channel of triangular cross section derive from Eq. (3–13) the hydraulic-jump equation

$$\frac{Q^2 y_s^2}{g y_1^5 B_s^2} = \frac{r^2(r^2 + r + 1)}{12(r + 1)}$$

where $r = y_2/y_1$. Verify that this equation is the special form of Eq. (3–12) for which $y' = my/b = \infty$.

A hydraulic jump forms in a channel of triangular section and apex angle of 90°. Upstream and downstream depths are 1 ft and 2 ft respectively; find the discharge in the channel.

3.32. Write down the equation which bears the same relation to the equation of the previous problem as Eq. (3–6) does to Eq. (3–5). Verify that this equation is true for the numerical example in the previous problem.

3.33. Derive from Eq. (3–12) an equation by which upstream conditions can be deduced from downstream conditions for the case of a hydraulic jump in a trapezoidal channel. Verify that this equation is true for the numerical example in Prob. 3.25.

3.34. Given the set of $y_2/y_1 - my_1/b$ curves obtained from Prob. C3.1 for the case of a hydraulic jump in a trapezoidal channel, show how contours of my_2/b and $Qm^{3/2}/y_2{}^2\sqrt{gy_2}$ may be drawn on the plane of the curves, and hence how upstream conditions can be deduced from downstream conditions. (Note: The technique developed in Figs. 2-15 and 2-16 by means of the family of broken lines on those figures affords a useful precedent.)

3.35. Derive Eq. (3–14) by adopting the viewpoint of the stationary observer, as in Fig. 3-7a; i.e., write the momentum equation between sections 1 and 2 for this situation. (Note: the equation must include not only the rate of momentum transfer across sections 1 and 2, but also the rate of momentum growth within the water contained between these sections.)

3.36. A river is flowing at a depth of 8 ft and a velocity of 3 ft/sec when it meets a tidal bore which abruptly increases the depth to 12 ft. Find the speed with which the bore moves upstream and the magnitude and direction of the velocity of the water behind the bore.

3.37. Water is flowing at a depth of 5 ft and a velocity of 3 ft/sec in a rectangular channel. The discharge at the upstream end is abruptly doubled; find the depth of the resulting surge and the speed with which it moves downstream.

3.38. Initial conditions are as in Prob. 3.37. The discharge at the downstream end is abruptly halved; find the depth of the resulting surge and the speed with which it moves upstream.

3.39. The same as Prob. 3.38, except that the channel is suddenly blocked completely at the downstream end.

3.40. Initial conditions are as in Prob. 3.37 and are produced by a sluice gate at the downstream end. The gate is abruptly dropped 6 in.; assuming a contraction coefficient of 0.61 for the flow under the gate, find the depth of the resulting surge and the speed with which it moves upstream. (Hint: for a number of trial values of y_2, the depth behind the gate, calculate q from the gate opening and from the surge characteristics. Proceed until these two values of q are found to be equal.)

Computer Programs

C3.1. Using Eq. (3–12), write and operate a program to compute simultaneous values of $r = y_2/y_1$ and $Z_{M_1} = Qm^{3/2}/y_1{}^2\sqrt{gy_1}$ for a hydraulic jump in a trapezoidal channel, and the following values of $1/y_1' = b/my_1$: 0, 0.5, 1, 2, 3, 4, 6, 8, 10, 12, 15, 20, 25, 30, 40, 50. From these results, plot curves of r vs. Z_{M_1}, using a logarithmic scale for Z_{M_1}, and compare them with those of Massey [1].

C3.2. Add features to the above program which will allow the downstream parameters y_2' and Z_{M_2} to be computed and listed along with each set of values of r, y_1', and Z_{M_1}. Compare these results with those given by the graphical treatment of Prob. 3.34.

C3.3. Write and operate a program to compute simultaneous values of M', y', and Q', the dimensionless numbers of Prob. 3.27 relating to flow in a circular culvert. Choose whatever range of values appears suitable, and plot curves of M' vs. y' for constant Q'.

C3.4. Write and operate a program to compute simultaneous values of y_2/y_1, y_1/D, and $Q/y_1^2 \sqrt{gy_1}$ for a hydraulic jump in a circular culvert. Two alternative methods can be used:

(a) Starting from the equation $M_1 = M_2$, where

$$M = \frac{Q^2}{Ag} + A\bar{y},$$

and using the result for $A\bar{y}$ given by Prob. 3.26, base the program on choosing values for two of the above numbers initially, and calculating the third therefrom. (Care should be taken to choose the easiest version of this system.)

(b) Write a program which in effect scans the M-y curves of the previous problem, and by interpolation between neighboring points on the curves, find values of y_1 and y_2 for which $M_1 = M_2$.

From these results plot curves of y_1/y_2 vs. $Q/y_1^2 \sqrt{gy_1}$, using a logarithmic scale for the latter, for $y_1/D = 0.1 \to 0.6$ by steps of 0.1, and compare the curves with those of Thiruvengadam [2].

C3.5. Superimpose contours of y_2/D and $Q/y_2^2 \sqrt{gy_2}$ on the curves of the previous problem, then add features to the computer programs which will cause these two parameters to be calculated and listed along with corresponding values of the upstream parameters. Compare these results with those of the graphical treatment above.

C3.6. Add features to the program C3.1 which will cause the head loss $\Delta E = E_1 - E_2$ across the hydraulic jump to be computed and listed, together with any dimensionless ratios deduced therefrom which will be useful in the solution of the hydraulic jump problem when Q, ΔE, and y_2 are known in advance. Plot the appropriate set of curves.

C3.7. Make the same extension to the program or programs of C3.4 that C3.6 makes to C3.1—i.e., aim at a presentation of results that will be useful when Q and ΔE are known in advance. In this case however, take it that y_2/D is specified equal to 0.75, and plot curves that will enable the designer to determine the culvert diameter when Q and ΔE are specified.

C3.8. Use the graphical results given by the above programs to rework Probs. 3.25, 3.26, 3.27, 3.31.

C3.9. Write a program to solve problems of the same kind as Prob. 3.40, and apply it to that particular problem.

Chapter 4

Flow Resistance

4.1 Introduction

The laws of flow resistance, already discussed in Sec. 1.8, are essentially the same in open channels as in closed pipes running full; in particular a resistance equation can be derived in each case by balancing the retarding shear force at the boundary against a propulsive force acting in the direction of flow. Whereas in pipe flow this propulsive force is supplied by a pressure gradient, in open channel flow it is supplied essentially by the weight of the flowing water resolved down a slope.

However, the boundary conditions in the two cases are somewhat different. Most pipes are of circular section, and in such pipes the shear stress restraining the fluid motion is uniformly distributed round the boundary of the cross section. In an open channel, on the other hand, there are two factors which tend to make the boundary shear nonuniform: the existence of a free surface (on which the shear stress is negligibly small), and the wide variety of possible cross-sectional shapes, each with its own distribution of shear stress round the solid boundary.

We might dispose of the first difficulty by arguing that the flow cross section is equivalent to that in one-half of a closed conduit, produced as shown in Fig. 4-1. In fact the equivalence is not exact, for in a closed conduit the point of maximum velocity would be at the center A', whereas in an open channel the maximum velocity often occurs not at the corresponding point A in the free surface, but at a point B somewhat below it. This depression of the point of maximum velocity is due to the action of secondary currents—i.e., circulating currents in the plane of the channel cross section, as shown by the arrows in Fig. 4-1a. These currents are often present in all types of channel, their presence being indicated by the tendency of floating objects to move from the banks towards the center of the stream. The action of the currents on the velocity distribution may be explained in the following way. First assume that a system of isotachs—i.e., lines of constant velocity, is drawn on the cross section, as shown by broken lines in Fig. 4-1a. Through the point B a chain-dotted line is drawn, which crosses all the isotachs at right angles and passes through the ends C, D, of the free surface. (The theory of orthogonal trajectories shows that it is in fact possible to draw such a line.)

We now consider the forces acting on a prism of water whose cross section is the lens-shaped figure $CADB$. By definition there can be no velocity gradient across the line CBD, so from the discussion of Sec. 1.8 it is clear that there can be no shear stress of any kind at this boundary. As the shear along the free surface CAD is negligible, there is no shear resistance to balance the weight of the prism resolved down the slope (as implied previously a slope of some kind must always be present).

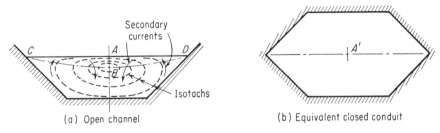

Figure 4-1.   *Comparison of Open Channel Flow with Closed-Conduit Flow*

This paradox can be resolved only by postulating secondary flows in the directions shown by the arrows in Fig. 4-1a. If these flows exist, the weight component is not really unbalanced, but performs instead the task of accelerating fluid which enters the region $CADB$ at low velocities near the banks, and leaves it at higher velocities nearer the center of the channel. It is important to note that the mechanism is still one of momentum exchange, and that one might describe this exchange as the equivalent of a shear resistance which balances the weight component. However this choice of terms would be rather unfortunate since it implies the existence of energy dissipation, which is not an essential part of the phenomenon.

As the above argument indicates, it is possible by making certain assumptions (Prob. 4.1), to establish an approximate relation between the strength of the secondary flow and the amount by which the point of maximum velocity is depressed. It must be realized that none of this argument explains why the secondary flows occur, and in fact no complete explanation has yet been found for the occurrence of these flows in straight channels. It is known that they frequently occur in such channels, not only in two "cells" as in Fig. 4-1a, but also (in the case of wide channels) in the form of many more than two cells, of alternating senses of rotation. At a channel bend there is usually only one cell of secondary flow, and in this case a simple explanation is possible; it is discussed in Sec. 7.3.

It appears from the preceding discussion that the shear-stress distribution on the boundaries of an open channel may be complicated not only by the asymmetry of the boundary but also by the presence of secondary flows. However, neither of these complications need deter us from writing a resistance equation in terms of a *mean* shear stress, even though the shear-stress distribution is unknown. The form of this equation will be essentially similar to that of the corresponding pipe flow equation, but its coefficient may depend on the cross-sectional shape as well as on other parameters such as the Reynolds number.

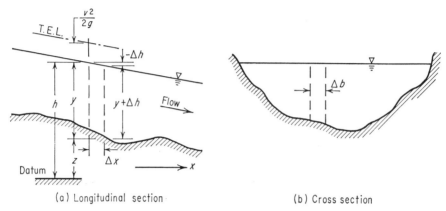

(a) Longitudinal section   (b) Cross section

Figure 4-2.   *Definition Sketch for the Resistance Equation*

4.2   The Resistance Equation

We now proceed to derive a suitable resistance equation. Consider the channel section shown in Fig. 4-2; on the small element shown, of dimensions y, Δx, and Δb, there will be a horizontal force due to the net hydrostatic thrust on the element. Assuming, as in Sec. 2.1, that the slopes are small and the pressure distribution hydrostatic, the pressure difference along any horizontal line drawn longitudinally through the element has a magnitude of $\gamma \Delta h$, where Δh is defined as the amount by which the water surface rises from the upstream to the downstream face of the element. The total horizontal hydrostatic thrust on the element, taken positive in the downstream direction, is therefore equal to $-\gamma y \Delta b \Delta h$, if $\Delta h / y$ and $\Delta z / y$ are small. The summation of this force over the whole section clearly gives the result $-\gamma A \Delta h$, where A is the cross-sectional area.

This force is resisted by a shear force equal to $\tau_0 P \Delta x$, where P is the wetted perimeter of the section and τ_0 is the mean longitudinal shear stress acting over this perimeter. The two forces are not quite parallel, but it is consistent with our assumption of small slopes to regard the two forces as parallel. The net force in the direction of flow is therefore equal to

$$-\gamma A \Delta h - \tau_0 P \Delta x \qquad\qquad (4\text{--}1)$$

We now consider the state of *uniform flow*, in which the channel slope and cross section, and the flow depth and mean velocity, remain constant as we move downstream. In this state there is no acceleration, and the net force on any element is zero. Hence from Eq. (4–1)

$$\tau_0 = \gamma R S_0 \tag{4–2}$$

where $R = A/P$ and is termed the "hydraulic mean radius," and S_0 is the bed slope, $-dz/dx$, which is equal to the water surface slope $-dh/dx$ in the case of uniform flow. Note that we define these slopes so as to be positive numbers when the surface concerned is dropping in the downstream direction.

Consider now the more general case in which the flow is nonuniform and the velocity may therefore be changing in the downstream direction. The force given by Eq. (4–1) is no longer zero, since the flow is accelerating. We consider steady flow, in which the only acceleration is convective, and equal to

$$v \frac{dv}{dx}$$

The force given by Eq. (4–1) applies to a mass $\rho A \Delta x$; therefore the equation of motion becomes

$$-\gamma A \Delta h - \tau_0 P \Delta x = \rho A v \frac{dv}{dx} \Delta x$$

i.e.,
$$\tau_0 = -\gamma R \left(\frac{dh}{dx} + \frac{v}{g} \frac{dv}{dx} \right)$$

$$= -\gamma R \frac{d}{dx} \left(h + \frac{v^2}{2g} \right)$$

$$= \gamma R S_f \tag{4–3}$$

where $S_f = -dH/dx$, the slope of the total energy line, and may be termed the "energy slope" or "friction slope." We see therefore that for any state of steady flow the shear stress τ_0 can be written as

$$\tau_0 = \gamma R S \tag{4–4}$$

provided that the slope S is properly defined. The definition $S = S_f$ fits the uniform flow case as well as nonuniform flow, since in uniform flow all three slopes—of bed, water surface, and total energy line—are equal. In other words, Eq. (4–2) is the special case of Eq. (4–3) when $S_0 = S_f$.

The Chézy Equation

In order to interpret Eq. (4–4), we need information about the magnitude of τ_0. Just as in the case of pipe flow, dimensional analysis leads to the result

$$\tau_0 = a \rho v^2 \tag{4–5}$$

where a is a dimensionless number (not necessarily constant) which may depend on the boundary roughness, on the Reynolds number, and on the cross-sectional shape. From Eqs. (4–4) and (4–5) we can readily obtain the equation

$$v = \sqrt{\frac{g}{a}} \, RS$$

and writing $\sqrt{g/a}$ as one constant C, we have the formula

$$v = C\sqrt{RS} \qquad\qquad (4\text{–}6)$$

known as the Chézy equation, since it was introduced by a French engineer of that name when in 1768 he was given the task of designing a canal for the Paris water supply. His reasoning in developing this equation was essentially that given above; i.e., that the resistance, varying as the square of the velocity, is to be balanced against a propulsive force varying directly as the slope. It is easily shown (Prob. 4.2) that this equation is essentially similar in form to the Darcy pipe-flow equation

$$h_f = \frac{fl}{D}\frac{v^2}{2g} \qquad\qquad (4\text{–}7)$$

since the diameter D in that equation arises in the same way as R in the Chézy equation, that is, by the division of the cross sectional area by the wetted perimeter.

Finally, it need hardly be pointed out that the Chézy equation, like Eq. (4–4), holds true for all cases of steady flow provided that the slope S is properly defined.

The Behavior of the Coefficient C

The Chézy coefficient C may be expected to depend, like the Darcy coefficient f, on the Reynolds number and the boundary roughness; it may also depend on the shape of the cross section. While the behavior of f in circular pipe flow has been thoroughly explored, notably by Nikuradse in 1932–35 and Colebrook and White in 1937–39 [1], a similarly complete investigation of the behavior of C has never been made, not only because of the extra variables involved in the open channel flow case, but also because of the extremely wide range of surface roughness sizes and types met in practice, and because of the difficulty in attaining steady, uniform, fully developed flow outside of laboratory installations.

These reasons are given in a study [1] of the question of friction factors in open channels sponsored by the American Society of Civil Engineers. From this and other sources it appears also that the effect of cross-sectional shape is small, at least within the limits of accuracy normally accepted in practice. This suggests that the behavior of C can be inferred directly from

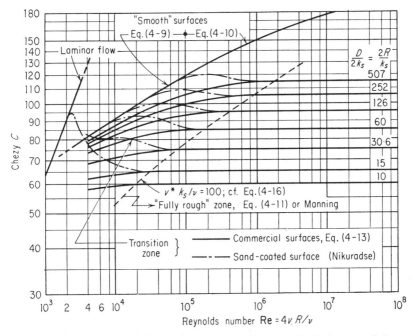

Figure 4-3.　*Modified Moody Diagram Showing the Behavior of the Chézy C*

the behavior of the Darcy f, using the relationship deduced in Prob. 4.2,

$$C = \sqrt{\frac{8g}{f}} \qquad (4\text{–}8)$$

and taking the pipe diameter D as equivalent to $4R$. This inference is allowable provided that the channel parameters are in the range normally found in pipe work and covered in pipe experiments; in practice this implies fairly small channels and fairly smooth surfaces—e.g., metal, concrete, or masonry.

Given the flow parameters, and a description of the surface roughness of the channel boundaries, a value for the resistance coefficient may be obtained from the well-known Moody diagram, which is based on the work of Nikuradse, and of Colebrook and White. Figure 4-3 shows the diagram plotted with C, instead of f, used as the ordinate. There are three possible types of turbulent flow, represented by three regions on the diagram. "Smooth" conditions occur when the boundaries are hydraulically smooth, and are represented by the uppermost line, having the equation

$$f = \frac{0.316}{\text{Re}^{1/4}} \qquad (4\text{–}9a)$$

or
$$C = 28.6\,\text{Re}^{1/8} \qquad (4\text{–}9b)$$

(the Blasius equation) for $Re < 10^5$; and the equation

$$\frac{1}{\sqrt{f}} = 2.0 \log_{10}\left(\frac{Re\sqrt{f}}{2.51}\right) \tag{4-10a}$$

or

$$C = 4\sqrt{2g}\, \log_{10}\left(\frac{Re\sqrt{8g}}{2.51C}\right) \tag{4-10b}$$

when $Re > 10^5$. In these and the following equations, Re is defined as vD/v, or $4vR/v$.

The flow conditions may be hydraulically "smooth" even when the surface is rough, provided that the roughness projections are small enough to be deeply buried within the laminar sublayer. As Re increases and this sublayer shrinks, the roughness projections become a significant factor and the flow enters a transitional stage, culminating when the projections break through the laminar sublayer and dominate the flow behavior, which is then termed *fully rough*; since the flow resistance is then entirely due to form drag on projections, the resistance coefficient is independent of Re, and dependent only on the ratio R/k_s, where k_s is a length parameter characteristic of the surface roughness. In Nikuradse's experiments on pipes uniformly coated with sand grains of uniform size, k_s was defined in the obvious way as the diameter of the sand grains; for other types of surface (e.g., of concrete or of commercial pipe), k_s is defined as the sand-grain diameter for a sand-coated surface having the same limiting value of f or C—i.e., the value reached at high values of Re. Many measurements have been made supplementing the original experiments of Nikuradse: the end result recommended [1] for fully rough flow is the equation

$$\frac{1}{\sqrt{f}} = \frac{C}{\sqrt{8g}} = 2 \log_{10}\left(\frac{12R}{k_s}\right) \tag{4-11}$$

where k_s has the values given in Table 4–1 for various types of surface.

Without direct field experience it may not be easy to determine accurate values of k_s from the surface descriptions given in Table 4–1; this need not cause as much concern as might at first appear, for the logarithmic relationship of Eq. (4–11) means that large errors in k_s give rise only to small errors in C.

The choice of a value of k_s for a certain surface implies an equivalence between that surface and one which is uniformly coated with sand grains of diameter k_s. This equivalence applies only to fully rough flow and does not extend to the transition region between smooth and fully rough flow; in this region f and C behave quite differently for surfaces like wood, metal, and concrete, than for sand-coated surfaces of the kind used by Nikuradse (Fig. 4-3). It follows that k_s alone does not fully describe the texture of a rough surface, and that a further length parameter is probably needed to indicate how the roughness projections are distributed—e.g., by giving the

TABLE 4-1 Values of k_s in Feet for Concrete and Masonry Surfaces

0.0005	Concrete class 4 (monolithic construction, cast against oiled steel forms with no surface irregularities).
0.001	Very smooth cement-plastered surfaces, all joints and seams hand-finished flush with surface.
0.0016	Concrete cast in lubricated steel molds, with carefully smoothed or pointed seams and joints.
0.002	Wood-stave pipes, planed-wood flumes, and concrete class 3 (cast against steel forms, or spun-precast pipe). Smooth troweled surfaces. Glazed sewer pipe.
0.005	Concrete class 2 (monolithic construction against rough forms or smooth-finished cement-gun surface, the latter often termed *gunite* or *shot concrete*). Glazed brickwork.
0.008	Short lengths of concrete pipe of small diameter without special facing of butt joints.
0.01	Concrete class 1 (precast pipes with mortar squeeze at the joints). Straight, uniform earth channels.
0.014	Roughly made concrete conduits.
0.02	Rubble masonry.
0.01 to 0.03	Untreated gunite.

average spacing between them. However it is not easy to achieve the required two-parameter description of roughness in a workable form, and no generally accepted method of doing so has yet been put forward. Moreover, such a description would serve little purpose in the case of the moderately smooth surfaces here being discussed, for all such surfaces behave in much the same way in the transition region. The behavior of these "commercial" surfaces (such as wood, metal or concrete) in the case of pipe flow is indicated by the following equation, due to Colebrook:

$$\frac{1}{\sqrt{f}} = \frac{C}{\sqrt{8g}} = -2\log_{10}\left(\frac{k_s}{14.83R} + \frac{2.52}{\mathrm{Re}\sqrt{f}}\right) \qquad (4\text{–}12)$$

and there is some evidence [1] in support of applying this equation to open channel flow with slightly modified coefficients:

$$\frac{C}{\sqrt{8g}} = -2\log_{10}\left(\frac{k_s}{12R} + \frac{2.5}{\mathrm{Re}\sqrt{f}}\right) \qquad (4\text{–}13)$$

using the values of k_s given by Table 4–1. However, further work would appear to be necessary before the matter is placed beyond doubt.

The three types of flow—smooth, transition, and fully rough—are distinguished from each other by the size of the dimensionless number $k_s v^*/v$, which is easily recognizable as a kind of Reynolds number. The quantity v^*, known as the *shear velocity* is defined as

$$v^* = \sqrt{\frac{\tau_0}{\rho}} = \sqrt{gRS_f} \qquad (4\text{–}14)$$

and will assume some importance in the treatment of sediment transport in Chap. 10. It has the dimensions of velocity, but is not identified with any physically real velocity. However it may be simply related to the mean velocity of flow v by combining Eqs. (4–6) and (4–14), leading to the result

$$\frac{v}{v^*} = \frac{C}{\sqrt{g}} = \sqrt{\frac{8}{f}} \qquad (4\text{–}15)$$

The transition region of flow is defined approximately by the limits

$$4 < \frac{v^* k_s}{v} < 100 \qquad (4\text{–}16)$$

the lower limit defining the end of the smooth region and the upper limit the beginning of the fully rough region. The ranges of applicability of Eqs. (4–9), (4–10), (4–11), and (4–13) are thus clearly defined.

It is desirable however to end this discussion by emphasizing the qualification to which it is subject—namely, that equations derived from pipe-flow experiments should be applied only to fairly small channels and conduits (of a few feet in cross-sectional dimensions) with fairly smooth surfaces. For large channels with very rough surfaces such as occur in nature, it is desirable to seek independent evidence, preferably based on direct field observations.

The Manning Equation

The empirical pipe-flow equations just discussed are largely based on experimental work done between 1930 and 1940. On the other hand, systematic observations on rivers and large channels had already been made by about the middle of the nineteenth century [1]. In 1869 Ganguillet and Kutter published a rather complicated formula for C, which achieved considerable popularity; however Gauckler in 1868 and Hagen in 1881 arrived independently at the conclusion that the data used by Ganguillet and Kutter were fitted just as well by a simpler formula stating that C varies as the sixth root of R. In 1891 the Frenchman Flamant wrongly attributed this conclusion to the Irishman R. Manning, and expressed it in the form

$$C = \frac{R^{1/6}}{n} \qquad (4\text{–}17)$$

or

$$v = \frac{R^{2/3} S^{1/2}}{n} \qquad (4\text{–}18)$$

where n is characteristic of the surface roughness alone, and the unit of length used is the meter. In 1911 Buckley converted this equation to foot-second units by inserting the factor $\sqrt[3]{3.28}$, or 1.486 (there being 3.28 ft in a meter). The end result

$$v = \frac{1.486 R^{2/3} S^{1/2}}{n} \qquad (4\text{–}19)$$

is known in the English-speaking world as the Manning equation, although on the continent of Europe it is sometimes known also as Strickler's equation. The coefficient 1.486, although arising from a perfectly reasonably conversion process, is far more precise than is necessary; in fact the precision with which n is known hardly warrants writing the coefficient more precisely than as 1.49 or even 1.5.

The Manning equation has proved most reliable in practice and extremely popular in most Western countries. However, in view of its completely different origins from those of Eqs. (4–11) and (4–13), it is desirable to make a comparison aimed at finding out whether there is a corresponding difference in the results it yields.

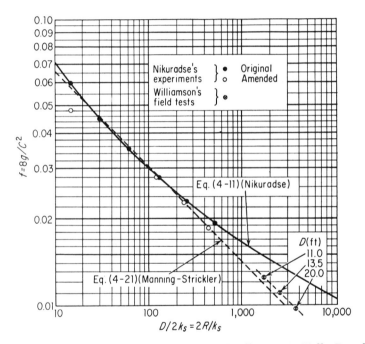

Figure 4-4. The Behavior of Resistance Coefficients in Fully Rough Flow

In the first place, one would expect the Manning equation to be applicable only in the fully rough zone, where Eq. (4–11) holds. The equation is plotted in Fig. 4-4 on logarithmic graph paper, and it is seen that the result is a curve which can be approximated to—rather roughly—by a straight line having a slope of 1:3, i.e., having the equation

$$f \propto \left(\frac{k_s}{R}\right)^{1/3} \propto \frac{g}{C^2},$$

Hence
$$C \propto \left(\frac{R}{k_s}\right)^{1/6} \qquad (4\text{–}20)$$

which confirms the form of the Manning equation and adds a further conclusion of interest, namely that

$$n \propto k_s^{1/6}$$

In 1951 Williamson [2] showed that some minor, but plausible, adjustments to Nikuradse's results made them approximate much more closely to Eq. (4–20). By correcting certain of Nikuradse's calculations, and increasing the assumed grain size to allow for a suitable thickness of the varnish used to stick the grains to the pipe walls, he found that points representing Nikuradse's results fell much closer than before to a straight line having a slope of 1:3; its equation was

$$f = 0.180\left(\frac{k_s}{D}\right)^{1/3} = 0.113\left(\frac{k_s}{R}\right)^{1/3} \qquad (4\text{–}21)$$

Further, he plotted some more observations made by himself on concrete pipelines up to 20 ft in diameter and found that, as shown at bottom right on Fig. 4-4, they fell on a line having the same slope; assignment of a suitable value of k_s would place the points on the line representing Eq. (4–21).

If we replace k_s with d, which will be used in a later chapter to indicate stone or gravel size on the channel bed, we can transpose Eq. (4–21) as follows:

$$f = \frac{8g}{C^2} = 0.113\left(\frac{d}{R}\right)^{1/3}$$

whence
$$C = \sqrt{\frac{8g}{0.113}}\left(\frac{R}{d}\right)^{1/6}$$

$$= \frac{1.49R^{1/6}}{0.031d^{1/6}}$$

which is identical with the Manning formula, provided that

$$n = 0.031d^{1/6} \qquad (4\text{–}22)$$

where d is measured in feet. Now in 1923 Strickler had independently produced an empirical equation relating n and d:

$$n = 0.034d^{1/6} \qquad (4\text{–}23)$$

which is very close to Eq. (4–22). The correspondence appears even closer when it is pointed out that Strickler's work was based on gravel-bed streams in which d was the *median* size of the bed material. However, the effective value of d from the resistance viewpoint is that of the larger size (two or three times the median) with which the bed tends to become armored. Hence to make a direct comparison between Eqs. (4–22) and (4–23) the coefficient of Eq. (4–23) should be reduced by a factor of the sixth root of a number between two and three—i.e., by between 10 and 20 percent. The effect is to make the agreement between the two equations even closer than it appears at first.

We conclude that there is a remarkably close correspondence between Eq. (4–11), based initially on quite small-scale pipe experiments (Nikuradse's largest pipe was $2\frac{1}{2}$ in. in diameter), and the Manning and Strickler equations, based on quite large-scale field observations. It follows that the Manning equation is suitable for all fully rough flow, although there will be a range of intermediate channel sizes for which Eq. (4–11) is equally suitable, within normally acceptable limits of accuracy. For transition flow, as described by Eq. (4–13), the Manning equation is no longer suitable, unless the coefficient n is recognized as dependent on Re, as in Fig. 4-5 (see notes on Table 4–2); the boundary between transition flow and fully rough flow is given by Eq. (4–16), and may conveniently be expressed in terms of the Manning equation parameters. Equations (4–14), (4–16), and (4–22) may be combined (Prob. 4.3) to give the result

$$n^6\sqrt{RS_f} \geq 1.9 \times 10^{-13} \qquad (4\text{–}24)$$

for fully rough flow. If this inequality is true the Manning equation is applicable.

Typical values of the coefficient n are listed in Table 4–2.

TABLE 4–2 Values of Manning's Roughness Coefficient n

Glass, plastic, machined metal	0.010
Dressed timber, joints flush	0.011
Sawn timber, joints uneven	0.014
Cement plaster	0.011
Concrete, steel troweled	0.012
Concrete, timber forms, unfinished	0.014
Untreated gunite	0.015–0.017
Brickwork or dressed masonry	0.014
Rubble set in cement	0.017
Earth, smooth, no weeds	0.020
Earth, some stones and weeds	0.025
Natural river channels:	
Clean and straight	0.025–0.030
Winding, with pools and shoals	0.033–0.040
Very weedy, winding and overgrown	0.075–0.150
Clean straight alluvial channels	$0.031d^{1/6}$
	(d=D-75 size in ft.)

Notes on Table 4.2

When a single value of n is given in the table, it is the mean value of a range of approximately ± 0.001. The categories such as "clean straight river channels" described at the end of the table clearly cover such a wide range of conditions that some field experience is desirable before a value of n can be estimated with reasonable confidence. However, the photographs given by Ven Te Chow [6] form a useful supplement to, or even substitute for, field experience.

The last entry in the table gives the result of Eq. (4–22), applicable mainly to alluvial channels of coarse noncohesive gravel or cobbles (known as *shingle* in British countries). The D-75 size may be taken as a good approximation to the value of d (larger than the median) with which the bed tends to become armored.

The reader will easily be able to verify that the values of k_s in Table 4–1 are generally consistent, via Eq. (4–22), with the above values of n.

When the channel bed and banks are thickly covered with vegetation an appreciable part of the flow takes place through the vegetation at low velocities. If the growth is of fine material such as grass the Reynolds number Re defined with respect to the stalk thickness will be low, and the resistance, and therefore the Manning n, will be dependent on Re. Since n will therefore depend on the velocity, it may possibly depend on Re defined with respect to the channel size as well as with respect to the stalk thickness. This has been shown to be true by the experiments of the U. S. Soil Conservation Service [3];

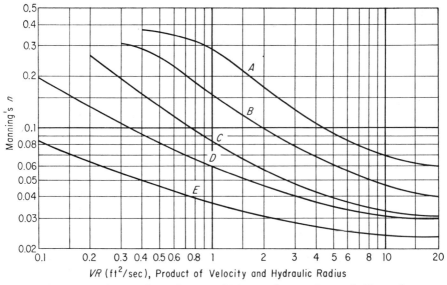

Figure 4-5. *The Behavior of Manning's n in Grassed Channels*

their results, for a number of North American grass species, are summarized in Fig. 4-5. The division into classes depends mainly on the length and the "stand"—i.e., the vigour and thickness of growth, according to the following table:

TABLE 4–3

Average length of grass	Class	
	Good stand	Fair stand
More than 30 in.	A	B
11–24 in.	B	C
6–10 in.	C	D
2–6 in.	D	D
less than 2 in.	E	E

Wide shallow grassed channels are a popular solution to the problem of passing large discharges down steep slopes without developing unduly high velocities.

4.3 Uniform Flow: Its Computation and Applications

Significance of Uniform Flow

Uniform flow has now been defined and a dynamic equation developed—the Manning equation—which adequately describes both uniform and non-uniform flow.

Uniform flow seldom occurs in nature, since natural channels are usually irregular. Even in artificial channels of uniform section, the occurrence of uniform flow may be relatively infrequent because of the existence of controls, such as weirs, sluice gates, etc., which dictate a depth-discharge relationship different from that appropriate to uniform flow.

However, uniform flow is a condition of such basic importance that it must be considered in all channel-design problems. For example, if it is proposed to instal certain controls in an irrigation canal it is necessary to compare their depth-discharge relation with those of uniform flow; as we shall see, the whole character of the flow in the canal will depend on the form this comparison takes. Again, if a canal is to be laid on a certain slope, is to have a lining of a certain coefficient n, and is to take a certain discharge, then the uniform-flow condition is the criterion governing the *minimum* cross-sectional area required. Other criteria may of course determine that the section must be greater than this minimum, but the section cannot conceivably be any smaller or the canal will be unable to take the required discharge.

Economical Design of a Channel Cross Section

A typical uniform-flow problem in the design of artificial canals is the economical proportioning of the cross section. A canal having a given Manning coefficient n and slope S_0 is to carry a certain discharge Q, and the designer's aim is to minimize the cross-sectional area A. Clearly if A is to be a minimum, the velocity v is to be a maximum; the Chézy and Manning formulas indicate, therefore, that the hydraulic radius $R = A/P$ must be a maximum. It can be shown that the problem is equivalent to that of minimizing P for a given constant value of A; the ideal cross section would, therefore, be a semicircle.

It can also be shown (Prob. 4.4) that the best trapezoidal shape is that which approximates most closely to a semicircle, in that a semicircle, having its center in the surface, can be inscribed in the trapezoid (Fig. 4-6). In the special case when the trapezoid is a rectangle, the best shape is that for which the width is twice the depth.

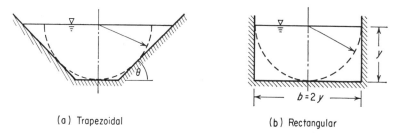

(a) Trapezoidal   (b) Rectangular

Figure 4-6.   *Channel Sections of "Optimum" Shape*

The above formulation of the problem is so oversimplified as to be somewhat artificial. In practice the economics of canal design is complicated by these factors:

Flow resistance is not the only important design consideration; there is also the possibility of scour in an erodible bed. This question is dealt with fully in Chap. 10.

The area A is only the waterway area; the total volume of excavation includes overburden as well, and a minimum value of A need not imply a minimum total excavation.

Even if attention is limited to the waterway area A, the minimum as given by the above criteria is a very flat one; the proportions of a canal section may vary widely without changing the required value of A significantly (Prob. 4.5).

If the canal has to be lined, the cost of the lining may well be comparable with excavation costs; the optimization problem must, therefore, be reformulated (Prob. 4.6).

For short channels in which the slope is not absolutely fixed by the local topography, the slope itself may be regarded as a variable in economic calculations. A reduced value of slope may require a larger waterway area, but less removal of overburden.

The cost of excavation is not solely dependent on the amount of material removed. Considerations such as ease of access and disposal may be more important than the volume of material excavated.

For all these reasons, it is only in the most restricted sense that the channel sections in Fig. 4-6 can be said to represent an "optimum" choice of section, although that term is commonly applied to them.

The Computation of Uniform Flow

The solution of the Manning equation presents no basic computational difficulties and in some cases—e.g., when S_0 or Q is the unknown, the solution is quite explicit. However when it is required to determine the channel cross section itself, or the uniform depth (normally written as y_0) a trial solution is needed except in the simple case (Probs. 4.10 and 4.15) where the channel is very wide (so that $P = B$, the surface width), and B is given by an exponential equation of the form

$$\frac{B}{B_s} = \left(\frac{y}{y_s}\right)^i \tag{2-37}$$

Examples of such a section would be the wide triangular and the wide parabolic; the latter might be approximated to by some river sections, but in general the uniform-flow problem arises mainly in artificial channels of finite width, not described by Eq. (2–37), in which y_0 can be obtained only by a trial solution.

When only a small number of cases are to be calculated, this solution can be obtained by a tabulation in which trial values of y lead to values of A, P, R, ... , etc., until the correct value of Q is obtained. Where a large number of calculations are to be made, it is convenient to introduce the concept of the "conveyance" of a channel, indicated by the symbol K and defined by the equation

$$Q = KS^{1/2} \tag{4-25}$$

so that

$$K = \frac{1.49AR^{2/3}}{n} \tag{4-26}$$

Either K alone or the product Kn can then be tabulated or plotted as a function of depth for any given channel section; the resulting tables or curves can then be used as a permanent reference, which will immediately yield values of depth for given values of Q, S, and n.

King [4] has introduced a form of the conveyance that is particularly suitable for rectangular and trapezoidal channels (which make up the vast majority of all artificial channels). It is indicated by K' and defined by the equation

$$Q = \frac{K'b^{8/3}S_0{}^{1/2}}{n} \tag{4-27}$$

where b is the base width of the channel. It follows (Prob. 4.11) that

$$K' = \frac{1.49(1 + my/b)^{5/3}}{[1 + 2\sqrt{1 + m^2}(y/b)]^{2/3}} \left(\frac{y}{b}\right)^{5/3} \tag{4-28}$$

where $m = \cot \theta$ and θ is the side-slope angle, as in Chap. 2. King lists values of K' for a wide range of values of m and of y/b; these tables are satisfactory for $y/b > 0.02$, but Posey and Pagay [5] have pointed out that when $y/b < 0.02$ interpolation in these tables is unreliable and it is sounder to treat the channel as a wide rectangular one in which $R = y$ and $A = by$. The result of this assumption is the equation

$$y_0 = 0.79 \left(\frac{Qn}{b\sqrt{S_0}} \right)^{0.6} \tag{4-29}$$

which is in error by amounts ranging from $+1$ percent for $m = 3$ to -1.6 percent for $m = 0$, when $y/b = 0.02$. It follows that this equation is perfectly satisfactory for use in trapezoidal channels when $y/b \leq 0.02$; when $y/b > 0.2$, King's tables may be used. Finally, it may be noted that in these tables m and y/b are treated as separate independent variables. It is clear from inspection of Eq. (4–28) that K' cannot be expressed as a function of my/b alone, as was the case with the corresponding functions in the energy and momentum problems of Chaps. 2 and 3.

The quantity K' may also be defined for the circular culvert, replacing b in Eq. (4–27) by the diameter D. As for the trapezoidal section, it can be computed and listed as a function of y/D (Prob. C4.1). Without access to such tables, trial methods must be used (Prob. 4.28). Curves may also be plotted showing the variation of K' with y/b or y/D (Probs. C4.1, C4.2) although for the trapezoidal section a separate curve must be drawn for each value of m; for the reason given above, the more compact presentation of Figs. 2-15 and 2-16 is not possible. However, it is possible to initiate the development plotted on these figures in broken lines, by which one can determine b or D, given Q and y (Prob. 4.29). This development is not, however, of great usefulness in uniform-flow computation; although it often happens in practice that b or D is the unknown to be calculated, the quantity known in advance (other than Q) is usually y_0/D (Prob. 4.27) or the velocity (Prob. 4.21), rather than y_0 alone.

Computer programs such as those of Probs. C4.1 and C4.2 make it unnecessary to resort to a trial process to determine y_0. However it makes a useful exercise to write a program which can reproduce this trial process in any particular case (Prob. C4.3).

From the results of Prob. C4.1 it will appear that K' has a maximum in the circular culvert when $y_0/D = 0.94$ approx. This maximum can give rise to instability when the culvert is running nearly full, for there will be a tendency for the culvert to run temporarily full at irregular intervals—a tendency reinforced by wave action. The underlying problems are of some interest to the theoretician, but for the practicing engineer the simplest way out of any resulting difficulty is to keep the ratio y_0/D at a safe distance below unity (say less than 0.75).

4.4 Nonuniform Flow

Uniform Flow as a Control

In Sec. 2.5 the concept of a control was introduced and discussed; it was defined as any feature which determines a relationship between depth and discharge. The discussion of uniform flow in Sec. 4.3 makes it clear that this state of flow may itself be thought of as a control, since from a resistance equation such as Manning's we may, given the depth, calculate the discharge.

However, uniform flow is not, of course, associated with particular localized features in the channel; it is the state which the flow tends to assume in a long uniform channel when no other controls are present. If there are other controls they tend to pull the flow away from the uniform condition, and there will be a transition—which may be gradual or abrupt—between the two states of flow.

Within this transition region, the flow will in general be nonuniform; as has already been pointed out, we can use resistance equations such as Chézy's to describe this state of flow provided that the slope S is interpreted as the slope of the total energy line, S_f. We now have to examine this form of the equation with a view to obtaining a complete solution, i.e., a complete description of the longitudinal flow variation within a nonuniform transition region. We can write

$$\frac{dH}{dx} = \frac{d}{dx}\left(z + y + \frac{v^2}{2g}\right) = -S_f = -\frac{v^2}{C^2 R}$$

whence

$$\frac{d}{dx}\left(y + \frac{v^2}{2g}\right) = -\frac{dz}{dx} - S_f$$

i.e.,

$$\frac{dE}{dx} = S_0 - S_f \tag{4-30}$$

Now it has been shown in Sec. 2.7 that for any shape of channel cross section

$$\frac{dE}{dy} = 1 - \mathrm{Fr}^2 \tag{2-23}$$

whence from Eq. (4-30)

$$\frac{dy}{dx}(1 - \mathrm{Fr}^2) = S_0 - S_f \tag{4-31}$$

which is a generalization of Eq. (2-15).

We may consider the resistance equation, in the form of either Eqs. (4-30) or (4-31), as a differential equation in $y, = f(x)$; it is not in general explicitly soluble, but many numerical methods have been developed for its solution.

These will be considered in the next chapter; meanwhile we consider certain general questions relating to the solution.

Mild, Steep and Critical Slopes

A *mild* slope is one on which uniform flow is subcritical; on a *steep* slope uniform flow is supercritical; on a *critical* slope uniform flow is critical. If as before we indicate uniform depth by y_0, then we can write

$$y_0 > y_c \text{ for mild slopes}$$

$$y_0 < y_c \text{ for steep slopes}$$

$$y_0 = y_c \text{ for critical slopes}$$

The classification of the slope will depend on the roughness, on the magnitude of the slope itself, and to a lesser extent on the discharge. The slightness of the discharge effect follows from the close similarity between the exponents of the equations

$$q \propto y_0^{5/3} \text{ (Manning equation, wide channel)}$$

and $\qquad q \propto y_c^{3/2}$

From equations of this sort relationships can readily be obtained (Prob. 4.12) between the discharge and the critical value of S_0, which can form an approximate guide even for channels of finite width.

The Occurrence of Critical Flow

An interesting result concerning the occurrence of critical flow can be derived from Eq. (4–31). Consider the special case $S_0 = S_f$. This means that either

$$\frac{dy}{dx} = 0$$

or $\qquad\qquad\qquad$ $\text{Fr} = 1$

The first alternative evidently implies uniform flow: the question arises whether the second alternative has any real physical meaning. Consider a long channel in two sections—one of mild slope upstream and one of steep slope downstream, joining at the point O (Fig. 4-7). The flow will gradually change from subcritical at a great distance upstream to supercritical at a great distance downstream, passing through critical at some intermediate point, as yet unknown.

In the transition region upstream of O the depth is less and the velocity is greater than in uniform flow. Since $S_f = v^2/C^2 R$ it is clear that $S_f > S_0$ in this region. Similarly, it follows that $S_f < S_0$ in the transition region downstream of O. If we consider the "point" O to be a short curve joining the two long slopes, there must, therefore, be some point on this curve (Fig. 4-7b) at which $S_f = S_0$. And since dy/dx is clearly not zero in this neighborhood, it follows that Fr must be unity and the flow critical.

It has been shown in Sec. 2.5 that critical flow occurs at the outflow from a lake into a steep channel or through a constriction in width; this last result shows that it will also occur at the head of a steep slope which is preceded by a mild slope, with resistance effects taken into account.

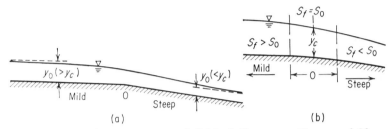

Figure 4-7.   *The Occurrence of Critical Flow at a Change of Slope*

These particular examples point to an underlying physical concept, which should be clearly understood by the reader. What these cases have in common is some sort of mechanism of *release*—water which has previously been restrained either by the banks of a lake or the roughness of a channel bed is suddenly released into a region where restraint is either nonexistent or is so small that the flow can no longer be forced into the subcritical condition. The flow is always critical at or near such a point of release.

Finally, it is worth reiterating here that a critical-flow section is a control. We shall see in the following paragraphs that such sections exert a most important influence on longitudinal flow profiles in nonuniform flow. Figure 4-7 shows one example of this influence.

4.5   Longitudinal Profiles

Classification of Longitudinal Profiles

It is clear that the profile shown in Fig. 4-7 is only one of a large number of cases that may occur when a control interferes with uniform flow, or when there is a transition from one state of uniform flow to another. Before embarking on the detailed numerical integration of Eq. (4–31), it is clearly desirable to develop some systematic process of counting up and classifying all the profiles that may conceivably occur; such an approach can be devised by first rewriting Eq. (4–31) as

$$\frac{dy}{dx} = \frac{S_0 - S_f}{1 - \mathrm{Fr}^2} \tag{4–32}$$

For a given value of Q, S_f and Fr are functions of the depth y. For a cross section of unrestricted shape S_f and Fr are not simple functions of y, so the equation is not in general explicitly soluble.

However, we are not, in this argument, concerned to obtain an explicit solution; we are concerned only to obtain a semiquantitative picture of the variation of y with x in various circumstances, just as in the curve-sketching exercises of elementary calculus. For this purpose we need only consider the signs of the numerator and denominator of Eq. (4–32), and how these signs depend on the magnitude of y. Now it follows from the Chézy equation that

$$S_f = \frac{v^2}{C^2 R} = \frac{Q^2}{C^2 A^2 R} = \frac{Q^2 P}{C^2 A^3}$$

and from Sec. 2.7 that

$$Fr^2 = \frac{v^2 B}{g A} = \frac{Q^2 B}{g A^3}$$

For a given Q, S_f and Fr^2 will vary in much the same way with depth, for P will not differ greatly from B, particularly in wide channels. Since both S_f and Fr^2 show a strong inverse dependence on A, we should expect them both to decrease as y increases for most normal channel sections. In fact they behave in this way for *all* channel sections; for if this behavior could be reversed it should be possible to derive a channel section for which Fr^2 will remain constant at all values of y. It can be shown (Prob. 4.7) that no such channel section can be derived. However if E rather than Q is kept constant it is possible to derive such a section, but this result is of no interest in the present discussion.

By definition, $S_f = S_0$ when $y = y_0$; we conclude therefore that

$$S_f \gtreqless S_0 \text{ according as } y \lesseqgtr y_0 \qquad\qquad \textbf{(4–33)}$$

$$Fr \gtreqless 1 \quad \text{according as } y \lesseqgtr y_c \qquad\qquad \textbf{(4–34)}$$

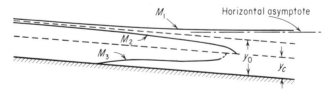

Figure 4-8. *Longitudinal Profiles on a Mild Slope*

With the aid of these inequalities we can now readily determine how the behavior of dy/dx is affected by the relative magnitudes of y, y_0, and y_c. For the first case consider a channel of mild slope, i.e., with $y_0 > y_c$. Such a channel is represented in Fig. 4-8, with broken lines representing depths of y_0 and y_c. Including the channel bed we therefore have three parallel lines dividing the whole region into three spaces, which are labeled 1, 2, and 3, starting at the top. From Eqs. (4–32) through (4–34) we can immediately state the following results:

Space 1, $y > y_0 > y_c$. $S_0 > S_f$, Fr < 1, and dy/dx is positive

Space 2, $y_0 > y > y_c$. $S_0 < S_f$, Fr < 1, and dy/dx is negative

Space 3, $y_0 > y_c > y$. $S_0 < S_f$, Fr > 1, and dy/dx is positive

Having established the sign of dy/dx in each zone, we need only determine its behavior at the boundaries of each region. We consider each one in turn.

1. $y > y_0 > y_c$. As $y \to y_0$, $S_0 \to S_f$ and $dy/dx \to 0$; i.e., the water surface is asymptotic to the line $y = y_0$, and:

As $y \to \infty$, Fr and $S_f \to 0$, i.e., $dy/dx \to S_0$ and the water surface is asymptotic to a horizontal line.

From these results a clear picture emerges of the circumstances in which this water surface profile would form, for it is simply the backwater curve that forms behind a dam. Following an obvious labeling system we designate this curve as M_1 (M for mild).

2. $y_0 > y > y_c$. As $y \to y_0$, $dy/dx \to 0$ as before; and:

As $y \to y_c$, then apparently $dy/dx \to \infty$, but this result is fictitious; under no circumstances will the water surface be at right angles to the stream bed. The situation in fact is essentially the same as the one analyzed in Sec. 4.4; $y = y_c$ and S_0 is momentarily equal to S_f, so that $y = y_0$. Then dy/dx tends to the indeterminate $0/0$, and, in fact, is finite; it is however, large enough to appear infinite on a longitudinal profile drawn with an exaggerated vertical scale, as in Fig. 4-8 and most of the subsequent figures.

Underlying this conclusion is the assumption that where $y \to y_c$, the channel terminates at a free overfall or at a transition to a steep slope. In fact the profile treated here, which we designate the M_2 curve, is the drawdown profile from uniform subcritical flow to a critical section such as an overfall.

3. $y_0 > y_c > y$. As $y \to 0$, both S_f and Fr tend to infinity, so that dy/dx tends to some positive finite limit, of magnitude dependent on the particular cross section. This result is of little practical interest, since zero depth never actually occurs. We postpone for the moment discussion of the case $y \to y_c$. The profile can be labeled M_3; clearly it will occur when supercritical flow is created by some control such as a sluice gate. The velocity is initially greater than the slope can maintain, and must, therefore decrease downstream, apparently until critical flow is reached. In the previous case (the M_2 curve) we were able to postulate a free overfall at the critical section, because the flow was subcritical and subject to influence by the overfall downstream—in fact the whole profile was *produced* by this channel feature. But in the present case the flow is supercritical and controlled from upstream—e.g., by a sluice gate. We have no right to assume that a free overfall will conveniently

occur just where the profile, plotted downstream from the sluice gate, calls for critical flow. And we cannot, therefore, assume that S_0 will momentarily be equal to S_f at the critical section making dy/dx finite; dy/dx will be infinite.

If the channel terminates in an overfall a short distance downstream from the sluice gate, critical flow is never attained and the difficulty does not arise. If the channel carries on for a great distance downstream, the difficulty is resolved by the occurrence of a hydraulic jump, making an abrupt change from the M_3 curve up to the uniform flow line or perhaps to an M_1 or M_2 curve. When this happens critical flow never actually occurs.

From the three cases so far dealt with, certain principles emerge that will be applicable to all cases. They are:

1. The sign of dy/dx can be readily determined from Eqs. (4–32) through (4–34).

2. When the water surface approaches the uniform depth line, it does so asymptotically.

3. When the water surface approaches the critical depth line, it meets this line at a fairly large finite angle.

4. If the curve includes a critical section, and if the flow is subcritical upstream, (as in the M_2 curve) then that critical section is produced by a feature such as a free overfall, which controls the subcritical flow upstream of it, and in fact determines and locates the whole profile. But if the flow is supercritical upstream (as in the M_3 curve) the control cannot come from the critical section, and indeed such a section will probably not occur in reality, but will be bypassed by a hydraulic jump. Analogous conclusions can be drawn when the profile being studied is downstream of the critical section. These are left as an exercise for the reader (Prob. 4.8).

5. Above all, every profile exemplifies the important principle that subcritical flow is controlled from downstream (e.g., the M_1 and M_2 curves) and supercritical flow from upstream (e.g., the M_3 curve). In fact these profiles owe their existence to the action of upstream or downstream controls.

Given the above conclusions, the reader will be able to verify (Prob. 4.8) that the profiles for steep, critical, horizontal, and adverse slopes are as

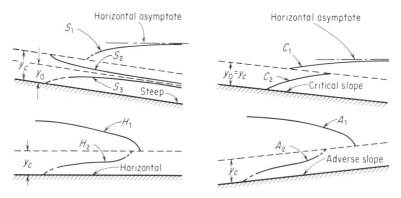

Figure 4-9. *Longitudinal Profiles on Various Types of Slope*

sketched in Fig. 4-9. He will also readily perceive that the situation of Fig. 4-7 may now be described as an M_2 curve followed by an S_2 curve. It is seen that the profiles for critical slopes are very nearly straight horizontal lines: they would be exactly so in the particular case where the Chézy C is constant and the channel is wide and rectangular in section (Prob. 4.9). For horizontal and adverse slopes y_0 is meaningless, but this gives rise to no difficulty; since S_0 is zero or negative, the numerator $(S_0 - S_f)$ of Eq. (4–32) will always be negative.

The Synthesis of Composite Profiles

Channels with a number of controls will have flow profiles that are compounded of many of the different types deduced in the previous section. The ability to sketch these composite profiles is in many cases necessary for understanding the flow in the channel or indeed for determining the discharge. In all cases it is necessary first to identify the controls operating in the channel, and then to trace the profiles upstream and downstream from these controls.

Two simple cases are shown in Fig. 4-10; in the first case the slope is mild, in the second it is steep. The curves for the mild slope are virtually self-explanatory, since they incorporate many of the features already discussed. For the steep slope we have already seen that critical flow must occur at the head of the slope—i.e., at the outflow from the lake; thereafter there must clearly be an S_2 curve tending towards the uniform depth line. There must be an S_1 curve behind the gate, and the transition from the S_2 to the S_1 curve must be via a hydraulic jump. Downstream of the sluice gate, the flow will tend to the uniform condition via an S_2 or S_3 curve, thence over the fall

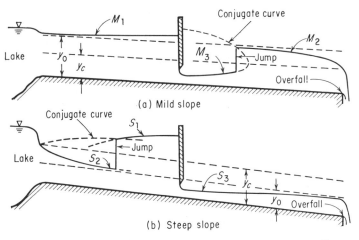

Figure 4-10. *Examples of Composite Longitudinal Profiles*

at the end of the slope. In this case there is nothing that impels the flow to seek the critical condition.

On Fig. 4-10 two profiles are drawn in broken lines above the M_3 and the S_2 curves. These are the loci of depths conjugate to the corresponding depths on the underlying real surface profiles, and are therefore known as "conjugate curves." Clearly a hydraulic jump will occur where such a curve intersects the real (subcritical) surface profile downstream; the conjugate curve therefore provides a convenient means of determining the location of a hydraulic jump. The shape of the conjugate curves in Fig. 4-10 clearly indicates the property shown also by the M-function curves (Fig. 3-4): that for a given discharge, as the depth increases the corresponding conjugate depth decreases.

A noteworthy deduction from the steep-slope profiles is that the sluice gate does not help to control the amount of the discharge, which is determined only by the conditions at the lake exit. The flow issuing from the lake is supercritical once the exit is passed, and does not know or care about the presence of the sluice gate further downstream. However, if the sluice gate were closed down further the water behind it would back up to a higher level, sending the hydraulic jump further upstream. If this process were continued the jump would eventually drown the lake outlet completely, establishing subcritical flow all the way from the lake to the gate, which would then be effectively controlling the amount of the discharge.

The implied conclusion—obvious once it is pointed out—is that if the sluice gate is to control the flow effectively it should be placed right at the lake outlet or a very short distance downstream.

The Discharge Problem

The mild-slope profiles suggest a question of much more general interest relating to the discharge. So far it has been assumed that the discharge must be known before longitudinal profiles can be sketched. But in many cases, such as that of Fig. 4-10a, the discharge will not be known initially, but could apparently be calculated, given the necessary information on lake levels, slopes, etc. Suppose that we assume a certain value of discharge; values of y_0 and y_c could then be calculated and the profiles plotted upstream and downstream from the controls. The profile plotted upstream from the sluice gate arrives at the lake with certain values of y and v, from which the specific energy E is obtained. But it does not follow that this value of E is the same as the specific energy available at the lake outlet, which is simply the height of the lake surface above the stream bed at that point. The two values of E must be equal if the flow is to be possible: if they are not it simply means that our trial value of discharge is wrong and that a new one must be chosen. The trial process goes on until the specific energy values at the lake outlet match up correctly.

The simplest form of this problem arises in the outflow from a lake of known surface level into a long uniform channel of mild slope. Since there is no downstream control (other than the channel roughness itself) there can be no M_1 or M_2 curve, and the flow becomes uniform immediately it enters the channel from the lake. The problem then simply amounts to finding a pair of values of Q and y which satisfy both the Manning equation and the specific energy equation, as shown graphically in Fig. 4-11. The solution is indicated by the intersection A of two Q-y curves—one representing uniform flow and the other constant specific energy, as in Fig. 2-5.

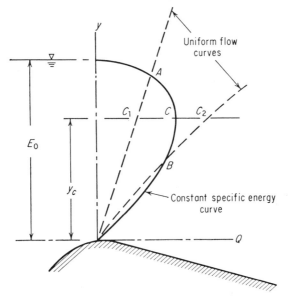

Figure 4-11. *Outflow from a Lake into a Long Channel*

Computation is quite simple, once it has been determined whether the channel slope is in fact mild. It is most important to notice that if the slope is steep, the discharge is no longer given by the intersection (B in this case) of the uniform-flow curve and the specific energy curve, because the flow does not become uniform right at the channel outlet; the arguments of Secs. 2.5 and 4.4 indicate that at this section the flow is critical, followed by an S_2 curve tending asymptotically to uniform flow. For if the flow did become uniform (and therefore supercritical) as soon as it entered the channel, this would imply a sudden jump from subcritical to supercritical flow, of the type discussed in Sec. 2.2 in connection with Fig. 2-3. It was shown there that a sudden jump from, say, A to B' in Fig. 2-3b is impossible, and that the transition must be made via the critical condition. The outflow section in Fig. 4-11 provides, logically, the constricted section at which one would expect this critical flow to occur. The magnitude of the discharge is of course indicated by the point C in that figure.

Given a certain magnitude of channel slope, it follows that the first task is to decide whether the slope is steep or mild. The equation of Prob. 4.12 gives a useful approximate guide even when the channel is of finite width, but to make an exact test it is necessary to determine whether the uniform flow curve in Fig. 4-11 cuts the specific energy curve on the subcritical limb, as at A, or on the supercritical limb, as at B. A convenient form of test is to determine the position (C_1 or C_2) at which the uniform flow curve cuts the line $y \doteq y_c$, and whether this point is to the right or left of C. The figures embodying this test can conveniently form the first line of the tabulation process by which the problem has to be solved when the slope is mild. The entire process is best illustrated by an example.

Example 4.1

A channel of rectangular section 10 ft wide, with $n = 0.014$ and $S_0 = 0.001$, leads from a lake whose surface level is 10 ft above the channel bed at the lake outlet. Find the discharge in the channel.

The maximum possible discharge would result from critical flow at the outlet, for which

$$y = y_c = \tfrac{2}{3} \times 10 \quad = 6\tfrac{2}{3} \text{ ft}$$

$$v_c = \sqrt{6\tfrac{2}{3}g} \quad = 14.64 \text{ ft/sec}$$

and

$$q = v_c y_c \quad = 97.7 \text{ cusecs/ft}$$

The criterion of Prob. 4.12 indicates a critical value of slope equal to

$$21.3 \times (0.014)^2 \times (97.7)^{-2/9} = 0.00151$$

which is too close to the actual value of 0.001 to be decisive, for this criterion is only very approximately applicable in this case. We proceed therefore to make the exact test by inserting $y_c = 6\tfrac{2}{3}$ ft into the Manning equation; since the test shows that the slope is mild, the tabulation proceeds by calculating uniform-flow conditions for a series of values of y until the resulting value of E becomes equal to the actual value, of 10 ft.

We use the constant

$$\frac{1.49\sqrt{.001}}{0.014} = 3.37$$

to calculate

$$v = 3.37 R^{2/3}$$

y, ft	A, sq. ft	P, ft	R, ft	$R^{2/3}$	v, ft/sec	Q, cusecs	$v^2/2g$, ft	E, ft	Remarks
$6\tfrac{2}{3}$	66.7	23.3	2.86	2.02	6.80	453	0.72	7.39	Test shows slope mild
9.2	92	28.4	3.24	2.19	7.36	677	0.84	10.04	
9.16	91.6	28.3	3.24	2.19	7.36	674	0.84	10.00	Correct discharge

The first line of the table shows that the slope is mild for the basic reason that the calculated velocity of 6.80 ft/sec is less than the critical velocity of 14.64 ft/sec; it follows that the uniform flow curve cuts the line $y = y_c$ to the left of C (Fig. 4-11). As an indicator of this fact, one may use either the velocity (6.80 < 14.64), the discharge (453 < 977) or the specific energy (7.39 < 10). The test calculation also indicates the order of magnitude of the velocity head, giving a guide to the first trial value of y.

If the slope had been steep, the discharge would have been given by the critical flow condition. In this case, some interest would attach to the calculation of the uniform depth, occurring further downstream; this depth can be obtained by exactly the same form of tabulation as above, aimed this time at finding the value of y which produces the already known value of discharge.

In this exercise, and in the discussion of Figs. 4-10 and 4-11, it has been assumed that there is no loss of energy as the water flows out of the lake. If such an exit loss exists, its magnitude can be found from the results given in Sec. 7.2; it is then easily incorporated into the tabulation of Example 4.1. This makes a convenient point at which to warn the reader against the elementary mistake of regarding the step down from lake level to channel water level in Fig. 4-10 as an energy loss only; this step represents an energy conversion rather than an energy loss, for its magnitude is equal to the velocity head *plus* the energy loss, if any.

4.6   Interaction of Local Features and Longitudinal Profiles

Introduction

In the treatment of the elementary discharge problem in Sec. 4.5 a difficulty arose which is characteristic of the whole subject of nonuniform flow. The difficulty was this: previously the channel slope had been discussed in a purely functional way—i.e., as if some outside arbiter had defined the slope as mild or steep before the discussion began. But in a real problem the engineer is given only a numerical value of slope, and must decide for himself whether it is mild or steep.

This difficulty was not a serious one, and could readily be dealt with as soon as it was clearly recognized. However it is worthy of explicit mention here because a more general form of this problem underlies much of the discussion in Sec. 4.5. It was shown in that section that controls are important as the points of origin for longitudinal profiles, but the controls were discussed as if they had all been nominated and their functioning described in advance. In practice this is never true; the engineer dealing with specific problems is given only the description of certain channel features and must decide whether and how they will act as controls.

In this respect it is important to observe that while any control present will influence and help to determine the whole flow profile, the profile in its turn may be said to influence the control, in the special sense that the form of a profile may determine whether a certain feature acts as a control or not. The most familiar example of this action is the "drowning" of a control, a possibility which has already been mentioned in the discussion of Fig. 4-10b, where it was pointed out that if the sluice gate part way down the steep slope were closed to a small enough opening, the S_1 curve behind it would fill all the upstream channel and drown the lake outlet, which would no longer be a critical section. The general principle which emerges is that a control may be drowned and deprived of its function by a stronger control downstream; a further example of this action is shown in Fig. 4-12. For the lowest profile

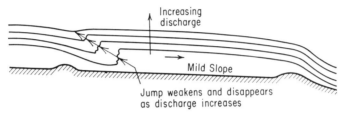

Figure 4-12.   *The Drowning of a Control by Increasing Discharge*

shown on this figure both weirs are acting as critical-flow controls; as the discharge increases the hydraulic jump moves upstream, becoming weaker as it does so, until finally it vanishes, the only remaining trace of it being a depression over the upstream weir. This weir is now drowned and flow over it is no longer critical. The action is in contrast to the case of Fig. 4-10b, where the drowning effect was produced by a change in the sluice-gate setting, not in the discharge.

The Effect of a Choke on the Flow Profile

The above discussion has dealt with the effect of a changing flow profile on a particular feature; it is also useful to consider this interaction in the converse way—i.e., to consider the effect on the flow profile of some feature as it is gradually converted into a control by some continuous change in the discharge or in the geometry of the feature. This latter type of change is unlikely to occur in physical reality, but a consideration of its effect makes a useful, if artificial, exercise for a designer seeking to determine a suitable size for some channel feature.

An example of great practical interest is provided by a local width contraction (e.g., bridge piers or a culvert) in a long uniform channel of mild slope. Suppose that initially the contraction is not a very severe one and the flow can be passed through it without requiring more specific energy than the upstream flow possesses; this means that the transition problem of Sec. 2.2 has a solution and the contraction does not act as a "choke." The flow within the contraction is therefore subcritical, as is the uniform flow for a great distance upstream and downstream (Fig. 4-13a). If now the contraction width is gradually reduced, a point is reached where the available specific energy is just sufficient to pass the flow through the contraction in the critical condition (Fig. 4-13b). This is the threshold of the choking condition, where the contraction becomes a control; it is determined according to the criteria discussed in Sec. 2.2. The present discussion is the one forecast in that section, where it was pointed out that the complete behavior of a choked contraction could be determined only by considering the channel as a whole, resistance effects included.

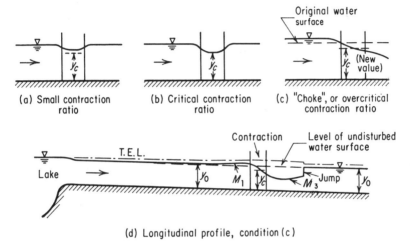

(a) Small contraction ratio

(b) Critical contraction ratio

(c) "Choke", or overcritical contraction ratio

(d) Longitudinal profile, condition (c)

Figure 4-13. *The Development of a Choke and its Effect on the Longitudinal Profile*

We now consider what happens if the contraction is narrowed even further. First, the flow within the contraction remains critical; this fact is of prime importance. Clearly there is no reason why this flow should return to sub-critical, for the condition which originally produced critical flow is now being pushed to even further extremes; on the other hand, the flow cannot pass to supercritical for the reasons given in Sec. 2.2. There is therefore no alternative to the maintenance of critical flow. Assuming for the moment that the discharge remains constant, it is seen that the discharge per unit width q within the contraction must increase, so that the critical depth y_c must also increase; it follows that the specific energy $E = 3y_c/2$ will increase, within the contraction and upstream, so that the upstream depth must increase and an M_1 curve will appear upstream (Fig. 4-13c). This behavior accords well with the intuitive notion that a severe constriction in the channel will cause the water to "back up" or "head up" so as to force the required discharge through the constriction.

We can now examine more critically the assumption that the discharge remains constant. Consider the channel as a whole, including the source of the flow, as in Fig. 4-13d. In this sketch is shown the whole extent of the M_1 curve of Fig. 4-13c; if it "runs out," as shown, before reaching the source, then the choking of the contraction has produced only a local disturbance which does not alter the discharge. On the other hand, if the contraction were made severe enough for the M_1 curve to reach right back to the lake, the discharge would be reduced somewhat.

In order to calculate the amount of this reduction it is necessary to calculate the shape of the M_1 curve; details of the necessary calculations must therefore be postponed until the subject of backwater computations is dealt with in Chap. 5. We are concerned here simply to discuss in general terms this phenomenon and certain other related ones; one further aspect that can profitably be discussed in this way is the behavior of the specific energy in the neighborhood of controls and similar features.

Specific Energy Changes Near Controls

First, it is important to see that the choking and backing up shown in Figs. 4-13c and d is independent of energy dissipation, and would occur even if the walls of the contraction were streamlined so as to eliminate energy loss. Nevertheless, energy concepts are useful in discussing certain consequences of the choking process. It is easily shown (Prob. 4.13) that in the M_1 curve the specific energy E increases in the downstream direction; it is this process which supplies the extra specific energy needed to pass the flow through the contraction. Further downstream, however, the flow must return to uniform, and to the appropriate value of E. The extra specific energy which was acquired upstream must therefore be given up, even if there is no energy loss in the contraction itself. And if there is no such energy loss, the required drop in E can occur only through the downstream development of super-critical flow (in which E decreases downstream) followed by a hydraulic jump. The total energy line therefore behaves as shown in Fig. 4-13d.

Similar reasoning would apply to a control such as a sluice gate, shown in Fig. 4-14. The argument is not of course limited to the case where the un-disturbed flow is uniform; in Fig. 4-14 the undisturbed flow (in the absence of the sluice gate) is an M_2 curve, and the process of departure from and return to this curve is essentially similar to that shown for uniform flow in Fig. 4-13d. It is noteworthy that in the present case the upstream profile produced by the sluice gate is itself an M_2 curve, although at a higher level than the original one. Reduction of the sluice-gate opening would raise the upstream profile even further until it represented uniform flow; further reduction in the opening would produce an M_1 curve. Downstream of the hydraulic jump the profile is, of course, unaltered by the presence of the sluice gate.

The preceding discussion has dealt with two types of control—the barrier type, such as a weir or sluice gate, and the choked-constriction type, in which critical flow occurs within the constriction. Both have the effect, when placed on a mild slope, of forcing a rise in the upstream water level and total energy line, and this can occur without any energy dissipation at the control itself.

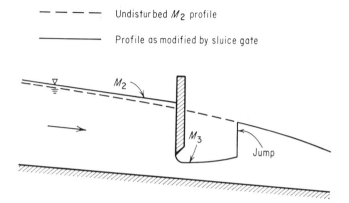

Figure 4-14.   *Modification of a Subcritical Profile by a Sluice Gate*

But it is conceivable that the same backing-up effect could arise simply from some feature which causes energy dissipation without acting as a control. Typical of such features are obstacles such as bridge piers which dissipate energy but present only a moderate degree of contraction to the flow. The behavior of bridge piers forms a subject of its own, which will be discussed fully in Chap. 7.

Control of upstream flow is the essence of the action here discussed; it is therefore mainly applicable on mild slopes where the undisturbed flow is subcritical. When the slope is steep the action of controls is to create an S_1 curve upstream, as in Fig. 4-10b, which may move upstream and drown the source. The end result may therefore be similar to that on a mild slope, although the details of the mechanism are different. The action of energy-dissipative features like bridge piers may take alternative forms, as will be discussed in Chap. 7.

References

1. Report, A.S.C.E. Task Force on Friction Factors in Open Channels, *Proc. Am. Soc. Civil Engrs.*, vol. 89, no. HY2 (March 1963), p. 97.
2. J. Williamson. "The Laws of Flow in Rough Pipes," *La Houille Blanche*, vol. 6, no. 5 (September-October 1951), p. 738.
3. V. J. Palmer and W. O. Ree. "Handbook of Channel Design for Soil and Water Conservation," *U.S. Dept. of Agriculture*, SCS-TP-61 (1954).

4. H. W. King. *Handbook of Hydraulics*, 4th ed. (New York: McGraw-Hill Book Company, Inc., 1954).
5. C. J. Posey and S. N. Pagay. "Manning's Formula for Wide Trapezoidal Channels," *Civil Engineering*, vol. 33, no. 3 (March 1963), p. 71.
6. Ven Te Chow. *Open-Channel Hydraulics* (New York: McGraw-Hill Book Company, Inc., 1959), Chap. 5.

Problems

4.1. A trapezoidal channel section like the one shown in Fig. 4-1 has a base width of 20 ft, side slopes of $1\frac{1}{2}$H : 1V, and a water depth of 6 ft. The channel is laid on a slope S_0 of 0.001, and the flow is uniform. The point of maximum velocity B is $1\frac{1}{2}$ ft below the water surface, and it may be assumed that the line CBD in Fig. 4-1 is a parabola. Water entering the region $CADB$ in the form of secondary flow has an average forward velocity of 5 ft/sec, and leaves this region with an average forward velocity of 7 ft/sec. Find the rate, in cusecs per foot length of channel, at which water circulates through the region $CADB$.

4.2. Prove for a circular pipe of diameter D, that $R = D/4$. Hence show that the Darcy equation for pipe flow is equivalent to a special case of the Chézy equation, provided that

$$C = \sqrt{\frac{8g}{f}} \qquad (4\text{--}8)$$

4.3. From Eqs. (4–16) and (4–22) show that the Manning formula is applicable provided that Eq. (4–24) holds true, assuming that ν for water $= 1.2 \times 10^{-5}$ ft²/sec.

4.4. Consider a trapezoidal channel section as shown in Fig. 4-6a. By eliminating b between the two equations for A and P, express P as a function of A, y, and θ. Hence show that if A and θ are fixed, P will be a minimum when:

$$A = y^2 (2 \operatorname{cosec} \theta - \cot \theta)$$

$$P = 2y (2 \operatorname{cosec} \theta - \cot \theta)$$

i.e., when $$R = \frac{y}{2}$$

and $$b = 2y \tan \frac{\theta}{2}$$

Hence show that if these equations hold true, it is possible to inscribe a semicircle in the channel section as shown in the figure.

4.5. A canal of rectangular section and $n = 0.014$ is to be laid on a slope of 0.001 and is to carry a discharge of 1,000 cusecs. Calculate the required waterway area for the following values of the ratio b/y : 1, 1.5, 2, 2.5, 3, 3.5, and plot the values of waterway area against b/y.

4.6. The canal described in the previous example is to be lined with concrete 6 in. thick; the cost of the concrete per cubic yard is 100 times the excavation cost per cubic yard. Either algebraically or by numerical methods, find the value of b/y for which the total cost is a minimum, and find relative values of total cost for this value of b/y, and for values equal to 0.75 and 1.5 of this one.

4.7. The problem is to derive a channel section, defined by a B-y equation, for which $\text{Fr}^2 = Q^2 B/gA^3$ is to remain constant at all values of y.

(a) Show that if Q is kept constant no such channel section can be derived, for the required section will have imaginary values of B.

(b) Show that if the specific energy E is kept constant, the required section will have the equation

$$\frac{B}{B_0} = \left(\frac{E}{E-y}\right)^{1+\text{Fr}^2/2}$$

where B_0 is the base width of the section.

4.8. From Eqs. (4–32) through (4–34), derive the profiles sketched in Fig. 4-9. Verify that the critical section of an S_2 curve must be produced by a channel feature, and that the critical section of an S_1 curve is normally bypassed by a hydraulic jump.

4.9. Show that if the Chézy C is constant and the section is wide rectangular, then Eq. (4–32) becomes

$$\frac{dy}{dx} = S_0 \frac{1 - \left(\dfrac{y_0}{y}\right)^3}{1 - \left(\dfrac{y_c}{y}\right)^3}$$

and hence that all profiles on a critical slope are straight horizontal lines.

4.10. For a channel section described by Eq. (2–37) and wide enough for P and B to be equal, derive the Q-y equation for uniform flow.

4.11. Prove Eq. (4–28). For $m = 2$, draw a graph of K' vs. y/b over the range $0.05 \leq y/b \leq 0.5$. From this graph determine the depth of uniform flow in a trapezoidal channel of base width 20 ft, $m = 2$, $S_0 = 0.001$, $n = 0.025$, $Q = 1{,}000$ cusecs.

4.12. Show that for a wide rectangular channel ($R = y$), the slope is steep or mild according as S_0 is greater or less than $21.3n^2 q^{-2/9}$.

4.13. For which of the profiles in Figs. 4-8 and 4-9 is the specific energy increasing in the downstream direction?

4.14. From Eqs. (4–32) through (4–34) prove that in the M_1 profile the water surface is falling in the downstream direction, and approaches its horizontal asymptote from above, whereas in the S_1 curve the reverse is true.

4.15. A natural river bed has a section which is approximately parabolic, such that the surface width is 150 ft when the center depth is 6 ft. If $n = 0.023$, find the center depth when the discharge is 1,500 cusecs and the slope is 4 ft per mile.

4.16. A sluice of triangular section is to be made from sawn timber ($n = 0.012$) of specified thickness, and is to carry a discharge of 1 cusec on a slope of 0.01. The cost of the sluice depends on the amount of wood used; neglecting any allowance for freeboard, show that for greatest economy the vertex angle of the triangle should be 90°, and find the depth and Froude number when the flow is uniform.

4.17. A trapezoidal channel has a base width of 20 ft, side slopes of $1\frac{1}{2}$H : 1V, and is lined with shot concrete, ($n = 0.016$). It is laid on a slope of 0.0006; find the discharge for a uniform depth of 8 ft.

4.18. Find the slope required to pass 200 cusecs through a rectangular channel 10 ft wide and lined with unfinished concrete, so that the Froude number shall not be less than 2.

4.19. For the same channel as in Prob. 4.18, and the same discharge, what would the uniform depth be if the slope were doubled?

4.20. Simultaneous flow gagings are made at two river sections A and B, 650 ft apart, with A upstream of B. When the flow is 17,000 cusecs, the following conditions exist at the two sections:

Section	Height of W.L. above datum, ft	Waterway Area, sq ft	Wetted Perimeter, ft
A	143.57	2,220	470
B	142.34	2,880	520

Choosing a suitable method of averaging the properties of the two sections, calculate the average value of n over the reach. Assume that form losses are negligible.

4.21. A trapezoidal channel of base width 30 ft and side slopes 2H : 1V is excavated in alluvial country, a sieve analysis of which gives a D-75 size of 2 in. The channel, which will be unlined, is to carry 1,250 cusecs under uniform flow conditions, and the velocity is not to exceed 5 ft/sec. Find the maximum allowable slope.

4.22. If the channel in Prob. 4.21 were laid on a slope of 0.0006, what would be the depth and Froude number of the flow?

4.23. The base width of the channel in Prob. 4.21 is not yet fixed, and the channel is to be laid on a slope of 0.0012. If the velocity is still not to exceed 5 ft/sec, find the minimum allowable width for the channel.

4.24. A trapezoidal channel of base width 20 ft, side slopes 2H : 1V, and $n = 0.022$, is laid on a slope of 0.0007 and carries a flow of 750 cusecs across a wide river terrace to a concrete chute of rectangular section which carries the flow down the steep face of the terrace to the river bed. The transition from the trapezoidal to the rectangular section is short but smooth, and is placed immediately upstream of the edge of the terrace. Find what the width of the chute must be so that there shall be no M_1 or M_2 curve from the upstream channel to the transition entrance.

4.25. A grassed channel is to take a flow of 600 cusecs, and is to be laid on a slope of 0.01. The grass height is expected to be 3–6 in., and the water velocity is not to exceed 8 ft/sec. Assuming that the channel is "wide" in the sense that $R = y$, find the minimum allowable width. (*Note*: Choose a trial width, hence VR, hence n, etc.)

4.26. A long rectangular channel 20 ft wide, with $n = 0.014$, is laid on a slope of 0.0004. At the downstream end of the channel is a sluice gate having a contraction coefficient of 0.8. When the vertical gate opening is 2.5 ft, find the discharge at which there will be no M_1 or M_2 curve behind the sluice gate.

4.27. A circular culvert is to carry a discharge of 450 cusecs on a slope of 0.0008, and is to run not more than half full. Circular pipes are available in spun concrete ($n = 0.011$) and internal diameters which are multiples of 1 ft. Choose a suitable culvert diameter.

4.28. At what depth would the culvert of the last problem take a discharge of 650 cusecs at uniform flow?

4.29. Devise a system, analogous to the one plotted in broken lines in Figs. 2-15 and 2-16, by which one can use the graphs of Probs. C4.1 and C4.2 to determine b or D, given Q and y_0.

4.30. A trapezoidal channel of base width 20 ft, side slopes $1\frac{1}{2}$H : 1V, $n = 0.023$, carries a discharge of 800 cusecs. Determine the critical value of slope.

4.31. A rectangular channel 20 ft wide, $n = 0.014$, forms the outflow from a lake whose surface level is 12 ft above the channel bed at the outlet. Determine the critical value of slope, and the discharge that occurs if this value of slope is exceeded.

4.32. Sketch all the longitudinal profiles that can occur for all possible values of the heights z_1 and z_2 in the figures shown.

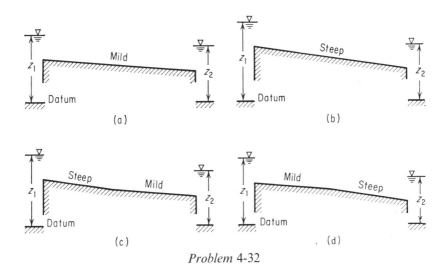

Problem 4-32

4.33. A long rectangular channel 20 ft wide, $n = 0.014$, is laid on a slope of 0.0003. Water is admitted to the channel by a sluice gate from a lake whose surface is 26.2 ft above the channel bed at this point. For a flow of 1,080 cusecs, sketch the longitudinal profile of the flow and discuss the formation and location of the hydraulic jump, if any.

4.34. The channel whose longitudinal section is shown in the figure is rectangular in section, 30 ft wide, with $n = 0.014$. Find the discharge in the channel, and sketch the longitudinal profile of the water surface. Determine whether a hydraulic jump occurs and if so whether it is upstream or downstream of the section A.

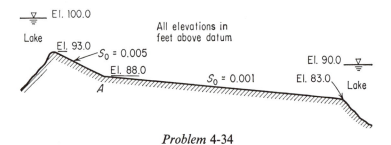

Problem 4-34

4.35.   If the channel of Prob. 4.31 has a slope of 0.0008, calculate the discharge and the uniform depth.

4.36.   If the channel of Prob. 4.31 has a slope of 0.005, calculate the discharge and the uniform depth.

4.37.   A relief drain from a large lake is to take the form of a rectangular section 20 ft wide and lined with concrete ($n = 0.014$). The channel will run 4,000 ft to a free overfall, where the channel bed level is to be set not less than 50.00 ft above datum. The channel is to discharge 800 cusecs, without allowing the lake level to rise more than 60.00 ft above datum; choose a suitable slope for the channel.

Computer Programs

C4.1.   Derive an equation corresponding to Eq. (4–28) giving K' as a function of y_0/D for uniform flow in a circular culvert. (There is no need to obtain K' as an *explicit* function of y_0/D; use the angle β, as defined in Prob. 3.26). Hence write and operate a program to compute simultaneous values of K' and y_0/D; plot the results on logarithmic graph paper, and verify that $K' = 0.465$ when $y_0/D = 1$, and has a maximum of about 0.50 when $y_0/D = 0.94$ approx.

C4.2.   Write and operate a program to compute simultaneous values of K' and y_0/b for $m = 0, 0.5, 1.0, 1.5, 2.0, 2.5, 3.0$, for the channel of trapezoidal section. Plot the results on the same sheet as those of the last problem.

C4.3.   Write a computer program to solve the following problem: given b, m, Q, S_0, and n, for a channel of trapezoidal section, determine y_0 by a trial process.

C4.4.   Use the above programs to rework Probs. 4.21, 4.22, 4.23, 4.24, 4.26, 4.27, and 4.28.

Chapter **5**

Flow Resistance—Nonuniform
Flow Computations

5.1   Introduction

The practicing engineer is faced with many situations in which he must be able to trace with reasonable accuracy the gradual variation of depth that takes place along a channel when the flow is nonuniform. His objectives in so tracing the longitudinal profile of the water surface may be many and various. One of them, mentioned in the previous chapter, is the determination of the discharge. Another is the assessment of the effect of a projected dam on upstream water levels: for what distance upstream does the dam produce a depth substantially different from uniform? Yet another is the tracing of flood levels upstream from a river mouth in order to assess the effect on these levels of certain proposed channel amendments.

Whatever the objectives may be, the problem is the integration of the equation of motion, either in the form

$$\frac{dE}{dx} = S_0 - S_f \qquad\qquad (4\text{–}30)$$

or the form

$$\frac{dy}{dx} = \frac{S_0 - S_f}{1 - \text{Fr}^2} \qquad\qquad (4\text{–}32)$$

In the previous chapter we have considered the general form that the solution might take: we now consider it in algebraic detail. It has already been remarked, in a discussion of Eq. (4–32), that it is not in general explicitly soluble; if the channel geometry and the form of the resistance equation are both particularly simple the equation may be solved explicitly, but in most cases we must seek a solution by numerical integration.

Now it is characteristic of numerical methods that the differing circumstances of particular problems may require differing methods of solution, although the basic equation itself remains unchanged. For example, it will be found that methods suitable for channels of uniform slope and section are not suitable for the irregular channels of natural rivers. However, all numerical solutions of the nonuniform flow equations have this in common

125

—that the calculation must start at a control, and proceed in the direction in which control is being exercised.

In this chapter detailed descriptions are given of methods that have been found suitable for problems occurring in engineering practice. The methods are grouped under two major headings—those suitable for application to uniform channels, and those suitable for irregular channels, such as in natural rivers.

Uniform Channels

5.2   Step Method—Distance Calculated from Depth

We consider Eq. (4–30) in the finite-difference form

$$\frac{\Delta E}{\Delta x} = \frac{\Delta\left(y + \dfrac{v^2}{2g}\right)}{\Delta x} = S_0 - S_f = S_0 - \frac{v^2}{C^2 R} \tag{5-1}$$

When the channel is of uniform slope and section, S_0 is a constant, so all the elements of the equation, except Δx, are dependent on y alone. This fact suggests a simple method of solution. We nominate a series cf values of y, and for each one E, v, C, and R can be immediately obtained. Hence Δx is calculated for each interval between successive values of y: the process is explicit, without the need for trial and error. The details are best shown by an example, the tabulation in which is almost self-explanatory.

Example 5.1

A trapezoidal channel of base width 20 ft and side slopes $1\frac{1}{2}$H : 1V is laid on a slope of 0.001 and carries a discharge of 1,000 cusecs. The channel terminates in a free overfall and it is desired to explore the "drawdown" to the overfall and the

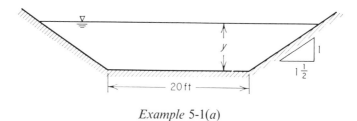

Example 5-1(a)

extent to which velocities are thereby increased. For this reason it is required to compute and plot the flow profile upstream from the overfall over the region where the velocity is at least 10 percent greater than at uniform flow. The Manning $n = 0.025$.

From the figure it is clear that

$$A = y(20 + 1\tfrac{1}{2}y)$$

$$P = 20 + 2\sqrt{3.25}\, y$$

$$= 20 + 3.60y$$

We obtain first the uniform depth y_0. At uniform flow we have

$$Q = \frac{1.49 A R^{2/3} S^{1/2}}{n} = \frac{1.49\sqrt{0.001}}{0.025} y_0(20 + 1.5y_0) \left\{ \frac{y_0(20 + 1.5y_0)}{20 + 3.6y_0} \right\}^{2/3} = 1{,}000 \text{ cusecs}$$

Whence by trial†

$$y_0 = 6.55 \text{ ft}$$

and
$$v_0 = 5.12 \text{ ft/sec}$$

Adding 10 percent to v_0, we obtain 5.63 ft/sec: at this velocity the depth will be about 10 percent less than y_0, i.e., in the neighborhood of 6 ft. We now obtain the critical depth y_c existing at the free overfall: at this point

$$v = \sqrt{g\frac{A}{B}}$$

i.e.,
$$Q^2 = g\frac{A^3}{B}$$

where B = the surface width = $20 + 3y$.
We therefore have

$$\frac{Q^2}{g} = \frac{A^3}{B} = \frac{y_c{}^3(20 + 1.5y_c)^3}{20 + 3y_c} = \frac{10^6}{g} = 31{,}050$$

i.e.,
$$y_c(20 + 1.5y_c) = 31.4\sqrt[3]{20 + 3y_c}$$

whence by trial†

$$y_c = 3.85 \text{ ft}$$

The range of depths to be covered therefore runs from 3.85 ft at the overfall to about 6 ft; the overfall itself is the control point from which the calculation must begin.

The depth intervals in the following tabulation represent a typical choice for this type of problem, in which a fairly close and accurate check is required on the exact amount of drawdown at any particular section, since the drawdown, with its attendant higher velocities, determines the need for protective lining to prevent scour. Also, Δy has been made larger near the overfall where the depth is changing rapidly. In other cases less precision is needed—e.g., in calculating flood profiles in a natural river, sections might be taken at intervals of a quarter-mile or more, instead of 200–300 ft as in this example.

† See Appendix on Mathematical Aids, Chap. 2.

EXAMPLE 5-1 Tabulation

1	2	3	4	5	6	7	8	9	10	11	12	13	14
y ft	A sq. ft	P ft	R ft	C^2 ft/sec^2	v ft/sec	$\dfrac{v^2}{2g}$ ft	E ft	$\dfrac{v^2}{C^2R}$	$\left(\dfrac{v^2}{C^2R}\right)_m$	$\left(S_0-\dfrac{v^2}{C^2R}\right)_m$	ΔE ft	Δx ft	$x=\Sigma\Delta x$ ft
3.85	99.5	33.9	2.93	5050	10.05	1.570	5.420	0.00683					0
									0.00572	−0.00472	+0.080	−17	
4.30	113.9	35.5	3.21	5210	8.78	1.200	5.500	0.00460					−17
									0.00412	−0.00312	+0.115	−37	
4.60	123.8	36.6	3.38	5300	8.08	1.015	5.615	0.00364					−54
									0.00327	−0.00227	+0.151	−66	
4.90	134.0	37.6	3.56	5400	7.46	0.866	5.766	0.00290					−120
									0.00262	−0.00162	+0.178	−110	
5.20	144.5	38.7	3.74	5480	6.92	0.744	5.944	0.00234					−230
									0.00213	−0.00113	+0.200	−177	
5.50	155.4	39.8	3.90	5560	6.44	0.644	6.144	0.00191					−407
									0.00177	−0.00077	+0.180	−234	
5.75	164.5	40.7	4.04	5620	6.08	0.574	6.324	0.00163					−641
									0.00151	−0.00051	+0.189	−370	
6.00	174.0	41.6	4.18	5690	5.75	0.513	6.513	0.00139					−1,011
									0.00134	−0.000340	+0.075	−220	
6.10	177.8	41.95	4.25	5720	5.60	0.488	6.588	0.00129					−1,231

We rewrite Eq. (5–1) in the form

$$\Delta x = \frac{\Delta E}{\left(S_0 - \dfrac{v^2}{C^2 R}\right)_m}$$

the subscript m indicating the mean value over an interval. Hence Column 13.

Column 5 is obtained from

$$C^2 = \left(\frac{1.49}{0.025}\right)^2 R^{1/3}$$

$$= 3530 R^{1/3}$$

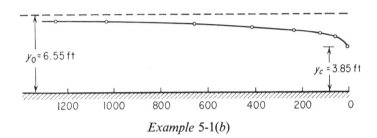

$y_0 = 6.55$ ft

$y_c = 3.85$ ft

| 1200 | 1000 | 800 | 600 | 400 | 200 | 0 |

Example 5-1(b)

Notes on the Tabulation in Example 5–1

1. The calculation of ΔE is inherently imprecise, since it is a small difference between two larger numbers; for this reason the velocity head must be calculated to three decimal places. A more precise method would be to use the equation

$$\Delta E = \Delta y(1 - \mathrm{Fr}^2)$$

implying the use of Eq. (2–23), although for nonrectangular channels this would require a further column for the surface width $B(\mathrm{Fr}^2 = v^2 B/gA)$. The value of Fr^2 used would be the mean over the interval.

2. As the depth approaches y_0, the difference $(S_0 - v^2/C^2 R)$ becomes small and less easy to calculate exactly. This is not a matter of great concern, since in practice we are usually not greatly interested in details of the asymptotic approach of y to y_0; we are more concerned with the surface profile in regions where a control has drawn the depth substantially away from y_0.

3. The signs in columns 11–14 are self explanatory but the reader should develop the habit of keeping careful track of these signs, since they provide to some extent a check against gross numerical errors. In this case we find that $S_f > S_0$ and $(S_0 - S_f)$ is negative, as one would expect when $y < y_0$. Since E increases down the table (E increases with y in subcritical flow) ΔE is always positive. Hence the quotient $\Delta x = \Delta E/(S_0 - S_f)$ is negative, which confirms that the calculation proceeds *upstream* from the overfall—i.e., in the direction in which control is being exercised.

4. We have seen that E decreases in the downstream direction. It is helpful to think of this result in the following way. The constancy of E in uniform flow is the

result of a balance between a gain in E due to a continual drop in the channel bed and a loss in total energy (and hence in E) due to flow resistance. This balance is expressed geometrically by the fact that the bed is parallel to the total energy line, since $S_0 = S_f$. Now if $y < y_0$, $S_f > S_0$ from Eq. (4–33), and the total energy line dips more steeply than the bed; E therefore decreases in the downstream direction. On the other hand if $y > y_0$, as in the M_1 profile, E will actually increase in the downstream direction.

The complete profile is plotted below the tabulation with the vertical scale exaggerated; by interpolation it is found that $v = 1.1\ v_0 = 5.63$ ft/sec at a distance of 1,200 ft from the overfall.

The method is useful for artificial canals, but alternative methods exist which are superior in some circumstances. These will be discussed in the next two sections.

5.3   Direct Integration Methods

We have seen that the flow equation

$$\frac{dy}{dx} = \frac{S_0 - S_f}{1 - \mathrm{Fr}^2} \tag{4–32}$$

is true for all forms of channel section, provided that the Froude number Fr is properly defined by the equation $\mathrm{Fr}^2 = v^2 B/gA = Q^2 B/gA^3$. We now re-write certain of the other elements of this equation with the aim of examining the possibility of a direct integration. It is convenient to use here the conveyance K, introduced in Sec. 4.3 and defined by the equation

$$Q = \frac{1.49AR^{2/3}S^{1/2}}{n} = KS^{1/2} \tag{4–25}$$

so that

$$K = \frac{1.49AR^{2/3}}{n} \tag{4–26}$$

It follows from Eq. (4–25) that for a given Q, $K^2 \propto 1/S$, so that the term $(S_0 - S_f)$ may be written as

$$S_0\left(1 - \frac{S_f}{S_0}\right) = S_0\left(1 - \frac{K_0^2}{K^2}\right) \tag{5–2}$$

where K_0 is the conveyance at uniform flow. Now K includes the geometrical term $AR^{2/3}$ which will vary with the depth y in some way dependent on the shape of the cross section; we assume that

$$(AR^{2/3})^2 \propto y^N \tag{5–3}$$

although strictly speaking this is true only for sections which are very wide and are described by the exponential equation

$$\frac{B}{B_s} = \left(\frac{y}{y_s}\right)^i \tag{2–37}$$

which applies to such sections as the triangular and parabolic. Comment on the applicability of Eq. (5–3) to such sections as the trapezoidal will be reserved until later. It follows from Eq. (5–3) that

$$\frac{K_0{}^2}{K^2} = \left(\frac{y_0}{y}\right)^N \tag{5–4}$$

We now consider $\mathrm{Fr}^2 = Q^2 B / g A^3$. Since this term equals unity at critical flow, then

$$\frac{Q^2}{g} = \frac{A_c{}^3}{B_c}$$

where A_c and B_c are the values of A and B at critical flow. Then

$$\mathrm{Fr}^2 = \frac{A_c{}^3}{B_c} \bigg/ \frac{A^3}{B} \tag{5–5}$$

and if we assume that the geometrical term A^3/B varies as y^M, (again implying that Eq. (2–37) is true) we have

$$\mathrm{Fr}^2 = \left(\frac{y_c}{y}\right)^M \tag{5–6}$$

Finally Eq. (4–32) may be written as

$$\frac{dy}{dx} = S_0 \frac{1 - \left(\dfrac{y_0}{y}\right)^N}{1 - \left(\dfrac{y_c}{y}\right)^M} \tag{5–7}$$

granted the geometrical assumptions implied in Eqs. (5–4) and (5–6). When the channel section is rectangular, Eq. (5–6) is true and $M = 3$; if the channel is also very wide, Eq. (5–4) is true and $N = 3\frac{1}{3}$ or 3, depending on whether the Manning n or the Chézy C is assumed to be constant. The latter is the case of Prob. 4.9.

The Bresse Function

Equation (5–7) is not in general integrable by elementary means. Of the few special cases in which it is integrable, by far the most important is the case of Prob. 4.9, $M = N = 3$. The integration, first undertaken by Bresse, proceeds as follows:

$$S_0 \frac{dx}{dy} = \frac{y^3 - y_c{}^3}{y^3 - y_0{}^3} = 1 - \frac{1 - \left(\dfrac{y_c}{y_0}\right)^3}{1 - \left(\dfrac{y}{y_0}\right)^3}$$

Integration of this equation then leads to the result

$$S_0 x = y - y_0 \left[1 - \left(\frac{y_c}{y_0} \right)^3 \right] \Phi \qquad (5\text{–}8)$$

where Φ is known as Bresse's function, and is equal to

$$\Phi = \int \frac{du}{1 - u^3} = \frac{1}{6} \log \frac{u^2 + u + 1}{(u - 1)^2} - \frac{1}{\sqrt{3}} \tan^{-1} \frac{\sqrt{3}}{2u + 1} + A_1 \qquad (5\text{–}9)$$

where $u = y/y_0$ and A_1 is a constant of integration. The term $(y_c/y_0)^3$ may also be written as $C^2 S_0/g$, a form that is more convenient for computation purposes. The Bresse function is tabulated in the Appendix to this chapter.

The Bakhmeteff Varied-Flow Function

The integral

$$F(u,N) = \int_0^u \frac{du}{1 - u^N} \qquad (5\text{–}10)$$

may be evaluated numerically even if N is not a whole number. Tables may thus be prepared covering a range of values of u and N, by means of which longitudinal profiles might be plotted for a number of different channel shapes. Preparing such tables is particularly simple nowadays when high speed computers are commonly available; however the first work along these lines, carried out by B. A. Bakhmeteff, was begun as early as 1912.

A difficulty is that when $N \neq M$, Eq. (5–7) does not lead to a result expressed wholly in terms of integrals such as the one in Eq. (5–10). Bakhmeteff overcame this difficulty by the following argument: we set $\mathrm{Fr}^2 = \beta S_f/S_0$, and then consider the behavior of β. It is equal to

$$\beta = \frac{\mathrm{Fr}^2 S_0}{S_f} = \frac{Q^2 B}{g A^3} \frac{S_0 A^2 C^2 R}{Q^2} = \frac{C^2 S_0}{g} \frac{B}{P} \qquad (5\text{–}11)$$

and will clearly not vary a great deal in moderately wide channels. Bakhmeteff treated β as a constant, allowing for its variation by dividing the channel length into sections, short enough so that β could be assumed constant over each section. Under this assumption, Eq. (4–32) becomes, via Eqs. (5–2) and (5–4),

$$S_0 \frac{dx}{dy} = \frac{1 - \beta \left(\dfrac{y_0}{y} \right)^N}{1 - \left(\dfrac{y_0}{y} \right)^N} = 1 - \frac{1 - \beta}{1 - \left(\dfrac{y}{y_0} \right)^N}$$

the integration of which involves only integrals of the form of Eq. (5–10). In the above equation β clearly corresponds to the term $(y_c/y_0)^3$ in Eq. (5–8).

Bakhmeteff examined also the assumption that

$$\frac{S_f}{S_0} = \left(\frac{y_0}{y}\right)^N$$

as in Eq. (5–4). He found that it was a fair approximation to the truth for small variations of depth in trapezoidal channels, and that N might therefore be assumed constant over channel reaches of moderate length. It was found that N varied from 2 for a deep narrow rectangular section (Prob. 5.6) to $5\frac{1}{3}$ for a triangular section (Prob. 5.7). We have already noted that $N = 3\frac{1}{3}$ for a wide rectangular section.

Bakhmeteff accordingly calculated the value of the integral of Eq. (5–10) for values of N covering the range from 2.0 to 5.5. The results of this classic work are embodied in tables [1], a few of which are given in the Appendix at the end of this chapter. However he did not explore in detail the large variations in N that occur with large variations in depth—e.g., the variation in N from $3\frac{1}{3}$ to $5\frac{1}{3}$ as the depth increases from zero to infinity in a trapezoidal channel, as indicated by the above results. This detailed exploration was left for a later study by another investigator.

Ven Te Chow's Analysis

Many investigators have suggested means of refining Bakhmeteff's original work; Ven Te Chow in particular has developed methods [2] which extend and consolidate Bakhmeteff's work while retaining the same form of varied-flow function. First, he deals with the case $N \neq M$ by the following argument.

Equation (5–7) may be rewritten as

$$dx = \frac{y_0}{S_0}\left[1 - \frac{1}{1 - u^N} + \left(\frac{y_c}{y_0}\right)^M \frac{u^{N-M}}{1 - u^N}\right] du \qquad (5\text{–}12)$$

where u, as before, is equal to y/y_0. The second term within the bracket leads directly to Bakhmeteff's varied flow function; the third can be transposed into this form by the substitutions $v = u^{N/J}$ and $J = N/(N - M + 1)$. Integration of this term then leads to

$$\int_0^u \frac{u^{N-M}}{1 - u^N}\, du = \frac{J}{N}\int_0^v \frac{dv}{1 - v^J} \qquad (5\text{–}13)$$

Thus the whole of the integral of Eq. (5–12) has been written in terms of the original varied-flow function without the use of any approximations. However, the use of Eq. (5–13) requires extensions to the tables of this function, since the argument J can take values much larger than Bakhmeteff's maximum of 5.5. Accordingly the tables have been extended by Ven Te Chow to cover values of the argument up to 9.8.

From Eqs. (5–12) and (5–13) we can finally write

$$x = \frac{y_0}{S_0}\left[u - F(u,N) + \left(\frac{y_c}{y_0}\right)^M \frac{J}{N} F(v,J)\right] + A_1 \qquad (5\text{–}14)$$

The same investigator examined the behavior of N and M for a number of sectional shapes and a number of width: depth ratios. Full details are given in Ref. [2] and results are summarized here. For any section it can be shown (Prob. 5.8) that

$$M = \frac{y}{A}\left(3B - \frac{A}{B}\frac{dB}{dy}\right) \qquad (5\text{–}15)$$

and that for trapezoidal channels this equation takes the form

$$M = \frac{3(1 + 2y')}{1 + y'} - \frac{2y'}{1 + 2y'}$$

$$= 3 + \frac{y'(1 + 4y')}{(1 + y')(1 + 2y')} \qquad (5\text{–}16)$$

where $y' = my/b$ as in Secs. 2–7 and 3–4. Similarly it can be shown (Prob. 5.9) that

$$N = \frac{2y}{3A}\left(5B - 2R\frac{dP}{dy}\right) \qquad (5\text{–}17)$$

which for trapezoidal channels takes the form

$$N = \frac{10(1 + 2my/b)}{3(1 + my/b)} - \frac{8\sqrt{1 + m^2}\,(y/b)}{3[1 + 2\sqrt{1 + m^2}\,(y/b)]} \qquad (5\text{–}18)$$

Here we find, as in Sec. 4.3, that equations containing P and R cannot be expressed wholly in terms of $y' = my/b$.

The Appendix to this chapter lists values of $F(u,N)$ for $N = 3$ (the Bresse function), $N = 3\frac{1}{3}$ and $N = 2.5$. The last two values are for the wide rectangular channel using the Manning equation; the latter of these is required for the third term of Eq. (5–14), since in this case

$$J = \frac{N}{N - M + 1} = \frac{3\frac{1}{3}}{3\frac{1}{3} - 3 + 1} = 2.5$$

The reader can compile tables for other values of N (Prob. C5.3).

We conclude this section with a worked-out example showing the use of the varied-flow function.

Example 5.2

A wide rectangular channel carries a discharge of 40 cusecs/ft width on a slope of 0.001; the Manning n is 0.025. Plot the flow profile for a distance of 2,000 ft upstream from a weir which increases the depth to 10 ft.

Now
$$q = 40 = \frac{1.49 y_0^{5/3} \sqrt{0.001}}{0.025} \; ; \quad y_0 = 6.25 \text{ ft}$$

whence
$$\frac{y_0}{S_0} = 6{,}250 \text{ ft}$$

Also,
$$y_c^3 = \frac{q^2}{g} = \frac{1{,}600}{g} = 49.7$$

$$\left(\frac{y_c}{y_0}\right)^M \frac{J}{N} = \frac{49.7}{6.25^3} \frac{2.5}{3\frac{1}{3}} = 0.153$$

Hence from Eq. (5–14)
$$x = 6{,}250[u - F(u,3\tfrac{1}{3}) + 0.153F(v,2.5)] + \text{const.}$$

where $v = u^{4/3}$ and $u = y/y_0$. Initially

$$u = \frac{10}{6.25} = 1.6$$

and this value is the starting point for the following tabulation, in which x is the distance measured downstream from some datum (see Note 1 below), and L is the distance measured upstream from the weir.

EXAMPLE 5–2 Tabulation

1	2	3	4	5	6	7	8
u	y (ft)	v	$F(u,3\tfrac{1}{3})$	$F(v,2.5)$	$0.153 \times$ col. 5	x (ft)	L (ft)
1.6	10.00	1.87	0.158	0.287	0.044	9,290	0
1.5	9.375	1.72	0.188	0.333	0.051	8,520	770
1.4	8.75	1.57	0.229	0.396	0.061	7,700	1,590
1.3	8.125	1.42	0.289	0.481	0.074	6,780	2,510

Notes on the Tabulation in Example 5.2

1. The function $F(u,N)$ may have any desired constant added to it. In the table given in the Appendix this constant is adjusted so that $F(0,N) = F(\infty,N) = 0$; clearly this is a convenient general-purpose adjustment of the constant. It means, incidentally, that the constant has differing values for $u > 1$ and $u < 1$, but this introduces no difficulty, since for all real profiles u is either greater or less than unity for the whole length of the profile.

A further adjustment to the constant is almost always needed to suit the requirements of a particular problem such as the above. Thus the tabulation gives $x = 9,290$ at the weir, which is the starting point of the problem. Recalling that x increases downstream, it is a simple matter to obtain values of L, the distance measured upstream from the weir.

2. Since the integration is direct, the successive values of x in the tabulation are independent of each other. One may think of the calculation proceeding in independent steps as indicated in Fig. 5-1a. On the other hand, the step method of Sec. 5.2 requires that each step is dependent on the preceding one, as indicated in Fig. 5-1b. Herein lies one of the main advantages of direct integration over stepwise numerical integration: thus if one only wished to find the section where $u = 1.3$, it would be unnecessary in the direct method to compute x for $u = 1.5$, 1.4, whereas in the step method these intermediate steps could not be avoided unless one were prepared to risk a loss of precision by making one step cover the whole range $1.3 \leq u \leq 1.6$.

Unfortunately this advantage of the direct method is seldom exploited, since one usually wishes to know all the intermediate values of x in order to plot a complete profile.

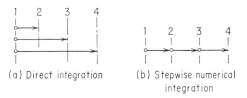

(a) Direct integration   (b) Stepwise numerical integration

Figure 5-1.   *Alternative Schemes of Computation for Longitudinal Profiles*

5.4   Step Method—Depth Calculated from Distance

The methods described so far have required the operator to nominate a series of depths, from which values of distance were to be calculated. While this procedure is often quite satisfactory, it is sometimes inconvenient. For instance we shall see in studying the discharge problem that it is usually necessary to determine the depth variation over a specified length of channel —e.g., from a lake outlet to an overfall. Attacking this problem by the method of Sec. 5.2, or even by the direct integration method, requires a profile to be plotted passing through the specified section, such as a lake outlet, and obtaining the depth at that section by interpolation. This procedure adds materially to the time and labor involved.

In such cases it would clearly be convenient to be able to nominate values of x and Δx in advance, and to calculate y from them. Since y is involved in so many of the elements of Eqs. (4–30) and (4–32), trial or graphical methods must be used, and of these two the latter is usually more convenient, the preliminary work of plotting the graph being repaid by the time saved in the subsequent calculation.

Many such methods have been put forward, based on varying arrangements of the flow equation. The method given here is a variant of an algebraic technique that has proved successful in flood routing problems.

Let two sections, separated by a distance Δx, be labeled 1 (downstream) and 2 (upstream) and let their parameters be subscripted accordingly. Then from Eq. (4–30)

$$E_1 - E_2 = (S_0 - \tfrac{1}{2}S_{f_1} - \tfrac{1}{2}S_{f_2})\Delta x \tag{5–19}$$

Suppose that the calculation is proceeding upstream, so that the conditions are known at 1 and unknown at 2. Then we arrange Eq. (5–19) in this way:

$$E_1 - \tfrac{1}{2}S_{f_1}\Delta x + (S_{f_1} - S_0)\Delta x = E_2 - \tfrac{1}{2}S_{f_2}\Delta x$$

i.e.,

$$U_2 - U_1 = (S_{f_1} - S_0)\Delta x \tag{5–20}$$

where

$$U = E - \tfrac{1}{2}S_f\Delta x \tag{5–21}$$

and is a function of y and Δx. Given Q and Δx (assumed constant) we can plot U against y and use the graph in the following scheme of computation: given x_1 and y_1, obtain U_1 from the graph, U_2 from Eq. (5–20), thence y_2 from the graph. It is desirable also to plot $(S_f - S_0)$ against y; this graph supplies values of $(S_{f_1} - S_0)$ to insert in Eq. (5–20).

If the calculation is proceeding downstream, so that conditions are known at 2 and unknown at 1, we arrange Eq. (5–19) in this way:

$$E_1 + \tfrac{1}{2}S_{f_1}\Delta x = (S_0 - S_{f_2})\Delta x + E_2 + \tfrac{1}{2}S_{f_2}\Delta x$$

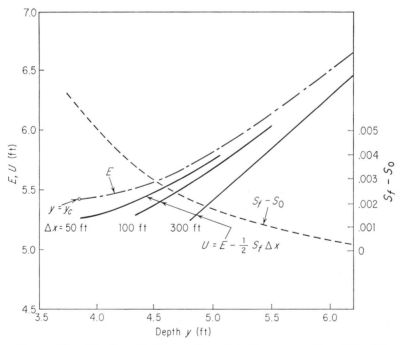

Figure 5-2.   The Specific Energy E, and the Parameters U and $(S_f–S_0)$, as Functions of y in Example 5-3

i.e.,
$$V_1 - V_2 = (S_0 - S_{f_2})\Delta x \qquad (5\text{--}22)$$

where
$$V = E + \tfrac{1}{2}S_f \Delta x \qquad (5\text{--}23)$$

Plotting V against y leads to a scheme of computation similar to that in the previous case; the reversals of sign must however be noted with care. The difference between the functions U and V is due to a difference in the sign of Δx, since a reversal of the direction of the calculation implies a reversal of the sign of Δx. In essence, therefore, U and V are the same function; however it is more convenient to regard them as different functions and to define Δx (as in the above argument) as the unsigned magnitude of the x interval, so that Δx is always positive.

We illustrate the method by working through the same example as was dealt with in Sec. 5.2. The rapid change in y near the overfall makes it desirable to use smaller values of Δx here than further upstream; however the magnitude of Δx can readily be changed part way through the calculation, as will be shown.

Since the calculation proceeds upstream, we use the U function. On Fig. 5-2, E and $(S_f - S_0)$ are plotted against y; U is also plotted for three different values of Δx—50, 100, and 300 ft. When Δx is changed, two separate lines are required in the tabulation. These lines show the same value of y, but different values of U.

EXAMPLE 5–3 Repetition of Example 5–1 by Graphical Method

x	Δx	y	U	$S_f - S_0$	$\Delta U = (S_{f_1} - S_0)\,\Delta x$
0		3.85	5.249	0.00583	
	50				0.292
-50		4.63	5.541	0.00257	
	50				0.128
-100		4.86	5.669	0.00201	
-100		4.86	5.593	0.00201	
	100				0.201
-200		5.15	5.794	0.00144	
	100				0.144
-300		5.35	5.938	0.00113	
-300		5.35	5.726	0.00113	
	300				0.339
-600		5.725	6.065	0.00065	
	300				0.195
-900		5.95	6.260	0.00044	
	300				0.132
-1200		6.09	6.392		

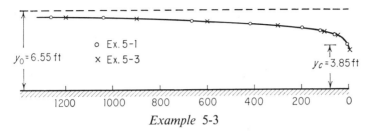

Example 5-3

The result of the calculation is plotted above, together with the result obtained by the method of Sec. 5.2. The two results are virtually indistinguishable.

The method may be converted into a wholly graphical one by plotting another curve at a (variable) distance $(S_f - S_0) \Delta x$ above the U-y curve, as shown in Fig. 5-3. It follows from Eq. (5–20) that the path between successive points 1 and 2 on the U-y curve may be traced via the upper curve in the way shown

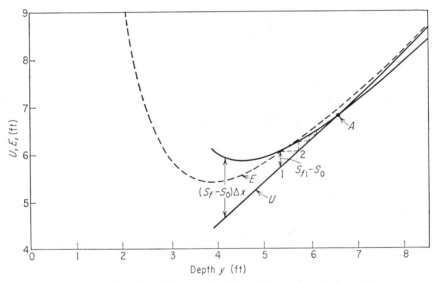

Figure 5-3.   *Graphical Scheme for Determining Depth from Distance in Longitudinal Profiles*

in the figure, and that all subsequent points may be found by tracing a sawtooth path between the two curves as shown. This path must terminate at the point A where $S_f = S_0$ and the two curves meet; this point therefore represents uniform flow. Since the intervals Δy become infinitesimally small as this point is approached, an infinite number of finite steps Δx must be taken, confirming that y approaches y_0 asymptotically.

The particular configuration drawn in Fig. 5-3 relates only to the M_2 curve; other arrangements will be required for other profiles. If the flow is supercritical the appropriate curves are based on the supercritical limb of the E-y curve, on the left of Fig. 5-3. In this case the two working curves would not intersect unless the slope were steep, and uniform flow therefore supercritical. The details are left as an exercise for the reader (Probs. 5.14 and 5.15.)

While the tabular and graphical methods described in this section are very quick and convenient once the curves of Figs. 5-2 and 5-3 are plotted, the preparation of these curves requires substantially as much work as does the tabulation of Sec. 5.2. Therefore the methods of this section are worthwhile only when a number of profiles are to be plotted for the same discharge—e.g., M_1 curves upstream of a sluice gate for a number of different gate openings. A method suitable for varying discharge is described in Sec. 5.7.

The methods described in the following sections have been developed particularly for use in irregular channels, but may of course be applied to uniform channels as well. Normally there is little point in so doing, but an exception is the modified Ezra method described in Sec. 5.7, which may be profitably applied to uniform as well as to irregular channels.

Irregular Channels

5.5   Step Method—Single Channels

In this and succeeding sections we shall deal with irregular channels such as occur in natural rivers. In such channels it often happens that the water overflows on to flood plains or berms, as described in Sec. 1.9. In this way two parallel systems of flow are formed which have to be considered separately. In this first section, however, we shall consider only those cases in which the water is contained within a single channel, forming a single system of flow.

Now in any numerical integration process carried out in natural river channels, it is necessary to work from chosen values of x, and to calculate the depth accordingly. One practical reason for this is that river-channel properties are usually measured only at certain fixed sections, as shown in Fig. 5-4. But even if the channel geometry were known exactly at every section along the length of a river, the irregular variation of channel properties with distance x would make it even more difficult to calculate x from a given y than to calculate y from a given x.

In plotting profiles in natural rivers we must therefore calculate y from x, and must accordingly use some sort of trial process. Actually the depth y seldom appears in the calculations; it is preferable to specify instead the height h of the water level above some fixed datum. This height is known as the *stage*, a term that occurs frequently in discussions of the behavior of natural rivers.

The most commonly used method is a straight trial process, the problem being formulated as follows: For a given discharge the river stage is known at one section (1) and it is required to calculate the stage at the adjacent section (2). At both sections a plan of the waterway is available, from which values of A, P, etc., may be determined for any given stage. The Manning n is also known.

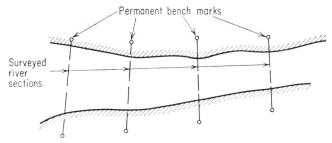

Figure 5-4.   *Layout of Surveyed Cross Sections in a Natural River Channel*

We take a trial value of the unknown stage h_2 and calculate A_2, v_2, $v_2^2/2g$, and hence H_2, the total energy. The friction slope $S_{f_2} = v_2^2/C_2^2R_2$ can also be calculated, and from the mean value $(S_{f_1} + S_{f_2})/2$ we obtain a total energy difference $H_2 - H_1$, and hence H_2. The test of the trial process is whether this value of H_2 is equal to the one previously calculated; if not, a further trial value must be taken.

Example 5.4 shows the process in tabular form for a typical problem, viz., the determination of flood levels in a natural river. The calculation proceeds upstream from a known depth at a certain section, identified in column 1 by its distance (36.55 miles) upstream from a section near the river mouth. This is the normal method of identifying river sections. All necessary data are given either at the head of the table or in the first line of the tabulation.

A column is included for eddy losses such as occur at sharp bends or sudden expansions. Unfortunately there is little generalized information on this topic, and it is common practice for river engineers to determine local values for loss coefficients from known flood levels,. In the last reach of the example, eddy losses due to a bend in the river are assumed equal to 0.1 times the mean of the velocity heads at the two ends of the reach.

The tabulation is almost self-explanatory. In column 5 is listed the value of total head H obtained simply from the assumed water-surface elevation and the velocity head calculated from the corresponding values of A and v. Columns 6–8 are self-explanatory; column 9 contains the friction slope S_f calculated direct from the velocity head by means of the factor worked out at the head of the table. The average S_f listed in column 10 is the average of the two values of S_f in column 9, relating to the current section and the

EXAMPLE 5-4 Longitudinal Profile in a Natural River by the Standard Step Method

$Q = 90{,}000$ cusecs, $n = 0.033$

$$\frac{2g \times 0.033^2}{1.49^2} = 0.0316 \qquad S_f = 0.0316\,\frac{v^2/2g}{R^{4/3}}$$

1	2	3	4	5	6	7	8	9	10	11	12	13	14
River mile	River stage, ft	Area, A sq ft	$\dfrac{v^2}{2g}$ ft	Total head, H ft	P ft	R ft	$R^{4/3}$	Friction slope, S_f	Average S_f over reach	Length of reach ft	h_f ft	h_e ft	Total head, H ft
36.55	65.21	11,950	0.88	66.09	623	19.2	51.4	0.00054	–	–	–	–	66.09
36.70	65.5	10,700	1.10	66.60	514	20.8	57.2	0.00061	0.00057	792	0.45	0.00	66.54
36.70	65.44	10,670	1.10	66.54	514	20.8	57.2	0.00061	0.00057	792	0.45	0.00	66.54
36.95	66.4	7,700	2.12	68.52	500	15.4	38.3	0.00175	0.00118	1320	1.56	0.00	68.10
36.95	65.96	7,450	2.26	68.22	497	15.0	37.0	0.00193	0.00127	1320	1.68	0.00	68.22
				Bridge, Rounded Piers, Opening: Span Ratio = 0.94									
36.97	66.4	7,700	2.12	68.52	500	15.4	38.3	0.00175	–	–	–	–	68.52
37.10	68.0	8,900	1.58	69.58	551	16.2	41.0	0.00122	0.00149	686	1.02	0.00	69.54
37.10	67.94	8,850	1.61	69.55	550	16.1	40.7	0.00125	0.00150	686	1.03	0.00	69.55
37.30	69.0	11,050	1.03	70.03	637	17.4	45.1	0.00072	0.000985	1056	1.04	0.13	70.72
37.30	69.75	11,500	0.95	70.70	645	17.8	46.5	0.00065	0.00095	1056	1.00	0.13	70.68

previous section (*not* the previous trial value of the current section). The resistance and eddy losses in columns 12 and 13 lead to the total head value in column 14; if this is not equal to the value in column 5 a line is ruled neatly through the whole line of calculation and a further trial water-surface level taken.

The water surface rise of 0.5 ft through the bridge piers is calculated by the methods given in Chap. 7. A distance of about 100 ft (0.02 mile) has been allowed for the width of the bridge and it has been assumed that the river section is the same immediately upstream of the bridge as it is immediately downstream.

It will be noticed that only two trial values of water-surface level have been taken at each section, and that the second trial is always successful. This has been accomplished by using the results of the first trial as a guide to the second trial, according to the following argument.

The aim of the process is to equalize the two values

$$H_2 = y_2 + z_2 + \frac{v_2{}^2}{2g}$$

and
$$H_2 = H_1 + \tfrac{1}{2}\Delta x(S_{f_1} + S_{f_2})$$

Let the difference between these two (i.e., the error) be denoted by H_E. The aim is to make this quantity vanish by changing the water-surface level— i.e., by changing y_2. (Clearly we cannot change z_2.) We are concerned therefore with the response of H_E to small changes in y_2, and this response is measured by the derivative dH_E/dy_2. Since z_2, H_1 and S_{f_1} are constants, then

$$\frac{dH_E}{dy_2} = \frac{d}{dy_2}\left(y_2 + \frac{v_2{}^2}{2g} - \tfrac{1}{2}\Delta x S_{f_2}\right)$$

$$= 1 - \mathrm{Fr}_2{}^2 - \tfrac{1}{2}\Delta x \frac{dS_{f_2}}{dy_2}$$

Since S_f varies approximately as the inverse cube of y,

$$\frac{dS_{f_2}}{dy_2} \approx -\frac{3S_{f_2}}{y_2} \approx -\frac{3S_{f_2}}{R_2}$$

and we have
$$\frac{dH_E}{dy_2} = 1 - \mathrm{Fr}_2{}^2 + \frac{3S_{f_2}\Delta x}{2R_2}$$

or
$$\Delta y_2 = \frac{H_E}{1 - \mathrm{Fr}_2{}^2 + \dfrac{3S_{f_2}\Delta x}{2R_2}} \tag{5-24}$$

where Δy_2 is the amount by which the water level must be changed in order to make a small error H_E vanish.

Since in a natural river the surface width B approximates closely to the wetted perimeter P, then

$$\text{Fr}^2 = \frac{v^2 B}{gA} \approx \frac{v^2}{gR} \approx \frac{v^2/2g}{R/2}$$

and is obtained from columns 4 and 7. If there are eddy losses, equal to

$$\tfrac{1}{2} C_L \left(\frac{v_1^2}{2g} + \frac{v_2^2}{2g} \right)$$

their effect can be allowed for by multiplying Fr_2^2 in Eq. (5–24) by the factor $(1 - \tfrac{1}{2} C_L)$.

Consider the application of Eq. (5–24) at river mile 36.95. After the first trial, $H_E = 68.52 - 68.10 = 0.42$. Also,

$$\text{Fr}^2 = \frac{2 \times 2.12}{15.4} = 0.275$$

and

$$\frac{3 S_f \Delta x}{2R} = \frac{1.5 \times 0.00175 \times 1320}{15.4} = 0.225$$

so that the required correction, Δy, is equal to

$$\frac{0.42}{1 - 0.275 + 0.225} = 0.44$$

and the sign of the correction will be negative, since it is clearly necessary to reduce the first (column 5) value of H. The next trial value of stage is accordingly 65.96, which leads to equal values of H in columns 5 and 14.

5.6 Step Method—Divided Channels

In this section we consider cases where the channel cross section is divided into distinct regions having distinct flow characteristics. The most common example of this situation is the one already mentioned, and shown in Fig. 5-5 —the case of overbank flow, when the flow over the berms has a different depth, and the surface possibly a different roughness, from those existing in the main channel.

If the channel is straight the water-surface level will remain substantially constant over the whole section of flow, since the hydrostatic pressure must remain constant along any horizontal line drawn across the section. However, the distinct regions of flow shown in Fig. 5-5 will almost certainly have different velocities and velocity heads; the problem then is to define a total head H applicable to the entire cross section. The solution is to use the energy coefficient α as defined in Sec. 1.9; the total head line then extends across the whole water surface, a distance of $\alpha v_m{}^2/2g$ above it, as shown in Fig. 5-5. This total head line, and any head losses deduced from it, are assumed to be applicable to the section as a whole, and also to each of the individual subsections.

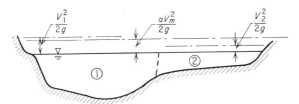

Figure 5-5. Example of Overbank Flow and the Use of a Velocity Coefficient to Define a Mean Velocity Head

This last assumption is not strictly true, since the separate subsections should properly be regarded as parallel systems each with its distinct head loss, as well as distinct velocity, depth, etc. It would however be extremely difficult to treat the problem in this way, for allowances would have to be made for the exchange of flow between the berms and the main channel between one section and the next, and in general the distribution of flow between the berms and the main channel could be determined only by a trial process. In fact the treatment would be more difficult than is warranted by the requirements of the river engineer, and the assumption of uniform total head across each section gives results that are quite accurate enough for practical purposes. This assumption, therefore, is normally adopted in practice.

The algebra required is conveniently handled by the use of the conveyance K. We have seen in Sec. 1.9 that the energy coefficient α is defined by the equation

$$\alpha = \frac{\Sigma(v_1{}^3 A_1)}{v_m{}^3 \Sigma A_1} = \frac{(\Sigma A_1)^2}{(\Sigma Q_1)^3} \Sigma\left(\frac{Q_1{}^3}{A_1{}^2}\right)$$

where subscripts 1, 2, 3 ... indicate the distinct subsections of the flow. Now since we are assuming that the same value of friction slope S_f applies to each subsection, it follows from the definition of the conveyance K in Eq. (4–25) that

$$\frac{Q_1}{K_1} = \frac{Q_2}{K_2} = \frac{Q_3}{K_3} = \cdots = \frac{Q_n}{K_n} = \frac{\Sigma Q_1}{\Sigma K_1} \tag{5–25}$$

and therefore that

$$\alpha = \frac{(\Sigma A_1)^2}{(\Sigma K_1)^3} \Sigma\left(\frac{K_1{}^3}{A_1{}^2}\right) \tag{5–26}$$

Since each element of Eq. (5–25) is equal to $S_f{}^{1/2}$, it follows that

$$S_f = \left(\frac{\Sigma Q_1}{\Sigma K_1}\right)^2 = \frac{Q^2}{(\Sigma K_1)^2} \tag{5–27}$$

EXAMPLE 5-5 Tabulation

$Q = 167,000$ cusecs.   River Sections Assumed Known, i.e., A and P Known Functions of Stage.

(One berm only, on left bank.)

1	2	3	4	5	6	7	8	9	10	11	12	13	14	15	16	17	18	19	20
River mile	Sub sec	Stage h, ft	A, sq ft	P, ft	R, ft	$R^{2/3}$	n	K $\times 10^{-6}$	K^3/A^2 $\times 10^{-10}$	α	v_m ft/sec	$\alpha \dfrac{v_m^2}{2g}$ ft	H, ft	$S_f = \left(\dfrac{Q}{\Sigma K}\right)^2$	Mean S_f	Δx, ft	h_f, ft	h_e, ft	H, ft
3.50	M.C.	22.9	9,860	665	14.8	6.04	0.03	2.96	26.7										
	L.B.		5,750	612	9.4	4.46	0.05	0.77	1.4										
	Total		15,610					3.75	28.7	1.32	10.7	2.35	25.25	0.00200					25.27
3.75	M.C.	26.3	14,750	805	18.32	6.93	0.03	5.07	60.0										
	L.B.		5,480	506	10.82	4.90	0.05	0.80	1.7										
	Total		20,230	1310				5.87	61.7	1.25	8.25	1.32	27.62	0.00081	0.00140	1320	1.84		27.11
	M.C.	25.75	14,340	805	17.8	6.80	0.03	4.85	55.3										
	L.B.		5,230	506	10.32	4.74	0.05	0.74	1.5										
	Total		19,570					5.59	56.8	1.25	8.54	1.42	27.17	0.00089	0.00145	1320	1.91	—	27.18
3.875	M.C.	26.4	11,960	671	17.8	6.80	0.03	4.04	46.1										
	L.B.		6,540	689	9.48	4.48	0.05	0.87	1.5										
	Total		18,500	1360				4.91	47.6	1.38	9.02	1.74	28.14	0.00116	0.001026	660	0.68		27.88
	M.C.	26.05	11,690	671	17.4	6.70	0.03	3.89	43.2										
	L.B.		6,330	689	9.19	4.38	0.05	0.83	1.4										
	Total		18,020					4.72	44.6	1.38	9.27	1.85	27.90	0.00125	0.001066	660	0.70	—	27.90
4.00	M.C.	26.5	10,470	626	16.7	6.52	0.03	3.39	35.5										
	L.B.		6,830	800	8.54	4.18	0.05	0.85	1.3										
	Total		17,300	1430				4.24	36.8	1.45	9.65	2.10	28.60	0.00155	0.00140	660	0.93		28.83
	M.C.	26.75	10,600	626	16.9	6.58	0.03	3.46	36.9										
	L.B.		6,990	800	8.74	4.25	0.05	0.89	1.4										
	Total		17,590					4.35	38.3	1.45	9.50	2.03	28.78	0.00148	0.00137	660	0.90	—	28.80

where Q is the total flow. Thus α and S_f, the two factors which are of critical importance in the tabulation, can be calculated without explicitly evaluating the discharges Q_1, Q_2, ... etc. The values of K which are to be inserted in Eqs. (5–26) and (5–27) are obtained from Eq. (4–26).

We now have all the information required to proceed with the computation. The necessary tabulation is shown in Example 5.5 for a typical flood-level computation similar to that of Example 5.4, but with river sections of the type shown in Fig. 5–5. It is seen that at each major section a separate line is devoted to each subsection. These lines are labeled M.C. (main channel) and L.B. (left berm) in column 2. Again it is assumed that a plan of each section is available, from which, given the water level, A and P can be determined for each subsection. The liquid interface between zones 1 and 2, shown as a broken line in Fig. 5-5, is not normally included in estimates of the wetted perimeter P because the shear stress along this interface is so much lower than at the solid boundaries. The calculation proceeds upstream from known conditions at river mile 3.50.

The scheme of tabulation is essentially similar to that of Example 5.4. The major complication in Example 5.5 is that α must be evaluated, via Eq. (5–26), before the first value of total head H, equal to stage plus $\alpha v^2/2g$, can be obtained and listed in column 14. The second value of H is obtained just as in Example 5.4 and listed in column 20. Note that the factors 10^{-6} and 10^{-10} at the heads of columns 9 and 10 are the reciprocals of the factors to be applied to the figures in the column. Thus the first value of the conveyance K, calculated from Eq. (4–26), is 2.96×10^6.

Equation (5–24) is still applicable as a guide to the second trial value of water level, provided that Fr_2^2 is multiplied by α and a suitable average value of R_2 is used. An appropriate value is obtained from the total values of A and P listed in columns 4 and 5 under the individual values; the resultant value of R is listed in column 6, also under the individual values.

We consider how the process works out at river mile 3.75. After the first trial, $H_E = 27.62 - 27.11 = 0.51$. Also the average R is equal to

$$\frac{20230}{1310} = 15.44$$

Hence

$$\alpha Fr_2^2 = \frac{2}{R}\left(\frac{\alpha v_m^2}{2g}\right) = \frac{2 \times 1.32}{15.44} = 0.171$$

and

$$\frac{3S_{f_2}\Delta x}{2R} = \frac{3 \times 0.00081 \times 1320}{2 \times 15.44} = 0.104$$

so that

$$\Delta y = \frac{0.51}{1 - 0.171 + 0.104} = 0.55$$

and the next trial value of stage is accordingly 25.75, which leads to almost equal values of H, 27.17 and 27.18.

When a river reach between two adjacent sections is on a curve, the length Δx may vary from one subsection to another. The easiest way of allowing for this effect is to vary the Manning n; thus if a berm has a smaller value of Δx than the main channel, we imagine its length to be increased to that of the main channel and the value of n reduced in proportion to $(\Delta x)^{1/2}$. Similarly, eddy losses may be treated by incorporating them into the resistance losses; since the Manning formula may be written

$$h_f = \frac{2gn^2\Delta x}{1.49^2R^{4/3}}\frac{v^2}{2g}$$

we allow for a loss coefficient C_L by raising the value of n sufficiently to increase the coefficient of $v^2/2g$ in the above equation by the amount C_L. The method is approximate, but sufficiently accurate in the light of the present imperfect knowledge of eddy losses.

It is often convenient to treat eddy losses in this way, since their presence can make the manipulation of the energy equation inconvenient. This is particularly true in the case of the method described in the next section.

5.7   The Ezra Method

In the treatment of irregular channels, it is possible, just as in the case of uniform channels, to replace a trial process by a graphical method based on plotting certain properties of the cross section against the water level, or stage. One of the most successful of these methods is due to A. A. Ezra [3].

The method is based on the following rearrangement of the energy equation, Eq. (5–19). We can write

$$h_2 + \alpha_2 \frac{v_2{}^2}{2g} = h_1 + \alpha_1 \frac{v_1{}^2}{2g} + \tfrac{1}{2}\Delta x(S_{f_1} + S_{f_2}) \qquad (5\text{–}28)$$

where, as before, section 2 is upstream of section 1. Eddy losses have been neglected on the assumption that if they are present they can be incorporated into the resistance losses. The above equation can be written

$$h_2 + F_B(h_2) = h_1 + F_A(h_1) \qquad (5\text{–}29)$$

where

$$F_A(h) = \alpha \frac{v^2}{2g} + \tfrac{1}{2}S_f\Delta x_u \qquad (5\text{–}30)$$

and

$$F_B(h) = \alpha \frac{v^2}{2g} - \tfrac{1}{2}S_f\Delta x_d \qquad (5\text{–}31)$$

corresponding to the U and V functions of Sec. 5.4. However the method of that section cannot conveniently be used in this case, since Δx varies from reach to reach. Ezra's method is based on plotting $h + F_A(h)$ and $h + F_B(h)$

against h on adjacent sheets, for every river section being considered. If the channel is divided, α is computed by Eq. (5–26) and S_f by Eq. (5–27).

Now the function $F_A(h)$ is to be applied to the section at the downstream end of a certain reach, and $F_B(h)$ to the one at the upstream end. Since each section will form, in turn, the upstream end of one reach and then the downstream end of the next one, $F_A(h)$ and $F_B(h)$ must be calculated and plotted for every section. However, it is most important to note that Δx is not the same for each function at the same section. For the reason given above, Δx in $F_A(h)$ is measured upstream from the section being considered, and in $F_B(h)$ it is measured downstream. Accordingly, Δx is subscripted u and d in Eqs. (5–30) and (5–31) respectively.

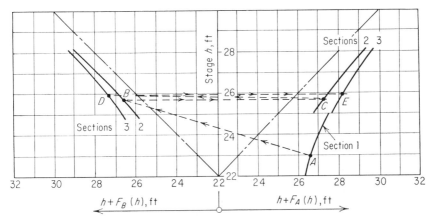

Figure 5-6. *Operation of the Ezra Method for Determination of Longitudinal Profiles*

The two sets of curves are plotted in Fig. 5-6 for the same example that was worked out in Sec. 5.6. Since the flow is subcritical in this case (as it almost always is in natural rivers) the calculation proceeds in the upstream direction. However the system would work in exactly the same way if the flow were supercritical and the calculation proceeded downstream from section 2 to section 1.

The flow at section 1 is represented by the point A, on the right hand, or F_A, diagram. The flow at section 2 will then be represented by the point B on the F_B diagram, having the same abscissa, 26.3, as A; Eq. (5–29) is therefore satisfied. To go from section 2 to section 3 we must first return to the F_A diagram via a horizontal line to the point C. Both of the points B and C represent section 2, so they must of course have the same ordinate, h_2. The need to switch from the F_B function to the F_A function before moving on to the next reach has been discussed previously. From the point C we transfer to D, representing section 3, and so on.

The graphical process may be made a little more convenient by plotting both sets of curves on one sheet, as in Fig. 5-7. Although this may make the curves a little crowded, the solution may then be very simply traced by the saw-tooth path shown.

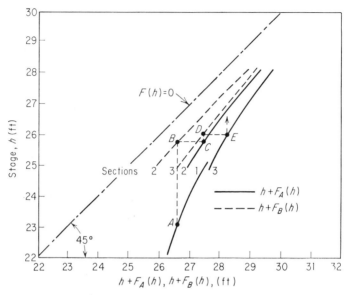

Figure 5-7.   Alternative Scheme for the Ezra Method

A further refinement is made possible by the fact that both the functions F_B and F_A vary as the square of the discharge Q. This suggests that one set of curves may be used for a number of different discharges by varying the scale of F_A and F_B; Fig. 5-8 shows how this result can be accomplished.

The ordinate is the stage h, as in Figs. 5-6 and 5-7. The abscissa is the variable k, defined by the equation

$$k = \frac{Q_0^2}{Q^2} F(h) \tag{5-32}$$

where Q_0 is some representative discharge chosen near the center of the range of discharges that is to be dealt with. Both sets of curves, k_A and k_B, are plotted in the first quadrant of the figure; we obtain the necessary values of k_A and k_B by calculating $F_A(h)$ and $F_B(h)$ for $Q = Q_0$. We use the same scale for k as for h.

We now consider how to trace the solution for the case $Q = Q_0$, when $k = F(h)$. The point A represents section 1 on the k_A curve: we move to the point B, representing section 2 on the k_B curve, along a line which is at 45° to each axis, since this line has the equation

$$h + F(h) = \text{constant}$$

and Eq. (5–29) is therefore satisfied. Thereafter we move from B to C along a horizontal line as in Figs. 5-6 and 5-7, then from C to D along a 45° line, and so on.

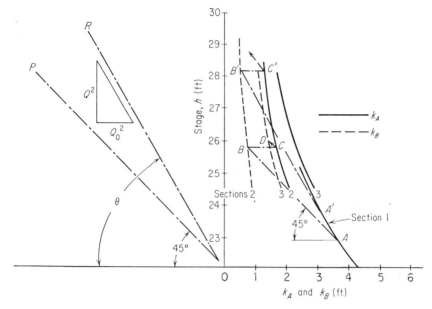

Figure 5-8. *Operation of the Modified Ezra Method, Allowing for Discharge Variation*

Thus far the result accomplished is the same one given by Figs. 5-6 and 5-7. However we can now trace a solution for any other value of Q, in this way: Let $r = Q^2/Q_0^2$, so that

$$F(h) = \frac{Q^2}{Q_0^2} k = rk \tag{5–33}$$

then to go from an initial point A' representing section 1, to a point B' representing section 2, we must move along a line having the equation

$$h + rk = \text{constant}$$

i.e., a line inclined at an angle $\tan^{-1}(-r)$ to the k axis. We may think of it as being parallel to a reference line OR drawn in the second quadrant at an angle θ to the k axis, where

$$\tan \theta = r = \frac{Q^2}{Q_0^2} \tag{5–34}$$

Any number of these reference lines may be drawn; also shown is the line OP for $Q = Q_0$, drawn at 45° to each axis. Thus the same two sets of k curves serve for any desired discharge.

The preparation of the curves in Figs. 5-6, 5-7, or 5-8 requires at least as much work as the tabulation procedures of Secs. 5.5 and 5.6. Therefore the basic Ezra method, as presented in Figs. 5-6 and 5-7, is worthwhile only when a number of profiles are to be plotted for the same discharge—e.g., backwater curves upstream of proposed dams of varying heights. The modified method of Fig. 5-8 is suitable when a number of discharges are to be considered; in this case the method is worthwhile even if only one type of channel feature, such as one dam of a certain height, is being considered.

This flexibility in the modified Ezra method makes it suitable for uniform channels as well. In this application a very simple representation is possible provided Δx is kept constant, so that k_A, k_B, and the depth are related in the same way for all sections. Every section is then represented, basically, by

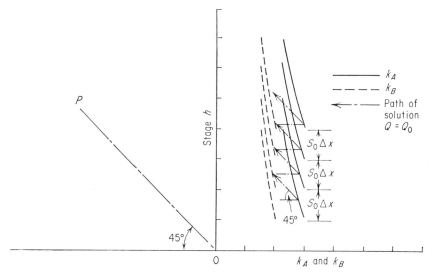

Figure 5-9.   *Adaptation of the Modified Ezra Method to Uniform Channels*

the same pair of curves, varying only in their displacement in the h direction on the h-k plane. Adjacent pairs are separated by a vertical distance $S_0\Delta x$, the difference between the bed levels of adjacent sections. The arrangement is shown in Fig. 5-9; the curves are crowded, but not so much as to make the method unworkable.

5.8   Grimm's Method

The methods applicable to natural rivers which have been discussed so far require full surveys of the river cross section to be made at close intervals along the river. The method described in this section, on the other hand, requires only data relating stage to discharge at each river section. These data may, of course, come from river records rather than from calculations, so that expensive surveys to determine the complete geometry of each section are not required. The method therefore has the advantage of low cost, but this advantage is offset by its limited accuracy and applicability.

First, it is assumed that the water-surface slope is the energy slope—that is, the velocity head is neglected. The method is particularly suited to obtaining the profiles that result after a new structure, such as a dam, is installed in the river. However, the method fails if alterations are made to the river channel itself, since it is assumed that stage-discharge data obtained from existing records will still be applicable in the new situation.

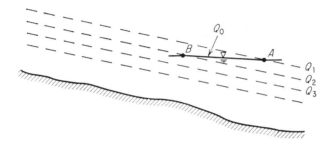

Figure 5-10. Grimm's Method of Determination of Backwater Profiles from Existing Records of Flow Profiles

The method itself is displayed in Fig. 5-10. It relies on the fact that for the same stage in the same channel

$$Q \propto S_w^{1/2} \tag{5–35}$$

where S_w is the water-surface slope. The broken lines in the figure represent water surface profiles at various discharges in the original channel; the solid line is the profile at a discharge Q_0 after the erection of a dam. The method of constructing this profile follows immediately from Eq. (5–35); if S_w is the slope of the Q_1 profile at A, and S_{w_0} the slope of the Q_0 profile, then

$$\frac{S_{w_0}}{S_{w_1}} = \left(\frac{Q_0}{Q_1}\right)^2 \tag{5–36}$$

The slope of the new profile is thence found at A, B, ... , etc., and the whole profile sketched in. In going from A to B, some judgment has to be exercised in choosing a mean slope which is an average of the slope calculated at A from Eq. (5–35) and the estimated slope which is expected at B but not yet calculated. With practice, however, it is possible to estimate this mean slope with little need for second trials.

According to Fig. 5-10 the basic step in the process is from one profile (e.g., Q_1) to the next, (e.g., Q_2), not from one channel section to the next. There is however no hard-and-fast rule about this, and either method can be used. If the latter is used, then further profiles must be interpolated—e.g., between the Q_1 and Q_2 profiles—so that each step does terminate on a profile, as at B.

The method is not only approximate, it is also somewhat limited in its application. It will be satisfactory only if records exist for stages high enough to approach those of the new profile—the solid line in Fig. 5-10. This will be true only if the new feature being introduced into the channel either lowers the water surface or raises it by only a small amount.

5.9   The Escoffier Method

It frequently happens that the Froude number of the flow is small enough to justify the assumptions made in the use of Grimm's method, that the velocity head is negligible and that the energy slope is equal to the water-surface slope. However, it often happens that profiles are required for river stages much higher than those observed in recorded floods, so that Grimm's method is unsuitable.

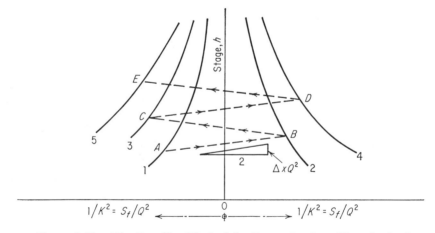

Figure 5-11.   *The Escoffier Method for Determination of Longitudinal Profiles*

In such cases the surface profile can be readily determined by a method due to F. F. Escoffier [4]. The method is shown, in slightly amended form, in Fig. 5-11. For each river section the quantity S_f/Q^2 (i.e., $1/K^2$), where K is the conveyance of the section, is plotted as abscissa against water level, or stage, as ordinate. The curves representing successive river sections are plotted alternately to the left and right of the vertical axis, as shown.

Given the stage at section 1, represented by the point A, we determine the stage at section 2 (the point B) by drawing a line AB at a slope equal to $\Delta x Q^2/2$, where Δx is the length of the reach between sections 1 and 2. For it follows from this construction that Δh, the stage difference between A and B, is equal to

$$\Delta h = \frac{S_{f_1} + S_{f_2}}{Q^2} \cdot \frac{\Delta x Q^2}{2} = \frac{S_{f_1} + S_{f_2}}{2} \Delta x \qquad (5\text{--}37)$$

satisfying the condition that the water surface slope is equal to the energy slope. The point C, representing the stage at section 3, is obtained by drawing a line BC at a slope equal to $-\Delta x Q^2/2$, and subsequent points, D, E, ... , are obtained by drawing lines CD, DE, ... of alternately positive and negative slope.

As with the modified Ezra method described in Sec. 5.7, only one set of curves is required for all discharges. Varying discharge is accommodated by varying the slope of the lines AB, BC, ... , etc.

5.10   The Discharge Problem

In Sec. 4.6 the interaction of local features and longitudinal profiles was dealt with. The problem which lends most point to the discussion in that section is that of calculating the discharge in a long channel. Typically, the argument proceeds from water levels which are known upstream and down-stream, or upstream alone. If the channel leading from the upstream source is on a steep slope, the problem is trivial, for the discharge is easily determined from the critical-flow condition at the lake outlet. The complications which arise, as in Sec. 4.6, from the presence of a control such as a choke, occur only when the channel slope is mild; the easiest way of dealing with them is to begin with the assumption of a long uniform downstream channel, as in Example 4.1, and then to determine whether any downstream controls exist that are strong enough to send their influence upstream to the source. This procedure is applicable whether the downstream controls are such as to decrease the flow (as in Fig. 4-13) or to increase the flow (as in the case of a free overfall further downstream).

In either case the numerical problem is to plot a backwater curve upstream from the control; for practical purposes this curve may be said to "run out" when the depth returns to within 1 percent of uniform depth, and the essence of the problem is to determine whether it does run out before it reaches right back to the source. If it does, the channel is "long" for practical purposes, as in Example 4.1. If it does not, the discharge will differ from its value in the long-channel case, and must be determined by trial. The details are best illustrated by an example.

Example 5.6

The situation is as in Example 4.1, i.e., a rectangular channel 10 ft wide, $n = 0.014$, runs on a slope of 0.001 from a lake whose surface is 10 ft above the channel bed at the lake outlet. But in this case the channel contracts to a short box culvert 6 ft wide, with a well-rounded entrance, 2,000 ft downstream of the lake outlet. Find the discharge in the channel.

We assume first that the contraction created by the culvert is not severe enough to reduce the flow—that is, we assume $Q = 674$ cusecs, as given by Example 4.1. Then within the culvert we have:

$$q = \frac{674}{6} = 112.3 \text{ cusecs/ft}$$

for this value of q,

$$y_c = \sqrt[3]{\frac{q^2}{g}} = 7.32 \text{ ft}$$

and

$$E_c = \tfrac{3}{2} y_c = 10.98 \text{ ft}$$

which is greater than the available specific energy if the upstream flow is uniform (in which case $E = 10$ ft, as at the lake outlet). It follows that the culvert acts as a choke, as in Fig. 4-13c. The question now is whether the resultant M_1 curve reaches right back to the lake, or stops short of it as in Fig. 4-13d. To answer this question we assume first that the discharge remains unaltered at 674 cusecs. Then we obtain the depth y immediately upstream of the culvert by noting that in that region $E = 10.98$ (since there is no energy loss at the culvert entry), and $q = 674/10 = 67.4$ cusecs/ft. It follows that

$$y + \frac{67.4^2}{2gy^2} = y + \frac{70.6}{y^2} = 10.98$$

whence by trial $y = 10.32$ ft, and we plot the M_1 profile upstream from this value. The tabulation proceeds as in Example 5.1; the summary given below contains only the more important columns.

(a) $Q = 674$ cusecs

y, ft	A, sq ft	$v^2/2g$, ft	E, ft	v^2/C^2R	$(S_0 - v^2/C^2R)_m$	ΔE, ft	Δx, ft	x, ft
10.32	103.2	0.66	10.98	0.000747				0
					0.000243	−0.100	−412	
10.2	102	0.680	10.880	0.000767				−412
					0.000222	−0.086	−386	
10.1	101	0.694	10.794	0.000789				−798
					0.000201	−0.086	−427	
10.0	100	0.708	10.708	0.000809				−1225
					0.000181	−0.085	−470	
9.9	99	0.723	10.623	0.000829				−1695
					0.000161	−0.086	−534	
9.8	98	0.737	10.537	0.000849				−2229

(b) $Q = 621$ cusecs

9.78	97.8	0.624	10.404	0.000719				
					0.000274	−0.069	−251	
9.7	97	0.635	10.335	0.000732				−251
					0.000256	−0.086	−336	
9.6	96	0.649	10.249	0.000756				−591
					0.000234	−0.086	−368	
9.5	95	0.663	10.163	0.000777				−959
					0.000214	−0.086	−402	
9.4	94	0.677	10.077	0.000795				−1361
					0.000195	−0.085	−435	
9.3	93	0.692	9.992	0.000815				−1796
					0.000174	−0.086	−495	
9.2	92	0.706	9.906	0.000837				−2291

Tabulation (a), based on $Q = 674$ cusecs and E at the culvert $= 10.98$, shows by interpolation that at $x = -2,000$ ft, i.e., at the lake outlet, $y = 9.843$ ft and $E = 10.574$ ft. This means that the M_1 curve has failed to "run out" before reaching the lake, since E has failed, by a margin of 0.574 ft, to fall to the available amount of 10.00 ft.

In choosing a new trial value of Q and plotting another M_1 profile one would not expect the change in E along the profile to differ very much from that occurring along the first profile. Therefore an obvious second trial value for Q is that which produces a value of E at the culvert which is 0.574 ft less than the previous one of 10.98 ft, say about 10.4 ft. Fixing the value of E leads to Q and to y upstream of the culvert, thus:
If $E_c = 10.4$ ft,

$$y_c = \tfrac{2}{3} \times 10.4 = 6.93 \text{ ft}$$

$$v_c = \sqrt{gy_c} = 14.94 \text{ ft/sec}$$

and these values of depth and velocity will obtain within the culvert, so that $Q = 6 \times 6.93 \times 14.94 = 621$ cusecs. Then, immediately upstream of the culvert

$$q = 621/10 = 62.1 \text{ cusecs/ft}$$

and
$$E = y + \frac{62.1^2}{2gy^2} = y + \frac{59.9}{y^2} = 10.4 \text{ ft}$$

whence by trial $y = 9.78$ ft. (Note that the immediate choice of a second trial value of y, instead of E, would have led to difficulties in determining Q and E.)

Given $Q = 621$ cusecs and $E = 10.4$ ft at the culvert, we plot another M_1 profile upstream; by completing tabulation (b), and interpolating, we obtain $E = 9.96$ ft at the lake outlet. This is too low, but is close enough to the correct value of 10.00 ft for a linear interpolation to yield a satisfactory result, thus:

E at Lake Outlet (ft)	Q (cusecs)	Remarks
10.574	674	1st trial
10.00	624	Correct answer
9.96	621	2nd trial

The numerical work in this example is quite formidable, and prompts one to enquire whether simpler methods might not be possible. Problems 5.1 and 5.3 have shown that an M_1 curve which is only a few thousand feet in length (as in Example 5.6) can be closely approximated by a calculation of one step only, and this suggests a trial process as follows.

Take a trial value of Q, and calculate E and y at the culvert. Also calculate y at the lake (where E must be equal to 10 ft). The friction slope $S_f = v^2/C^2R$ is then calculated as in the tabulation of Example 5.6 at each end of the channel, and Δx obtained from Eq. (5–1). If this value of Δx is not equal to the known interval of 2,000 ft, a further trial value of Q must be taken. The reader should have no difficulty in devising a suitable scheme of tabulation (Prob. 5.30).

The numerical methods of Secs. 5.4 and 5.7 may appear to offer more simple processes of solution, but this is not the case. The method of Sec. 5.4 requires the preparation of curves, as in Figs. 5-2 and 5-3, based on a known discharge or series of discharges; this is not suited to a problem in which discharge is the unknown, for a set of these curves would have to be drawn for each trial value of discharge. The Ezra method, on the other hand, can be modified, as in Fig. 5-8, to accommodate any value of discharge, and as in Fig. 5-9 to apply to channels of uniform section. Figure 5-9 could be used in the single-step method described above and suggested for Prob. 5.30, but it would still be necessary to calculate the depth at each end of the channel, for each trial value of Q, before using Fig. 5-9 as a shortcut to solve Eq. (5–1). It is doubtful whether there is any real saving in time, when it is remembered that the curves of Fig. 5-9 must first be calculated and plotted.

Of course, any shortcut numerical method usually depends on some preparatory work by which tables or graphs are set up for future reference. But in the discharge problem the best way in which to formulate this preparatory work is to consider, for a given discharge, the mutual dependence of the water levels at the two ends of the channel reach. This leads to what may be termed the two-lake problem.

The Two-Lake Problem

This term implies a more complete formulation of the discharge problem than has been considered so far. Consider the two lakes, as in Fig. 5-12, connected by a channel reach of mild slope, and in general short enough for M_1 or M_2 curves originating at the lower lake to extend upstream as far as the upper lake. As the flow enters the channel from the upper lake the water surface drops by an amount equal to the velocity head plus losses, but at the downstream end the water flows into the lower lake without a corresponding change in water surface level, for in general the velocity head in the channel is not recovered as the flow expands abruptly into the lake (cf. flow from a pipe into a large reservoir). However if the water level in the lower lake falls below the level corresponding to critical depth in the channel, the water level in the channel cannot match itself to the water level in the lake, for this would imply a sudden change from subcritical to supercritical flow; what happens is that the water after leaving the channel forms a kind of outflow "dome" spreading out over the lake surface, as in Fig. 5-12a. The flow within the channel remains critical.

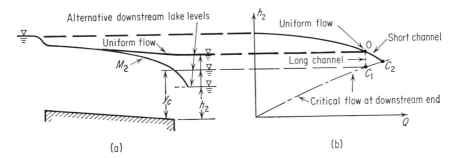

Figure 5-12.   *The Two-Lake Problem—Fixed Upstream Lake Level*

From the above discussion, and from Example 5.6, it can be seen that if the upper lake level is held fixed, the discharge will vary with the lower lake level h_2 somewhat as shown in Fig. 5-12b. As h_2 falls, Q rises until the flow in the channel is uniform, as in a very long channel. As h_2 falls further, an M_2 profile forms and the behavior of Q depends on how the length of this profile compares with the channel length. If the channel is long, so that the M_2 profile does not reach back to the upper lake, Q will remain unaltered at its uniform flow value, which is therefore a maximum (along the line OC_1 in Fig. 5-12b); if the channel is short, this value will be exceeded and Q will continue to increase as h_2 falls, along the line OC_2 in Fig. 5-12b. The points C_1 and C_2 each represent critical flow in the channel, at differing discharges; they therefore lie on a curve having the equation $Q = By\sqrt{gy}$ in the case of a rectangular channel. When the lower lake level falls below C_1 or C_2 the discharge remains unchanged at the maximum value of Q_1 or Q_2.

As the channel is shortened further, the maximum Q_2 increases until the channel is short enough to become steep; control then shifts to the upstream end of the channel, where the flow becomes critical. Unless the channel is very short, Q_2 is not much greater than Q_1 (e.g., Prob. 5.27), and for channels longer than a thousand feet or so, Q_1 can be taken as the maximum capacity for all practical purposes (as in Prob. 4.37).

The best way of storing data about a channel reach, to serve as a basis for discharge calculations, is in the form of Q contours on the h_1-h_2 plane, where h_1 and h_2 are the upper and lower lake levels measured upwards from the upper and lower channel bed levels respectively, as in Fig. 5-13a. Since we are dealing only with the case where the upper lake level exceeds that of the

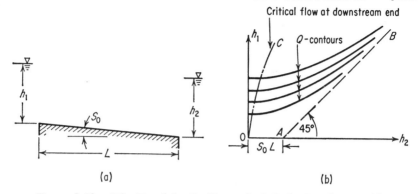

Figure 5-13.   *The Two-lake Problem—both Lake Levels Variable*

lower lake, all the curves must lie to the left of a line AB drawn on the h_1-h_2 plane at an angle of 45° to the h_2 axis and with an h_2 intercept of $S_0 L$. All the Q contours must in the limit converge towards this line, representing the state of affairs when both lake levels are very high, and tending to the same absolute level. The line OC represents critical flow at the downstream end of the channel; to the left of this line all the Q contours are straight lines parallel to the h_2 axis, as suggested by the discussion of Fig. 5-12, provided that the channel is long.

Computation of coordinates for these Q contours is lengthy but straightforward, since no trial process is required. It makes a good exercise in the use of the high-speed computer (Prob. C5.8). Once the curves are plotted they readily yield all the design information needed in practice.

If the channel is very short (i.e., only a few hundred feet long), a useful first approximation can be achieved by neglecting resistance effects altogether. The only head losses then are the velocity head dissipated at the outflow into the lower lake, and the losses in any hydraulic jump which may form in the

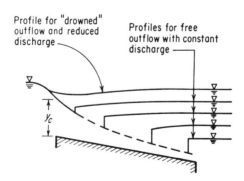

Figure 5-14.   *The Two-lake Problem—effect of Rising Tailwater in a Steep Channel*

channel, which must now be regarded as steep. The problem becomes an exercise in the specific energy and momentum concepts of Chaps. 2 and 3 (Prob. 5.32).

If the channel is steep even when resistance is taken into account the problem remains essentially the same; control resides in the critical section at the upper end of the channel until the lower lake level becomes high enough to drown the upper lake outflow (Fig. 5-14). If resistance is taken into account the downstream subcritical profile is an S_1 curve; if resistance is neglected, this profile derives simply from the specific energy-depth curve. In the former case the use of a high-speed computer is warranted (Prob. C5.9).

The Choice of Channel Parameters

In most of the discussion in this and the preceding chapter it has been assumed that the leading dimensions of the channel have been fixed in advance. This sort of treatment is necessary in order to elucidate basic principles, but it should not obscure the fact that in practice the design engineer must often regard these leading dimensions as variables, of which he must choose the best or most economical values.

For example, the bed slope of a channel may often be fixed by considerations outside the designer's control—as in the case of a natural river or of a long artificial channel to be laid along a fixed line; but there are many cases of artificial channels in which the slope must be set by the engineer to match his design requirements. An example has already been seen in Prob. 4.37; the reader can explore further aspects of this case in Prob. 5.33 by considering how the control section at the downstream end increases in importance as the slope reduces to zero and then becomes adverse. Problems 5.34 and 5.35 exemplify the type of problem controlled essentially by water levels at the two ends of a reach, and show that there may be many different bed levels and slopes which satisfy a given specification of water surface levels.

This last statement revives, by implication, the problem of the economical channel section introduced in outline in Sec. 4.3, for it suggests the query: which bed level and slope will be best? This is turn suggests another query: which *combination* of cross section, bed level, and slope will be best? It would be pointless to attempt here the exploration of all the economic complexities met by the practicing engineer, but Probs. 5.36 and 5.37 give some hint of the kind of considerations governing the design of a typical artificial channel of mild slope. Problems 5.38 and 5.39 contain a similar but less complex version of the economic problem for a steep slope. Problem 5.40 raises another common practical problem—the choice between deepening and widening a channel.

Clearly, there will an endless variety of practical problems of this sort, and it is difficult to lay down general rules. However, one paramount rule suggests itself. In considering any proposed channel, decide first what factors are the most important in fixing the discharge: control sections upstream or downstream, given water surface slopes, or bed slopes. Once this decision is made, the designer can see clearly those respects in which he has a free choice, and those in which he has a restricted choice. He can then take the final step of exploring the ways in which his free choices turn out to be restricted by economic considerations.

5.11   The High-Speed Computer

The use of the high-speed computer has already been considered in Chap. 2 and subsequently; the problems introduced in this chapter warrant a more complete discussion of the use and potentiality of the computer.

All the laborsaving methods treated so far in this chapter are devised for the convenience of the human operator, who can quickly and easily look up tables or trace solutions on a graph, but who finds it more lengthy and difficult to do arithmetical calculations. Also he prefers his calculations to be direct and explicit, and not to involve the tedious repetition of trial-and-error processes. Indeed many of the methods described in this chapter are aimed at lifting the problem out of the trial-and-error category, which is the worst, into the tabular or graphical category, which is the best.

The high-speed computer takes quite a different view of the matter. It can do calculations with great speed, and is not deterred by the repetitions required in trial processes. Also, the human programmer directing the machine can easily and quickly write the instructions governing these trial processes. The computer can look up tables just as readily as it can perform calculations, but tables are generally regarded as a nuisance in computer work, for two reasons: they take up valuable storage space in the machine, and the programmer must transfer them *manually* to tape or cards and then load them into the machine. The second reason is not applicable in the case of standard functions (e.g., the trigonometric and expónential functions), for which listings on tape or cards would be held on file, but the first reason still holds. On this account the standard practice on most machines is to compute sine functions, etc., by subroutines rather than to look them up in tables.

In using graphs their ordinates would be listed in the machine as tables, and the scanning of a graph is simulated by looking up the table. Most programming systems have a table-look-up instruction which includes all the necessary scanning and interpolation procedures; if this instruction is not used the scanning of a graph by a computer becomes an awkward but interesting problem, as the reader will have previously found (Prob. C3.4b).

It appears therefore that the computer's view of the labor problem almost exactly reverses the view of the human operator. It does not follow, however, that an engineer with access to a computer need scrap the procedures described in this chapter. For quick spot checks or for small schemes, the

engineer will always have a use for methods that can be operated at the desk
or in the drawing office, and that will yield solutions as quickly and casually
as does the slide rule. For comprehensive reviews of large schemes the computer comes into its own, but until it becomes as cheap, as small, and as
ubiquitous as the slide rule it will not completely displace the methods given
in this chapter.

At the present stage of development it is profitable to discuss how the
computer may possibly be combined with manual methods. To put the whole
matter in perspective consider the elementary operations chart of Fig. 5-15;

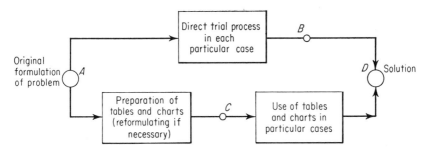

Figure 5-15.   *Alternative Methods of Scheduling Computations*

as shown in this chart, there are in general two alternative paths to a solution:
that of direct calculation, including a trial process if required, and that in
which the problem is reformulated to suit the preparation of tables or charts,
which are applied thereafter in particular cases. Examples of both methods
are found in this chapter, the second being resorted to when the first becomes
awkward for the human operator. As we have seen, the computer is not
worried by the task of straight calculation, even when it involves trial solutions; for instance, to determine y from x and Δx in a uniform channel it
would be better to have the computer carry out the necessary trial process
rather than make it work through the special method of Sec. 5.4. In terms of
Fig. 5-15, the path ABD is best suited to the computer, and the path ACD
to the human operator.

However, a third process is possible and economically viable. As discussed
previously, the manual use of tables and charts (the step CD in Fig. 5-15)
still has a part to play even when a computer is available. In such cases it
becomes worthwhile to use the computer for the step AC—i.e., the preparation
of tables and charts for later use by human operators. The computer problems
at the end of this chapter include problems of this AC type, as well as of the
ABD type. In many cases it will be found possible to try both approaches on
the same problem; in these cases care must be taken to avoid hasty judgments
on the relative speed of the two methods, for it must be remembered that
effort expended on the AC step is a form of capital expenditure which should

not be charged to one problem alone, but is to be spread over many possible future problems.

Before working the computer problems at the end of this chapter, the reader will find it helpful to consult a recent study by Pickard [5] in the application of the computer to longitudinal profiles.

References

1. B. A. Bakhmeteff. *Hydraulics of Open Channels* (New York: McGraw-Hill Book Company, Inc., 1932), Chap. 8.
2. Ven Te Chow. *Open-Channel Hydraulics* (New York: McGraw-Hill Book Company, Inc., 1959), Chap. 10.
3. A. A. Ezra. "A Direct Step Method for Computing Water-Surface Profiles," *Trans. Am. Soc. Civil Engrs.*, vol. 119 (1954), pp. 453–462.
4. F. F. Escoffier. "Graphic Calculation of Backwater Eliminates Solution by Trial," *Engineering News-Record*, vol. 136, no. 26 (June 27, 1946), p. 71.
5. W. F. Pickard. "Solving the Equations of Uniform Flow," *Proc. A.S.C.E.*, vol. 89, no. HY4, part I (July 1963), p. 23.
6. V. B. Dul'nev. "Universal Formula for Computation of Nonuniform Flow in Open Channels" (in Russian), *Gidrotekh Stroit*, vol. 33, no. 9 (September 1963), p. 47.

Appendix to Chapter 5

TABLE OF THE VARIED-FLOW FUNCTION

$$F(u,N) = \int_0^u \frac{du}{1-u^N} \text{ with constant of integration}$$

adjusted so that $F(0,N) = 0$ and $F(\infty,N) = 0$

u \ N	$2\frac{1}{2}$	3	$3\frac{1}{3}$	u \ N	$2\frac{1}{2}$	3	$3\frac{1}{3}$
0.00	0.000	0.000	0.000	0.60	0.658	0.637	0.628
0.02	0.020	0.020	0.020	0.61	0.673	0.650	0.641
0.04	0.040	0.040	0.040	0.62	0.686	0.663	0.653
0.06	0.060	0.060	0.060	0.63	0.700	0.676	0.666
0.08	0.080	0.080	0.080	0.64	0.716	0.690	0.679
0.10	0.100	0.100	0.100	0.65	0.731	0.703	0.692
0.12	0.120	0.120	0.120	0.66	0.746	0.717	0.705
0.14	0.140	0.140	0.140	0.67	0.762	0.731	0.718
0.16	0.161	0.160	0.160	0.68	0.777	0.746	0.732
0.18	0.181	0.180	0.180	0.69	0.795	0.761	0.746
0.20	0.201	0.200	0.200	0.70	0.811	0.776	0.760
0.22	0.222	0.221	0.220	0.71	0.828	0.791	0.775
0.24	0.243	0.241	0.240	0.72	0.845	0.807	0.790
0.26	0.263	0.261	0.261	0.73	0.863	0.823	0.805
0.28	0.284	0.282	0.281	0.74	0.881	0.840	0.821
0.30	0.305	0.302	0.301	0.75	0.900	0.857	0.837
0.32	0.326	0.323	0.322	0.76	0.919	0.874	0.853
0.34	0.347	0.343	0.342	0.77	0.940	0.892	0.870
0.36	0.368	0.364	0.363	0.78	0.962	0.911	0.887
0.38	0.391	0.385	0.383	0.79	0.985	0.930	0.905
0.40	0.413	0.407	0.404	0.80	1.008	0.950	0.924
0.42	0.435	0.428	0.425	0.81	1.032	0.971	0.943
0.44	0.458	0.450	0.447	0.82	1.057	0.993	0.963
0.46	0.481	0.472	0.469	0.83	1.083	1.016	0.985
0.48	0.504	0.494	0.490	0.84	1.110	1.040	1.007
0.50	0.528	0.517	0.512	0.85	1.139	1.065	1.030
0.52	0.553	0.540	0.535	0.86	1.171	1.092	1.055
0.54	0.578	0.563	0.557	0.87	1.205	1.120	1.081
0.56	0.604	0.587	0.580	0.88	1.241	1.151	1.109
0.58	0.631	0.612	0.604	0.89	1.279	1.183	1.139

Table of the Varied-Flow Function—(continued)

u \ N	$2\frac{1}{2}$	3	$3\frac{1}{3}$	u \ N	$2\frac{1}{2}$	3	$3\frac{1}{3}$
0.90	1.319	1.218	1.172	1.26	0.633	0.410	0.320
0.91	1.362	1.257	1.206	1.28	0.609	0.391	0.303
0.92	1.400	1.300	1.246	1.30	0.587	0.373	0.289
0.93	1.455	1.348	1.290	1.32	0.568	0.357	0.275
0.94	1.520	1.403	1.340	1.34	0.549	0.342	0.262
0.950	1.605	1.467	1.398	1.36	0.531	0.329	0.251
0.960	1.703	1.545	1.468	1.38	0.513	0.316	0.239
0.970	1.823	1.644	1.559	1.40	0.496	0.304	0.229
0.975	1.899	1.707	1.615	1.42	0.481	0.293	0.220
0.980	1.996	1.783	1.684	1.44	0.467	0.282	0.211
0.985	2.111	1.880	1.772	1.46	0.455	0.272	0.203
0.990	2.273	2.017	1.895	1.48	0.444	0.263	0.196
0.995	2.550	2.250	2.106	1.50	0.432	0.255	0.188
0.999	3.195	2.788	2.590	1.55	0.405	0.235	0.172
1.000	∞	∞	∞	1.60	0.380	0.218	0.158
1.001	2.786	2.184	1.907	1.65	0.359	0.203	0.145
1.005	2.144	1.649	1.425	1.70	0.340	0.189	0.135
1.010	1.867	1.419	1.218	1.75	0.322	0.177	0.125
1.015	1.705	1.286	1.099	1.80	0.308	0.166	0.116
1.020	1.602	1.191	1.014	1.85	0.293	0.156	0.108
1.03	1.436	1.060	0.896	1.90	0.279	0.147	0.102
1.04	1.321	0.967	0.813	1.95	0.268	0.139	0.095
1.05	1.242	0.896	0.749	2.00	0.257	0.132	0.089
1.06	1.166	0.838	0.697	2.10	0.238	0.119	0.079
1.07	1.111	0.790	0.651	2.20	0.220	0.107	0.071
1.08	1.059	0.749	0.618	2.3	0.204	0.098	0.064
1.09	1.012	0.713	0.586	2.4	0.190	0.089	0.057
1.10	0.973	0.681	0.558	2.5	0.179	0.082	0.052
1.11	0.939	0.652	0.532	2.6	0.169	0.076	0.048
1.12	0.907	0.626	0.509	2.7	0.160	0.070	0.043
1.13	0.878	0.602	0.488	2.8	0.150	0.065	0.040
1.14	0.851	0.581	0.479	2.9	0.142	0.060	0.037
1.15	0.824	0.561	0.452	3.0	0.135	0.056	0.034
1.16	0.802	0.542	0.436	3.5	0.106	0.041	0.024
1.17	0.782	0.525	0.421	4.0	0.087	0.031	0.017
1.18	0.760	0.509	0.406	4.5	0.072	0.025	0.013
1.19	0.740	0.494	0.393	5.0	0.062	0.019	0.010
1.20	0.723	0.480	0.381	6.0	0.048	0.014	0.007
1.22	0.692	0.454	0.358	7.0	0.038	0.010	0.005
1.24	0.662	0.431	0.338	8.0	0.031	0.008	0.004
				9.0	0.027	0.006	0.003
				10.0	0.022	0.005	0.002
				20.0	0.015	0.002	0.001

Problems

5.1. For the same channel and the same discharge (1,000 cusecs) as in Example 5.1, compute and plot the surface profile upstream from a sluice gate which raises the depth to 10 ft, to a section where the depth is 20 percent greater than uniform depth. Use the method of Sec. 5.2, and let the depth change in five equal steps.

5.2. Recalculate Example 5.1, with larger increments of depth, (a) by omitting every second step in the existing tabulation, (b) by taking only two steps in all: from $y = 3.85$ to $y = 5.20$, thence to $y = 6.10$. Compare the results so obtained with the results of Example 5.1.

5.3. Recalculate Prob. 5.1, covering the whole range of depth in one step. Compare the result with that of Prob. 5.1.

5.4. A rectangular channel 20 ft wide, $n = 0.014$, is laid on a slope of 0.001 and terminates in a free overfall. Upstream 1,000 ft from the overfall is a sluice gate which produces a depth of 1.55 ft immediately downstream. Using the method of Sec. 5.2, and the "conjugate curves" introduced in Sec. 4.5, compute and plot the profile between the sluice gate and the overfall for a discharge of 600 cusecs.

5.5. Upstream of the sluice gate in Prob. 5.4, the channel has the same properties as it did downstream, except that the slope is 0.005. Upstream 1,000 ft from the sluice gate the channel begins with a free outfall from a lake. For the same discharge as in Prob. 5.4, and using the same method, compute and plot the profile between the lake and the sluice gate.

5.6. Show that for a deep narrow rectangular section, the exponent N in Eq. (5–3) is equal to 2.

5.7. Show that for a wide-channel section described by Eq. (2–37), the exponents N and M in Eqs. (5–4) and (5–6) are equal to $(2i + 3\frac{1}{3})$ and $(2i + 3)$ respectively. Hence determine N and M for (a) wide parabolic and (b) wide triangular sections. For which of the exponents, N or M, is it unnecessary to assume a wide channel in order to obtain a simple result?

5.8. From Eq. (5–6) prove Eqs. (5–15) and (5–16).

5.9. From Eq. (5–3) prove Eqs. (5–17) and (5–18).
(*Note*: in each of Probs. 5.8 and 5.9, begin by differentiating each side of the equation with respect to y.)

5.10. In the situation of Prob. 4.33, compute enough of the profile to locate the hydraulic jump. Use any method.

5.11. In the situation of Prob. 4.34, compute the whole profile, including the location of the hydraulic jump. Use any method.

5.12. A very wide rectangular channel carries a discharge of 100 cusecs per foot width. $S_0 = 0.001$, $n = 0.026$. A spillway structure produces a depth of 2.5 ft immediately downstream; 2,000 ft downstream of the spillway the slope of the channel changes and becomes steep. Use the varied-flow function to compute and plot the profile between the spillway and the head of the steep slope.

5.13. The channel of Prob. 5.12 has the same properties upstream of the spillway as it has downstream. Assuming that there is no energy loss as the flow goes over the spillway, compute and plot the profile for a distance of 10,000 ft upstream of the spillway.

5.14. Verify that the procedure outlined in Fig. 5-3 can be generalized as shown in the accompanying figure, with the various types of mild-slope and steep-slope profiles traced out in their appropriate regions as shown. Can the M_3 and S_1 curves be traced on this figure?

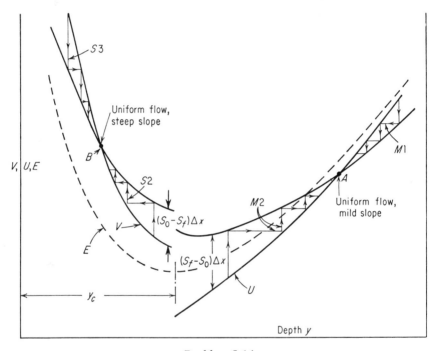

Problem 5-14

5.15. Construct a figure similar to that accompanying Prob. 5.14, showing the regions appropriate to the flow profiles occurring on adverse and horizontal slopes.

5.16. Construct curves, as in Fig. 5-3, suitable for the case of Prob. 5.1, with $\Delta x = 400$ ft. Recalculate this problem for each of the following initial values of depth: 10 ft, 10.5 ft, 11.0 ft, 11.5 ft, 12.0 ft.

5.17. Recalculate Prob. 5.4 by the method of Fig. 5-3, choosing suitable values for Δx.

5.18. Recalculate Prob. 5.5 by the method of Fig. 5-3, choosing suitable values for Δx.

5.19. The following table shows how the wetted perimeter P (feet) and the cross sectional area A (square feet) varies with stage at three river sections, of which I is furthest downstream.

Stage, ft	I		II		III	
	P	A	P	A	P	A
107	264	1,150	224	400		
108	298	1,428	325	675	0	0
109	332	1,743	395	1,035	105	70
110	355	2,086	422	1,443	141	193
111			444	1,876	170	348
112			466	2,330	190	528
113			486	2,806	212	729
114			507	3,303	230	950
115					248	1,189
116					266	1,446
117					282	1,720
118					300	2,011

The Manning n is 0.035, and none of the sections have berms. The distance I–II is 2,000 ft, and II–III is 2,300 ft. When the discharge is 7,500 cusecs, the river stage at I is 108 ft; find the stage at the other two sections, using the method of Sec. 5.5.

5.20. The longitudinal profile is to be computed for a flow of 100,000 cusecs in a natural river channel. Along the reach being considered there is a distinct berm as well as the main channel. Each of the two channels so formed is

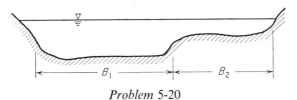

Problem 5-20

approximately rectangular in section, as shown in the figure, and wide enough for the width to be taken equal to the wetted perimeter P. The properties of each section are listed below:

River mile	Widths, ft		Elevation of bed above datum, ft		Manning n	
	B_1	B_2	M.C.	Berm	M.C.	Berm
13.25	380	350	49.1	54.7	0.03	0.05
13.50	490	300	49.6	57.1	0.03	0.05
13.75	400	480	51.6	59.6	0.03	0.05
14.00	600	280	55.9	61.4	0.03	0.05

Between river mile 13.25 and 13.75 the channel is straight and there are no eddy losses. Between 13.75 and 14.00 the channel is curved so that the berm is on the inside of the curve, and the length of the berm section is only 960 ft. The main channel section is the full length of 1,320 ft. Over this reach the curve also introduces eddy losses equal to 0.2 times the mean of the velocity heads at the ends of the reach.

At river mile 13.25 the river stage is 64.5 ft when the discharge is 100,000 cusecs; calculate the stage at the other three sections by the method of Sec. 5.6.

5.21. Using the modified Ezra method of Fig. 5-8, recalculate Prob. 5.19 for discharges of 5,000, 7,500, and 10,000 cusecs, the stages at section I being 107, 108, and 109 ft respectively.

5.22. Using the modified Ezra method of Fig. 5-8, recalculate Prob. 5.20 for discharges of 75,000, 100,000 and 150,000 cusecs, the stages at river mile 13.25 being 62.0, 64.5, and 68.5 ft respectively.

5.23. Recalculate Prob. 5.21 by the Escoffier method, and compare the results with those of Prob. 5.21.

5.24. Recalculate Prob. 5.22 by the Escoffier method, and compare the results with those of Prob. 5.22.

5.25. In the modified Ezra method as shown in Fig. 5-9, show that the horizontal distance between corresponding k_A and k_B curves is equal to $S_f \Delta x Q_0^2 / Q^2$. Hence verify that y increases or decreases *in the direction in which the solution is being traced*, according as S_f is, respectively, greater or less than S_0. Show that this conclusion is consistent with all the results of Sec. 4.5 concerning the behavior of dy/dx, whether they relate to subcritical or supercritical flow.

5.26. For the same channel as in Prob. 5.1, use the modified Ezra method of Fig. 5-9 to plot the surface profile for a distance of 2,000 ft upstream from the sluice gate, using four equal steps of 500 ft. This is to be done for discharges of 500, 750, and 1,000 cusecs, with corresponding depths at the sluice gate of 8, 9, and 10 ft respectively.

5.27. As in Example 4.1, a rectangular channel 10 ft wide, $n = 0.014$, runs on a slope of 0.001 from a lake whose surface level is 10 ft above the channel bed at the lake outlet. A free overfall is to be located at some downstream section; how far should it be from the lake outlet so as to make the depth at the outlet 1 per cent less than it would have been if no overfall were present? If the overfall were in fact placed at half this limiting distance from the lake, what would the discharge be?

5.28. A trapezoidal channel of base width 15 ft, side slopes $1\frac{1}{2}$H : 1V, $n = 0.023$, runs on a slope of 0.0005 from a lake which is to provide pondage as part of a flood-control scheme. The maximum lake level expected is 10 ft above the channel bed at the lake outlet; calculate the discharge capacity of the channel, assuming that it has the same slope and cross section for a great distance downstream.

5.29. In the situation of Prob. 5.28, the channel is to be crossed by a road at a section 3,000 ft downstream from the lake outlet. It is proposed to use a circular culvert for the road crossing instead of a bridge. Assuming that the culvert entry is well rounded and that its invert is level with the channel bed, determine the limiting culvert diameters which must be exceeded if the culvert (a) is not to act as a choke; (b) is not to reduce the discharge from the lake.

5.30.   In the situation of Prob. 5.29, the culvert diameter is to be 10 ft. Determine the discharge capacity of the channel when the lake level is at its maximum, calculating any M_1 curve between the culvert and the lake in one step.

5.31.   The object in this problem is to verify by actual calculation the general picture of events shown in Fig. 4-12. Consider a very wide $(R = y)$ channel with $n = 0.025$, $S_0 = 0.0001$, and with two short well-rounded humps A and B, 1.5 ft high. A is 3,000 ft upstream of B, and downstream of B there is a steep slope. Determine and plot the whole profile (including the hydraulic jump, if any) between A and B and for a few hundred feet upstream of A, for discharges of 5, 10, 20, 30, 40, 60, and 80 cusecs/ft width. Note in each case whether the profile immediately upstream of B is of M_2 type, and whether a hydraulic jump forms or not. Prove that for a jump to form it is a necessary (but not sufficient) condition that the profile be of M_1 type, and verify this conclusion from your numerical results.

5.32.   A channel of rectangular section 20 ft wide is 500 ft long and is laid on a uniform slope of 0.010. It drains a lake whose maximum surface level is 10 ft above the channel bed at the lake outlet, and discharges into another lake downstream. Neglecting resistance, draw a suitable set of Q contours on the h_1-h_2 plane as defined in Fig. 5-13.

5.33.   In the situation of Prob. 4.37, what is the maximum possible flow that could be delivered by the channel under the specified upstream lake level and downstream bed level? What would the corresponding bed slope be? What bed slope would be necessary to make the channel deliver half the above maximum flow?

5.34.   A trapezoidal channel, of base width 30 ft and side slopes 2H : 1V, has a bed slope of 0.001, $n = 0.027$, and is to deliver 2,500 cusecs to a lake, whose level may fluctuate. The channel bed at its outflow into the lake is 10 ft above datum; determine the water level at a section one mile upstream from the outflow when the lake level is (a) 14 ft, (b) 21 ft above datum.

5.35.   Rework the previous problem for the following two alternative designs of channel: (a) the bed of the channel is lowered by 3 ft along its whole length; (b) the bed level at the outflow is kept at its original level of 10 ft above datum, and the bed slope is reduced to 0.0005.

5.36.   As part of a flood-control scheme, a relief channel with a capacity of 10,000 cusecs is to be cut through a seaside resort town, discharging direct into the sea. High real estate values in the locality dictate a rectangular concrete-lined section for the channel in order to keep the top width down to a minimum. The highest tide level is 5 ft above mean sea level (M.S.L.) and it is required that the water level in the channel shall not rise more than 40 ft above M.S.L. at a section 1.5 miles upstream from the mouth of the channel. Choose a bed slope and bed levels so as to minimize the excavation required, assuming that the width of the channel is to be (a) 100 ft, (b) 50 ft. Take $n = 0.014$.

5.37. Find some real locality to which Prob. 5.36 would be applicable, and from that locality assemble data on real estate values, and on excavation, concreting, and bridging costs. Make an economic study comparing the cost of a 100-ft-wide channel with that of a 50-ft-wide channel, assuming that the total length of the channel is to be 2 miles and that it is to be crossed by three two-lane bridges, all suitable for heavy commercial traffic. Assume that any "working strip" of land that has to be acquired, as well as the land for the channel itself, will have the same width whatever the width of the channel may be.

5.38. A concrete lined ($n = 0.014$) channel of rectangular section is to discharge 3,000 cusecs from one lake to another; the lakes are one mile apart and their surface levels differ by 15 ft. Choose a bed slope and bed levels so as to minimize the excavation required, assuming that the width of the channel is to be (a) 40 ft, (b) 20 ft.

5.39. In the situation of Prob. 5.38, the mean ground level along the line of the proposed channel is 3 ft above the line joining the lake surfaces. The cost of finished concrete per cubic yard is 100 times the excavation cost per cubic yard. The channel walls and bed are to be 1 ft thick, and the freeboard allowance is to be 2 ft. The cost of the land is negligible. By numerical trial, find the most economical width.

5.40. Determine whether it is more economical to increase the capacity of a channel by increasing its width, or its depth, in the following cases: (a) a natural stream near its mouth; (b) a steep channel leading from a lake; (c) a long channel of mild slope leading from a lake. In all cases assume that the channel is approximately rectangular in section, and that the water levels in the channel are to remain unchanged at the increased discharge.

Computer Programs

(*Note*: Unless specific values of the known variables are given, as for example in C5.2, these programs are to be written in a form which is as general as possible— i.e., allowing the future operator of the program to insert any desired value of base width, bed slope, etc., or in the case of an irregular channel a listing of A and P against stage. Remember also that it is good practice to provide for listing all this input data at the output stage of the program.)

C5.1. Write a computer program which will determine longitudinal profiles in a channel of uniform trapezoidal section, by the method of Sec. 5.2. Incorporate means of computing and listing the uniform depth.

C5.2. Write and operate a computer program to calculate M from Eq. (5–16) for values of $y' = my/b$ from 0 to 2 in steps of 0.1, and N from Eq. (5–18) for $m = 0, 1.0, 1.5, 2.0$, and y/b running from 0 to 2 in steps of 0.1.

C5.3. Write and operate a computer program to calculate the varied-flow function for the same values of $u = y/y_0$ as in the Appendix, and for values of N equal to the mean values of N, M, and J over the channel reach in Prob. 5.1. Rework this problem using these results and those of Prob. C5.2, and compare the solution with the one obtained as in Sec. 5.2.

C5.4. Incorporate features in the program of Prob. C5.1 which will enable y to be calculated and listed for initial specified values of x and Δx.

C5.5. Write a computer program which will determine longitudinal profiles in an irregular river channel without berms—i.e., the velocity coefficient $\alpha = 1$.

C5.6. Incorporate features in the program of Prob. C5.5 allowing for berm flow on each side of the channel. This means that the channel is to be divided into three parts, and that A and P must be listed as a function of stage for each part.

C5.7. Write a computer program which, given the data required by the program of Prob. C5.6, will calculate and list the functions required for the operation of the Ezra method.

C5.8. Write a computer program which, given any channel of trapezoidal section, any length and any mild slope, will calculate and list data from which the curves of Fig. 5-13b may be plotted. Operate the program for the channel of Example 5.6, and plot the resultant set of curves. Plot another curve representing the Q-h_2 relation dictated by the condition of critical flow within the culvert, taking h_2 as the depth immediately upstream of the culvert. From the intersections of this curve with the Q contours, solve Example 5.6 and compare the result with that previously obtained.

C5.9. Write a computer program which, given any channel of trapezoidal section, any length and any steep slope, will calculate and list data from which curves corresponding to those of Fig. 5-13b may be plotted. Operate the program for the case of a rectangular channel 20 ft wide, 2,000 ft long with $n = 0.014$ and a bed slope of 0.0025. Plot the resulting curves and compare them with those of Prob. 5.32; is any agreement between them to be expected?

C5.10. Write a computer program which, given any channel of trapezoidal section, any length and any slope, will test whether the slope is mild or steep and choose the program of C5.8 or C5.9 accordingly. Operate the program for the case of a rectangular channel as in Prob. C5.9, but with a slope of 0.0018.

C5.11. Apply the above programs to whichever of the other problems and examples are appropriate, and compare the results so obtained with those derived without the use of a computer.

Chapter **6**

Channel Controls

6.1 Introduction

A control has previously been defined as any channel feature, natural or man-made, which fixes a relationship between depth and discharge in its neighborhood. We now consider just what form this depth-discharge relationship may take, and how it may depend on the physical nature of the feature—e.g., a weir or a sluice gate, which forms the control. The engineer's interest in the problem is twofold: first, he is concerned with the functioning of the control itself (such as the ability of a spillway to discharge floodwaters at the required rate); second, he wishes to know the extent to which controls may interfere with or even dominate the shape of the longitudinal profile of the water surface, as discussed in Chap. 5..

Also, it is convenient to include in this chapter the treatment of certain appurtenances—for example, energy dissipators, which are not controls but usually occur in association with controls in engineering works.

6.2 Sharp-Crested Weirs

A device of this kind normally consists of a vertical plate mounted at right angles to the flow and having a sharp-edged crest, as in Fig. 6-1. Such weirs are commonly used as a means of flow measurement, but are of some fundamental interest as well because their theory forms a basis for the design of spillways. Because the edge is sharp, opportunities for boundary-layer development are limited to the vertical face of the weir, where velocities are low; we may therefore expect the flow to be substantially free from viscous effects and the resultant energy dissipation.

We consider first the simplest form of weir, consisting of a plate set perpendicular to the flow in a rectangular channel, its horizontal upper edge running the full width of the channel. This last feature means that the flow is essentially two-dimensional, without lateral contraction effects; since these effects are suppressed by the channel sidewalls, this type of weir is sometimes termed the "suppressed" weir.

Figure 6-1 shows a longitudinal section of flow over such a weir; an elementary analysis can be made by assuming that the flow does not contract as it passes over the weir, and that the pressure is atmospheric across the

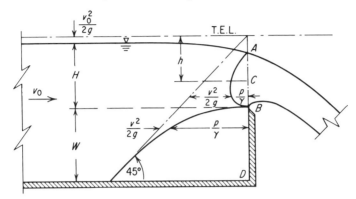

Figure 6-1.   *The Sharp-Crested Weir*

whole section AB. Under these assumptions the velocity at any point such as C is equal to $\sqrt{2gh}$, and the discharge q per unit width accordingly equal to

$$\int_{v_0^2/2g}^{H+v_0^2/2g} \sqrt{2gh}\; dh = \tfrac{2}{3}\sqrt{2g}\left[\left(\frac{v_0^2}{2g}+H\right)^{3/2} - \left(\frac{v_0^2}{2g}\right)^{3/2}\right]$$

the depth h being measured downwards from the total energy line, and not from the upstream water surface.

The effect of the flow contraction may be expressed by a contraction coefficient C_c, leading finally to the result

$$q = \tfrac{2}{3}C_c\sqrt{2g}\left[\left(\frac{v_0^2}{2g}+H\right)^{3/2} - \left(\frac{v_0^2}{2g}\right)^{3/2}\right] \tag{6-1}$$

We can make this expression more compact by introducing a discharge coefficient C_d; Eq. (6–1) then becomes

$$q = \tfrac{2}{3}C_d\sqrt{2g}\; H^{3/2} \tag{6-2}$$

where

$$C_d = C_c\left[\left(1+\frac{v_0^2}{2gH}\right)^{3/2} - \left(\frac{v_0^2}{2gH}\right)^{3/2}\right] \tag{6-3}$$

We should expect both C_c and the ratio $v_0^2/2gH$ to be dependent on the boundary geometry alone, in particular on the ratio H/W; it follows that C_d should be a function of H/W alone. Such was indeed found to be the case in the early experimental work of Rehbock [1]; his results are closely approximated by the formula

$$C_d = 0.611 + 0.08H/W \tag{6-4}$$

according to which C_d becomes equal to 0.611 as W becomes very large.

Since in this case $v_0^2/2gH$ will become negligibly small, Eq. (6–3) shows that C_c also will be equal to 0.611. Now this happens to be the numerical value of $\pi/(\pi + 2)$, shown last century by Kirchoff to be the contraction coefficient of a jet issuing without energy loss, and with negligible deflection by gravity, from a long rectangular slot in a large tank—the two-dimensional problem shown in Fig. 6-2a. The solution was later extended by von Mises [2] to the case of the finite-size tank shown in Fig. 6-2b.

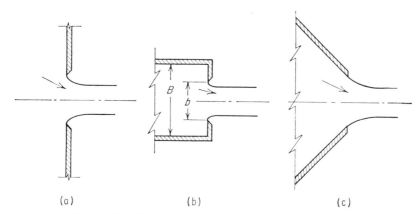

(a)   (b)   (c)

Figure 6-2.   *Examples of Jets Uninfluenced by Gravity*

This two-dimensional problem in ideal fluid flow is the only case for which explicit theoretical solutions have been obtained for the contraction coefficient of flow from a sharp-edged opening. The problem remains intractable for those cases where the jet is three-dimensional, or deflected by gravity. However Lauck [3], by using a numerical integration process, obtained a theoretical result confirming the above experimental value of 0.611 for a high weir, and Southwell and Vaisey [4] later used the relaxation method—a means of determining the flow pattern by successive trials—on the case where $W = 0$, i.e., the free overfall. In recent years the advent of high-speed computers has enabled more attention to be given to numerical methods, which are constantly being improved, e.g., as described by Birkhoff [5]; however an explicit solution still remains elusive.

We have seen that C_c has the same numerical value for the high weir and for the slot shown in Fig. 6-2a; this might suggest a similar correspondence between the values of C_c for the low weir and for the slot in a finite tank, Fig. 6-2b, with b/B in this figure corresponding to $H/(H + W)$ in Fig. 6-1. However this is not the case; as b/B tends to unity, so does C_c, but as $H/(H + W)$ tends to unity (the free overfall) then C_c, as we shall see, remains substantially less than unity.

We now reconsider Eq. (6–4); clearly this formula cannot be true when the ratio H/W becomes large: in fact experimental work has shown that it is true only for values of H/W up to approximately 5; for $5 < H/W < 10$, C_d begins to diverge from the value given by the formula, reaching a value of 1.135 when $H/W = 10$. If W vanishes completely, so that H/W becomes infinite, we have the case of the free overfall, discussed in a preliminary way in Chaps. 2 and 4. In this case H becomes equal to y_c, the critical depth, and q is determined accordingly; in fact it has been shown experimentally that critical flow also occurs just upstream of a very low weir (better described as a sill in this case) i.e., in the range $H/W > 20$. We can therefore write

$$q = \sqrt{g}\, y_c^{3/2} = \sqrt{g}\, (H + W)^{3/2} \qquad (6\text{–}5)$$

and if this result is substituted into Eq. (6–2) we obtain

$$C_d = 1.06\left(1 + \frac{W}{H}\right)^{3/2} \qquad (6\text{–}6)$$

applicable to very low weirs, or sills. The range $10 < H/W < 20$ has not yet been completely explored.

From Eqs. (6–3) and (6–6) it is readily shown (Prob. 6.1) that $C_c = 0.715$ for the completely free overfall ($W = 0$). The existence of a contraction coefficient in this situation is a matter that will be more fully discussed in Sec. 6.4.

Figure 6-1 shows two important features of the pressure distribution down the vertical section ABD: first, that it is nonhydrostatic, because of distinct curvature of the streamlines in the vertical plane; second, that contrary to the assumptions made in the elementary analysis the pressure is above atmospheric within the section AB. This last feature is consistent with the contraction and acceleration that takes place downstream of AB; clearly a pressure force is required along AB in order to bring about this acceleration into a region of atmospheric pressure.

It has been assumed throughout that the pressure is atmospheric along the lower surface of the jet, or "nappe," as well as along the upper surface. Care must therefore be taken to see that this condition is fulfilled if the above equations are to be used as a basis for flow measurement. If the nappe is contained within parallel walls downstream of the weir, it may well enclose the air between itself, walls, and floor; this air will be gradually removed by the flow, the pressure in this region reduced, and the discharge for a given head increased. Means should therefore be provided for ventilating the region below the nappe.

Any type of weir other than the one just described will involve a three-dimensional flow problem, in that there will be contraction from the sides as well as in the vertical plane. Typical examples are the "contracted" rectangular weir, and the triangular weir, both shown in Fig. 6-3. Such weirs are commonly used for flow measurement, normally in tanks large enough to be effectively infinite, so that C_c has its minimum value. In this case general equations are available relating discharge to head; however if the weirs are used in smaller tanks then each particular setup must be calibrated individually, because of the obvious difficulties in obtaining general formulas corresponding to Eq. (6–4).

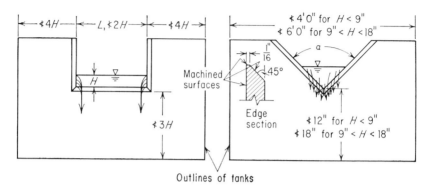

Figure 6-3. Weir and Tank Dimensions Recommended by British Standard Specifications [7]

Francis [6] found experimentally that the amount of lateral contraction at each end of the contracted rectangular weir was equal to one-tenth of the head H, provided that the length L of the weir was greater than $3H$. On the basis of this result, it is commonly accepted that the discharge Q is given by the equation

$$Q = \tfrac{2}{3}C_c(L - 0.2H)\sqrt{2g}\,H^{3/2} \tag{6–7}$$

and that C_c is 0.611, as for the suppressed weir.

The triangular weir, or V notch, can be analyzed in the same elementary way as the suppressed rectangular weir (Prob. 6.2) leading to the result

$$Q = \frac{8}{15}\,C_c \tan\frac{\alpha}{2}\sqrt{2g}\,H^{5/2} \tag{6–8}$$

The most commonly used value of the notch angle α is 90°; for this case C_c is found to be around 0.585, somewhat less than for the rectangular weir. Given these values Eq. (6–8) becomes

$$Q = 2.50H^{2.50} \tag{6–9}$$

although precise measurements have tended to indicate that the equation

$$Q = 2.48H^{2.48} \tag{6–10}$$

is more accurate. The "half-90°" notch is also commonly used; its width at any level is one half that of the 90° notch, i.e., $\alpha = 2\tan^{-1}\tfrac{1}{2}$.

Formulas for all three of these weirs are also to be found in the British Standard Specifications [7]. While these specifications have no legal force outside the British Commonwealth, they are of interest in that they offer firm guidance as to the exact form of the weir edge, the required tank dimensions (Fig. 6-3) and the degree of accuracy to be expected in the result. The formula given for the suppressed weir is the equivalent of:

$$q = \tfrac{2}{3}\sqrt{2g}\,(0.604 + 0.081H/W)(H + 0.0034)^{1.5} \qquad \textbf{(6–11)}$$

where all quantities are in foot-second units. This formula is virtually identical with that of Rehbock, as in Eqs. (6–2) and (6–4). The stated margin of error of this formula is ± 1.5 percent.

For the contracted rectangular weir, the formula is the equivalent of

$$Q = \tfrac{2}{3}\sqrt{2g} \times 0.615(L - 0.1H)H^{1.5} \qquad \textbf{(6–12)}$$

with a margin of ± 2 percent applicable for $L/H \geq 2$. For the 90° and half-90° V notches:

$$Q = 2.48H^{2.48} \qquad (6\text{–}10)$$

and

$$Q = 1.24H^{2.48} \qquad \textbf{(6–13)}$$

respectively, with a margin of ± 1.5 percent. In this case the recommended range for H is 3 in.–15 in.

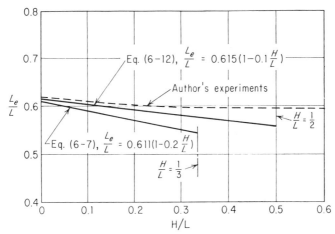

Figure 6-4. *The Effective Width of a Contracted Rectangular Weir*

The difference between the effective widths in Eqs. (6–7) and (6–12) is noteworthy. It is displayed in Fig. 6-4, in which the ratio L_e/L is plotted against H/L, the "effective width" L_e being defined as $Q/\tfrac{2}{3}\sqrt{2g}\,H^{1.5}$. The difference is seen to be quite substantial. The results of some experiments performed by the author are indicated by a broken line in the same figure, and are seen to lie somewhat above the line representing Eq. (6–12). At higher values of H/L the broken line turns upwards again; this is reasonable since one would not expect the effective width L_e to keep decreasing indefinitely. The ultimate tendency must be for the overall contraction coefficient to approach a constant value not greatly different from 0.6. However this may be, it appears that further experiments would be desirable at lower values of H/L in order to resolve the discrepancy between Eqs. (6–7) and (6–12).

The finish of the edge and upstream surface of a weir is important, since roughness of the surface or rounding of the edge tends to suppress the lateral components of flow, increase C_c, and hence increase the discharge.

6.3   The Overflow Spillway

This type of spillway, illustrated in Figs. 6-5 through 6-8, is the most common of all types. It divides naturally into three zones—the crest, the face, and the toe—each with its separate problems. It will be seen that some of these problems, like those involved in the sharp-crested weir, are not completely tractable by existing theory, and that recourse must be had to experiment.

The Spillway Crest

Normally the crest is shaped so as to conform to the lower surface of the nappe from a sharp-crested weir, as shown in Fig. 6-5b. The pressure on the crest will then be atmospheric, provided that the resistance of the solid surface to flow does not induce a material change in the pressure distribution. This could happen only if the boundary layer over the crest were very thick; and it is easily shown (Prob. 6.3) that the boundary layer, which will grow effectively only from the neighborhood of the point A, is in fact a very small fraction of the head over the crest. Therefore we may expect the pressure over the crest to be atmospheric, and in this fact lies the virtue of this crest shape;

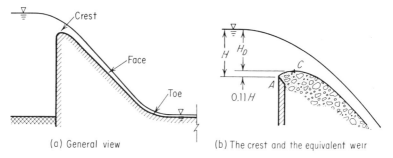

(a) General view   (b) The crest and the equivalent weir

Figure 6-5.   The Overflow Spillway

pressures above atmospheric will reduce the discharge, and pressures below atmospheric will increase the discharge, but at the risk of introducing instability and cavitation.

We consider the case of the high spillway. For the equivalent weir, $H/W = 0$ and the substitution of Eq. (6–4) into Eq. (6–2) then leads to the result

$$q = 3.27H^{3/2} \tag{6–14}$$

Experiments show that the rise from weir crest to the high point of the nappe (the spillway crest) is $0.11H$, as in Fig. 6-5. Using this fact we can express Eq. (6–14) in terms of H_D, the head over the spillway crest. We obtain

$$q = 3.97H_D{}^{3/2} \tag{6–15}$$

where H_D may be termed the *design head*; as we have seen, operation at this head will make the pressure over the crest atmospheric. However the spillway will also have to operate at lower heads, and possibly higher heads as well. The former will evidently result in above-atmospheric pressures on the crest, and a lower discharge coefficient; the latter in the reverse of these effects.

In the latter case, the dangers inherent in low pressures are not as great as might at first appear. The matter is clarified by the experiments of Rouse and Reid [8] and of Dillman [9], plotted in Fig. 6-6. It is seen from this figure that the actual head H_A may safely exceed the design head by at least 50 percent, with a 10 percent increase in the discharge coefficient, provided of course that the local pressure does not fall below the cavitation level. The

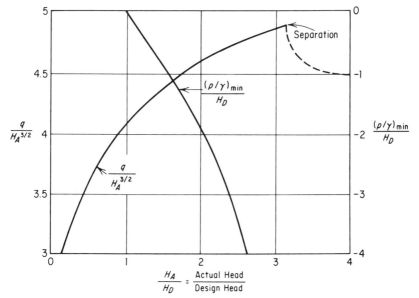

Figure 6-6. *Spillway Crest Characteristics, after H. Rouse and L. Reid* [8] *and O. Dillman* [9]

magnitude of this local pressure (at the crest) was measured by the same investigators, and is also plotted in Fig. 6-6.

As to the details of the crest shape, extensive experiments by the U. S. Bureau of Reclamation have resulted in the development by the U. S. Army Corps of Engineers of curves which can be described by simple equations, yet approximate closely to the nappe profiles measured in the U.S.B.R. experiments. The profile for a vertical upstream face is shown in Fig. 6-7; others were also developed for various angles of the upstream face to the vertical.

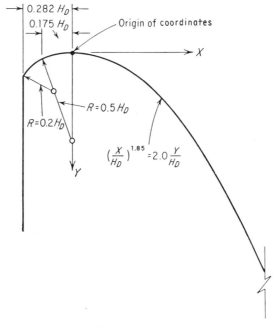

Figure 6-7. Standard Spillway Crest (U.S. Army Engineers Waterways Experiment Station)

The above discussion has been confined to the case of the high spillway without convergence in plan. Complications arise when the spillway is of finite height, converges in plan, or is divided by piers into sections. These problems have been the subject of extensive studies by the two agencies referred to above, and a comprehensive account of their results is given by Ven Te Chow [10].

The Spillway Face

Flow down the steep face of the spillway, normally at about 45° to the horizontal, has a rather special character which makes the methods of Chaps. 4 and 5 unsuitable for its treatment. In this case acceleration and boundary layer development are both taking place during much of the journey down the spillway face, as shown in Fig. 6-8. Turbulence does not become fully developed until the boundary layer fills the whole cross section of the flow, at the point marked C. Downstream of this point the flow might be expected to conform to the S_2 profile developed in Chap. 4, but the extreme steepness of the slope introduces more complications, chiefly the phenomenon of air entrainment, or "insufflation," of which an example is shown in Fig. 6-9.

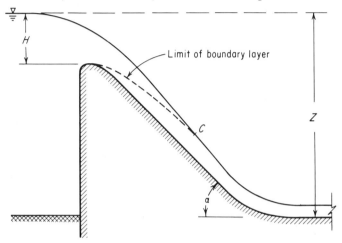

Figure 6-8.   *Boundary Layer Development on Spillway Face*

It is now generally agreed that insufflation begins at this very point C, where the boundary layer meets the water surface. The resulting mixture of air and water, containing an ever-increasing proportion of air, continues to accelerate until uniform flow occurs, or the base of the spillway is reached.

Clearly the designer will wish to know the velocity reached at the base, or toe, of the spillway, but the above remarks make it clear that the computation of this velocity would be tedious and difficult, even if one were certain of the correct assumptions to adopt concerning the nature of the flow. Considerable work has been done on this problem at the U. S. Bureau of Reclamation, and the results have been presented by Bradley and Peterka [11] in the form of a chart (Fig. 6-10) from which this velocity may be estimated. In the authors' words, this chart "represents a composite of experience, computation, and a limited amount of experimental information obtained from prototype tests on Shasta and Grand Coulee Dams. There is much to be desired in the way of experimental confirmation; however, it is felt that this chart is sufficiently accurate for preliminary design."

The "theoretical velocity" v used in Fig. 6-10 is defined as

$$v_t = \sqrt{2g(Z - H/2)}$$

where Z and H are as shown in Fig. 6-8. The curves are said to be applicable

Figure 6-9.   Air entrainment on the Face of a Spillway

[Courtesy *The Auckland Star*]

to steep slopes, from 0.8H : 1V to 0.6H : 1V. It is clear that the general form
of the curves is correct, for as the head H rises and the discharge increases, the
actual velocity v_a will more nearly approach v_t.

At the other extreme, when the head H and the discharge are small, the
depth of flow will be small and the effect of resistance will be marked. We
should therefore expect the flow to become uniform before it has fallen very
far below the spillway crest, and in fact it can readily be shown (Prob. 6.20)
that the appropriate portions of the curves in Fig. 6-10 have a form consistent
with this hypothesis.

The mechanism of air entrainment is not yet completely understood, and
reliable field data on the concentration of entrained air are extremely sparse.
However, some important laboratory experiments carried out by Straub and
Anderson [12] have yielded many useful data. The flume used was 50 ft long,
with its bed roughened by particles having a mean diameter (k_s) of 0.028 in.
The sidewalls were smooth and the width-to-depth ratio was always large
enough for sidewall effects to be small—i.e., air entrainment was substantially
due to boundary layer growth from the bed, not from the sidewalls. The
longitudinal slope was varied from 7.5 to 75°, and the depth was adjusted
by an inlet gate so as to produce flow at the downstream end that was uniform
in velocity, depth, and distribution of entrained air.

The air concentration c was defined as the volumetric ratio air: (air plus water). The mean concentration $\bar{c}$ was found to be a function of $S/q^{1/5}$ alone, where $S = \sin \alpha$ (Fig. 6-8) and q is the discharge per unit width. A Task Committee set up by the American Society of Civil Engineers to study air

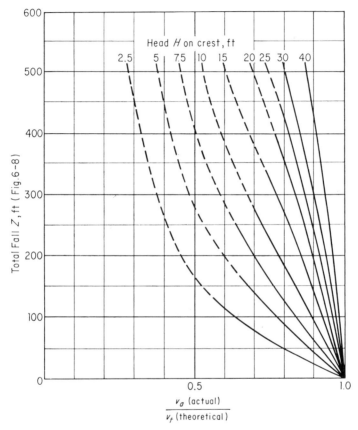

Figure 6-10.   *Velocity at the Foot of a Spillway, after J. N. Bradley and A. J. Peterka* [11]

entrainment found [13] that all available results, including Straub and Anderson's, fitted the following equation

$$\bar{c} = 0.743 \log_{10}(S/q^{1/5}) + 0.876 \tag{6-16}$$

with a standard error of 0.061. The Task Committee also recommended that Eq. (6–16) be applied to open, i.e., ungated, spillways (in which flow at the base might not be uniform), although in Straub and Anderson's experiments uniform flow obtained in all cases, being produced by upstream gate control. Further work is needed to clear this matter up.

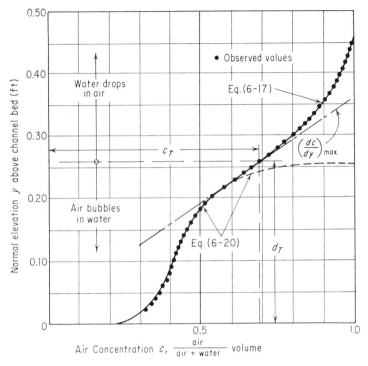

Figure 6-11. *Typical Distribution of Entrained Air, after L. G. Straub and A. G. Anderson* [12]

The work of Straub and Anderson was also notable for its clarification of the detailed structure of aerated uniform flow. These authors drew a distinction between a lower region of flow, consisting essentially of water impregnated with air bubbles, and an upper region consisting of spray, or air containing drops of water. The two regions are shown in Fig. 6-11, in which air concentration during a typical experimental run is plotted against normal elevation y above the channel bed. The depth d_T marked on this figure is the transition depth marking the division between the two regions of flow. (In this discussion all water depths normal to the bed will be indicated by d, as noted in Sec. 2.1. We define y as above, i.e., as distance normal to the bed; the difference between this definition and the usual one need not cause any confusion.)

Straub and Anderson developed theoretical analyses of the variation of the air concentration in each region. For the upper region, the argument was as follows: drops of water are projected upwards into this region by turbulent velocity fluctuations at the water surface $(d = d_T)$, and it is assumed that these fluctuations follow a Gaussian probability distribution. Given this assumption it can readily be shown that the proportion of water particles reaching or passing through a level y_1 above the transition depth is equal to

$$\frac{2}{h\sqrt{\pi}} \int_{y_1}^{\infty} e^{-(y_1/h)^2} \, dy_1$$

where h is a measure of the mean distance the particles are projected above d_T. It follows that the air concentration c at the level y_1 is related to the concentration c_T at d_T by the equation

$$\frac{1-c}{1-c_T} = \frac{2}{h\sqrt{\pi}} \int_{y_1}^{\infty} e^{-(y_1/h)^2} \, dy_1 \qquad (6\text{--}17)$$

which fits closely to the plotted observations in Fig. 6-11. This good agreement is of course an agreement in the form of the function only; the theory cannot predict values of c_T and h, which must be deduced (Prob. 6.21) from the observed magnitudes of c and dc/dy at $d = d_T$. Given these values, it can then be shown that Eq. (6–17) is a good fit to the observations in the range $d > d_T$, or $y_1 > 0$.

The distribution of air bubbles in the lower region can be investigated by a well known theory which is commonly used to examine the distribution of suspended sediment in stream flow. The two problems are of course essentially similar although air bubbles tend to rise and sediment tends to fall. In each case a systematic drift in one direction is balanced by transport in the other direction due to the interaction of fluid turbulence and a continuous change in concentration in the y-direction.

A full discussion of this theory will be postponed until sediment transport is dealt with in Chap. 10, but an outline of the theory will be given here. As pointed out in Sec. 1.8, the essential action of turbulence is the continuous interchange of material between neighboring regions of fluid. If these regions have differing forward velocities, the result is a shear force acting between them, induced by an exchange of momentum; if they have differing concentrations of some quality or substance—e.g., temperature, salinity, or air content, the result is a movement of that quality or substance from the region of high concentration to that of low concentration. The rate of this movement will therefore depend on the concentration gradient and on a coefficient ε_b measuring the intensity of the turbulence; this rate is balanced by the steady upward drift of the air bubbles at a velocity v_b. The resulting equation is

$$-cv_b + \varepsilon_b \frac{dc}{dy} = 0 \qquad (6\text{--}18)$$

and its integration will depend on the behavior of the mixing coefficient ε_b. If, anticipating the argument of Sec. 10.4, we assume a parabolic distribution of ε_b,

$$\varepsilon_b = av^* y \left(\frac{d_T - y}{d_T}\right) \qquad (6\text{--}19)$$

then Eq. (6–18) can be integrated to yield

$$c = c_1 \left(\frac{y}{d_T - y} \right)^{v_b/av*} \tag{6–20}$$

where $v*$ is the shear velocity $\sqrt{\tau_0/\rho}$, as in Sec. 4.2, and c_1 is the concentration at $y = d_T/2$. Again, Fig. 6-11 shows good agreement between plotted observations and theory, subject to the same limitation that two constants (c_1 and $v_b/av*$ in this case) have to be determined from the experimental results. Granted this limitation, there is close agreement in form between Eq. (6–20) and the experimental results.

There remains the question of flow resistance. Straub and Anderson found that uniform nonaerated flow in their laboratory channel followed the Chézy equation very closely, with the coefficient C equal to 90.5, i.e.,

$$q = 90.5 d_m^{3/2} S^{1/2} \tag{6–21}$$

where d_m is the uniform depth. Uniform aerated flow followed a similar relationship, and for this reason d_m provides a convenient reference depth to which depths representative of the aerated flow can be related. Two further depths were defined: d_u, the depth at which $c = 0.99$, and above which there is a negligible amount of water; and d_w, defined by the equation

$$d_w = \int_0^\infty (1 - c)\, dy \tag{6–22}$$

and therefore equal to the effective depth of water only, q/v_m, where v_m is the mean velocity. The results of Straub and Anderson's experiments can then be summed up in this way. Given q and S, d_m is calculated from Eq. (6–21); d_T and d_w can then be obtained from it by means of the following empirical relationships, true in the range $0.25 < \bar{c} < 0.75$:

$$\frac{d_T}{d_m} = 1 + 2(\bar{c} - 0.25)^2 \tag{6–23}$$

$$\frac{d_w}{d_m} = 1 - 1.3(\bar{c} - 0.25)^2 \tag{6–24}$$

When $\bar{c} < 0.25$ the upper, spray-filled, region is nonexistent and d_m, d_w, and d_T are identical. By definition, the upper limiting depth d_u is given by the equation

$$d_u = \frac{d_w}{1 - \bar{c}} \tag{6–25}$$

Equation (6–21) is of course true only for the particular channel and surface roughness used in Straub and Anderson's experiments, but pending further investigation it is suggested that Eqs. (6–23) and (6–24) should form a useful guide for the designer even when applied to different surfaces and resistance equations, as in Prob. 6.22.

The Spillway Toe

When the flow reaches the end of the inclined face of the spillway it is deflected through a vertical curve into the horizontal or into an upward direction, Fig. 6-12. In the latter case we have the ski-jump and the bucket-type energy dissipators, to be discussed in a later section.

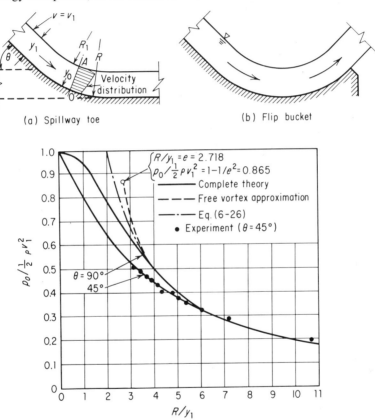

(a) Spillway toe   (b) Flip bucket

(c) Maximum pressure — theory and experiment

Figure 6-12.   *Flow at the Spillway Toe, after F. M. Henderson and D. G. Tierney* [14]

In either case, centrifugal pressures will be developed which can set up a severe thrust on the spillway sidewalls. These pressures cannot be accurately calculated by elementary means, but certain approximations suggest themselves; e.g., if one assumed that the depth y_0 at the center of the curve (Fig. 6-12a) is equal to the depth y_1 of the approaching flow, then the centrifugal pressure at the point O will be equal to

$$p_0 = \frac{\rho v_1{}^2 y_1}{R} \tag{6-26}$$

where v_1 and R are also defined in Fig. 6-12a. This result can only be an approximation, for a pressure rise along AO must, by the Bernoulli equation, be accompanied by a fall in velocity, so that the velocity profile will be somewhat as shown in Fig. 6-12a. The average velocity will then be less than v_1, and the depth y_0 greater than y_1, so that Eq. (6–26) will not be correct.

A better approximation can be made by assuming that the streamlines crossing OA form parts of concentric circles, and that the velocity distribution along this line is accordingly the same as that in the free, or irrotational, vortex, i.e.,

$$v = \frac{C}{r}$$

where C is a constant and r is the radius of any streamline. Since the streamlines are concentric circles, r is also a measure of distance along AO from A to O. If R_1 is the radius of the streamline at A, then $C = v_1 R_1$. The discharge q across AO is given by

$$q = v_1 y_1 = \int_{R_1}^{R} v\, dr = v_1 R_1 \int_{R_1}^{R} r^{-1}\, dr$$

$$= v_1 R_1 \log \frac{R}{R_1}$$

i.e.,

$$\frac{y_1}{R_1} = \log \frac{R}{R_1}$$

and

$$\frac{y_1}{R} = \frac{R_1}{R} \log \frac{R}{R_1} \tag{6-27}$$

Since y_1 and R are known in advance, R_1 can be obtained by trial from this equation. Given R_1/R, we can obtain p_0, the pressure at O, from the condition

$$p_0 + \tfrac{1}{2}\rho v_0{}^2 = \tfrac{1}{2}\rho v_1{}^2$$

i.e.,

$$\frac{p_0}{\tfrac{1}{2}\rho v_1{}^2} = 1 - \left(\frac{v_0}{v_1}\right)^2 = 1 - \left(\frac{R_1}{R}\right)^2 \tag{6-28}$$

assuming no energy dissipation between A and O; this assumption appears to be justified by experiment.

The "free-vortex" method leads to results that are quite accurate within a certain range, but it suffers from a curious limitation, arising from the fact that the function $(\log_e x)/x$ has a maximum value of $1/e$, which occurs when $x = e$, the base of the natural logarithms (Prob. 6.4). Applying this result to Eq. (6–27) we see that the ratio R/y_1 has a minimum value of e when $R/R_1 = e$, even though R/y_1 is by the nature of the problem an independent variable, which may in practice assume any value at all. The effect of this curious result is that the theory cannot be applied when $R/y_1 < e$, and a curve displaying the results of the theory must, as in Fig. 6-12c, terminate at the point where $R/y_1 = e$, although lower values of R/y_1 are quite possible. The corresponding terminal value of $p_0/\frac{1}{2}\rho v_1{}^2$ will, from Eq. (6–28), be equal to $1 - 1/e^2$.

A complete solution of the problem requires the use of the mathematical theory of irrotational flow. This has been done by Henderson and Tierney [14] for the case where there is an open sluice through the spillway, as shown by broken lines in Fig. 6-12a, and for the more usual case discussed above, where the toe is a curved solid surface. Theoretical and experimental results for the latter case are shown in Fig. 6-12c, which displays the behavior of p_0, the pressure at O, for angles θ (Fig. 6-12a) of 45° and 90°. It is seen that the free vortex method gives results approximating closely to those of the complete theory when $R/y_1 > 6$, as does the elementary result of Eq. (6–26). However, the latter fails to predict any thickening of the jet and therefore seriously underestimates the total thrust and bending moment on the sidewalls. For complete details the reader is referred to the original papers.

All the above discussion implies the assumption that the flow is irrotational. This assumption is a reasonable one, since losses must be small over the short length of spillway involved, and the highly turbulent approaching flow must have a transverse velocity distribution very close to the uniform distribution which is characteristic of irrotational flow of a perfect fluid. Also, there is little risk of separation when a solid boundary is continually curving *into* the flow, as in this case. Pressure distributions should therefore be close to those in the irrotational flow of a perfect fluid, and this conclusion is confirmed by the good agreement between theory and experiment shown in Fig. 6-12.

Further, the effect of gravity has been ignored, so that the pressures derived are purely those due to centrifugal action. We take gravity into account simply by adding hydrostatic pressure, calculated as in Sec. 2.1, to the pressures obtained above. This additional pressure may be substantial in the case of the bucket-type energy dissipator, in which a structure like that of Fig. 6-12b is deeply drowned under a turbulent but stationary eddy, as shown in Fig. 6-36.

6.4　The Free Overfall

In this situation, shown in Fig. 6-13, flow takes place over a drop which is sharp enough for the lowermost streamline to part company with the channel bed. It has been previously mentioned as a special case ($W = 0$) of the sharp-crested weir, but it is of enough importance to warrant individual treatment.

Clearly, an important feature of the flow is the strong departure from hydrostatic pressure distribution which must exist near the brink, induced by strong vertical components of acceleration in the neighborhood. The form of this pressure distribution at the brink *B* will evidently be somewhat as shown in Fig. 6-13, with a mean pressure considerably less than hydrostatic. It should also be clear that at some section *A*, quite a short distance back from the brink, the vertical accelerations will be small and the pressure will be hydrostatic. Experiment confirms the conclusion suggested by intuition—that from *A* to *B* there is pronounced acceleration and reduction in depth, as in Fig. 6-13.

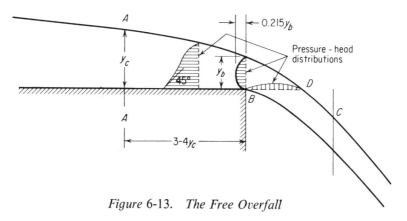

Figure 6-13. The Free Overfall

If the upstream channel is steep, the flow at *A* will be supercritical and determined by upstream conditions. If on the other hand the channel slope is mild, horizontal, or adverse, the flow at *A* will be critical. This is readily seen to be true by recalling the argument of Sec. 4.4, according to which flow is critical at the transition from a mild (or horizontal, or adverse) slope to a steep slope. Imagine now that in this case (shown in Fig. 4-7) the steep slope is gradually made even steeper, until the lower streamline separates and the overfall condition is reached. The critical section cannot disappear; it simply retreats upstream into the region of hydrostatic pressure—i.e., to *A* in Fig. 6-13.

The local effects of the brink are therefore confined to the region *AB*; experiment shows this section to be quite short, of the order of 3–4 times the depth. Upstream of *A* the profile will be one of the normal types determined by channel slope and roughness, and discussed in Chap. 4; if our interest is confined to longitudinal profiles the local effect of the brink may be neglected because *AB* is so short compared with the channel lengths normally considered in profile computations.

However, our interest may center on the overfall itself, because of its use either as a form of spillway or as a means of flow measurement—the latter arising from the unique relationship between brink depth and the discharge. Apart from these matters of practical interest, the problem, like that of the sharp-crested weir, continues to attract the exasperated interest of theoreticians who find it difficult to believe that a complete theoretical solution can really be as elusive as it has so far proved to be.

In the following discussion it is convenient to subdivide the flow into two regions of interest—first, the brink itself, and the falling jet, which we may call the "head" of the overfall; and second, the base of the overfall where the jet strikes some lower bed level and proceeds downstream after the dissipation of some energy.

The Head of the Overfall

The simplest case is that of a rectangular channel with sidewalls continuing downstream on either side of the free jet, so that the atmosphere has access only to the upper and lower streamlines, not to the sides. This is a two-dimensional case and it is only in this form of the problem that serious attempts have been made at a complete theoretical solution.

Consider section C (Fig. 6-13), a vertical section through the jet far enough downstream for the pressure throughout the jet to be atmospheric, and the horizontal velocity to be constant. If we simplify the problem further by assuming a horizontal channel bed with no resistance, and apply the momentum equation to sections A and C, it is easily shown (Prob. 6.5) that

$$\frac{y_2}{y_1} = \frac{2Fr_1^2}{1 + 2Fr_1^2} \qquad (6\text{–}29)$$

where the subscripts 1 and 2 characterize sections A and C respectively; if the flow is critical at section A the above equation becomes

$$\frac{y_2}{y_c} = \frac{2}{3} \qquad (6\text{–}30)$$

which sets a lower limit on the brink depth y_b; since there is some residual pressure at the brink, y_b must be greater than y_2. It follows that

$$\frac{2}{3} < \frac{y_b}{y_c} < 1 \qquad (6\text{–}31)$$

Actually there is no part of the jet, however far downstream, where the pressure is completely atmospheric; if there were the streamlines would all become parabolas, and these curves cannot exhibit the property of asymptotic convergence which the streamlines actually possess. However, the point is a somewhat academic one, for it can be shown (Probs. 6.6 and 6.7) that the internal pressure in the jet tends to zero much faster than does the width of the jet, as the jet moves further downstream. In the limit, when the jet has fallen infinitely far below the brink, Eq. (6–30) will be true and the horizontal velocity will be equal to $3v_c/2$. From this last fact it can be seen that the internal pressure of the jet plays a decisive role in developing the ultimate form of the jet, for the horizontal velocity on the lower streamline is originally equal to $\sqrt{2g(3y_c/2)} = \sqrt{3}v_c$ (at B) and that on the upper streamline to v_c (at A). The horizontal forces required to bring each of these to the ultimate value of $3v_c/2$ are supplied by the pressure gradients at either end of pressure profiles such as that shown on the horizontal section BD.

The form of the pressure distribution at B has already been referred to; the pressure profiles just upstream must be of the form indicated in Fig. 6-13, with inflexions as shown. These are necessary in order to return the pressure distribution to hydrostatic at the bed, where the vertical acceleration must be zero. One consequence of this property of the profiles is that the pressure on the bed must remain finite very close to the brink; the longitudinal pressure gradient there must therefore be infinite, and the same is true of the lateral pressure gradient, as indicated by the pressure profile at B. It follows that the radius of the curvature of the lower streamline must momentarily be zero just downstream of B. This is a well-recognized property of all free jets.

The foregoing discussion, although of some general interest, does not lead to specific conclusions. For these we depend on experiment and on approximate analysis. The experiments of Rouse [15] showed that the brink section has a depth of $0.715\,y_c$. Rouse also pointed out that combination of the weir Eq. (6–1) with the critical flow equation

$$q = v_c y_c = y_c \sqrt{g y_c}$$

setting $H = y_c$ and $v_0 = v_c$, led to the result $y_b = 0.715\,y_c$, although there might be some doubt about the physical significance of the result.

More recently Replogle [16] has carried out further experiments with substantially the same results, and has also measured the brink pressure profile, with the result shown in Fig 6-13. The brink depth and pressure profile are approximately consistent with each other (Prob. 6.5), although the pressure is somewhat smaller than it should be for complete consistency. Bed resistance over the length AB can account for only about half of this discrepancy (Prob. 6.8), but the matter is of little significance because the discrepancy represents a very much smaller percentage error in the brink depth than in the brink pressure. Replogle [16] discusses the relationship between corresponding errors in the various parameters and shows, among other things, that the effect of velocity variation at the brink, expressed by a momentum coefficient β, is quite negligible.

Although no complete theoretical solution has yet been obtained, a number of solutions based on trial or approximate methods have been advanced, most of them offering remarkably close confirmation of Rouse's result

$y_b = 0.715 \, y_c$. Southwell and Vaisey [4] used relaxation methods to plot the complete flow pattern, finding in the process a value of y_b of approximately $0.705 \, y_c$. Jaeger [17] and Roy [18] used ingenious approximations to obtain near-complete solutions in the neighborhood of the brink; each found that $y_b = 0.72 \, y_c$. Fraser [19] used an iterative method due to Woods [20] to trace the upper and lower streamlines, concluding that $y_b = 0.71 \, y_c$. Hay and Markland [44] used the electrical analogy to determine an experimental solution in an electrolytic plotting tank. The profile they deduced was very close to Southwell and Vaisey's except near the brink, where they found $y_b = 0.676 \, y_c$.

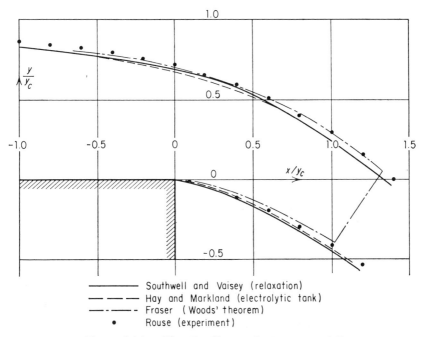

Figure 6-14.   *Flow Profiles at the Free Overfall*

The conclusion suggested by all this work, summarized in Fig. 6-14, is that a brink depth $y_b = 0.715 \, y_c$ can safely be used for flow measurement, with a likely error of only 1 or 2 percent.

The preceding discussion has concentrated particularly on the two-dimensional case with a horizontal bed and no resistance. The theoretical methods described assume a perfect fluid; if the bed is smooth and the upstream flow fully turbulent, this assumption should create little error in the analysis, for the bed resistance between sections A and B, Fig. 6-13, will have little effect on the brink depth (Prob. 6.8). No analysis has yet been attempted of the case where the slope and resistance are large, but comprehensive experimental results have been obtained by Delleur and others [21]. They are illustrated in Fig. 6-15.

Experimental results have also been obtained by Diskin [22] for trapezoidal channels and by a number of investigators for circular pipes. In these cases the whole periphery of the flow at the brink is exposed to atmosphere, so one would expect the average static pressure at this section to be appreciably less than in the two-dimensional case, and y_b to be correspondingly closer to a value calculated from the momentum equation by neglecting the hydrostatic thrust at the brink.

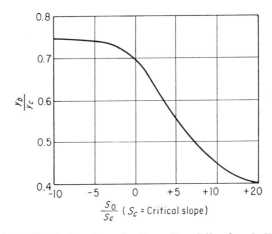

Figure 6-15. *Brink Depth at the Free Overfall, after J. W. Delleur et al.* [21]

For trapezoidal channels of side slopes $mH:1V$, Diskin [22] found that the result of this momentum argument could be conveniently expressed in terms of the dimensionless number $y' = my/b$, used in Chap. 2 for the determination of critical depth in trapezoidal channels. If, as before, we indicate brink conditions and critical conditions by the subscripts b and c respectively, the momentum argument yields (Prob. 6.9)

$$y'_b = \tfrac{1}{2}[\sqrt{1 + 4T_c} - 1] \tag{6–32}$$

where

$$T = \frac{6y'(1 + y')^3}{9 + 20y' + 10y'^2}$$

From these equations, brink conditions may be calculated from critical conditions. Diskin's experiments showed very good agreement with Eq. (6–32) for $m = 2$, but for $m = 1.5$ the measured value of y_b was about 4–5 percent greater than the calculated value. In these experiments the ratio y/b went up to 0.8 approximately.

It would seem reasonable to expect better agreement with theory for the flatter side slopes, since in this case the atmosphere has access to a greater perimeter per unit cross-sectional area, and the mean pressure over the brink section would accordingly be less. Whatever the reason may be, it appears that further experiments are needed to settle the matter; meanwhile Diskin's results have established that Eq. (6–32) is accurate enough for the design of drop structures, if not for systems of flow measurement.

Many investigators have made measurements of brink depth in circular pipes running full; the original aim of these investigations was to develop a system of flow measurement, which became known as the California pipe method. This title became attached to the first empirical equation developed, which can be written in the following dimensionless form

$$\frac{Q}{D^2\sqrt{gD}} = 1.55\left(\frac{y_b}{D}\right)^{1.88} \tag{6–33}$$

for pipe diameters between 3 and 12 in. However, other observers have made measurements of discharge differing by up to 10 percent from those given by Eq. (6–33). Accordingly this equation can only be recommended for approximate measurements; for more precise results the arrangement should be calibrated in situ.

Another feature of the experimental results is that the extent of the agreement with the results of the simplified momentum argument varies from one investigator to another, and from one pipe size to another. These wide variations appear to arise from difficulties of measurement peculiar to the circular pipe, e.g., the existence of cross waves and of a hump in the transverse water surface near the brink; they have led Diskin [23] to suggest that a section some distance back from the brink would be a more satisfactory measuring station on which to base a system of flow measurement. He found that by choosing such a station rather arbitrarily at 5.2 times the diameter upstream of the brink, a consistent empirical relation could be found between $Q/D^2\sqrt{gD}$ and y_s/D, where y_s is the depth at the chosen station. It was

$$\frac{Q}{D^2\sqrt{gD}} = 0.88\left(\frac{y_s}{D}\right)^{2.23} \tag{6–34}$$

and this equation fitted closely (with a mean deviation of 2 percent) the results of three series of experiments on 6-in., 8-in., and 10-in. pipe. The same three series had failed to disclose an equally consistent relationship among the discharge, diameter, and the brink depth.

It appears therefore that Eq. (6–34) forms a satisfactory basis for flow measurement, at least within the range of pipe diameters tested by Diskin, and within his stated range of error. A limitation on the method is that it requires the pipe to run part full for some distance back from the brink. It is pointed out by Smith [23] that at higher values of y_c/D the pipe runs full up to a distance L_f from the brink equal to a few pipe diameters. The figures given by Smith are:

$Q/D^2\sqrt{gD}$	0.65	0.54	0.47
y_c/D	0.82	0.75	0.70
L_f/D	0.25	0.65	2.6

from which it appears that y_c/D would have to be less than 0.6 for a free surface to exist over a substantial distance back from the brink. Vennard [23] suggests that the flow should be ventilated by holes drilled along the top of the pipe; this would have the effect of increasing L_f. Of course, no measures of this sort would be effective unless the slope and roughness of the pipe are such that uniform flow is possible when the pipe runs part full.

The Base of the Overfall

The situation is illustrated in Fig. 6-16, which shows a complete "drop structure" such as is installed at intervals in steep channels in order to dissipate energy without scouring the channel. We are concerned here with the events occurring where the jet strikes the floor and turns downstream at section 1.

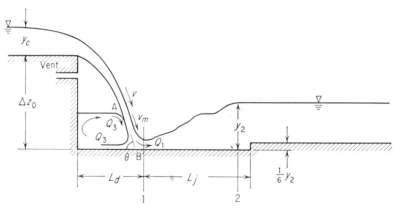

Figure 6-16.   *The Drop Structure*

In this region there will be a great deal of energy loss because of circulation induced by the jet in the pool which forms beneath the nappe. The function of this pool is to supply the horizontal thrust required to turn the jet into the horizontal direction. The amount of the energy loss has been determined by the experiments of Moore [24], results of which are plotted in Fig. 6-17. Interesting comment on these results is provided by the analysis of White, who in a discussion of Ref. [24] assumed the following mechanism by which the jet sets up circulation in the pool: near the point A, a thin layer of water, having negligible momentum, is entrained into the jet; it mixes with the jet, the two streams merging into a single one having a uniform velocity v_m. The effect is that the jet becomes thicker and slower moving; then when the jet strikes the floor at B it divides into a main stream moving forward with velocity $v_1 = v_m$, and into a smaller stream which returns to the pool, where it dissipates the momentum which it acquired from the jet. The discharge rate Q_3 of this smaller stream is of course equal to the rate of entrainment at A.

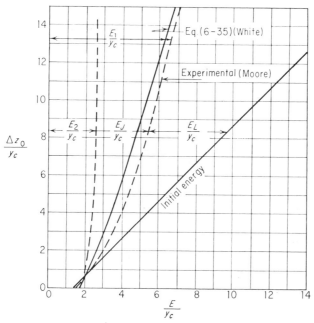

Figure 6-17.   *Energy Dissipation at the Base of the Free Overfall*

Application of the momentum equation to this situation leads to the result

$$\frac{y_1}{y_c} = \frac{\sqrt{2}}{1.06 + \sqrt{\dfrac{\Delta z_0}{y_c} + \dfrac{3}{2}}} \tag{6–35}$$

by means of which the specific energy at section 1 is readily obtained from the equation

$$\frac{E_1}{y_c} = \frac{y_1}{y_c} + \frac{y_c^2}{2y_1^2} \tag{6–36}$$

which is a restatement of Eq. (2–17). A curve combining the results of Eqs. (6–35) and (6–36) is plotted in Fig. 6-17, and it is seen that the agreement with experiment is remarkably good considering the approximations that must be inherent in this formulation of the problem.

This analysis by White is commended to the reader as a good example of the way in which skilful formulation and a grasp of fundamentals can yield surprisingly exact solutions of problems which may at first glance appear intractable by theoretical methods. The details can be worked out by the reader as an exercise (Probs. 6.10 and 6.11).

The Drop Structure

Figure 6-17 shows that the energy loss E_L at the base of an overfall may be 50 percent or more of the initial energy, referred to the basin floor as datum. If, as in Fig. 6-16, there is a hydraulic jump downstream of section 1 dissipating further energy, the energy loss in the entire "drop structure" may be very substantial. The loss due to the hydraulic jump is readily calculated (Prob. 6.12) in terms of the parameters of Fig. 6-17, and a curve is plotted on that figure displaced to the left of the E_1/y_c curve by the amount E_J/y_c, where E_J is the loss in the jump. This left-hand curve then indicates the remaining specific energy E_2 downstream of the jump. It is seen that the ratio E_2/y_c does not vary greatly with $\Delta z_0/y_c$; this suggests that a value of 2.5 for E_2/y_c may form a satisfactory basis for a preliminary design (Probs. 6.13 and 6.14).

Rand [25] assembled the results of experimental measurements made by himself, by Moore [24], and others, and from them obtained the following exponential equations, which fit the data with errors of 5 percent or less:

$$\frac{y_1}{\Delta z_0} = 0.54\left(\frac{y_c}{\Delta z_0}\right)^{1.275} \tag{6–37a}$$

or

$$\frac{y_1}{y_c} = 0.54\left(\frac{y_c}{\Delta z_0}\right)^{0.275} \tag{6–37b}$$

$$\frac{y_2}{\Delta z_0} = 1.66\left(\frac{y_c}{\Delta z_0}\right)^{0.81} \tag{6–38}$$

$$\frac{L_d}{\Delta z_0} = 4.30\left(\frac{y_c}{\Delta z_0}\right)^{0.09} \tag{6–39}$$

$$L_j = 6.9(y_2 - y_1) \tag{6–40}$$

where L_d and L_j are the horizontal distances covered by the jet and the hydraulic jump respectively, as shown in Fig. 6-16. With the help of these equations the designer can proportion the simple drop structure completely. The upward step of $y_2/6$ at the end of the structure, shown in Fig. 6-16, is a standard design feature which helps to localize the jump immediately below the overfall.

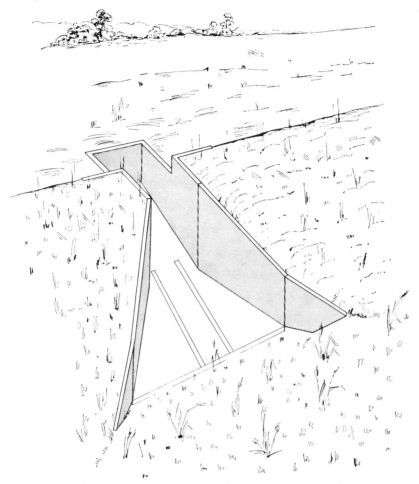

Figure 6-18.   The Box Inlet Drop Structure, after F. W. Blaisdell and C. A. Donnelly [26]

It has been assumed in the preceding discussion that the flow at the brink of the overfall is critical, i.e. that the upstream slope is mild. This is often true of the drop structure, whose very purpose is usually to allow the main channel to be laid on a mild slope. However, steep upstream slopes sometimes occur, leading to supercritical flow at the brink. Rouse [24] gives details of the jet shape and behavior for this type of situation.

The drop structure described above constitutes a two-dimensional problem and is the simplest type in use. Many variants of this design are used in practice; e.g., a basin may be used that is shorter than Eqs. (6–39) and (6–40) would indicate, embodying special means of forming and locating the jump, as in Sec. 6.7. A common type is the box-inlet drop structure, shown in Fig. 6-18, which has the advantage of dissipating more energy by making three streams meet at the foot of the drop; it is discussed in detail by Blaisdell and Donnelly [26].

6.5 Underflow Gates

The vertical sluice gate shown in Fig. 6-19a has already been discussed in Chaps. 2 and 3; it forms one of a general class of *underflow gates*, of which the radial or Tainter gate, and the drum gate, Fig. 6-19b and c, are also typical. Such gates may be used for a variety of purposes—e.g., as controls

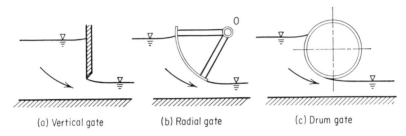

(a) Vertical gate   (b) Radial gate   (c) Drum gate

Figure 6-19. *Typical Underflow Gates*

at the crest of an overflow spillway, or at the outlet from a lake to a river or irrigation canal. The choice of one type or another in a particular case may depend on a variety of factors, and each type has its own advantages. For example, the vertical gate has the disadvantage of requiring a costly roller-and-track assembly by which to transmit its thrust to the sidewalls; the Tainter gate is more economical in this respect but may incur extra structural costs through having its thrust concentrated at the hinge O.

Underflow gates can conveniently be discussed under the headings of free outflow and submerged outflow.

Free Outflow

The outflow is said to be free when, as in Fig. 6-20, the issuing jet of supercritical flow is open to atmosphere and is not overlaid, or submerged, by tailwater of excessive depth. The analysis is elementary provided that the contraction coefficient C_c is known, and provided that we are concerned only with the regions of substantially uniform depth upstream of section 1 and downstream of section 2. We can write, from the energy equation

$$y_1 + \frac{q^2}{2gy_1{}^2} = y_2 + \frac{q^2}{2gy_2{}^2}$$

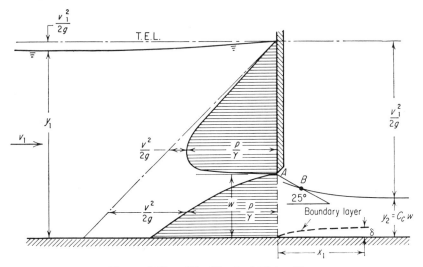

Figure 6-20.   *The Vertical Sluice Gate*

and from a simple rearrangement

$$q = y_1 y_2 \sqrt{\frac{2g}{y_1 + y_2}} \qquad (6\text{–}41)$$

It is customary to express this discharge as the product of the depth w, the velocity $\sqrt{2gy_1}$, and a discharge coefficient C_d. Accordingly Eq. (6–41) is rewritten as follows (noting that $y_2 = C_c w$)

$$q = C_c w \sqrt{2g y_1 \frac{y_1}{y_1 + y_2}}$$

$$= C_d w \sqrt{2g y_1} \qquad (6\text{–}42)$$

where

$$C_d = \sqrt{\frac{C_c}{1 + C_c w / y_1}} \qquad (6\text{–}43)$$

and since C_c is dependent on the boundary geometry, i.e., on w/y_1, C_d will be wholly dependent on w/y_1. The velocity $\sqrt{2gy_1}$ does not of course occur anywhere in the system; it simply makes a convenient reference velocity on which to base a standard equation such as Eq. (6–42).

The determination of C_c for a given value of w/y_1 involves the exploration of the region of rapidly varying flow close to the gate; like the free overfall problem, it is a question of theoretical hydrodynamics that has not so far yielded a complete solution. Southwell and Vaisey [4] have applied relaxation methods to the case where $y_2/y_1 = 0.321$, finding $C_c = 0.608$. Much more comprehensive results were obtained by Benjamin [27] through an analysis in which the flow was divided into two parts by the section B (Fig. 6-20), chosen rather arbitrarily as the section where the tangent to the surface makes an angle of 25° with the bed. It was assumed that downstream of B the curvature d^2y/dx^2 of the water surface was small, and that all higher derivatives of y were of rapidly diminishing order. These assumptions imply, among other things, that the pressure distribution is hydrostatic. With their help an approximate analysis was developed, whose solution was the same as that of the well-known solitary wave (Chap. 8).

In the region AB these assumptions are no longer valid; here an approximate solution was found by taking von Mises' [2] solution for the no-gravity case, and superimposing on it an allowance for the variation of surface velocity between A and B. The two solutions were fitted together at the section B. The results are given in the following table.

TABLE 6-1

w/E_1	0	0.1	0.2	0.3	0.4	0.5
C_c	0.611	0.606	0.602	0.600	0.598	0.598

The lack of variation in C_c is remarkable: for the corresponding no-gravity case treated by von Mises C_c increases quite rapidly as the ratio of opening width to tank width increases. In the present case the constancy of C_c is clearly due to the depression by gravity of the downstream free surface well below the level it would reach in the no-gravity case.

Benjamin also determined the criterion for the establishment of uniform flow downstream, free of waves. The existence of this condition is normally taken for granted by engineers, but it was shown by Benjamin that a necessary condition for its existence is that $Fr_2 \geq 1.25$, i.e., $Fr_1 \leq 0.792$. If $Fr_2 \leq 1.25$ a system of standing waves will be set up downstream; however, this condition would hardly occur in normal sluice gate operation since it implies that the gate opening w is nearly equal to the upstream depth y_1.

Experimental results obtained by Benjamin are plotted in Fig. 6-21, together with the values listed in Table 6–1. Two sets of experimental results are plotted, each for a fixed value of the opening w. The marked discrepancies between the different experimental results, and between experiment and theory, are explained by the existence of boundary-layer growth approximately as indicated in Fig. 6-20 (and in Fig. 1-9b). Comparing systems having the same Froude number at corresponding parts of each system, and being therefore substantially similar (e.g., x_1/w in Fig. 6-20 = a constant) then if δ = boundary-layer thickness, we have

$$\frac{\delta}{w} \propto Re^{-1/2} \propto \left(\frac{\nu}{vw}\right)^{1/2}$$

since the smooth bed and rapidly converging flow will produce a laminar rather than a turbulent boundary layer. Since Fr is a constant

$$v \propto w^{1/2}$$

i.e.,
$$\frac{\delta}{w} \propto \left(\frac{1}{w^{3/2}}\right)^{1/2} \propto w^{-3/4} \qquad (6\text{--}44)$$

and δ/w represents the proportional increase in C_c due to boundary layer growth. Now it is easily verified from Fig. 6-21 that the ratio of departures

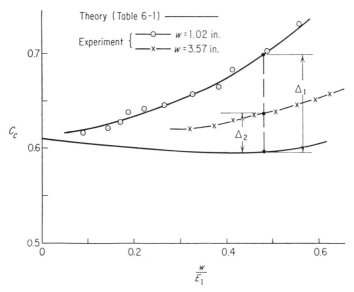

Figure 6-21.   *The Contraction Coefficient of the Vertical Sluice Gate, after T. Brooke Benjamin* [27]

Δ_1 and Δ_2 of the experimental lines from the theoretical line is very close to the value

$$\frac{\Delta_1}{\Delta_2} = \left(\frac{w_2}{w_1}\right)^{3/4} = \left(\frac{3.57}{1.02}\right)^{3/4} = 2.56 \qquad (6\text{--}45)$$

offering good confirmation of Eq. (6–44). Direct measurements of boundary-layer thickness also indicated that it was of the right order of magnitude to explain the observed discrepancies.

This example of boundary-layer growth has limited interest for the engineer concerned with field installations, for in these cases boundary-layer thickness will be small and it will be sufficiently accurate to take C_c as being constant and equal to 0.61. However, it is of some interest for the laboratory worker, who may be inclined to overlook the effects of viscosity on small-scale experimental systems.

The contraction coefficient C_c will clearly be greater than 0.61 when the gate surface is inclined to the vertical at the gate lip A, as in the case of the Tainter gate, Fig. 6-22. The discharge characteristics of this gate will depend on ratios between the four lengths y_1, w, a, and r; a lengthy program of experiment or analysis would therefore be needed to cover completely a wide

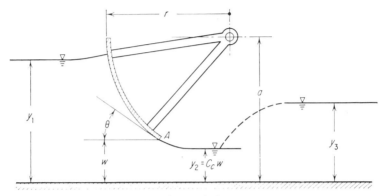

Figure 6-22.   *The Radial or Tainter Gate*

range of values of these ratios. The situation would be difficult to analyze by theoretical means, and no serious attempt has been made to do so; however, the experiments of Toch [28] covered a substantial range of values of the independent variables, both for free and drowned outflow. His results are plotted in Fig. 6-23.

An interesting feature of Toch's results is that the contraction coefficient C_c was determined very largely by the angle θ, Fig. 6-22, and to a much lesser extent by the ratio w/y_1. The effect of w/y_1 was not made completely clear by the experiments, but for a given value of θ, C_c did not depart by more than 6–7 percent from the value computed by von Mises [2] for the non-gravitational case of a two-dimensional orifice with inclined sidewalls, Fig. 6-2c. The comparative closeness of this agreement is interesting in view of the substantial differences between the two systems.

This relationship, interesting though it is, does not offer a very precise guide for the designer; for his purposes it is necessary to use the curves of Fig. 6-23 if they are applicable. However, for preliminary estimates where great precision is not required, it may be convenient to use the following equation, obtained by fitting a parabola on the $C_c - \theta$ plane as closely as possible to both von Mises' and Toch's results:

$$C_c = 1 - 0.75\theta + 0.36\theta^2 \tag{6–46}$$

where the unit of θ is taken as 90°. This equation gives results which are accurate to within ± 5 percent, provided that $\theta \leq 1$.

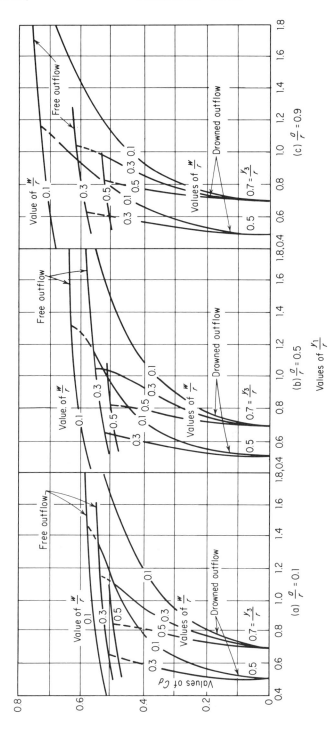

Figure 6-23.   The Discharge Coefficient of the Radial Gate (Fig. 6-22), after A. Toch [28]

Drowned Outflow

Consider the longitudinal section of flow shown in Fig. 6-24. The depth y_2 is produced by the gate, and the depth y_3 is produced by some downstream control. If y_3 is greater than the depth conjugate to y_2—i.e., the depth needed to form a hydraulic jump with y_2, then the gate outlet must become "drowned" as shown in the figure. The effect is that the jet of water issuing from beneath the gate is overlaid by a mass of water which, although strongly turbulent, has no net motion in any direction.

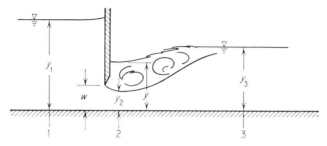

Figure 6-24.   Drowned Outflow from a Sluice Gate

An approximate analysis can therefore be made by treating the case as one of "divided flow," already discussed in Chaps. 2 and 3, in which part of the flow section is occupied by moving water, and part by stagnant water. While there will be some energy loss between sections 1 and 2, a much greater proportion of the loss will occur in the expanding flow between sections 2 and 3. We therefore assume as an approximation that all the loss occurs between 2 and 3—i.e., that $E_1 = E_2$:

$$y_1 + \frac{q^2}{2gy_1^2} = y + \frac{q^2}{2gy_2^2} \qquad (6\text{–}47)$$

Note that the piezometric head term at 2 is equal to the total depth y, not the jet depth y_2. Between 2 and 3 we can use the momentum equation, $M_2 = M_3$:

$$\frac{q^2}{gy_2} + \frac{y^2}{2} = \frac{q^2}{gy_3} + \frac{y_3^2}{2} \qquad (6\text{–}48)$$

noting that at 2 the hydrostatic thrust term is based on y, not y_2.

In the normal situation occurring in practice, y_1, y_2, and y_3 are known and it is required to calculate q; the second unknown y will also emerge from the calculation. The solution is elementary, for elimination of q^2/g leads to a quadratic in y.

While the case treated above is that of a rectangular channel, other forms of channel geometry (e.g., a circular culvert discharging into a trapezoidal channel) can easily be dealt with by the methods given in Chap. 3. However, numerical methods will be needed to deal with the resulting equations.

It remains to be seen whether the above approximate treatment is confirmed by experiment; in the event, the confirmation is remarkably good. The experimental results of Henry [29] are plotted in Fig. 6-25, in the form of curves showing the variation of C_d [defined by Eq. (6–42)] vs. y_1/w. Solid lines indicate experimental results and the broken line for $y_3/w = 5$ indicates

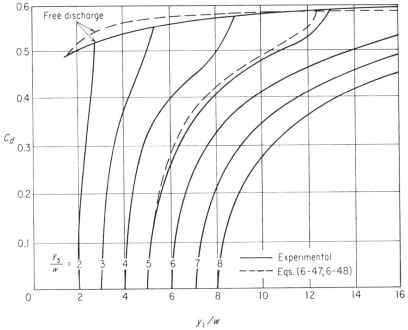

Figure 6-25. *The Discharge Coefficient of the Vertical Gate* (*Fig.* 6-24), *after H. R. Henry* [29]

the result calculated from Eqs. (6–47) and (6–48), assuming $C_c = 0.6$. Over most of the range the calculated value of C_d is greater than the measured by a margin of about 0.016, representing a proportional increase of 3–5 percent. Only when the free outflow condition is approached at higher values of C_d does the divergence between theory and experiment become more marked, presumably because of the pronounced departure from hydrostatic conditions at section 2.

The experiments of Crausse and others [30] were carried out in a flume 20 cm wide, with $w = 4$ cm. The variation of q with y_3 for $y_1 = 20.5$ cm, and the variation of y_1 with y_3 for $q = 2.57$ liters/sec/dm, showed very close agreement between theory and experiment, with differences no greater than 2 percent.

The more recent experiments of Rao and Rajaratnam [31] also showed good agreement between theory and experiment for the case of submerged flow from a culvert. This situation is equivalent to the downstream part of the case shown in Fig. 6-24, and offers independent confirmation of Eq. (6–48), in that separate measurements of y and q fitted this equation with very small error.

The evidence therefore is generally in favour of Eqs. (6–47) and (6–48), at least within a margin of 5 percent error. It is interesting to note that the boundary-layer effect observed by Benjamin [27] would have little effect on laboratory discharge measurements on submerged outflow, for the boundary layer would simply lift the emerging jet a distance equal to its displacement thickness; this lateral movement of the jet would not change in any way the application of Eqs. (6–47) and (6–48), which require only that the jet be buried somewhere within the total depth y, not necessarily at the very bottom.

Toch's experimental results, which are plotted in the same form as Henry's, C_d vs. y_1/w, showed general agreement with Eqs. (6–47) and (6–48) although in a more irregular way, with discrepancies of up to 10 percent. Some of this error may however be attributable to uncertainties about the behavior of the contraction coefficient C_c.

6.6   Critical Depth Meters

It was pointed out in Chap. 2 that the fixed relationship between depth and discharge that marks critical flow makes this type of flow a convenient basis for discharge measurement. In order to apply this principle it is necessary to create some device or use some feature that sets up critical flow at a known section in its vicinity; then measurement of the depth at this section enables the discharge to be calculated.

If such devices can be constructed, they may be free of two important disadvantages that are characteristic of weirs: first, fairly large head loss, when available head may be at a premium; and second, the existence of a dead-water region behind the weir where silt can accumulate and greatly change the head-discharge relationship of the weir.

Before considering any particular device, we first consider certain general principles. It has been shown in Chap. 4 that critical flow will occur at the change of grade B in Fig. 6-26, provided that the pressure distribution is hydrostatic. This will be true if the downstream slope, although steep, is not excessively so—say of the order of 0.01. In this case, Section B would constitute an ideal *critical depth meter*; the depth here is definitely critical, and is not changing so rapidly that slight errors in locating the depth-measuring device would give rise to serious errors in estimates of the depth.

However, the long downstream slope in Fig. 6-26 is not usually available in practice. The problem then is to incorporate the essential features of this flow situation into some comparatively short structure that can be fitted into a channel of specified (usually mild) slope. The necessary shortness of such a unit would mean that the downstream slope would have to be very steep, and the flow would have some of the character of flow over an overfall—in which, as we have seen, the critical section retreats upstream to some ill-defined location;

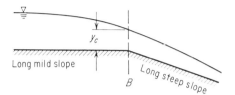

Figure 6-26.   Critical Flow at a Change of Slope

at the brink the depth is a well-defined fraction of the critical depth but the rapid variation in depth calls for precise location of the depth-measuring instrument. These facts are the basis of certain practical difficulties in the design and use of critical depth meters; these difficulties, however, are by no means insuperable.

The Broad-Crested Weir

If a weir has a crest broad enough to maintain hydrostatic pressure distribution in the flow across it, the flow will apparently be critical, as in Fig. 6-27a, and the discharge is readily determined from the upstream head H:

$$q = \tfrac{2}{3}H\sqrt{\tfrac{2}{3}gH} \qquad (6\text{–}49)$$

where H is the height of the upstream total energy line above the weir crest. If the velocity of approach is appreciable, Eq. (6–49) will have to be used with successive approximations which yield H as well as q.

The picture of events shown in Fig. 6-27a is, however, an oversimplified one which never actually occurs in practice. If the weir is short, as in Fig. 6-27b, there will be no clearly defined region of critical flow because the whole weir length will be occupied by the regions of rapidly changing depth produced by the two ends of the weir. If the weir is lengthened, as in Fig. 6-27c, resistance effects become appreciable. In both these cases Eq. (6–49) will not necessarily be true; the most satisfactory course of action in practice seems to be to use a long weir, and either use Eq. (6–49) with a correction for resistance, or base the flow measurement on the brink depth $y_b = 0.715\ y_c$. In this latter case

$$q = \frac{y_b}{0.715}\sqrt{\frac{gy_b}{0.715}} = 1.65 y_b\sqrt{gy_b} \qquad (6\text{–}50)$$

Correcting Eq. (6–49) for resistance amounts to reducing the head H by an amount δ^*, the maximum displacement thickness of the boundary layer. The discharge q given by Eq. (6–49) is then multiplied by a coefficient $C, = (1 - \delta^*/H)^{3/2}$. The analysis of Hall [32] gives the following value for C, applicable to a *square-edged* weir crest (not rounded as in Fig. 6-27):

$$1 - C = 0.069(L/H - 1 + 2.84\,\mathrm{Re}^{0.25})^{0.8}\,\mathrm{Re}^{-0.2} \qquad (6\text{–}51)$$

where $\mathrm{Re} = v_1 H/v$, the Reynolds number of the flow related to the theoretical velocity $v_1 = \sqrt{2gH/3}$. The coefficient 2.84 within the bracket relates to a weir of infinite height W (defined as in Fig. 6-1), with a vertical upstream face. When H/W is appreciable, or when the upstream face is inclined to the vertical, this coefficient has lower values. For details the reader is referred to the original paper [32].

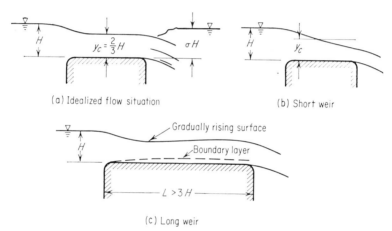

(a) Idealized flow situation   (b) Short weir

(c) Long weir

Figure 6-27.   *The Broad-Crested Weir*

Values of C calculated from Eq. (6–51) are within 1.5 percent of those measured in the early experiments of Bazin [33], in which L/H ranged from 3 to 34. When the weir is of finite width B, the boundary-layer growth along the sides will introduce another coefficient equal to

$$C_B = 1 - \frac{2\delta^*}{B} = 1 - \frac{4H(1 - C)}{3B} \qquad (6\text{–}52)$$

since $C = 1 - 3\delta^*/2H$ approximately. Experimental confirmation of this equation is not yet available.

A rounded upstream edge on the weir, as in Fig 6-27, would presumably have some influence on boundary-layer growth, producing coefficients different from those in Eqs. (6–51) and (6–52). It is not known whether the difference would be appreciable, as the matter still awaits investigation.

To sum up: investigations made so far on the broad-crested weir favor the use of the long weir, obtaining the discharge either from the upstream head H

(Fig. 6-27a) via Eqs. (6–49), (6–51), and (6–52) or from the brink depth y_b via Eq. (6–50). In the former case the weir crest is sufficiently long if $L/H > 3$; in the latter case a longer crest is needed to develop the full drawdown profile which makes Eq. (6–50) true, but the exact length needed is not yet known with certainty. The use of the brink depth is probably best confined to those cases where a long mild slope upstream of an overfall happens to be available, as discussed in Sec. 6.4.

The one question remaining is that of submergence. If the submergence factor σ (Fig. 6-27) is greater than a value of about 0.83—0.85, the flow will be affected by downstream conditions and the above equations will no longer be applicable. A conservative value of σ would be 0.80; below this value the weir will certainly discharge freely, with no submergence effects.

The Parshall Flume

As the submergence criterion indicates, a broad-crested weir can operate under quite a small head difference; however, it has the disadvantage of having a dead water region upstream in which silt and debris can accumulate. A system without this disadvantage would be provided by an open flume with a substantially level bed and a constriction in width, having the general effect of a Venturi meter. The difference in water levels Δh between the constricted throat and the upstream flow would then be measured and used as in the pipe flow case to determine the discharge.

Such "Venturi flumes" have been tried but have been found rather unsatisfactory because of the very small values of Δh which occur at low Froude numbers. The logical extension of the Venturi flume is the "standing wave flume," in which the throat geometry is so arranged as to force the occurrence of critical flow there, followed by a short length of supercritical flow and a hydraulic jump. The result is a critical depth meter which has no dead water region although it has the usual difficulties concerning the location and measurement of the critical depth.

The best-known flume of this type is of the design introduced by Parshall [34] and named after him. Detailed designs for this flume have been developed for a wide range of discharges, and the difficulties mentioned above readily disposed of by suitable choice of a standard section (not necessarily the critical one) at which the depth is measured, and by a comprehensive program of calibration, from which reliable empirical formulas have been established. Figure 6-28 shows the standard design of Parshall flume for throat widths B of 1 to 8 ft; it will be clear from the discussion of critical flow in Chap. 2 that this combination of sidewall convergence and sudden dip in the bed is well designed to promote the formation of critical flow near the section A. For the above range of throat widths the discharge is given by the formula

$$Q = 4BH_a{}^{1.522B^{0.026}} \tag{6–53}$$

where H_a is the depth measured at the upstream location shown in Fig. 6-28, and all quantities are in foot-second units. This formula is true for values of the submergence ratio H_b/H_a up to 0.7; above that value the discharge is less than the formula indicates.

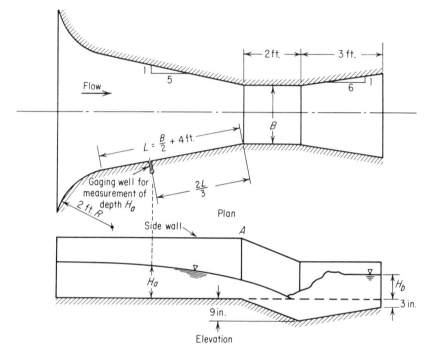

Figure 6-28.   *Parshall Flume Dimensions for Widths B of* 1 *ft to* 8 *ft*

For throat widths less than 1 ft and greater than 8 ft other designs and other formulas are given by Parshall [35,36] covering a range of widths from 3 in. to 50 ft.

6.7   Energy Dissipators

In the design of a control structure there is often a need to provide for the dissipation of excess kinetic energy possessed by the downstream flow. The result is that devices known as *energy dissipators* are a common feature of control structures. The need for them may arise from the occasional discharge of flood waters, as in the spillway of a dam, or from some other factor. Indeed the primary function of some control structures is to act as energy dissipators; an example is the drop structure, discussed in Sec. 6.4.

The study of these devices forms a large and important subject, to which numerous technical papers in the literature have been devoted (see

References). In this section we shall be concerned mainly to set out the basic principles of their operation, although some empirical data are included for the sake of completeness.

In general two methods are in common use to dissipate the energy of the flow. First, there are abrupt transitions or other features which induce severe turbulence: in this class we can include sudden changes in direction (such as result from the impact at the base of a free overfall); and sudden expansions (such as in the hydraulic jump). In the second class are methods based on throwing the water a long distance as a free jet, in which form it will readily break up into small drops which are very substantially retarded before they reach any vulnerable surface. In the first class, the free overfall has already been dealt with in Sec. 6.4; in this section will be considered the hydraulic jump and the *stilling basin*, whose design is based on the action of the hydraulic jump. Also the "ski jump" energy dissipator, representative of the second class, will receive some attention. In the first instance, a review will be made of the important basic properties of the hydraulic jump.

The Hydraulic Jump

The discussion of this phenomenon in Chap. 3 has established the following points: that it may be analyzed by the momentum principle, leading to the equation

$$\text{Fr}_1{}^2 = \frac{1}{2}\frac{y_2}{y_1}\left(\frac{y_2}{y_1}+1\right) \tag{6-54}$$

and that the energy loss across the jump is equal to

$$\Delta E = \frac{(y_2 - y_1)^3}{4y_1y_2} \tag{6-55}$$

We may think of the depth ratio y_2/y_1 as being a measure of the "strength" of the jump; by Eq. (6-54), $\text{Fr}_1{}^2$ is also a measure of this property, as one would expect from previous conclusions about the role of the Froude number as a general indicator of the state of affairs in open channel flow. The cube term in Eq. (6-55) shows that the energy loss increases very sharply with the strength of the jump.

Hydraulic jumps can in fact be classified into a number of different types, depending on the size of Fr_1; these types are illustrated in Fig. 6-29. When $\text{Fr}_1 < 1.7$, the energy difference in Eq. (6-55) is so small that a broken wave front does not form and there is instead a train of unbroken standing waves —the so-called "undular jump." This phenomenon raises interesting questions of principle, which have been explored by Benjamin and Lighthill [37]. The question of most basic interest lies in the fact that the energy difference ΔE is not necessarily dissipated into heat and lost: part of it may be "radiated" downstream through the stationary wave train. This occurrence arises from

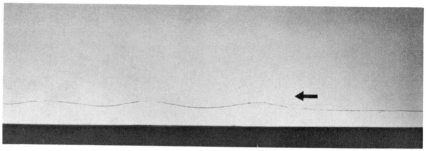

(a) *Undular jump on smooth bed.* $Fr_1 = 1.25$

(b) *Broken jump on smooth bed.* $Fr_1 = 1.55$

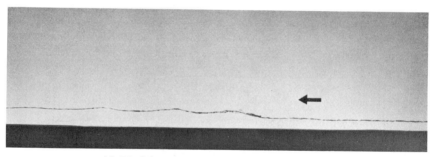

(c) *Undular jump on rough bed.* $Fr_1 = 1.55$

Figure 6-29 (a-c).   Various Forms of the Hydraulic Jump

the fact, discussed more fully in Chap. 8, that in a train of waves energy is transmitted at a "group velocity" which is in general less than the velocity of the waves themselves. Accordingly, if the waves are regarded as stationary, energy is transmitted in a direction opposite to that in which the waves are moving relative to the water—i.e., downstream in Fig. 6-29.

Benjamin and Lighthill [37] showed by theoretical arguments that if radiation alone were to account for all of the energy difference ΔE, it could do so only when Fr_1 is less than about 1.25; above this value the waves required to transmit the energy would be steep enough to break. But in practice, energy

(*d*) *Weak broken jump.* $Fr_1 = 2.4$

(*e*) *Stronger broken jump, some instability.* $Fr_1 = 3.4$

(*f*) *Strong and stable broken jump.* $Fr_1 = 5.5$

Figure 6-29 (d-f). Various Forms of the Hydraulic Jump

dissipation due to viscous resistance makes an appreciable contribution to ΔE, so that Fr_1 can rise to values of about 1.7 before the waves break and form a turbulent front of the type shown in Figs. 6-29*d, e,* and *f.* We would expect therefore that in the range $1.25 < Fr_1 < 1.7$ the bed roughness would have some part to play, and this is shown to be the case by Figs. 6-29*b* and *c,* in each of which Fr_1 has the same value of about 1.55. In Fig. 6-29*b* the bed is smooth and the front of the jump is broken, whereas in Fig. 6-29*c,* where the bed has been roughened with sandpaper, the front is wavy and unbroken because ΔE is largely accounted for by viscous resistance, and little energy has been radiated through the wave train. In Fig. 6-29a, $Fr_1 = 1.25$ and the front is wavy and unbroken even though the bed is smooth.

In the range $1.7 < Fr_1 < 2.5$ (Fig. 6-29d) the front, although broken, is comparatively quiet and marked by surface turbulence only, but when $2.5 < Fr_1 < 4.5$ (Fig. 6-29e), the upstream flow penetrates the turbulent front in the form of an oscillating jet which gives rise to irregular surface waves; these may persist for a great distance downstream. When $4.5 < Fr_1 < 9.0$, (Fig. 6-29f), the jump is strong but stable in form and free from waves, although the distinction between the cases $Fr_1 < 4.5$ and $Fr_1 > 4.5$ may not be apparent in the photographs of Fig. 6-29. When $Fr_1 > 9.0$ the jump is effective but becomes increasingly rough as Fr_1 rises. When $Fr_1 > 13$ the jump is so rough as to make conditions rather difficult and stilling basin design expensive.

When the jump is formed on a substantially horizontal bed, as in Fig. 6-29, the length L also varies with Fr_1 but it remains substantially constant at about $6y_2$ over the range $4.5 < Fr_1 < 13$, being somewhat smaller outside this range. The downstream end of the jump is somewhat ill-defined and any definition of length must accordingly be somewhat arbitrary; however, the adoption of the above figure results in satisfactory designs of aprons and stilling basins, so the question of definition need not be pursued any further.

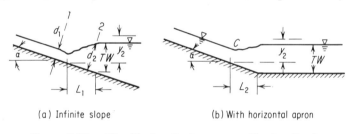

(a) Infinite slope   (b) With horizontal apron

Figure 6-30.   *The Hydraulic Jump on a Sloping Surface*

However, when the jump forms on a steep slope (Fig. 6-30) the question of length becomes important, for the momentum equation must now include the weight (resolved down the slope) of the water contained in the jump itself. The writing of the equation forms a useful exercise in the application of the momentum principle, which is left to the reader (Prob. 6.15); but in the end result recourse must be had to experiment, which alone can determine the profile of the jump and hence the weight of water between sections 1 and 2 in Fig. 6-30a.

Much experimental work has been done on this question, and perhaps the most comprehensive statement of the results of this work is that of Bradley and Peterka [11]. From this statement it appears that there are basically two different types of jump—Case A, Fig. 6-30a, in which the slope extends a long way downstream; and Case B, Fig. 6-30b, in which the slope is terminated by a horizontal apron. In both cases the water surface downstream tends to the horizontal, for in Case A the downstream condition can only be a level pool or an S_1 profile.

It is convenient to express results in terms of y_2, the downstream depth required to form a jump on a horizontal apron. For case A, the downstream end of the jump was defined by Bradley and Peterka as the section where the high-velocity jet begins to lift from the floor, or a point on the level tail-water surface immediately downstream from the surface roller, whichever occurred farthest downstream. The length L_1 consistent with this definition was found to be greater than on a horizontal bed, and substantially constant in the range $4.5 < \mathrm{Fr}_1 < 13$, as for the horizontal bed. The results are approximated closely by the equation

$$\frac{L_1}{y_2} = 6.1 + 4S_0 \tag{6-56}$$

where the slope S_0 is defined as the tangent of the angle α. Outside the above range of Fr_1 the value of L_1 is less than in Eq. (6–56).

The actual downstream depth we denote by TW (for tailwater); it is measured vertically at the "downstream end" of the jump as defined above. For all values of S_0 and Fr_1, it is approximated by the equation

$$\frac{TW}{y_2} = 1 + 11.2S_0{}^{3/2} \tag{6-57}$$

We may think of Case B in this way; a jump, initially formed on the horizontal apron, is made to encroach on the upstream slope by an increase in the tailwater depth TW above the amount y_2. Suppose that the ratio TW/y_2 is steadily increased from unity; the point C advances up the slope until when $TW/y_2 = 1.3$, the length L_2 is given by the empirical equation

$$\frac{L_2}{y_2} = 0.82S_0{}^{-0.78} \tag{6-58}$$

Any further increases in TW are matched by exactly equal rises in the level of the point C, and thus by horizontal movements equal to those amounts divided by S_0. Accordingly we can write, for $TW/y_2 \geq 1.3$:

$$\frac{L_2}{y_2} = 0.82S_0{}^{-0.78} + \frac{(TW/y_2) - 1.3}{S_0} \tag{6-59}$$

The above description gives the main points of Bradley and Peterka's results; for details the reader should consult the original paper [11]. One point that emerges clearly from the results is this: although TW is always greater than y_2, the downstream water level is always lower than it would be if a horizontal apron extended downstream from the toe of the jump, along the broken lines in Fig. 6-30. The reader should have no difficulty in finding the physical reason for this result.

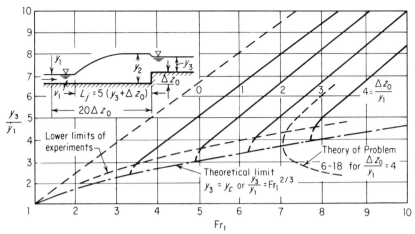

Figure 6-31.   *The Hydraulic Jump at an Abrupt Rise, after J. W.*
Forster and R. A. Skrinde [38]

When the jump is formed on a horizontal apron made of a smooth material like concrete, there is little variation in y_1 or y_2 upstream or downstream of the jump; consequently, as mentioned in Sec. 3.2, the jump may easily drift upstream or downstream unless it is held in place by some special device. These devices may take many forms; e.g., the intersection of a steep slope and a horizontal apron (Case *B* above) will have the desired stabilizing effect if *TW* is large enough. Other forms are furnished by local features having high resistance to flow; for example a sudden rise (Fig. 6-31) or a sudden drop (Fig. 6-32) in bed level.

The former case has been investigated by Forster and Skrinde [38]; their experimental results (full lines in Fig. 6-31) indicate values of y_3 substantially less than those obtained by applying the momentum equation between sections 1 and 3 and assuming a thrust on the vertical face determined by the depth y_2 (broken line in Fig. 6-31). The reason for this discrepancy must be that the high velocity upstream jet does not completely lose its identity before reaching the step; the impact of the residual jet on the step then gives rise to a thrust greater than the one calculated on the above assumption. The thrust need only be slightly greater (Prob. 6.18) in order to account for the observed difference between analysis and observation. The jump was set to the length $L_j = 5(\Delta z_0 + y_3)$ in all experiments.

The case of the sudden drop in bed level is a little more tractable by elementary theory, although two distinct situations are possible (Cases *A* and *B* in Fig. 6-32), the thrust on the step being determined by the downstream depth in Case *A* and the upstream depth in Case *B*. The former (Prob. 6.16) leads to the equation

$$\mathrm{Fr}_1{}^2 = \frac{y_2}{2(y_2 - y_1)}\left[\left(\frac{y_2}{y_1} - \frac{\Delta z_0}{y_1}\right)^2 - 1\right] \qquad (6\text{--}60)$$

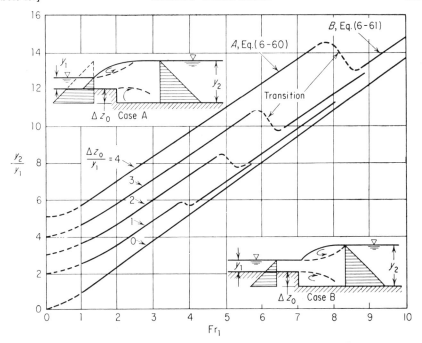

Figure 6-32.   *The Hydraulic Jump at an Abrupt Drop, after E.-Y. Hsu* [38]

and the latter (Prob. 6.17) to the equation

$$\text{Fr}_1{}^2 = \frac{y_2}{2(y_2 - y_1)}\left[\left(\frac{y_2}{y_1}\right)^2 - \left(\frac{\Delta z_0}{y_1} + 1\right)^2\right] \qquad (6\text{–}61)$$

The experiments of Hsu, described in a discussion of Ref. [38], gave results (Fig. 6-32) following each of these equations in turn; they also disclosed the existence of a transition from one state to the other. The occurrence of this transition, during which the jump has an unstable undular form, cannot be predicted by theory.

Stilling Basins

A stilling basin is a short length of paved channel placed at the foot of a spillway or any other source of supercritical flow. The aim of the designer is to make a hydraulic jump form within the basin, so that the flow is converted to subcritical before it reaches the exposed and unpaved riverbed downstream. Desirable features of the stilling basin are those that tend to promote the formation of the jump, to make it stable in one position, and to make it as short as possible.

Some preliminary considerations have been dealt with in Chap. 3. Problems

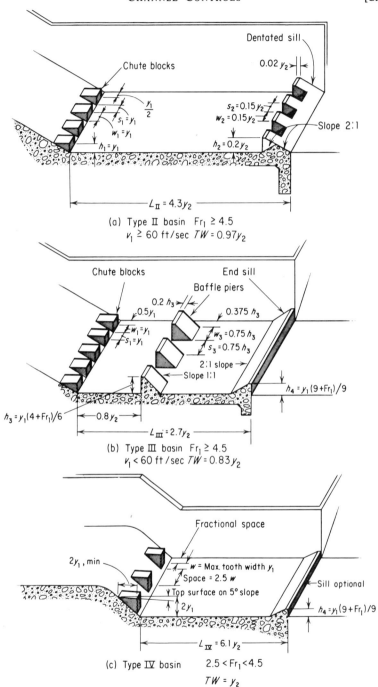

Figure 6-33. Recommended Proportions for U. S. Bureau of Reclamation Stilling Basins, after J. N. Bradley and A. J. Peterka [11]

3.21 through 3.24 and 3.29 have shown that for a jump to form unaided, the floor of the stilling basin must be placed a substantial distance below tail-water level; the required excavation may of course make the basin very expensive. The above problems have also shown that excessive depth of excavation can be avoided by widening the basin, by the installation of baffles to increase resistance to the flow, or by a raised sill at the end of the basin. This last expedient, however, may have the disadvantage of giving rise to supercritical flow immediately downstream of the sill; this would in many cases defeat the whole purpose of the stilling basin.

The abrupt rise and abrupt drop already discussed will of course introduce extra resistance that tends to promote jump formation as well as to localize the jump when it is formed. The action of increased stilling basin width suggests not only that the basin should be as wide as possible but also that if it were tapered in plan (the width increasing downstream) then the jump would remain stable in one position for given values of upstream and downstream depth.

All the above remarks derive from general principles, and it might be expected that in the hands of an experienced designer they would form a satisfactory basis for stilling basin design. This is in fact the case, but the experience and experiment required are considerable. Model studies are often necessary, although in recent years the need for them has been reduced some-what by the experimental development of general-purpose designs, notably by the U. S. Bureau of Reclamation.

The U.S.B.R. designs, as reported by Bradley and Peterka [11], are shown in Figs. 6-33a through c. Basins II and III are for $Fr_1 > 4.5$, the former for high spillways in which v_1 exceeds 60 ft/sec, the latter for $v_1 < 60$ ft/sec. In both cases the flow entering the basin is split and aerated by "chute blocks" —triangular blocks mounted at the base of the slope—and at least some part of the flow leaving the basin is directed upward and away from the unpaved

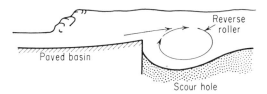

Figure 6-34. *The Effect of Upward Deflection of Flow by an End Sill*

riverbed by the end sill arrangement. The effect of this latter device is to protect the bed from the direct action of the current, and possibly to set in motion a reverse roller, Fig. 6-34, which by directing bed material back towards the basin prevents undermining of the structure.

The major difference between the two designs is that in Basin III lower velocities allow the installation of "baffle piers" downstream of the chute blocks; at higher velocities such piers would be liable to severe damage. The added resistance offered by the piers allows the use not only of a lower tail-water level, but also of a shorter basin. The basin lengths given in Fig. 6-33 are maxima, applicable in the range $6 < Fr_1 < 14$; outside this range slightly lower values may be used. As before, we use as a standard the theoretical depth y_2, required to form a plain jump on a horizontal floor without special devices or appurtenances (described as a Type I basin by Bradley and Peterka). Since Basin II requires a tailwater depth almost equal to y_2, this design accomplishes only the localizing and shortening of the jump; the baffle piers in Basin III accomplish a distinct lowering of the required tail-water, as shown in Fig. 6-33, as well as a further shortening of the basin.

Basin IV, Fig. 6-33c, is designed for the special purpose of suppressing at their source the waves which travel downstream when $2.5 < Fr_1 < 4.5$. Basin length and tailwater depth are the same as for the unaided jump, so the sole function of this design (apart from localizing the jump) is wave suppression. Other means of accomplishing the same end are described by Bradley and Peterka [11]; they include a drop structure in which the water falls through a grille, and underpass structures which damp out waves by skimming the surface of the water.

Another well-known standard design has been developed at the St. Anthony Falls Hydraulic Laboratory, University of Minnesota, for the U. S. Soil Conservation Service, and reported by Blaisdell [39,40]. It is intended for much the same use as the U.S.B.R. Basin III, i.e., on low-head structures, but is designed for a greater range of upstream Froude numbers, viz. $1.7 < Fr_1 < 17$. The general dimensions of the basin, which is usually given the abbreviated title of the SAF Stilling Basin, are shown in Fig. 6-35. The length L_B and the tailwater depth TW are as given in Table 6-2.

TABLE 6-2 Design Parameters for St. Anthony Falls Stilling Basin

Fr_1	1.7–5.5	5.5–11	11–17
L_B/y_2	$4.5/Fr_1^{0.76}$	$4.5/Fr_1^{0.76}$	$4.5/Fr_1^{0.76}$
TW/y_2	$1.1-Fr_1^2/120$	0.85	$1-Fr_1^2/800$

The sidewalls may be parallel or diverging in plan. If the latter, then the baffle piers, or "floor blocks" (as opposed to chute blocks) must have their width and spacing increased in proportion to the amount of divergence, so that they continue to intercept the flow coming from between the chute blocks, as they do when the sidewalls are parallel. The same values of L_B and TW must be used whether the sidewalls are parallel or diverging, although it would appear that diverging walls would give better performance.

Noteworthy is the extreme shortness of the SAF basin; when $Fr_1 = 9$, $L_B/y_2 = 0.865$, compared with 2.7 for the U.S.B.R. Basin III.

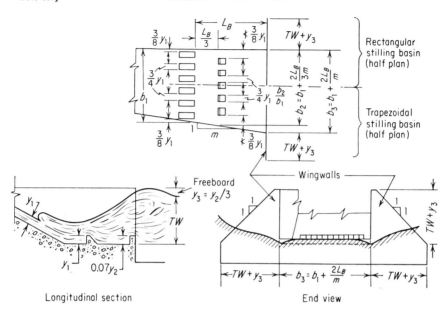

Figure 6-35.   *Recommended Proportions for the St. Anthony Falls*
Stilling Basin, after F. W. Blaisdell [39,40]

For all the standard designs described above, it is recommended that no allowance for air entrainment need be made. This recommendation is based on experience which indicates that prototype stilling basins in which air entrainment is present perform at least as well as model basins, in which there is no air entrainment. Just why this should be so is not immediately obvious; while modified momentum equations can be written allowing for the effects of air entrainment (Prob. 6.19), their application to this problem depends on how the residual air concentration at the end of the jump compares with the initial value upstream (Prob. 6.23). However, an analysis by Rajaratnam [41], based on certain reasonable assumptions, leads to results showing that the final effect of air entrainment on the required tailwater level is small.

Although air entrainment makes no difference to the tailwater level to be supplied downstream of the basin, its existence within the basin calls for a generous freeboard allowance, such as is supplied in the SAF basin.

The standard designs here described may not be suitable for every project and they do not by any means exhaust the possibilities open to the designer. Many of the features discussed here—sudden drops, end sills, blocks, tapered sidewalls, to name only a few—may be combined with advantage in particular cases in ways depending on the economics dictated by the special circumstances of each project. However, model studies are essential to test such "custom-made" designs.

Bucket-Type and Ski-Jump Energy Dissipators

The bucket-type energy dissipator illustrated in Fig. 6-36 is normally cheaper than the stilling basin but is remarkably effective. Essentially it is a means of deflecting the flow upwards at a considerable angle to the horizontal, preferably about 45°, and this is achieved simply by means of a concave profile of large radius. The purpose of the upward deflection is to promote the formation of a reverse roller, as shown in Fig. 6-36, which returns bed

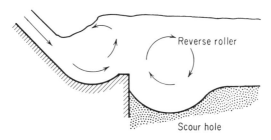

Figure 6-36.   *The Bucket-Type Energy Dissipator*

material towards the structure just as in Fig. 6-34. The riverbed profile shown in Fig. 6-36 is typical of the stable bed profiles developed by the action of the bucket.

Tests on this type of energy dissipator show that for satisfactory operation it needs a tailwater depth at least equal to the depth needed to form a hydraulic jump; for lesser depths the jet leaving the bucket lip tends to burst through the surface, with considerable formation of spray.

The bucket-type dissipator is not, therefore, a remedy for inadequate tail-water depth. If the tailwater level is in fact insufficient for hydraulic jump formation, as it often is in the case of high spillways, the ski-jump dissipator

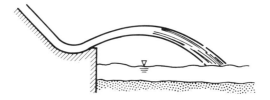

Figure 6-37.   *The Ski-Jump, or Flip-Bucket, Energy Dissipator*

may be used. Figure 6-37 illustrates a common form of this device, often called a trajectory or flip bucket. In this figure the bucket is at the foot of an open spillway; it may equally well be placed at the end of a tunnel spillway. In either case the objective is the same: to produce a long jet which will lose some of its energy in flight and will land at a safe distance downstream. Unless the riverbed material is hard rock, model tests are usually required to determine what constitutes a safe distance.

More recent data on the flip-bucket type are given by Rhone and Peterka [42]. Study of the leading dimensions of actual designs reveals a wide variety of choice exercised by designers, as to values of the angle of projection, the elevation of the bucket, and so on. This is not surprising, for the single governing criterion is that the jet should land at a safe distance downstream; this criterion can be met by a wide variety of bucket arrangements and locations.

The material in this section has dealt with the subject only in brief outline. In order to fill in the details, the reader is referred to the extensive summary and bibliography given in the report [43] of a Task Force set up for the purpose by the American Society of Civil Engineers, and to the comprehensive design recommendations of the U. S. Bureau of Reclamation [45].

References

1. T. Rehbock. Discussion of "Precise Weir Measurements" by E. W. Schoder and K. B. Turner, *Trans. Am. Soc. Civil Engrs.*, vol. 93 (1929), p. 1143.
2. R. von Mises. "Berechnung von Ausfluss und Überfällzahlen," *Z. ver. Deuts. Ing.*, vol. 61 (1917), p. 447.
3. A. Lauck. "Ueberfall über ein Wehr," *Z. angew. Math. Mech.*, vol. 5 (1925), p. 1.
4. R. V. Southwell and G. Vaisey. "Relaxation Methods Applied to Engineering Problems: XII, Fluid Motions Characterized by 'free' Streamlines," *Phil. Trans. Roy. Soc.* (London), A, vol. 240 (1946), p. 117.
5. G. Birkhoff. "Calculation of Potential Flows with Free Streamlines," *Proc. Am. Soc. Civil Engrs.*, vol. 87, no. HY6 (November 1961), p. 17.
6. J. B. Francis. *Lowell Hydraulic Experiments*, 4th ed. (New York: D. Van Nostrand Company, Inc., 1883).
7. "Pump Tests," *Brit. Stand. Spec. No. 599* (1939, amended 1945).
8. H. Rouse and L. Reid. "Model Research on Spillway Crests," *Civil Engineering*, vol. 5 (January 1935), p. 10.
9. O. Dillman. "Untersuchungen an Überfällen," *Mitt. des Hyd. Inst.*, Munich, no. 7 (1933).
10. Ven Te Chow. *Open-Channel Hydraulics* (New York: McGraw-Hill Book Company, Inc., 1959), Chap. 14.
11. J. N. Bradley and A. J. Peterka. "The Hydraulic Design of Stilling Basins," *Proc. Am. Soc. Civil Engrs.*, vol. 83, no. HY5, Proc. Papers 1401-6 (October 1957).
12. L. G. Straub and A. G. Anderson. "Experiments on Self-Aerated Flow in Open Channels," *Trans. Am. Soc. Civil Engrs.*, vol. 125 (1960), p. 456.
13. A.S.C.E. Task Committee. "Aerated Flow in Open Channels," *Proc. Am. Soc. Civil Engrs.*, vol. 87, no. HY3 (May 1961), p. 73.
14. F. M. Henderson. "Flow at the toe of a Spillway: I—The 'Open-Toe' Spillway: II (with D. G. Tierney) The 'Solid-Toe' Spillway," *La Houille Blanche*,

vol. 17, no. 6 (November 1962), p. 728, and vol. 18, no. 1 (January-February 1963), p. 42.

15. H. Rouse. "Discharge Characteristics of the Free Overfall," *Civil Engineering*, vol. 6 (April 1936), p. 257.

16. J. A. Replogle. Discussion of Ref. (22), *Proc. Am. Soc. Civil Engrs.*, vol. 88, no. HY2 (March 1962), p. 161.

17. C. Jaeger. "Hauteur d'eau à l'extremité d'un long déversoir," *La Houille Blanche*, vol. 3, part 6 (1948), p. 518.

18. S. K. Roy. "Note on the Stream Depth at the Edge of a Free Overfall," *La Houille Blanche*, vol. 4, part 6 (1949), p. 832.

19. W. B. Fraser. "Gravity-deflected Jets in Two-Dimensional Flow," M. E. Thesis, University of Canterbury, N.Z. (May 1961).

20. L. C. Woods. "Compressible Subsonic Flow in Two-Dimensional Channels with Mixed Boundary Conditions," *Quart. J. Mech. and App. Math.* vol. 7, part 3, (1945), p. 263.

21. J. W. Delleur, J. C. I. Dooge, and K. W. Gent. "Influence of Slope and Roughness on the Free Overfall," *Proc. Am. Soc. Civil Engrs.*, vol. 82, no. HY4 (August 1956), p. 1038.

22. M. H. Diskin. "End Depth at a Drop in Trapezoidal Channels," *Proc. Am. Soc. Civil Engrs.*, vol. 87, no. HY4 (July 1961), p. 11.

23. C. D. Smith. "Brink Depth for Circular Channels," *Proc. Am. Soc. Civil Engrs.*, vol. 88, no. HY6 (November 1962), p. 125, with discussions by M. H. Diskin (March 1963), p. 203, and J. K. Vennard (May 1963), p. 403, and others.

24. W. L. Moore. "Energy Loss at the Base of a Free Overfall," *Trans. Am. Soc. Civil Engrs.*, vol. 108 (1943), p. 1343, with discussion by M. P. White, p. 1361, H. Rouse, p. 1381, and others.

25. W. Rand. "Flow Geometry at Straight Drop Spillways," *Proc. Am. Soc. Civil Engrs.*, vol. 81 (September 1955), no. 791.

26. F. W. Blaisdell and C. A. Donnelly. "The Box Inlet Drop Spillway and its Outlet," *Trans. Am. Soc. Civil Engrs.*, vol. 121 (1956), p. 955.

27. T. Brooke Benjamin. "On the Flow in Channels When Rigid Obstacles Are Placed in the Stream," *J. Fluid Mech.*, vol. 1, part 2 (July 1956), p. 227.

28. A. Toch. "Discharge Characteristics of Tainter Gates," *Trans. Am. Soc. Civil Engrs.*, vol. 120 (1955), p. 290.

29. H. R. Henry. Discussion on "Diffusion of Submerged Jets," by M. L. Albertson et al., *Trans. Am. Soc. Civil Engrs.*, vol. 115 (1950), p. 687.

30. E. Crausse, G. Pouzens, and P. Cachon. "Contribution à l'étude des vannes de fond," *Compt. rend. de l'Acad. Fr.*, vol. 234 (June 1952), p. 2521.

31. Govinda Rao and N. Rajaratnam. "The Submerged Hydraulic Jump," *Proc. Am. Soc. Civil Engrs.*, vol. 89, no. HY1 (January 1963), p. 139.

32. G. W. Hall. "Discharge Characteristics of Broad-Crested Weirs using Boundary Layer Theory," *Proc. Inst. C.E.* (London), vol. 22 (June 1962), p. 177.

33. H. E. Bazin. "Expériences nouvelles sur l'écoulement en déversoir," *Annales des Ponts et Chaussées*, vol. 7, series 7 (1896), p. 249.

34. R. L. Parshall. "The Improved Venturi Flume," *Trans. Am. Soc. Civil Engrs.*, vol. 89 (1926), p. 841.

35. R. L. Parshall. "Parshall Flumes of Large Size," *Colorado Agricultural Experiment Station Bulletin*, No. 426A (March 1953).

36. R. L. Parshall. "Measuring Water in Irrigation Channels with Parshall Flumes and Small Weirs," *U. S. Soil Conservation Service, Circular* 843 (May 1950).

37. T. Brooke Benjamin and M. J. Lighthill. "On Cnoidal Waves and Bores," *Proc. Roy. Soc.* (London) A, vol. 224 (1954), p. 448.
38. J. W. Forster and R. A. Skrinde. "Control of the Hydraulic Jump by Sills," *Trans. Am. Soc. Civil Engrs.*, vol. 115 (1950), p. 973, with discussion by E.-Y. Hsu, p. 988.
39. F. W. Blaisdell. "Development and Hydraulic Design, Saint Anthony Falls Stilling Basin," *Trans. Am. Soc. Civil Engrs.*, vol. 113 (1948), p. 483.
40. F. W. Blaisdell. "The SAF Stilling Basin," *U. S. Soil Conservation Service*, Report SCS-TP-79 (May 1949).
41. N. Rajaratnam. "Effect of Air-Entrainment on Stilling Basin Performance," *J. Irrig. and Power*, (India), vol. 19, no. 5 (May 1962), p. 334.
42. T. J. Rhone and A. J. Peterka. "Improved Tunnel-Spillway Flip Buckets," *Trans. Am. Soc. Civil Engrs.*, vol. 126, part 1 (1961), p. 1270.
43. A.S.C.E. Task Force. "Energy Dissipators for Spillways and Outlet Works," *Proc. Am. Soc. Civil Engrs.*, vol. 90, no. HY1 (January 1964), p. 121.
44. N. Hay and E. Markland. "The Determination of the Discharge over Weirs by the Electrolytic Tank," *Proc. I.C.E.* (London) vol. 10 (May 1958), p. 59.
45. "Hydraulic Design of Stilling Basins and Energy Dissipators," U. S. Bureau of Reclamation Engineering Monograph (1964).

Problems

6.1.　If the approach flow in Eq. (6–3) is critical, prove from this equation and from Eq. (6–6) that $C_e = 0.715$ for a completely free overfall.

6.2.　Prove Eq. (6–8).

6.3.　Assuming that the velocity along the spillway crest AC in Fig. 6-5b is equal to $\sqrt{(2gH)}$, that the distance AC is approximately $0.2H$, and that the boundary layer thickness δ at A is negligible, show that at the crest C

$$\frac{\delta}{H} = 0.0072H^{-0.3}$$

assuming a turbulent boundary layer, $\nu = 1.2 \times 10^{-5}$ ft²/sec, and H measured in feet.

6.4.　Prove that the function $(\log_e x)/x$ has a maximum value of $1/e$, occurring when $x = e$.

6.5.　Prove Eq. (6–29) by applying the momentum equation between sections A and C in Fig. 6-13. Also, by applying this equation between sections A and B, and assuming that the flow at A is critical, prove that the pressure distribution at B measured by Replogle (Fig. 6-13) is consistent with a brink depth of $0.7y_c$. Similarly, show that a brink depth of $0.715y_c$ is consistent with a maximum pressure head at the brink of about $0.3y_b$. (Assume that the mean pressure head at section B is two-thirds of the maximum).

6.6.　Consider a horizontal section A_1B_1 through a falling jet of liquid, the section being at a depth h below the total energy line. Let u and v be the horizontal and vertical velocity components respectively; assume that the jet is so nearly vertical that v approximates closely to the resultant velocity, and that the internal pressure head of the jet is a small fraction of h. Use the continuity equation

$$\frac{\partial u}{\partial x} + \frac{\partial v}{\partial y} = 0$$

to prove that the horizontal velocity difference Δu across $A_1 B_1$ is approximately equal to $q/2h$, where q is the discharge per unit width, and that the horizontal width Δx of the jet is approximately given by

$$\frac{\Delta x}{y_c} = \left(\frac{y_c}{2h}\right)^{1/2}$$

6.7. In the case of Prob. 6.6, consider a further horizontal section $A_2 B_2$ just below $A_1 B_1$. Assume that the change in u as a fluid element goes from A_1 to A_2 is one-half the change that occurs in Δu (defined in Prob. 6.6) as the flow passes from $A_1 B_1$ to $A_2 B_2$. Hence show from the Euler equation that the horizontal pressure head gradient at A is equal to

$$\frac{d(p/\gamma)}{dx} = \left(\frac{y_c}{2h}\right)^{3/2}$$

Hence, assuming that the pressure distribution across AB is parabolic, prove that the maximum internal pressure head p_m/γ is given by

$$\frac{p_m/\gamma}{y_c} = \frac{1}{4}\left(\frac{y_c}{2h}\right)^2$$

For the free overfall at the end of a horizontal channel as in Fig. 6-13, consider a horizontal section at a vertical distance $3y_c/2$ below the brink B. Show that the horizontal width of such a section is approximately 60 times the maximum internal pressure head at that section.

6.8. For the free overfall shown in Fig. 6-13, assume that the distance $AB = 4y_c$, and that the surface drag coefficient over the length AB is equal to 0.005 (assumed applicable to the mean of the velocities v_c and v_b). Taking $\text{Fr}_1 = 1$, recalculate Prob. 6.5 allowing for the resistance over AB, and show that this resistance accounts for about half the discrepancy indicated by Prob. 6.5, i.e. for an error of about 1 percent in the brink depth.

6.9. Prove Eq. (6–32).

6.10. Consider the impact of the jet at the base of an overfall, as in Fig. 6-16. Let Q_1 equal the discharge over the brink, and Q_3 the discharge entrained from the standing pool and then detached again at the floor, as in the figure. Assuming that the two streams Q_1 and Q_3 have the same velocity immediately after they separate, prove from the momentum equation that

$$\frac{Q_1}{Q_3} = \frac{1 + \cos\theta}{1 - \cos\theta}$$

Assuming that the velocity head in this region is equal to the entire depth below the total energy line, prove that

$$\cos\theta = \frac{1.5v_c}{\sqrt{2g(\Delta z_0 + 3y_c/2)}} = \frac{1.06}{\sqrt{\Delta z/y_c + 3/2}}$$

6.11. Consider the entraining of the stream Q_3 back into the main stream at the top of the standing pool in Fig. 6-16. Assuming that this stream has negligible momentum before its entrainment, show from the results of Prob. 6.10 that the velocity of the combined stream after entrainment is equal to

$$v_m = v_1 = \frac{v}{2}(1 + \cos \theta)$$

where v is the velocity of the free-falling jet, assumed equal to $\sqrt{2g(\Delta z_0 + 3y_c/2)}$. Hence prove Eq. (6–35).

6.12. Using the result given in the Appendix to Chap. 3 for the energy loss in a hydraulic jump, calculate the loss for the three cases $E_1/y_c = 4, 5$, and 6, and verify that the three corresponding values of E_2/y_c give points lying on the appropriate curve of Fig. 6-17.

6.13. Assuming an average value of 2.5 for E_2/y_c as in Fig. 6-17 for the drop structure of Fig. 6-16, prove that the energy dissipated in the structure is $(\Delta z_0 - y_c)$, that $Fr_2 = 0.27$ approximately, and that the height of the step at the end of the structure, normally made equal to $y_2/6$, will therefore be equal to $0.4y_c$.

6.14. An irrigation channel carrying a flow of 80 cusecs is to be laid across country along a line having an average slope of 0.005, but the slope of the channel is to be limited to 0.001 to avoid the danger of erosion of the channel bed. The extra fall is to be absorbed by drop structures of the type shown in Fig. 6-16 with a cross-channel width of 10 ft. Determine how many such structures should be installed in a length of 10 miles if the depth Δz_0 is to be (a) 6 ft, (b) 8 ft.

6.15. Resolving all forces parallel to the bed, write down the momentum equation governing the hydraulic jump in Fig. 6-30a. Neglect resistance and write the volume of water per unit width between sections 1 and 2 as $\frac{1}{2}KL(d_1 + d_2)$. Hence show that

$$\frac{d_2}{d_1} = \frac{1}{2}(\sqrt{1 + 8G_1{}^2} - 1)$$

where
$$G_1 = \frac{Fr_1}{\sqrt{\cos \alpha - \dfrac{KL \sin \alpha}{d_2 - d_1}}}$$

6.16. For Case A in Fig. 6-32, prove Eq. (6–60).

6.17. For Case B in Fig. 6-32, prove Eq. (6–61).

6.18. For the case where $\Delta z_0/y_1 = 4$ in the situation of Fig. 6-31, assume that the thrust on the upward step is hydrostatic and determined by the depth y_2. Verify that when $Fr_1 = 8$, the ratio y_3/y_1 has the value given by the dotted curve in Fig. 6-31. Determine the actual thrust on the step, as a multiple of $\gamma y_1{}^2$, required to give y_3/y_1 its experimental value of 5, and calculate the depth y_0, the effective head on the step required to produce this thrust. Verify that y_0 is very much closer to y_2 (assumed given by the normal hydraulic jump equation) than it is to E_1, which would be the value of y_0 if the upstream jet were striking the step unimpeded.

6.19. Aerated water is flowing with a mean air concentration $\bar{c}$, defined as the volumetric ratio air : (air plus water). Assuming that the air is uniformly distributed through the mixture, prove that the momentum function defined for a unit width of a rectangular channel is equal to

$$\frac{q^2}{gy_w} + \frac{y_w^2}{2(1-\bar{c})} \quad \text{or} \quad \frac{q^2}{gy_t(1-\bar{c})} + \frac{y_t^2(1-\bar{c})}{2}$$

where y_t is the total depth of the mixture, and y_w is the equivalent depth of the water alone.

Assuming, as an approximation to the distribution of Fig. 6-11, that the density of the mixture increases linearly from zero at the surface to a maximum at the bed, show that the factor $1/2$ in the hydrostatic thrust terms above is then replaced by the factor $2/3$.

6.20. Show that for sufficiently high values of the total head Z defined as in Fig. 6-8, the curves in Fig. 6-10 indicate uniform flow approximately.

6.21. Fit Eq. (6–17) to the upper curve of Fig. 6-11 by determining c_T and h from the observed values of c and dc/dy at $d = d_T$. Find some tables of the "error function" $\int e^{-x^2}\, dx$, for example in a text on statistical theory, and use them to evaluate c as a function of y_1 in Eq. (6–17). Verify that the values so calculated are close to the experimental values of Fig. 6-11.

6.22. At the foot of a spillway inclined at $45°$ to the horizontal, find the values of q for which the mean air concentration $\bar{c}$ would be equal to (a) 0.25, (b) 0.50, (c) 0.75. In each case calculate the depths d_m, d_w, and d_T, assuming that Manning's $n = 0.012$, and that the spillway is wide enough to make $R = d_m$, and high enough for the flow to be uniform.

6.23. A spillway has $Z = 200$ ft and $H = 20$ ft, the symbols being defined as in Fig. 6-8. At the foot of the spillway the discharge q per unit width is 500 cusecs per foot. Using Fig. 6-10, Eq. (6–16), and the results of Prob. 6.19, calculate the downstream depth required to form a hydraulic jump, assuming that the mean downstream air concentration is equal to (a) zero, (b) 50 percent, (c) 100 percent of the upstream air concentration. Also, (d), calculate upstream and downstream depth if the upstream velocity is given by Fig. 6-10 but there is no entrained air present.

6.24. The outflow from a lake is regulated by a radial gate mounted between vertical walls, the bed of the channel so formed being horizontal. Lake level is 10 ft above the channel bed, and with the gate open and discharging the depth in the downstream channel is 7 ft. The gate radius is 20 ft, and the hinge is 10 ft above the bed. Using Eq. (6–46), determine the gate opening at which the hydraulic jump is just kept clear of the gate, and calculate the corresponding rate of discharge.

6.25. With the lake level and the gate opening as in Prob. 6.24, calculate the discharge if the downstream depth is raised to (a) 8 ft; (b) 9 ft. Compare these values with those obtained direct from Fig. 6-23.

6.26. Water is flowing at the rate of 300 cusecs and a depth of 5 ft in a trapezoidal channel of base width 10 ft and side slopes 1H : 1V. At the end of this section of channel a sluice gate leads into a rectangular section 9 ft wide.

After a short distance there is an abrupt transition back to the original trapezoidal shape. A hydraulic jump forms just downstream of this transition;
find the depth in the rectangular section and in the downstream trapezoidal
section.

The downstream water level is now increased so that the jump moves
upstream into the rectangular channel section, producing a depth there of
4.5 ft. The discharge and gate opening remain unchanged; calculate the new
depth in the upstream trapezoidal section.

6.27. Water discharges from a lake through a short horizontal 5-ft diameter culvert into a trapezoidal channel of base width 6 ft and side slopes $1\frac{1}{2}$H : 1V.
The culvert invert is 10 ft below lake level, and level with the channel bed.
The depth in the channel a short distance downstream is 8 ft; find the discharge. Assume that the head loss at the entrance to the culvert is 0.4 times
the velocity head in the culvert.

6.28. A broad-crested weir is proposed as a means of measuring discharges up to
100 cusecs in a channel of rectangular section 10 ft wide. Determine a suitable
length of crest, if the results of Hall [32] are to be applicable. Find the discharge for a weir head H of 1.5 ft and a weir-crest height W of 3 ft above the
stream bed, determining whether the approach velocity will make an appreciable correction to H necessary. For the above value of H/W, replace
2.84 by 2.70 in Eq. (6–51).

6.29. A Parshall flume is to be installed in an irrigation channel of trapezoidal
section having a base width of 6 ft and side slopes of $1\frac{1}{2}$H : 1V, excavated on
a slope of 0.001 in alluvial country mainly composed of gravel having a
D-75 size of $1\frac{1}{2}$ in. The maximum discharge is to be 200 cusecs.

Assuming that the flume may be raised above the general line of slope
followed by the bed in order to keep the submergence ratio H_b/H_a of Fig. 6-28
below 0.7, determine a suitable throat width for the flume and the amount,
if any, by which the flume must be raised. Also find the amount by which the
depth just upstream of the flume exceeds the uniform depth at the maximum
discharge.

Is there a unique solution to this problem? If not, work out a number of
representative solutions and determine what considerations, if any, set a
limit to the process of working out solutions.

6.30. It is proposed to measure the outflow from a small Pelton wheel installation
by the California pipe method or some variation of it. The maximum discharge is 2 cusecs, and the maximum brink depth is to be about one-half the
pipe diameter. Choose a suitable pipe diameter.

6.31. For the free-vortex approximation to flow at the toe of a spillway, investigate
the case where $R/y_1 = 5$. By trial from Eq. (6–27), find R_1 and hence the
maximum depth y_0 at AO as multiples of y_1. Now integrate the pressure
along a strip of unit width lying along AO, obtaining the thrust on this strip
as a multiple of $\frac{1}{2}\rho v_1^2 y_1$. By a further integration obtain the moment of this
thrust about O, as a multiple of $\frac{1}{2}\rho v_1^2 y_1^2$. Compare the results with those of
the complete theory of Henderson and Tierney [14].

6.32. Water flowing at a depth of 2 ft and a velocity of 50 ft/sec arrives at the foot
of a 30° slope at the entry to a horizontal stilling basin, as in Fig. 6-30b.

Downstream conditions produce a depth of 25 ft in the stilling basin; locate the upstream limit of the hydraulic jump.

6.33. Consider the spillway of Prob. 6.23, and neglect air entrainment. What tailwater level would be required for successful operation of the U.S.B.R. Type II Basin? How could the design be modified so as to reduce the required tailwater level?

6.34. The lake level behind a spillway is 50 ft above tailwater level when the maximum flood discharge of 10,000 cusecs is flowing. The head over the spillway crest is 10 ft. Determine suitable stilling basin designs (including basin elevation) of the U.S.B.R. Type III for basin widths of (a) 40 ft, (b) 50 ft, (c) 60 ft.

6.35. In the situation of Prob. 6.24, determine the horizontal thrust on the gate. It is required to calculate the total thrust on the gate hinge (O in Fig. 6-19b). For this calculation, what is the most conservative design assumption to make about the position of the line of action of the horizontal thrust? Adopting this assumption, calculate the magnitude and direction of the total thrust.

Chapter 7

Channel Transitions

7.1 Introduction

A transition may be defined as a change either in the direction, slope, or cross section of the channel that produces a change in the state of the flow. Most transitions produce a permanent change in the flow, but some (e.g., channel bends) produce only transient changes, the flow eventually returning to its original state. Practically all transitions of engineering interest are comparatively short features, although they may affect the flow for a great distance upstream or downstream.

The class of transitions includes the class of controls; all controls are transitions, although not all transitions are controls. However, it is convenient to treat the two topics separately, as is done in this and the preceding chapter; it so happens that channel transitions may be readily divided into those that *always* act as controls (treated in Chap. 6), and those that may in some circumstances act as controls, which are treated in this chapter. Some transitions, such as bridge piers, act as a kind of partial control—i.e., the depth-discharge relationship at the piers is determined jointly by the piers themselves and by the channel conditions upstream or downstream.

In the treatment of transitions, as of every other topic in open channel flow, the distinction between subcritical and supercritical flow is of prime importance. It will be seen that the design and performance of many transitions are critically dependent on which one of these two flow regimes is operative.

7.2 Expansions and Contractions

These features are often required in artificial channels for a variety of practical purposes. They have been discussed in Chap. 2 in an elementary way, i.e., assuming no energy loss and neglecting the possibility of any other complications. As we shall see, supercritical flow in particular brings about certain complex flow phenomena which make the simplified viewpoint of Chap. 2 quite inadequate.

As implied by this last remark, the behavior of expansions and contractions depends on whether the flow is subcritical or supercritical. The following treatment is subdivided accordingly.

Subcritical Flow

If for the time being we postpone consideration of wave formation at changes of channel section, this type of flow raises no problems which are not already implicit in the theory of Chaps. 1, 2, and 3. The problem which still requires explicit consideration is that of energy loss when the expansion or contraction is abrupt, and we should expect this problem to be tractable by methods similar to those used in the study of pipe flow. For example, consider the abrupt expansion in width of a rectangular channel shown in plan view in Fig. 7-1; by analogy with the pipe-flow case we would treat this case

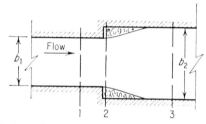

Figure 7-1.   Plan View of Abrupt Channel Expansion

by setting $E_1 = E_2$ and $M_2 = M_3$, assuming (a) that the depth across section 2 is constant and equal to the depth at section 1; (b) that the width of the jet of moving water at 2 is equal to b_1.

Manipulation of the resulting equations is much more awkward than in the pipe-flow case, but if it is assumed that Fr_1 is small enough for Fr_1^4 and higher powers to be neglected, it can be shown (Probs. 7.1 and 7.2) that the energy loss between sections 1 and 3 is equal to

$$E_1 - E_3 = \Delta E = \frac{v_1^2}{2g}\left[\left(1 - \frac{b_1}{b_2}\right)^2 + \frac{2\mathrm{Fr}_1^2 b_1^3(b_2 - b_1)}{b_2^4}\right] \tag{7–1}$$

The last term inside the bracket is the open channel flow term, which vanishes as Fr_1 tends to zero. In this case $y_1 = y_2 = y_3$, and the situation is equivalent to closed-conduit flow. Accordingly, Eq. (7–1) reduces to the form applicable to closed conduits, i.e.,

$$\Delta E = \frac{(v_1 - v_3)^2}{2g} \tag{7–2}$$

The term containing Fr_1^2 in Eq. (7–1) does not contribute a great deal to the total head loss unless $\mathrm{Fr}_1 > 0.5$, or $b_2/b_1 < 1.5$. The former condition is not often fulfilled, and the latter would, if true, make the total head loss very small, in which case little interest would attach to the relative size of its components. Equation (7–2) can therefore be recommended as safe for most normal circumstances; in fact the experiments of Formica [1] have indicated a head loss for sudden expansions some 10 percent less than the value given by this equation.

Just as in the pipe-flow case, the head loss is reduced by tapering the side walls; when the taper of the line joining tangent points is 1:4, as in the broken lines in Fig. 7-2a, the head loss is only about one-third of the value given in Eq. (7–1); it is given by some authorities as

$$\Delta E = 0.3 \frac{(v_1 - v_3)^2}{2g} \tag{7–3}$$

and by others as $\qquad \Delta E = 0.1\left(\frac{v_1{}^2}{2g} - \frac{v_3{}^2}{2g}\right) \tag{7–4}$

The former of these is to be preferred, but over the range $1.5 < v_1/v_3 < 2.5$ the two equations do not give greatly different results. In any event, a more gradual taper does not usually make savings in head commensurate with the extra expense, so the amount of 1:4 is the one normally recommended for channel contractions in subcritical flow. Given that this angle of divergence is to be used, the exact form of the sidewalls is not a matter of great importance, provided that they follow reasonably smooth curves without sharp corners, as in the two cases shown in Fig. 7-2. In the first of these, both upstream and downstream sections are rectangular and the sidewalls are generated by vertical lines; in the second case a warped transition is required to transfer from a trapezoidal to a rectangular channel.

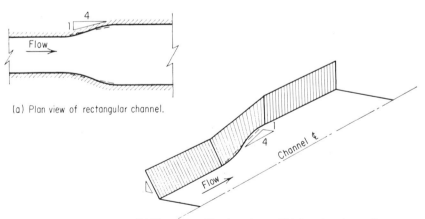

(a) Plan view of rectangular channel.

(b) Warped transition from trapezoidal to rectangular section.

Figure 7-2.   Channel Expansions for Subcritical Flow

Head losses through contractions are smaller than through expansions, just as in the case of pipe flow. An equation could be derived analogous to Eq. (7–1) with section 2 taken at the vena contracta just downstream of the entrance to the narrower channel, and section 3 where the flow has become uniform again downstream. However, direct experimental measurements provide a better approach, for experiment would be needed in any case to determine the contraction coefficient. The results of Formica [1] indicate head losses of up to 0.23 $v_3^2/2g$ for square-edged contractions in rectangular channels and up to 0.11 $v_3^2/2g$ when the edge is rounded—e.g., in the cylinder-quadrant type shown in Fig. 7-3. The results of Yarnell [14] obtained in

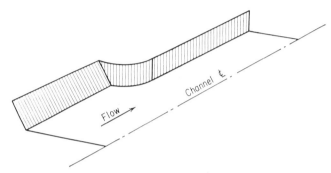

Figure 7-3.   Cylinder-Quadrant Contraction for Subcritical Flow

connection with an investigation into bridge piers indicated larger coefficients —up to 0.35 and 0.18 for square and rounded edges respectively. Formica's results showed that the coefficients increased with the ratio y_3/b_2, reaching the above maximum values when this ratio reached a value of about 1.3. When $y_3/b_2 \leq 1$, these coefficients reduced to about 0.1 and 0.04. Yarnell did not report values of depth : width ratio.

The formation of waves at the channel transition may modify somewhat the way in which the flow passes from one state to another, but if, as recommended above, the change in section is smooth and gradual, wave action should not be so severe as to modify the conclusions expressed in Eqs. (7–1) through (7–4). Nevertheless wave action can be quite pronounced, particularly when the flow approaches critical; an elementary example of this has previously been shown in Fig. 2-9. The matter of wave action will be discussed more fully in the next subsection.

The only other matter requiring attention in the design of a contraction is the possibility that the contraction might "choke" because the available specific energy was insufficient to pass the increased discharge per unit width, as discussed in Chap. 2. The remedy is to lower the channel floor far enough to ensure critical flow at the downstream end, or even subcritical flow in order to have a certain margin of safety (Prob. 7.3). A similar adjustment of level may be made simply to avoid unduly large departure from uniform flow; in this case the upstream and downstream lengths of channel form the context within which the problem must be considered (Prob. 7.4).

Supercritical Flow

In the preceding discussion on subcritical flow it was assumed that the velocity and depth remained the same across every section; this assumption is approximately correct, subject only to the qualification discussed in Sec. 2.6, viz., that within a contraction the velocity may be higher near the sidewalls than it is in midstream. However, when the flow is supercritical there

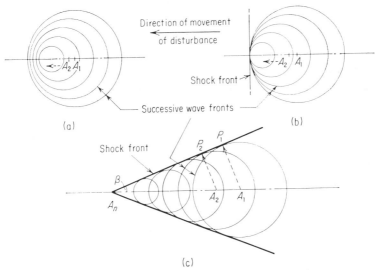

Figure 7-4.   Movement of a Small Disturbance at a Speed (a) Less than (b) Equal to (c) Greater than the Natural Wave Velocity

is a further complication in the form of wave motion. This is not confined to supercritical flow, but assumes particular importance when the flow is in that condition. What happens is that any obstacle in the path of the flow generates a surface wave which moves across the flow and is at the same time carried downstream; the end result is an oblique standing wave precisely analogous to the Mach waves characteristic of supersonic flow.

The formation of such waves is illustrated in Fig. 7-4. Consider a mass of stationary fluid, with a solid particle moving through it at a speed v comparable with the natural wave speed c, i.e., the speed with which a disturbance propagates itself through the fluid. When the particle is at A_1, it initiates a disturbance which travels outwards at the same velocity in all directions—

i.e., at any subsequent instant there is a circular wave front centered at A_1. Similar wave fronts are initiated when the particle passes through points A_2, A_3, etc. When $v < c$, as in Fig. 7-4a, the particle lags behind the wave fronts; when $v = c$, as in Fig. 7-4b, the particle moves at the same speed as, and in the same position as, a shock front formed from the accumulated wave fronts generated during the previous motion of the particle. But when $v > c$, as in Fig. 7-4c, the particle outstrips the wave fronts. When it reaches A_n the wave fronts have reached positions such that they can all be enveloped by a common tangent A_nP_1, which will itself form a distinct wave front. Since a disturbance travels from A_1 to P_1 in the same time as the particle travels from A_1 to A_n, it follows that

$$\frac{A_1P_1}{A_1A_n} = \sin \beta = \frac{c}{v} = \frac{1}{\text{Ma}}, \quad \text{or} \quad \frac{1}{\text{Fr}} \qquad (7\text{--}5)$$

The alternatives in this equation arise because the argument could equally well apply to compression waves in a compressible gas or surface waves in a liquid. Accordingly the ratio v/c could equally well be regarded as the Mach number Ma, appropriate to the former case, or the Froude number Fr, appropriate to the latter.

When the fluid is moving and the particle is stationary—e.g., is a slight irregularity on the wall of a flume, the oblique wave front will be stationary; indeed any flume walls no matter how smooth, will set up a system of inter-locking oblique waves as shown in Fig. 7-5. But, for the following reasons, care is needed in applying Eq. (7–5) to the situation shown in this figure.

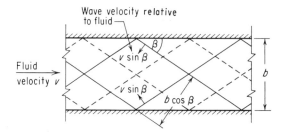

Figure 7-5. Oblique Standing Wave Patterns in a Uniform Channel

Equation (7–5) is true without qualification provided that c is in fact equal to $\sqrt{gy}$; this is true for long waves of small amplitude. But it has been pointed out in Chap. 3 that a large disturbance travels at a velocity greater than $\sqrt{gy}$; a further qualification arises from the equation

$$c^2 = \frac{gL}{2\pi} \tanh \frac{2\pi y}{L} \qquad (2\text{--}11)$$

from which it was shown in Sec. 2.3 that $c^2 = gy$ when the wavelength L is large. But L may be finite in the situation of Fig. 7-5; in fact it is simply related to the channel width b. There are therefore two factors—finite amplitude and finite wavelength—which may require a modification of Eq. (7–5), or at least a reinterpretation of the ratio v/c. The former will be dealt with first.

A convenient example of a large disturbance is the deflection of a vertical channel wall through a finite angle $\Delta\theta$, as in Fig. 7-6. The oblique wave front

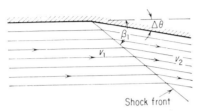

Figure 7-6. Plan View of Inclined Shock Front in Supercritical Flow

then formed will bring about a finite change in depth Δy, and it is unlikely that the total deflection angle β_1 will be given by Eq. (7–5). However we can readily analyze the situation by treating the wave front as a hydraulic jump on which a certain velocity component has been superimposed parallel to the front of the jump; clearly this component must be the same on both sides of the front, for the change in depth Δy does not bring about any force directed *along* the front of the jump. We can therefore write, using the terms defined in Fig. 7-6,

$$v_1 \cos \beta_1 = v_2 \cos(\beta_1 - \Delta\theta) \qquad (7\text{–}6)$$

Considering now the velocity components normal to the wave front, the continuity equation becomes

$$v_1 y_1 \sin \beta_1 = v_2 y_2 \sin(\beta_1 - \Delta\theta) \qquad (7\text{–}7)$$

and the momentum equation must clearly lead to the result

$$\frac{v_1^2 \sin^2 \beta_1}{g y_1} = \frac{1}{2} \frac{y_2}{y_1} \left(\frac{y_2}{y_1} + 1 \right) \qquad (7\text{–}8)$$

which differs from the ordinary hydraulic-jump equation, Eq. (3–4), only in that v_1 is replaced by $v_1 \sin \beta_1$. It follows that

$$\sin \beta_1 = \frac{1}{\mathrm{Fr}_1} \sqrt{\frac{1}{2} \frac{y_2}{y_1} \left(\frac{y_2}{y_1} + 1 \right)} \qquad (7\text{–}9)$$

which reduces to Eq. (7–5) when the disturbance is small and y_2/y_1 tends to unity.

The special case of the small disturbance can be investigated further by eliminating v_1/v_2 between Eqs. (7–6) and (7–7), leading to the result

$$\frac{y_2}{y_1} = \frac{\tan \beta_1}{\tan(\beta_1 - \Delta\theta)} \qquad (7\text{–}10)$$

Setting $y_2 = y_1 + \Delta y$, and letting $\Delta\theta$ tend to zero, we obtain

$$\frac{dy}{d\theta} = \frac{y}{\sin \beta \cos \beta} = \frac{v^2}{g} \tan \beta \qquad (7\text{–}11)$$

dropping all subscripts. This equation indicates how the depth would increase continuously along a curved wall (Fig. 7-7); each value of θ determines a

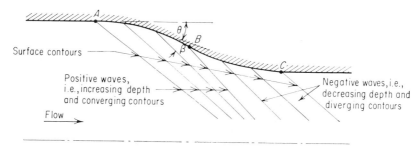

Figure 7-7. *Wave Patterns due to Flow along a Curved Boundary*

value of y not only at the wall but also along a line radiating from the wall as in the figure. We may think of this line as representing one of a series of small shocks or wavelets, each originated by a small change in θ, although in fact there is a continuous change in depth rather than a series of shocks. To be truly consistent with the angle β_1 defined in Fig. 7-6, β must be defined as the angle between the boundary tangent and the wave front, as in Fig. 7-7, since the fluid which is about to cross any wave front at any instant is moving parallel to the boundary tangent where that wave front originates; this conclusion is a logical generalization of the picture of events shown in Fig. 7-6. Granted the above definition of β, Eq. (7–5) is true, and the second step in Eq. (7–11) is justified.

The integration of Eq. (7–11) depends on the assumptions that are made about energy dissipation. A continuous change in depth of this sort would be accompanied by little loss of energy, so it would be reasonable to assume constant specific energy. This assumption gives the following result for the integration of Eq. (7–11):

$$\theta = \sqrt{3} \tan^{-1} \frac{\sqrt{3}}{\sqrt{Fr^2 - 1}} - \tan^{-1} \frac{1}{\sqrt{Fr^2 - 1}} - \theta_1 \qquad (7\text{–}12)$$

where θ_1 is the constant of integration obtained by substituting the initial value Fr_1 into the other terms on the right hand side of the equation. This implies that $\theta = 0$ when $y = y_1$. If Eq. (7–12) is plotted, as in Fig. 7-8, it is convenient to plot $(\theta + \theta_1)$ against Fr; the user then enters this diagram at whatever point is indicated by the initial value Fr_1, and assumes $\theta = 0$ at this point,

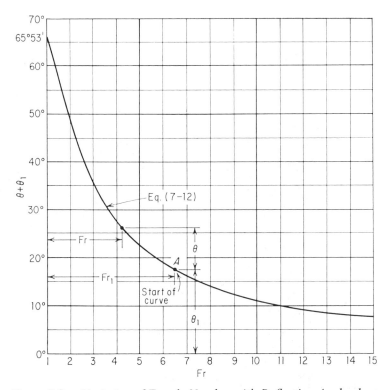

Figure 7-8.   Variation of Froude Number with Deflection Angle along a Curved Boundary in Supercritical Flow

e.g., the point A. Then θ is given by differences between subsequent values of the ordinate and the initial one at A, as shown in the figure. The depth y is simply obtained from Fr by recalling that the specific energy, assumed constant in this case, is equal to $y(1 + Fr^2/2)$.

This theory of the oblique shock wave is largely due to Ippen [2]; the experiments of this investigator on flow near reversed-curve boundaries of the type shown in Fig. 7-7 indicated close agreement with Eq. (7–12) except that the point of maximum depth occurred somewhat downstream of the inflection point B (Fig. 7-7) and not right at B, as the theory would indicate (Prob. 7.5). If it is assumed that the velocity rather than the specific energy

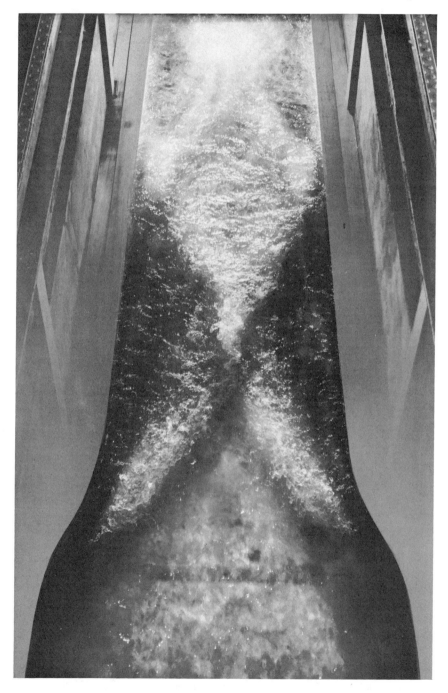

Figure 7-9. Oblique Standing Waves in a Channel Contraction

remains constant—a reasonable assumption if Fr is fairly large—we obtain instead of Eq. (7–12) a simpler result (Prob. 7.5) which gives values of y only slightly lower than those given by Eq. (7–12).

From the above discussion it is clear how Eq. (7–12) or Fig. 7-8 can be used to determine completely the flow pattern through a contraction such as Fig. 7-7. The depth y is known as a function of θ along the curved boundary and along the wavelets (which are surface contours) drawn from each point at the angle β, defined as in Fig. 7-7 and given by Eq. (7–5). Along the concave wall section AB, y is increasing, Fr is decreasing and β is increasing. However β is not increasing as fast as θ itself; the result is that the disturbance lines are converging and would eventually intersect, if the channel were wide enough, with the formation of an abrupt surge front. Along the convex wall section BC the depth is decreasing and the disturbance lines, which are now diverging, may be thought of as negative wavelets. At points where disturbance lines from opposite sides of the channel intersect, the net change of depth is the algebraic sum of the changes in depth indicated by the individual disturbance lines. This summation process, combined with reflections backwards and forwards between opposite walls, produces a diamond-shaped pattern of standing waves persisting well into the narrower downstream channel as in Fig. 7-9.

A standing wave system of this kind is clearly not a satisfactory operating condition, since it requires higher side walls than if the flow were uniform. In fact uniform flow can be produced downstream, but only by using a contraction with straight side walls, as in Fig. 7-10a. If the angle θ is properly chosen, the shock waves reflected from the junction B will arrive at the side walls exactly at the intersections C, C', and there will be no further reflections. If, however, these shock waves strike the side walls at other points such as D, D', as in Fig. 7-10b, a complicated system of reflections again results in a diamond-shaped pattern of standing waves.

For the design of such a transition it is necessary to refer back to Eqs. (7–6) through (7–10), dealing with finite values of the wall deflection angle $\Delta\theta$. The waves AB, BC, are also of finite height and the energy loss will therefore be appreciable. Various schemes are possible for manipulating the equations, depending on which parameters are assumed known and which are to be calculated. A convenient scheme is outlined in Prob. 7.7; for complete details the reader is referred to the graphical treatment given by Ippen [2]. An objection to this type of transition is that the desired system of waves exists only at one value of Fr_1; above or below this value the reflected waves do not strike the points C, C'. However, this design should give good results at the downstream end of chutes which are long enough for the flow to be uniform; in such cases, as the Manning equation shows, the Froude number will remain approximately constant over a fair range of discharges.

We consider now the effect of finite wave length, referring back to Fig. 7-5 for the purpose. Consider a symmetrical standing wave pattern like that of

Fig. 7-9, shown schematically by the solid diagonal lines in Fig. 7-5. The flow situation is the summation of a uniform flow with velocity v, and two oblique symmetrical wave motions proceeding as shown, each with velocity $v \sin \beta$

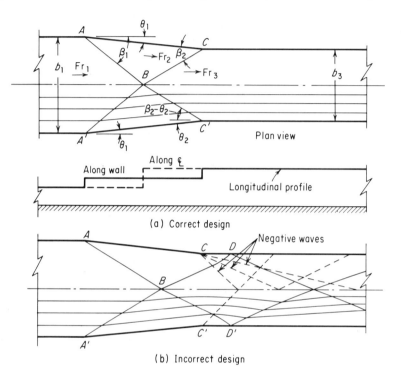

Figure 7-10. *Supercritical Wave Patterns in a Straight-Walled Contraction, after A. T. Ippen* [2]

and wave length $b \cos \beta$. A generalization of Eq. (7–5) is now simply obtained by substituting these values into Eq. (2–11); the result is

$$v^2 \sin^2 \beta = \frac{g\, b \cos \beta}{2\pi} \tanh \frac{2\pi y}{b \cos \beta}$$

or
$$\sin \beta = \frac{1}{\mathrm{Fr}} \sqrt{\frac{\tanh(2\pi y / b \cos \beta)}{2\pi y / b \cos \beta}} \qquad (7\text{–}13)$$

This result is due to Engelund and Munch-Petersen [23,24], who also pointed out that Eq. (7–13) is not a unique solution to the problem. The wave pattern would still fulfil all the boundary conditions if wave fronts represented by the broken lines in Fig. 7-5 were added to the system; in this case the wavelength would be equal to $b \cos \beta/2$, b would be replaced by $b/2$ in Eq. (7–13), and β would take a new value. In fact it is clear that the wavelength may be set equal to $b \cos \beta/m$, where m is any integer; to each value of m there will then correspond a distinct value of β.

The resulting wave pattern may therefore be compounded of a number of standing waves, each with its own wavelength and deflection angle β. The analysis cannot tell us anything about the relative heights of these waves, which will presumably be determined in some way by the boundary conditions; thus, in the contraction of Fig. 7-9, it is clear that only "first order" waves, corresponding to the full lines in Fig. 7-5, are present to any significant degree. The experiments of Harrison [25] have shown that small disturbances in a straight trapezoidal channel give rise only to first order waves, with measured values of β corresponding very closely to those calculated from Eq. (7–13) by the substitutions $b = B$, $y = \sqrt{A/B}$, and $\mathrm{Fr} = v/\sqrt{gA/B}$. The experiments of Engelund and Munch-Petersen [23] detected only first order waves at contractions in both subcritical and supercritical flow, but expansions in supercritical flow developed wave patterns in which first, second, and third order waves, i.e. those with $m = 1$, 2 and 3, were present in appreciable heights. A Fourier analysis of the water surface elevation showed that the heights of these three wave components were in the approximate ratio $1:\frac{1}{2}:\frac{1}{4}$, in the order named above. However, much more work will be necessary before it is possible to make systematic predictions of relative wave heights.

Indeed there is nothing in Eq. (7–13) that tells us anything about wave height at all. Even if it were known that only first order waves were present in a particular case, their height could not be determined from Eq. (7–13) alone; more information would have to be supplied concerning the changes in channel geometry that give rise to the waves. In this respect Ippen's theory [2] is more complete, for it successfully relates wave height to channel geometry. Its limitation is that by treating waves as surges it becomes applicable only to supercritical flow, whereas wave formation is by no means negligible in subcritical flow, particularly near the critical condition. Referring again to Fig. 2-9, the reader will be able, by measuring the wavelength and water depth and substituting in Eq. (2–11), to verify that $\mathrm{Fr} < 1$ in this case. Equation (7–13) enables us to make corresponding calculations about oblique waves, although not about their height.

Clearly there is a need for a general theory which combines features of both those discussed so far, making it possible to predict finite heights of waves having finite length. At present, the most direct usefulness of Eq. (7–13) lies in those situations where the successful design of a transition depends on the designer's ability to predict wave deflection angles with reasonable accuracy. A case of this sort is that of flow round channel bends, to be discussed in Sec. 7.3.

Reverting to more general considerations in the design of contractions, we have seen that in the case of subcritical flow it is necessary to guard against the possibility that the contraction will choke because the downstream section is too narrow. In supercritical flow there is another complication in the form of a second possible choking mechanism, in addition to the one based on considerations of specific energy and discussed in Chap. 2. Both mechanisms are illustrated in Fig. 7-11, which shows two possible forms of

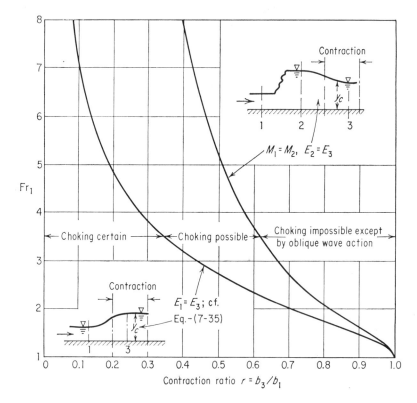

Figure 7-11.   *The Choking of a Width Contraction in Supercritical Flow*

limiting condition. The first is clearly possible and can be determined by the methods of Chap. 2; the second, in which there is a hydraulic jump immediately upstream of the contraction, is also a clear physical possibility and is readily determined from the equations $M_1 = M_2$, $E_2 = E_3$.

There are thus two distinct criteria which will not in general yield the same numerical results. The conclusion must be of the kind indicated by Fig. 7-11, that for a given contraction ratio there is an intermediate range of values of Fr_1 for which choking may or may not occur. (It will of course be clear to the reader that criteria of this sort depend only on Fr_1 and on the channel geometry.) The construction of Fig. 7-11 can be verified by Prob. 7.6, and in Prob. 7.8 the reader can construct a similar figure for the case of an upward step in the channel bed. Within the intermediate range referred to either choked or unchoked flow will be stable, and either can be set up in a small laboratory flume just by using a paddle to sweep the flow upstream or downstream. However, in the case of the contraction in width the waves set up by the converging side walls will tend to promote the choked condition, and for design purposes Fr_1 should be kept high enough, or the contraction ratio low enough, to make choking impossible. If this is done the effect is to keep Fr_3 (Fig. 7-10a) well above unity, as is recommended [2] for the straight-sidewall design shown in this figure.

If the effects of finite wavelength are neglected it should be possible to design expansions in supercritical flow by following the theory discussed above and summarized in Eq. (7–5) through (7–12). For example, the application of Eq. (7–12) to the curves AB, BC in Fig. 7-7 is not essentially different from its application to the expansion problem, where a curve such as BC, generating negative disturbance lines, is followed by a curve such as AB, generating positive disturbance lines.

The argument has been developed in a completely general way by Rouse, Bhoota, and Hsu [3] who first studied, analytically and experimentally, the way in which supercritical flow issues from the end of a rectangular channel and spreads out over a level floor of infinite extent. This situation is, in effect, the most abrupt channel enlargement possible. The second step was to design sidewalls corresponding to streamlines enclosing 90 percent of the flow in

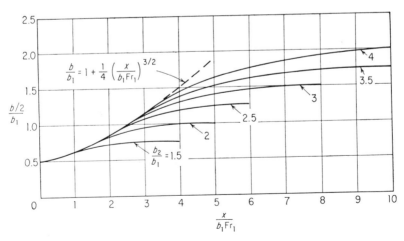

Figure 7-12. Generalised Channel-Expansion Curves for Supercritical Flow, after H. Rouse, B. V. Bhoota, and E.-Y. Hsu [3]

the above situation; the third step was to develop reverse curves whose effect was to direct the flow into a finite width downstream with the minimum of wave disturbance. The result of this work is shown in Fig. 7-12 in the form of a family of generalized expansion curves giving the sidewall shape for any value of Fr_1, and a wide range of values of the width ration b_2/b_1. Strictly speaking, the analysis should have allowed for the effects of finite wavelength, but in fact any approximations in the theory were made up for at the experimental stage. In the end result, successful designs for channel expansions were produced.

7.3   Changes of Direction

It is often necessary for the engineer to understand and control the behavior of the flow around bends in rivers and canals. In natural rivers the development of successive bends, or meanders, is an essential part of the process by which the river channel is molded; if these bends become too tight they may give rise to large head losses and unduly high flood levels upstream. When this happens the engineer must take remedial action; similarly, he will avoid large head losses in artificial canals by giving the bends sufficiently large radii.

However, the question of head losses is not the most important one in the consideration of this problem. A bend may introduce only small losses, but may at the same time set up a disturbance in the flow which persists for some distance downstream, perhaps damaging the banks of the river or canal, or perhaps overtopping the banks through wave action of some kind. Again, there may be, downstream of the bend, some critically important structure such as a spillway, whose satisfactory action depends on whether the incoming flow is regular and free from disturbance. We therefore consider first the general way in which a bend may modify the flow.

Basic Character of Flow Around a Bend

Consider the channel cross section shown in Fig. 7-13a. In the following analysis it will be assumed that any energy loss incurred in flow around the bend is small and may be neglected; this assumption is very close to the truth in many cases, and in all cases it is accurate enough for the results of the analysis to be qualitatively useful.

All the stream filaments across the width will have the same total energy H_0; if h is the height of the water surface above datum, then

$$h + \frac{v^2}{2g} = H_0 \qquad (7\text{--}14)$$

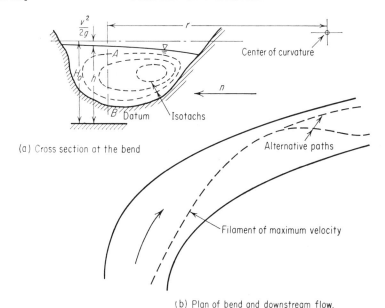

(a) Cross section at the bend

(b) Plan of bend and downstream flow.

Figure 7-13. *Flow Behavior at a Channel Bend and Further Downstream*

at any vertical section such as *AB*, where v is the mean velocity down the section. Let r be the radius of curvature (drawn in the horizontal plane) of the streamlines at any section *AB*, and n the distance measured outwards across the width of the section. If all the streamlines are concentric, n and r can be made identical by measuring n outwards from the common center, but this condition will not in general be fulfilled.

We now refer back to the Euler equation, Eq. (1–7). If we take the s direction inwards across the width of the section—i.e., in the negative n direction, then $a_s = v^2/r$ by elementary dynamics. If the pressure distribution is hydrostatic then $\partial(p + \gamma z)/\partial s = -\gamma \partial h/\partial n$, since at all points on *AB*, $p + \gamma z = \gamma h$. Equation (1–7) then becomes

$$\frac{dh}{dn} = \frac{v^2}{gr} \tag{7–15}$$

By differentiation from Eq. (7–14) we obtain

$$\frac{dh}{dn} + \frac{v}{g}\frac{dv}{dn} = 0 \tag{7–16}$$

Combining Eqs. (7–15) and (7–16) we obtain

$$\frac{dv}{dn} + \frac{v}{r} = 0 \tag{7–17}$$

a general equation which is true for flow in closed conduits as well as in free surface flow. This equation, and Eq. (7–15), indicate clearly that since v/r and v^2/gr are always positive, v decreases and h increases from the inner to the outer bank. The effects of resistance modify this behavior of v somewhat; the maximum velocity is not right at the inner bank, but at some distance from it, as in Fig. 7-13a. Nevertheless the bend has the distinct effect of moving the point of maximum velocity well towards the inner bank.

Figure 7-13b shows how the filament of maximum velocity moves across to the outer boundary immediately downstream of the bend; in some cases it may subsequently oscillate from side to side of the channel as shown. In an artificial channel in erodible material, the persistence of this high-velocity filament may have damaging effects on the channel bed.

Secondary currents, already discussed in Chap. 1, always occur at bends. The explanation is a simple one: because the pressure distribution is very nearly hydrostatic, the transverse pressure gradient supplied by the water surface gradient dh/dn is nearly the same at all points on a vertical section AB. However the velocity v is by no means constant, being much lower near the bed. Consequently the balance implied by Eq. (7–15) does not exist at all points on AB; near the bed the inward pressure gradient preponderates, and sets up an inward flow along the bed. This in its turn is compensated for by an outward flow at the upper levels, the final effect being a circulatory pattern like that shown in Fig. 7-14. In natural rivers this flow pattern is a most

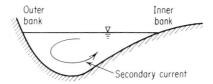

Figure 7-14. *Typical Section at a Natural River Bend*

important influence, tending to scour sediment from the outside of the curve and deposit it on the inside, forming a cross section of the general shape indicated in Fig. 7-14.

An interesting consequence of this scouring mechanism is that if a branch channel is taken off from the outside of a curve, the sediment concentration in the branch will be much less than in the main channel. This fact is of some importance in the layout of irrigation canals, which readily become choked if they are required to carry too much sediment; it is therefore desirable to connect them to the outside rather than the inside of curves. This fact, if not the explanation for it, seems to have been known even in ancient times; the remains of irrigation systems built by ancient Mediterranean civilizations show that branch channels were always connected either to straight sections or to the outside of curves, never to the inside of curves.

The above descriptions neglect the existence of waves, which will in fact occur as discussed in Sec. 7.2, particularly when the flow is supercritical. The plan view of a typical curve in a rectangular channel is shown in Fig. 7-15,

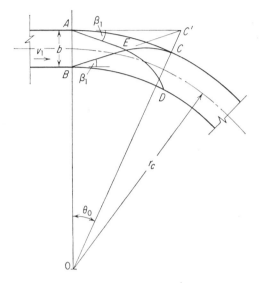

Figure 7-15.   Schematic Wave Pattern at a Circular Curve

together with the positive and negative disturbance lines AED, BEC, that are generated at the tangent points A, B, at an angle β_1, given by Eq. (7–13). The zone AEB is undisturbed. It is clear from the general form of the disturbance lines in Fig. 7-15 that the depth increases along AC to a maximum at the point C, where the first negative disturbance reaches this wall, and similarly that a minimum depth is reached at the point D on the inner wall. If we assume that AC is approximately equal to $AC' = b/\tan \beta_1$, the central angle θ_0 is given by

$$\tan \theta_0 = \frac{b}{(r_c + b/2)\tan \beta_1} \tag{7–18}$$

Detailed tracing of the wave pattern shows that further maximum depths occur on the outside wall at intervals of $2\theta_0$, and that there is the same interval between maxima on the inside wall, but with a maximum on one wall occurring opposite a minimum on the other wall. The general appearance of the flow pattern then suggests that the flow rocks backwards and forwards between successive maxima on opposite sides of the channel; this pattern persists beyond the curve itself (maintaining the same "wavelength"), and for a great distance downstream.

The experiments of Ippen and Knapp [4,5] showed that in supercritical flow the maximum transverse surface slope was about twice the amount indicated by Eq. (7–15), and that corresponding maximum and minimum surface levels were symmetrically placed about the upstream surface level. The water surface therefore ranges between limits $v^2 b/rg$ above and below the upstream surface.

The Design of Channel Bends

All of the preceding discussion relates to channel bends in which no special attempt has been made to suppress the disturbances arising in the bend and persisting downstream with possible harmful effects. The question arises whether special design features could be introduced in order to suppress these disturbances.

The most obvious feature to try is "banking" or superelevation analogous to that of highway and railroad curves. A less obvious expedient is to aim at suppressing the waves shown in Fig. 7-15 by shaping the channel sidewalls in some way. Dealing with superelevation first, Eq. (7–15) makes it clear that it is cross slope of the water surface, not of the bed, that provides the necessary centripetal force. If any cross slope is built into the bed, it should be with the intention of producing a certain desired cross slope in the water surface. The relationship between the two cross slopes will depend on the distribution of q, the discharge per unit width, across the channel, and it should therefore be possible for the designer to produce any desired q distribution by using a suitable cross slope on the bed.

As to the choice of q distribution: it is clear that the disturbance created by bends in the way shown in Fig. 7-13 amounts simply to an undue concentration of discharge near the inner bank; if this irregularity could be removed, by keeping q constant across the section, the flow would return more quickly to normal after passing through the bend. It is clear therefore that the designer should aim to develop a transverse bed slope that will produce a constant q over the section. We now consider how to achieve this result for a channel of rectangular section.

If the desired result is achieved the streamlines viewed from above will become concentric circles, for each element of fluid will remain at the same lateral distance from each sidewall as it passes round the curve. This means that n and r in Eq. (7–17) are identical, and that this equation can be written

$$\frac{dv}{v} + \frac{dr}{r} = 0 \qquad (7–19)$$

leading to the well-known free-vortex equation

$$vr = C, \text{ a constant} \qquad (7–20)$$

Since $q = vy$ is a constant, it follows that

$$\frac{y}{r} = \frac{q}{C}, \text{ a constant} \qquad (7\text{-}21)$$

i.e., that

$$\frac{dy}{dr} = \frac{y}{r} \qquad (7\text{-}22)$$

Now the total energy H_0 is assumed constant over the section, and since q also is constant, it follows that Eq. (2–15), relating changes in depth to changes in bed level, is applicable; and this equation holds true whether the

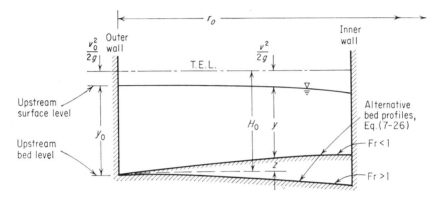

Figure 7-16. *Transverse Bed Profiles Required at a Channel Bend for Constant q across the Section*

changes considered are taking place in the downstream or cross-stream direction. Adapting the equation to the present purpose, we can write it as

$$\frac{dy}{dr} = -\frac{dz}{dr}\frac{1}{1 - \mathrm{Fr}^2} \qquad (7\text{-}23)$$

and it becomes clear that if Fr < 1 the bed should fall towards the outside of the curve, as shown in Fig. 7-16. From Eqs. (7–22) and (7–23) we have

$$\frac{dz}{dr} = (\mathrm{Fr}^2 - 1)\frac{dy}{dr} = (\mathrm{Fr}^2 - 1)\frac{y}{r} \qquad (7\text{-}24)$$

so that

$$\frac{dh}{dr} = \frac{dz}{dr} + \frac{dy}{dr} = \mathrm{Fr}^2 \frac{dy}{dr} = \mathrm{Fr}^2 \frac{y}{r} = \frac{v^2}{gr} \qquad (7\text{-}25)$$

a result which verifies Eq. (7–15). Finally we can obtain an explicit z-r relationship by integrating Eq. (7–24); this process is dependent on the choice of constants of integration implied in the choice of a bed level at some part of the section. Since the water surface is at its highest at the outer wall it is convenient to set the depth and bed level here at their original upstream values; the bed level falls or rises from this point according as Fr is greater or less than unity. According to the above theory the surface level should not then rise above the upstream level at any point.

We define z as the height of bed above the upstream bed level, y_0 and v_0 as the upstream values of y and v, r_o as the radius of the outer wall and H_0 as the total energy referred to upstream bed level. Then

$$z + y + \frac{v^2}{2g} = H_0 = y_0 + \frac{v_0{}^2}{2g}$$

and from Eq. (7–21), $y = rq/C$ and $v = C/r$. Also $C = v_0 r_o$; hence the above equation becomes

$$z + \frac{qr}{v_0 r_o} + \frac{v_0{}^2 r_o{}^2}{2gr^2} = H_0 \qquad (7\text{–}26)$$

which defines the transverse bed profile, and is in effect the integral of Eq. (7–24). If the radius of the curve is large, average values of Fr, y, and r may be used in Eq. (7–24) to give a straight-line bed profile.

When the flow is subcritical, a possible complication arises because within the curve the Froude number Fr is everywhere greater than the upstream Froude number Fr_0. Therefore the flow may become critical at the inner wall; however it is easy to determine the criterion governing this possibility and it is left as an exercise for the reader (Prob. 7.9).

A bend designed in accordance with Eq. (7–26) should suppress those disturbances which are due to wandering of the filament of maximum velocity, as in Fig. 7-13. To some extent this design should also help to suppress wave disturbances, for waves must be materially influenced by redistribution of the flow, such as this design achieves. However, complete suppression of wave disturbance requires a more positive measure, such as that proposed and tested by Knapp and Ippen [6]. Their method was to add transition curves of radius $2r_c$ upstream and downstream of the main curve, of radius r_c. The aim was to produce a counter wave of directly opposite phase to the positive wave AED in Fig. 7-15. The length θ_0 of the transition curve was obtained by replacing r_c by $2r_c$ in Eq. (7–18). The effect is that a negative wave like BEC in Fig. 7-15 grows from the start of the transition curve, arriving at the outer wall just at the point where the main curve begins, and thereby helping to suppress the positive wave originating at this point.

For this design, it is clearly desirable to have a precise estimate of β_1, so Eq. (7–13) is likely to be more satisfactory than Eq. (7–5), which would be correct only if the channel were very wide. In fact, Harrison [25] has obtained good results by the application of Eq. (7–13) to the design of a curve in a trapezoidal channel, with Fr = 1.6. Knapp and Ippen [6], applying Eq. (7–5) to the design of curves for supercritical flow, found it necessary to combine the transition curves with banking of the bed in order to achieve nearly complete suppression of wave action. In these experiments the bed was set on a transverse slope equal to v^2/gr, which is not strictly in accord with the above theory; however at high values of Fr it becomes an acceptable approximation, for as Eqs. (7–24) and (7–25) show, dz/dr and dh/dr tend to the same values as Fr increases.

In subcritical flow, Eq. (7–13) indicates very short lengths for the transition curves; in fact the length required would be determined simply by the need for reasonably gentle longitudinal slopes leading into and away from the banking in the bed. It is probably most convenient to make the transition radius $2r_c$, as in the supercritical case.

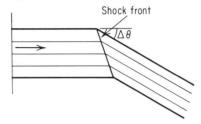

Figure 7-17.   *Change of Channel Direction for Supercritical Flow*

Another possibility in supercritical flow is to use the inclined shock wave as a basis for the design of a combined contraction and change of direction, as in Fig. 7-17. The usefulness of this design is limited by the fact that for every value of Fr_1 there is a maximum possible value of the deflection angle $\Delta\theta$. When $\Delta\theta$ exceeds this value, supercritical flow is impossible and the flow at the transition becomes subcritical.

Head Losses in Channel Bends

Experimental data on this subject are far from complete because of the large number of independent variables which must influence the magnitude of the loss coefficient C_L, the ratio between head loss and velocity head. These variables include Fr_1, y/b, r_c/b, Re and θ, the total angle of deflection. The results presented by Mockmore [7] and by those discussing his paper indicated a surprisingly close agreement with the simple relationship

$$C_L = 2\frac{b}{r_c} \tag{7-27}$$

for a number of data taken from artificial channels and natural rivers, with θ ranging between 90° and 180°. On the other hand, A. Shukry [8] has presented a large number of experimental results covering a fair range of values of the above variables, and of the Reynolds number up to a value of 7.5×10^4, in smooth channels of rectangular section. In these experiments C_L showed substantial variations with all the parameters listed above, but was in all cases less than one-third as great as Eq. (7–27) indicates. Further experiments by Allen and Chee [9] for the particular case $\theta = 180°$, $r_c/b = 1.57$, gave values of C_L approximately the same as in Shukry's experiments, but varying in a rather different manner with the ratio y/b.

In Shukry's experiments the y/b ratio was never less than 0.6. His results showed that C_L increased materially with decreasing y/b, and if much lower values of y/b had been used—e.g., those appropriate to natural rivers, it is conceivable that values of C_L approaching those given by Eq. (7–27) would have been measured. This tentative conclusion is not, however, borne out by Allen and Chee's work, in which C_L diminished as y/b was reduced from 0.5 to 0.2.

Further experimental work is needed to clear the matter up. There are certain indications that the value of C_L is critically dependent on the strength of the secondary flow, shown in Fig. 7-14; this in turn will be dependent on whether any secondary currents are present in the flow approaching the bend. For instance in Shukry's experiments the approach flow happened to possess secondary currents circulating in the opposite sense to those induced by the bend itself. The effect must have been to reduce the strength of the secondary flow existing in the bend, and it was suggested in the discussion of Shukry's paper [8] that this effect may well have been responsible for the low values of C_L measured by Shukry.

It may be that slight differences in the approach conditions can make large differences in the strength of the secondary currents, and hence in the magnitude of C_L. Indeed some explanation of this sort does appear necessary to account for the wide disparities between measured values of C_L obtained by different investigators in apparently similar laboratory setups. The same explanation fits Shukry's observation that C_L increased with decreasing y/b, for it is well known that the strength of the secondary flow within the bend increases substantially as y/b decreases.

However the details may work out, a complete solution must await further investigation; meanwhile it is suggested, as a conservative practice in river and canal work, that Eq. (7–27) be used, even though it may in some cases give values of C_L that are three or four times too large.

It is possible that channels designed as in Fig. 7-16 would have C_L values less than in channels of rectangular section, but experimental work would be required to establish this.

7.4   Culverts

In engineering practice the term "culvert" is commonly applied to any large underground pipe, particularly when used in short lengths where a channel crosses a highway or railroad. From the hydraulic viewpoint, however, the defining feature of a culvert is that it may or may not run full, and this distinction is often of great practical importance. For example if the culvert has an unsubmerged outlet and an appreciable fall from inlet to outlet, this fall will augment the total available head if the culvert runs full; otherwise it will not.

Whether a culvert does in fact run full depends on a number of factors—the diameter, length and roughness of the culvert, as well as the headwater and tailwater levels. If the culvert entry is of conventional type, length is one of the most important of these factors, and for this reason a culvert is sometimes called "hydraulically long" if it runs full and "hydraulically short" if it does not. However it will be shown subsequently that recent work on the design of culvert entries has rather diminished the importance of culvert length as a factor in making the culvert run full.

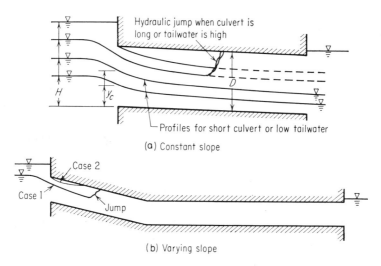

Figure 7-18.   *Typical Culvert Flow Profiles*

Figure 7-18 illustrates the operation of the various factors governing culvert performance. The upstream head H is defined as the height of pool level above the "invert"—i.e., the low point of the culvert periphery. The maximum vertical dimension of the culvert section is D. The three lowermost profiles in Fig. 7-18a are the type that would occur when the culvert is short and steep; in these cases the discharge is determined by the culvert entry and the flow at the entry will be critical. It is noteworthy that in one of these cases H is greater than D, but not so much greater as to make the water surface touch the "soffit"—i.e., the high point of the culvert periphery.

If the tailwater level is high, as indicated in the same figure, the culvert will fill at the downstream end of the culvert and a hydraulic jump will move upstream; if the headwater and tailwater levels are high enough, or if the culvert resistance is high enough, the jump will move right up to the entrance. In this case the entrance is completely drowned, the culvert is running completely full, and its discharge is determined by its resistance and the headwater and tailwater levels.

A slightly different situation in shown in Fig. 7-18b. Here there is a short steep upstream length, and a long flat downstream length, with tailwater above the soffit level. Clearly the downstream length will run full, as will some of the upstream length. The full portion extends far enough up the steep length to supply the head needed for the downstream length, as shown in the figure. At low discharges therefore it will extend only a short distance and the culvert entry will be running free and undrowned—case 1 in Fig. 7-18b. As the discharge increases the hydraulic jump which terminates the full part of the culvert will move further and further upstream until finally the entry is drowned and the whole culvert is running full (case 2). In this condition the discharge is determined by the culvert resistance, whereas in case 1 it is fixed only by the entry conditions. Whether the flow is of case 1 or case 2 type is decided simply by calculating and comparing the two appropriate discharges (Probs. 7.22 and 7.23).

The two situations shown in Fig. 7-18 do not however exhaust all the possibilities. A high tailwater level is not necessarily required in order to make a culvert run full; this can also occur if the culvert is very long and of gentle slope. For even if there were initially free surface flow at the upstream end, retardation of the flow would eventually cause the depth to increase to the point where the culvert could only run full.

The foregoing discussion concerning the effect of tailwater level, resistance, etc., implies that to some extent each case must be treated on its merits, as in Probs. 7.22 through 7.24. Generalized data have been prepared which give approximate guidance in deciding whether a culvert will run full or not, but much of this work has been rendered out of date by more recent work on special types of inlet. This is discussed in the next paragraph.

The Circular Culvert and the Hood Inlet

The developmental work of Blaisdell [10] has shown that even a short steep culvert can be made to run full at low heads if the inlet is formed by cutting the culvert at an angle so that the soffit projects beyond the invert, as in Fig. 7-19. He described this design as the "hood inlet."

Blaisdell's experiments, which related only to circular culverts, established that the optimum hood length L was equal to $\frac{3}{4}D$, and that for this hood length the culvert would "prime" and run full if H/D exceeded 1.25, for slopes up to 0.361; here the slope is defined as the tangent of the slope-angle. Essential to the performance of the inlet is a vortex inhibitor such as the vertical splitter shown in Fig. 7-19; an alternative device is the square or circular plate shown in the same figure. Without a vortex inhibitor the inlet

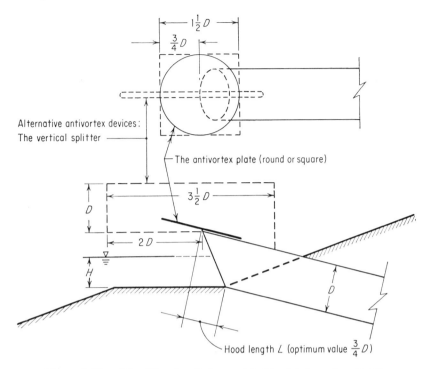

Figure 7-19. *The Circular Culvert with Hood Inlet, after F. W.*
Blaisdell [10]

would suck in air at low degrees of submergence (i.e., when H/D does not greatly exceed 1), reducing the discharge by as much as two-thirds in some cases.

The experiments also included extensive calibration of the inlet when running in the unsubmerged condition, i.e., with $H/D < 1.25$, and with steep culvert slopes. In this condition flow at the entry should be critical, and in fact we might expect the circular-culvert line on Fig. 2-16 to give the discharge, for the specific energy E_c on that figure is identical with the head H defined above. We should, however, expect the actual discharge to be somewhat less because of side contractions in the entering flow; these would have been particularly pronounced in the case of Blaisdell's experiments, in which the culvert entry projected into the headpool to form a reentrant mouthpiece.

The effect of these contractions did in fact appear, in the form of reduced discharge, but their effect became less pronounced at higher slopes, becoming almost negligible at the maximum slope, S_0, of 0.361. Over the range $0.025 < S_0 < 0.361$ the slight but appreciable effect of slope can be expressed by multiplying the theoretical discharge (Fig. 2-16) by the factor $(S_0/0.4)^{0.05}$, which may be thought of as a discharge coefficient. (When the slope is mild

or horizontal control shifts to the downstream end, and the inlet characteristics depend on the culvert resistance.) Combining the above result with
the exponential equations fitted in Sec. 2.7 to the curve of Fig. 2-16, we can
write for the range $0 < H/D < 0.8$:

$$\frac{Q}{D^2\sqrt{gD}} = 0.48\left(\frac{S_0}{0.4}\right)^{0.05}\left(\frac{H}{D}\right)^{1.9} \tag{7-28}$$

and for the range $0.8 < H/D < 1.2$:

$$\frac{Q}{D^2\sqrt{gD}} = 0.44\left(\frac{S_0}{0.4}\right)^{0.05}\left(\frac{H}{D}\right)^{1.5} \tag{7-29}$$

These equations give results within 2 or 3 percent of Blaisdell's experimental results for the hood inlet circular culvert. It is of interest to make a
comparison with the earlier experimental work of Mavis [11], performed on
circular culverts with square-edged entries—i.e., with each culvert leading
from a plane face at right angles to the culvert centerline. The comparison is
made in Fig. 7-20, from which it is seen that Mavis' results plot as a curve

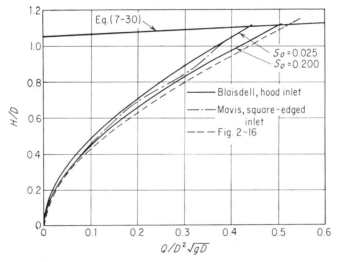

Figure 7-20. *Discharge Characteristics of Circular Culverts with
Unsubmerged Inlets*

contained between Blaisdell's two curves for $S_0 = 0.025$ and 0.20, representing discharge coefficients of between 0.87 and 0.97; it would seem that a
conservative figure for the square-edged entry would be 0.90. Mavis' results
indicated no effect of slope.

Also plotted on Fig. 7-20 is the theoretical curve, taken from Fig. 2-16.

The curves for the hood inlet terminate on a line having the equation

$$\frac{H}{D} = 1.05 + 0.142 \frac{Q}{D^2\sqrt{gD}} \tag{7-30}$$

representing the start of the priming process, during which there is "slug flow," and water-air mixture flow, the former of these terms representing flow of alternate "slugs" of water and air. Above this line the culvert flows full and its behavior is described by the normal equation of pipe flow; the only difference made by the hood inlet is to raise the inlet loss coefficient from its normal value of about 0.5 to one of about 1.0.

The Box Culvert

Culverts of square or rectangular section are quite common, and are known as box culverts. Presumably the hood-inlet principle could be applied to this type as well as to the circular culvert, but the relevant experimental studies have not been made. Meanwhile it is recommended that the circular culvert with hood inlet be used where it is especially desirable to make the culvert run full. There may still be occasions where box culverts are desirable for structural or other reasons, and for these cases the following data are available on the flow characteristics of short steep box culverts that do not run full.

When $H/D < 1.2$ approximately, the water surface at the entry does not touch the soffit, and the flow there is critical. The discharge is therefore equal to

$$Q = \tfrac{2}{3}C_B BH\sqrt{\tfrac{2}{3}gH} \tag{7-31}$$

where B is the width and C_B is a coefficient expressing the effect of width contraction in the flow. If the vertical edges are rounded to a radius of $0.1B$ or more there is no side contraction and $C_B = 1$. If the vertical edges are left square, $C_B = 0.9$. It appears that there is never any appreciable contraction from the invert itself, for there are no detectable differences among results obtained with inverts level with the headpool bed, and with square-edged or rounded inverts mounted well above the headpool bed.

When $H/D > 1.2$ approximately, the water surface touches the soffit, and for this and higher values of H the culvert entry acts essentially as a sluice gate. Experimental results show that the combined effect of vertical and horizontal contractions can be expressed as a single coefficient of contraction C_h in the vertical plane, which for rounded soffit and vertical edges is 0.8, and for square edges is 0.6. Discharges calculated on these assumptions, i.e., from the equation

$$Q = C_h BD\sqrt{2g(H - C_h D)} \tag{7-32}$$

are within 2 percent of measured discharges when $H/D > 1.2$. Equation (7-31) is true within a similar margin of error when $H/D < 1.2$. Figure 7-21 shows that curves representing the two equations intersect near the point where $H/D = 1.2$, with only a slight discontinuity in slope. A curve representing measured discharges does in fact pass through this region with only a slight rounding off being required as it passes from the curve representing Eq. (7–31) to that representing Eq. (7–32).

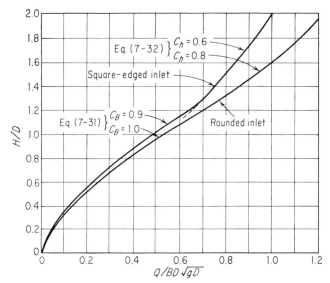

Figure 7-21.   *Discharge Characteristics of Short Box Culverts, Flow Controlled by Inlet Conditions*

7.5   Bridge Piers

Bridge piers form obstructions to the flow, and there must therefore be some mutual interference between the piers and the flow. The thrust of the flow on the piers is quite small compared with the other loads on the piers, so it is of small engineering interest. The possibility of scour around the pier foundations is of much more interest, and will be dealt with in a later chapter. Apart from scour, the most important problem is that of the "backwater" effect of the piers on the flow—the increase in upstream depth, Δy, shown in Fig. 7-22. The flow shown in this figure is subcritical, for the essence of the phenomenon is an influence transmitted by the piers *in the upstream direction*. The subsequent discussion will therefore be confined to subcritical flow unless it is otherwise indicated.

The importance of the backwater effect derives from the economic significance of river levels. In many localities large sums are spent in protecting property from river floodwaters, and the cost of the protective work is critically dependent on the predicted flood level. The building of a bridge

raises the upstream water levels and can set in train disputes about liability
for the consequent increased costs of flood protection. See, for example, the
discussions of Nagler's [12] pioneering work on the subject. It is therefore
highly desirable that the engineer should be able to help resolve these disputes
by predicting with reasonable accuracy the amount of backwater caused by
any specified bridge piers in any specified flow situation.

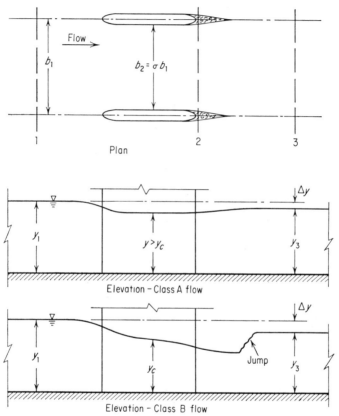

Figure 7-22.   Definition Sketch, Flow between Bridge Piers

A preliminary discussion can be centered on Fig. 7-22, which shows the
basic situation in its simplest form, with the piers set parallel to the direction
of flow. In any subsequent analysis the bed slope and resistance can be
neglected, for the aim is to consider the effect of the piers in isolation from all
other effects. The backwater Δy is properly defined as the difference caused
in the water level at section 1 by introducing the bridge piers into the stream,
and the definition implied in Fig. 7-22 will be equivalent to this if bed slope
and resistance between sections 1 and 3 are neglected.

The most elementary approach would be to set $E_1 = E_2$, $M_2 = M_3$. As
explained in Sec. 2.6 this approach would be wholly inadequate unless the

contraction ratio $\sigma = b_2/b_1$ were quite small, say 0.5 or less. The approach fails because of nonuniform velocity distribution across the space between the piers, and because surface drag along the pier faces may introduce differences between E_1 and E_2 that are comparable with the total amount of the backwater Δy itself.

A more realistic approach is to bypass section 2, considering only the equation

$$M_1 - M_3 = P_f \qquad (7\text{–}33)$$

where P_f is the drag force on the pier. This method should give good results if the drag force P_f can be accurately estimated, but such an estimate must contain some element of guesswork. Thus we might argue that since the piers are "blunt" bodies in what is essentially two-dimensional flow, the drag coefficient C_D should be between 1.5 and 2.0—cf. the flat plate held normal to the flow, whose C_D is 1.9. Surface drag along the faces of the piers might increase our estimate of C_D; so might the possibility of radiating energy downstream through a stationary wave train in the wake of the piers, as already discussed in connection with the hydraulic jump in Sec. 6.7. These considerations would lead to an estimated C_D of between 2.0 and 2.5, the latter value being assumed in order to make the estimate conservative.

However, estimates of this sort can never be exact; for example, there is no mention in the above argument of the effect of pier shape. In the last analysis recourse must be had to experiment, although Eq. (7–33) still forms the basis of a useful theoretical exercise. As such, it is elaborated in Probs. 7.10 and 7.11 for the reader who wishes to follow it through in detail.

The classic experimental work is that of Yarnell [13,14] who obtained results for a number of different pier types and Froude numbers, and for values of $\alpha = 1 - \sigma$, the ratio pier-width-to-span, equal to 11.7, 23.3, 35.0, and 50.0 percent. These values represent contractions much more severe than those in many modern bridges, in which pier widths as low as 5 or 6 percent of the span are quite common. Yarnell's results, however, showed such a systematic trend with varying σ that interpolation within all intervals, including the interval $0 < (1 - \sigma) < 0.117$, would seem to be entirely reliable.

In presenting any results on this problem it is desirable to take the independent variables from the downstream section 3 rather than from section 1, since conditions at section 3 will be initially known by the designer (e.g., by plotting a backwater profile from some control further downstream). We should then expect the ratio $\Delta y/y_3$ to be a function of σ, of the pier shape, and of the Froude number Fr_3. This is in fact true of Yarnell's results, to which he fitted the following empirical equation

$$\frac{\Delta y}{y_3} = K\mathrm{Fr}_3{}^2(K + 5\mathrm{Fr}_3{}^2 - 0.6)(\alpha + 15\alpha^4) \qquad (7\text{–}34)$$

where K characterizes the pier shape according to the following table:

TABLE 7-1

Pier Shape	K
Semicircular nose and tail	0.9
Lens-shaped nose and tail	0.9
Twin-cylinder piers with connecting diaphragm	0.95
Twin-cylinder piers without diaphragm	1.05
90° triangular nose and tail	1.05
Square nose and tail	1.25

A "lens-shaped" nose is formed from two circular curves each of radius equal to twice the pier width, and each tangential to a pier face.

The above coefficients relate to piers of length four times the width. Further tests were made on piers of length-width ratio 7 and 13, and the backwater Δy was found to be from 83 to 96 percent of the values given by Eq. (7–34) and Table 7-1. Equation (7–34) is therefore recommended as giving a conservative figure for all normal values of length-width ratio.

The Limiting Contraction Ratio

All of the above results apply to the case where σ is not small enough to choke the flow and set up critical conditions in the space between the piers. The backwater Δy as given above is therefore simply due to energy dissipation between sections 1 and 3. If however the flow does become choked there can be a substantial increase in Δy arising from the need for a higher local value of specific energy, as discussed in Secs. 2.2. and 4.6.

The first problem is to determine the limiting value of σ, which we should expect to be a function of the Froude number. This limiting value will normally represent quite a severe contraction, so it will be assumed that the velocity is uniform across section 2, Fig. 7-22, and that the equations $E_1 = E_2$, $M_2 = M_3$, will be at least approximately true. Assuming critical flow at section 2, the first of these leads (Prob. 7.12) to the equation

$$\sigma^2 = \frac{27\mathrm{Fr}_1{}^2}{(2 + \mathrm{Fr}_1{}^2)^3} \tag{7–35}$$

and the second leads (Prob. 7.13) to the equation

$$\sigma = \frac{(2+1/\sigma)^3 \mathrm{Fr}_3{}^4}{(1 + 2\mathrm{Fr}_3{}^2)^3} \tag{7–36}$$

Equation (7–35) was used by Yarnell to distinguish between "Class A" or unchoked flow, and "Class B" or choked flow. In fact, Eq. (7–36) is more likely to be correct because it does not depend on any assumptions about energy conservation, and in any case it is more useful because its independent variables derive from section 3, and are therefore known initially.

When the flow is of Class B there will be a hydraulic jump downstream, as shown in Fig. 7-22. This need not affect the computation of the backwater Δy, which simply proceeds upstream from the critical flow section to section 1, with a suitable allowance for energy loss between the two sections. This energy loss was expressed by Yarnell as a multiple of $v_1{}^2/2g$, the ratio between the two depending on the contraction ratio α and on the pier shape; however his results can also be expressed in this way:

$$E_1 - E_2 = C_L \frac{v_2{}^2}{2g} \tag{7-37}$$

the coefficient C_L depending on the pier shape alone. For square-ended piers $C_L = 0.35$ and for rounded ends $C_L = 0.18$. These results have already been referred to in Sec. 7.2 in a discussion on contractions, of which bridge piers are a special case.

Given the amount of energy loss the upstream depth y_1 can be calculated, and from it the backwater $\Delta y = y_1 - y_3$. The above values of energy loss are for a pier length-width ratio of 4; longer piers increase the loss, giving rise to an increase in Δy over the value for the 4:1 pier. The amount of the increase in Δy was approximately 5 percent and 10 percent for 7:1 and 13:1 piers respectively, with rounded ends. For the square-ended piers the backwater was slightly less for longer piers than for the 4:1 pier.

Piers Not Parallel to the Flow

In normal design practice care is taken to align the piers with the direction of flood flow. If doubts are felt about what this direction is, allowance should be made for possible misalignment. Yarnell made a few tests on 4:1 piers at angles of 10° and 20° to the flow; he found that a 10° angle of skew gave values of Δy very little greater than unskewed piers, but that a 20° angle produced values of Δy about 2.3 times as great as did unskewed piers. Since a 4:1 pier skewed at 20° has about 2.3 times the "frontal area," i.e., the projected area normal to the flow, of an unskewed pier, it appears that for angles up to 20° the effect of skew may be conservatively allowed for by assuming that the effective width of the pier is equal to its width projected normal to the flow.

7.6   Lateral Inflow and Outflow

This phenomenon, which occurs in a variety of circumstances, provides useful exercises in the proper use of the energy and momentum equations, although it can lead to problems of some difficulty. Consider the cases in which flow is being continuously added or subtracted along the length of the channel. The former of these is exemplified by the "side-channel spillway" used in a few major works, notably the Hoover Dam; in this case an open spillway discharges into a channel running along the foot of the spillway, and carrying the flow away in a direction generally parallel to the spillway crest, as shown in Fig. 7-23. The latter is exemplified by the "side weir" illustrated in Fig. 7-24, which is sometimes used to withdraw surplus flow from channels or drains. Another example, shown in Fig. 7-25, is a channel with a slotted bed through which water drops from this channel to another.

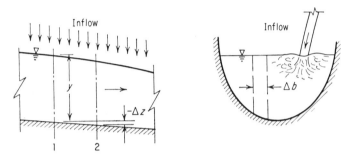

Figure 7-23.   Definition Sketch for the Lateral Inflow Problem

In the side-channel of Fig. 7-23 there may be considerable energy dissipation as the arriving flow mixes with the water already in the channel. The energy equation is not therefore available in a usable form. On the other hand the arriving flow has no momentum in the direction of the channel flow, and the momentum equation is therefore available in a simple usable form. If the channel bed is horizontal and bed resistance is neglected, then the momentum function

$$M = \frac{Q^2}{gA} + A\bar{y} \qquad (7\text{–}38)$$

remains constant along the channel, i.e.,

$$\frac{dM}{dx} = 0 \qquad (7\text{–}39)$$

and *this conclusion is unaffected by the variation of Q with x,* which merely introduces the term dQ/dx when dM/dx is evaluated; beyond this it does not complicate the issue any further.

If bed slope and bed resistance exist, further terms must be added to Eq. (7–39). Now M is equal to force divided by specific weight γ; we consider the forces acting on the section 1–2 in Fig. 7-23. The slope adds a force $-\Sigma(\gamma y \Delta z)\Delta b = -\gamma A \Delta z$ acting downstream, and the resistance adds a force

$\tau_0 P \Delta x$ acting upstream, to those already acting. We can therefore write

$$\gamma \Delta M + \gamma A \Delta z + \tau_0 P \Delta x = 0$$

i.e.,
$$\frac{dM}{dx} + A\frac{dz}{dx} + \frac{\tau_0 P}{\gamma} = 0$$

or
$$\frac{dM}{dx} = A(S_0 - S_f) \tag{7-40}$$

since $S_0 = -dz/dx$ and $\tau_0 = \gamma R S_f$. The terms of Eqs. (7–38) and (7–40) are as defined in Chaps. 3 and 4.

When flow is being drawn from the channel as in Figs. 7-24 and 7-25, energy losses in the overflow process will be small, and in the absence of bed slope or resistance, the specific energy may be assumed constant along the channel. (We recall that the Bernoulli expression represents energy *per unit weight*, a property of the fluid which remains unaltered by flow division.) On the other hand the momentum equation will not be readily applicable, for the overflow will have a momentum component along the channel. We should therefore be able to write

$$\frac{dE}{dx} = 0 \tag{7-41}$$

or if there is bed slope and resistance, this equation becomes

$$\frac{dE}{dx} = S_0 - S_f \tag{7-42}$$

which is a restatement of Eq. (4–30). For the reason given above the division of the flow does not alter the form of this equation.

The degree of difficulty in solving these equations depends very much on the particular conditions of each problem. In the inflow case Q will be a known (usually linear) function of x, and this simplifies the analysis. In fact if Q varies directly with x, and bed slope and resistance are negligible, Eq. (7–40) is explicitly soluble for the case of a rectangular channel, provided that there is a downstream control determining the downstream depth, as there must always be in this case. Details are left as an exercise for the reader (Prob. 7.14).

When S_0 and S_f are appreciable, Eq. (7–40) will yield to numerical methods similar to those of Chap. 5, although if y is to be determined from previously set values of x and Δx it is necessary to resort to a trial method such as the one given in Sec. 5.4 (Prob. 7.15). However there is no real difficulty provided that there is a readily identifiable control from which to start the calculation. This is always true when the channel slope is mild, but when it is steep enough there will be a critical section at some unknown point, which must be determined before the computation begins.

The criterion by which this critical section is identified can be developed from Eqs. (7–38) and (7–40). From these equations we obtain, noting that $d(A\bar{y})/dy = A$,

$$\frac{dM}{dx} = \frac{2Q}{gA}\frac{dQ}{dx} - \frac{Q^2}{gA^2}\frac{dA}{dx} + A\frac{dy}{dx}$$

$$= A(S_0 - S_f)$$

whence
$$\frac{dy}{dx} = \frac{S_0 - S_f - \dfrac{2Q}{gA^2}\dfrac{dQ}{dx}}{1 - \mathrm{Fr}^2} \qquad (7\text{–}43)$$

noting that $dA/dy = B$, the surface width, and that $\mathrm{Fr}^2 = Q^2 B/gA^3$ (Sec. 2.7). As in the argument of Sec. 4.4, critical flow occurs or $dy/dx = 0$ when the numerator of the fraction in Eq. (7–43) is zero. We assume that dQ/dx is constant (as it is in most cases of practical interest) and we write it as $Q_x = Q/x$. The above condition then becomes

$$S_0 = S_f + \frac{2Q^2}{gA^2 x}$$

$$= \frac{v^2 P}{C^2 A} + \frac{2Q^2}{gA^2 x}$$

$$= \mathrm{Fr}^2\left(\frac{gP}{C^2 B} + \frac{2A}{Bx}\right) \qquad (7\text{–}44)$$

where, as in Chaps. 2 and 4, B is the surface width and C is the Chézy coefficient. We now exclude from consideration the case $dy/dx = 0$, and assume therefore that if Eq. (7–44) is satisfied, $dy/dx \neq 0$ and $\mathrm{Fr} = 1$. Elimination of A between Eq. (7–44) and the equation $\mathrm{Fr}^2 = Q^2 B/gA^3 = 1$ then leads to the result

$$S_0 = \frac{gP}{C^2 B} + \frac{2}{B}\left(\frac{Q_x^2 B}{gx}\right)^{1/3} \qquad (7\text{–}45)$$

whence
$$x = \frac{8Q_x^2}{gB^2\left(S_0 - \dfrac{gP}{C^2 B}\right)^3} \qquad (7\text{–}46)$$

which gives the location of the critical-flow section; such a section will exist only if the value of x given by Eq. (7–46) is not greater than L, the total length of the channel, i.e., if S_0 is greater than the value obtained by substituting $x = L$ into Eq. (7–45). If this condition is not satisfied, flow will be subcritical along the whole channel and subject to a control at the downstream end. (This control may, of course, take the form of an overfall beyond the region of inflow; near such an overfall flow will be critical.) If a critical-flow

section does exist within the channel, flow will of course be subcritical upstream of this section and supercritical downstream. If, in this situation,
there is a control at the downstream end of the channel, a hydraulic jump will
occur if the controlled depth is large enough, and may even move upstream
and drown out the critical section (cf. Fig. 5-14).

The development leading to Eqs. (7–44) through (7–46) is a statement in
general form of an argument originally due to Keulegan [15]. When the
channel is of finite width or C is not constant, Eqs. (7–45) and (7–46) must
be solved by trial, as in Prob. 7.25. Explicit solutions are possible, and were
given by Keulegan, for the special case of a wide rectangular channel with
C constant. In this case the alternative Eqs. (7–44) and (7–45) become,
respectively,

$$S_0 = \frac{g}{C^2} + \frac{2y_c}{x} \tag{7-47}$$

and

$$S_0 = \frac{g}{C^2} + 2\left(\frac{q_x^2}{gx}\right)^{1/3} \tag{7-48}$$

where $q_x = Q_x/B$. Similarly, Eq. (7–46) becomes

$$x = \frac{8q_x^2}{g\left(S_0 - \frac{g}{C^2}\right)^3} \tag{7-49}$$

No arguments have been presented justifying the exclusion of the possibility
that $dy/dx = 0$ and $Fr \neq 1$ if Eq. (7–44) is satisfied. In fact no such arguments
appear to be possible, but the experiments of Beij [16] on short (30 ft long)
steep, $(0.005 < S_0 < 0.03)$ roof gutters indicated clearly that when Eq. (7–44)
was satisfied the flow was critical and dy/dx was positive, as it was over the
whole length of the channel in every experiment. It does seem physically very
likely that if the channel is steep enough for Eq. (7–44) to be satisfied, dy/dx
will never be zero for constantly increasing q; nevertheless it remains an
interesting theoretical possibility that dy/dx may be zero in some special cases.

There are indications that the turbulent mixing of the lateral inflow with
the main body of the channel flow makes the flow resistance greater than
normal; e.g., the experiments of Beij on smooth metal gutters indicated that
the equivalent Darcy f was given not by the Blasius law but by the equation

$$f = \frac{1280}{Re} \tag{7-50}$$

where $Re = 4vR/v$, as in Chap. 4.

Li [17] has made an interesting study of the case in which resistance is
negligible, approximated in practice by short channels of large cross section,
e.g. wash-water troughs. In this case, the retention of the slope S_0 in the
equation of motion means that it is no longer directly integrable, as in Prob.

7.14, and numerical methods must be used. Solutions have been fully worked out by Li [17] and the reader is referred to the original paper for details. The classification of the flow was found to depend on relationships between Fr_L and G, where $G = S_0 L/y_L$, and the subscript L indicates parameters fixed by a downstream control at the downstream end of the channel. The criterion for the drowning-out of the critical section by the controlled downstream depth is, for a channel of rectangular section

$$G < 1 + Fr_L, \text{ approximately} \tag{7-51}$$

If the critical section is not drowned, its location is given by Eq. (7–46), setting $C = \infty$.

We turn now to the lateral outflow case. It can easily be shown (Prob. 7.16) that the expansion of Eq. (7–42) leads to the result

$$\frac{dy}{dx} = \frac{S_0 - S_f - \dfrac{Q}{gA^2}\dfrac{dQ}{dx}}{1 - Fr^2} \tag{7-52}$$

which is similar to Eq. (7–43) except for the coefficient of the last term in the numerator. In the outflow case of Fig. 7-24, to which Eq. (7–52) applies, Q is not a known function of x because the magnitude of the overflow is itself dependent on the depth of flow in the main channel. Reliable data are not yet available on the outflow-head relationship for side weirs, although the present evidence is that an equation of the usual form

$$q = CH^{3/2} \tag{7-53}$$

is at least approximately true, H being the head over the weir crest, and the coefficient C diminishing as the Froude number in the main channel increases and the longitudinal component of flow over the weir becomes more pronounced. Ackers' [18] analysis of Coleman and Smith's [19] experiments indicates that $C = 4.10$. This is consistent with an increase in Francis' coefficient of 3.33 due to the normal drawdown by a factor of 0.87 at the weir crest, since $4.10 = 3.33/(0.87)^{3/2}$. (In the present case H must be measured at or near the weir crest, not some distance upstream as with the normal weir.) However, Frazer's [20] experiments indicate that $C = 4.10$ only when the Froude number in the channel is very low; his results indicated that

$$C = 4.15 - \frac{1.81 y_c}{y} - \frac{0.14 y_c}{L} \tag{7-54}$$

where L is the length of the weir and no allowance is made for end contractions. Ackers' result of 4.10 implied an allowance of $0.2H$ for contraction at each end of the weir, i.e., an effective length of $L - 0.4H$. Despite the reservations expressed about this practice in Sec. 6.2, it seems allowable in this case because the ratio L/H is usually high.

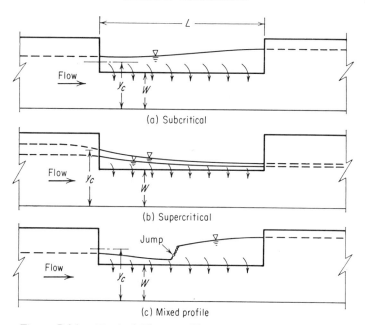

Figure 7-24.   *Typical Flow Profiles at Side-Discharge Weirs*

Figure 7-24 indicates the three main types of flow that are possible—wholly subcritical, wholly supercritical, or with the two regimes present and separated by a hydraulic jump. It is easily verified (Prob. 7.17) that the water surface in the channel should rise or fall in the manner shown in Fig. 7-24. Solutions of Eq. (7–52) are made more difficult by lack of knowledge of how Q varies with x; on the other hand, since side weirs are usually not very long, it may be assumed that $S_0 = S_f = 0$, i.e., that E is a constant over the length of the weir. By assuming also a rectangular channel, and a constant C in the weir equation, Eq. (7–53), de Marchi [21] obtained the following explicit solution.

The discharge over the weir per unit length of weir is equal to

$$-\frac{dQ}{dx} = C_1\sqrt{2g}(y - W)^{3/2} \qquad (7\text{–}55)$$

where Q is the discharge in the main channel, and C_1 is $\sqrt{2g}$ times smaller than the coefficient C in Eq. (7–53). At any section Q is given by

$$Q = by\sqrt{2g(E - y)} \qquad (7\text{–}56)$$

where b is the width. Substituting these results into Eq. (7–52), and setting $S_0 = S_f = 0$, leads to the equation

$$\frac{dy}{dx} = \frac{2C_1}{b}\frac{\sqrt{(E - y)(y - W)^3}}{(3y - 2E)} \qquad (7\text{–}57)$$

which was integrated by de Marchi, giving the result

$$\frac{xC_1}{b} = \frac{2E - 3W}{E - W}\sqrt{\frac{E - y}{y - W}} - 3\sin^{-1}\sqrt{\frac{E - y}{y - W}} + \text{constant} \qquad (7\text{--}58)$$

In applying this equation difficulties arise in the fixing of the boundary conditions. For example, in the subcritical case calculations cannot proceed initially from the downstream section because Q is not known at that section; a trial procedure is needed (Probs. 7.18 and 7.19). It is clear from Eq. (7–52) that if $S_0 = S_f = 0$, flow cannot be critical anywhere along the weir, since $dQ/dx \neq 0$. However, critical depth could occur just at the beginning of the weir (but not of course at the end) and when this happens a "drawdown" like that of the free overfall is usually observed upstream of the weir, as in Fig. 7-24b. This drawdown is evidence of non-hydrostatic pressure distribution, which raises doubts about Eq. (7–56) and de Marchi's analysis generally.

These complications raise problems which are not yet fully resolved; for details the reader is referred to their recent discussion in the symposium of which Refs. [18] and [20] form a part.

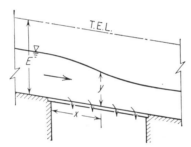

Figure 7-25.   *Withdrawal of Flow through a Bottom Rack*

When the outflow takes place through openings in the bed, as in Fig. 7-25, the outflow velocity head is equal to the specific energy when the openings are uninterrupted and longitudinal, and to the overlying water depth when the openings are in a perforated screen, for in the latter case the change in flow direction brings about an energy loss equal to the velocity head in the channel. The discharge coefficient varies, in the former case from 0.44 to 0.50, and in the latter case from 0.75 to 0.80, as the bed slope varies from 0.2 to zero. These conclusions are due to Mostkow [22].

The longitudinal profiles which can occur are substantially similar to those in Fig. 7-24. Equations analogous to de Marchi's can be deduced for the two cases described above; their derivation is left as an exercise for the reader (Probs. 7.20 and 7.21). In the application of these equations, as of de Marchi's equation, the boundary conditions can present difficulties (Probs. 7.30 and 7.31).

References

1. G. Formica. "Esperienze preliminari sulle perdite di carico nei canali dovute a cambiamenti di sezione" (Preliminary tests on head losses in channels due to cross-sectional changes), *L'Energia elletrica*, Milan, vol. 32, no. 7 (July 1955), p. 554.

2. A. T. Ippen. "Channel Transitions and Controls," in H. Rouse (ed.), *Engineering Hydraulics* (New York: John Wiley & Sons, Inc., 1950), Chap. 8.

3. H. Rouse, B. V. Bhoota, and E.-Y. Hsu. "Design of Channel Expansions," *Trans. Am. Soc. Civil Engrs.*, vol. 116 (1951), p. 347.

4. A. T. Ippen and R. T. Knapp. "Experimental Investigations of Flow in Curved Channels," U. S. Engineer Office, Los Angeles (1938).

5. R. T. Knapp. "Design of Channel Curves for Supercritical Flow," *Trans. Am. Soc. Civil Engrs.*, vol. 116 (1951), p. 296.

6. R. T. Knapp and A. T. Ippen. "Curvilinear Flow of Liquids with Free Surface at Velocities above that of Wave Propagation," *Proceedings, 5th International Congress of Applied Mechanics* (New York: John Wiley & Sons, Inc., 1939), p. 531.

7. C. E. Mockmore. "Flow Round Bends in Stable Channels," *Trans. Am. Soc. Civil Engrs.*, vol. 109 (1944), p. 593.

8. A. Shukry. "Flow Round Bends in an Open Flume," *Trans. Am. Soc. Civil Engrs.*, vol. 115 (1950), p. 751.

9. J. Allen and Sek Por Chee. "The Resistance to the Flow of Water Round a Smooth Circular Bend in an Open Channel," *Proc. Inst. C.E.* (London), vol. 23 (November 1962), p. 423.

10. F. W. Blaisdell. "Hood Inlet for Closed Conduit Spillways," *Proc. Am. Soc. Civil Engrs.*, vol. 86, no. HY5 (May 1960), p. 7.

11. F. T. Mavis. "The Hydraulics of Culverts," *Pennsylvania State College Eng. Exp. Stn. Bulletin 56* (October 1, 1942).

12. F. A. Nagler. "Obstruction of Bridge Piers to the Flow of Water," *Trans. Am. Soc. Civil Engrs.*, vol. 82 (1918), p. 334.

13. D. L. Yarnell. "Pile Trestles as Channel Obstructions," U. S. Department of Agriculture, Tech. Bull. no. 429 (July 1934).

14. D. L. Yarnell. "Bridge Piers as Channel Obstructions," U. S. Department of Agriculture, Tech. Bull. no. 442 (November 1934).

15. G. H. Keulegan. "Determination of Critical Depth in Spatially Variable Flow," *Proceedings 2nd Midwestern Conference on Fluid Mechanics*, Ohio State University (1952).

16. K. H. Beij. "Flow in Roof Gutters," *J. Res. U. S. Nat. Bur. Stand.*, vol. 12 (1934), p. 193.

17. W.-H. Li. "Open Channels with Non-Uniform Discharge," *Trans. Am. Soc. Civil Engrs.*, vol. 120 (1955), p. 255.

18. P. Ackers. "A Theoretical Consideration of Side Weirs as Stormwater Overflows," *Proc. Inst. C.E.* (London), vol. 6 (February 1957), p. 250.

19. G. S. Coleman and D. Smith, "The Discharging Capacity of Side Weirs," *Proc. Inst. C.E.* (London), Selected Engineering Paper No. 6 (1923).

20. W. Frazer. "The Behaviour of Side Weirs in Prismatic Rectangular Channels," *Proc. Inst. C.E.* (London), vol. 6 (February 1957), p. 305.

21. G. de Marchi. "Saggio di teoria del funzionamento degli stramazzi laterali" (Essay on the Performance of Lateral Weirs), *L'Energia elletrica*, Milan, vol. 11, no. 11 (November 1934), p. 849.

22. M. A. Mostkow. "Sur le calcul des grilles de prise d'eau," *La Houille Blanche*, vol. 12, no. 4 (September 1957), p. 570.

23. F. A. Engelund and J. Munch-Petersen. "Steady Flow in Contracted and Expanded Rectangular Channels," *La Houille Blanche*, vol. 8, no. 4 (August–September 1953), p. 464.

24. F. A. Engelund. Basic Research Progress Report No. 6, Tech. Univ. of Denmark, Copenhagen (June 1964).

25. A. J. M. Harrison. D.S.I.R. Hydraulics Research Station, Wallingford, Berks., personal communication.

Problems

7.1. For the abrupt expansion in a rectangular channel shown in Fig. 7-1, prove from the momentum equation that

$$\mathrm{Fr}_1{}^2 + \frac{r}{2} = \frac{\mathrm{Fr}_1{}^2}{rs} + \frac{rs^2}{2}$$

and hence that

$$\Delta E = E_1 - E_3 = \frac{v_1{}^2}{2g}\left(1 + \frac{2}{\mathrm{Fr}_1{}^2} - \frac{1}{r^2 s^2} - \frac{2s}{\mathrm{Fr}_1{}^2}\right)$$

where $r = b_2/b_1$ and $s = y_3/y_1$.

7.2. From the first result in Prob. 7.1, prove that if Fr_1 is small enough for $\mathrm{Fr}_1{}^4$ and higher powers to be neglected, then

$$s = 1 + \frac{\mathrm{Fr}_1{}^2(r-1)}{r^2}$$

Hence prove Eq. (7–1).

7.3. Water flows at a velocity of 5 ft/sec and a depth of 3 ft in a rectangular channel 10 ft wide. There is then a contraction to a width of 7 ft; by how much should the bed be lowered through the contraction so that the downstream Froude number will not exceed 0.8? Assume that the transition is rounded, so that the head loss is 0.04 times the downstream velocity head.

7.4. The situation is the same as in Prob. 4.24, except that for convenience in construction the rectangular chute is to be made 20 ft wide. It is still required that there be no M_1 or M_2 curve in the trapezoidal channel; determine the amount by which the channel bed must rise through the transition from the trapezoidal to the rectangular section. The head loss in the transition can be assumed to be negligible.

7.5. Experiments by Ippen yielded the following figures for the variation of the depth y, expressed as a multiple of the initial depth y_1, for supercritical flow along a boundary consisting of two reversed circular arcs. $\mathrm{Fr}_1 = 4$.

Converging boundary:

θ	0°	2°	4°	6°	8°	10°	12°	14°	16°
y/y_1	1.09	1.13	1.18	1.35	1.50	1.82	2.09	2.28	2.33

Diverging boundary:

θ	16°	14°	12°	10°	8°	6°	4°	2°	0°
y/y_1	2.33	2.40	2.32	2.18	1.89	1.59	1.45	1.25	1.22

On the same graph, plot the above results, together with two curves, obtained (a) from Eq. (7–12) or Fig. 7-8; (b) by proving the following equation on the assumption that the velocity is constant along the wall

$$\theta = \beta - \beta_1 + \sin(\beta - \beta_1)\cos(\beta + \beta_1)$$

and using it with Eq. (7–5) to calculate simultaneous values of y/y_1 and θ.

7.6.   Water flows at a depth y_1 in a rectangular channel immediately upstream of a width contraction, within which the flow is critical. If r is the ratio of contracted to upstream width, and $y' = y_1/y_c$, where y_c is the critical depth appropriate to the *contracted* section, prove that

$$y' + \frac{r^2}{2y'^2} = \frac{3}{2}$$

if there is no energy loss. From this result and the hydraulic jump equation, derive the equations governing the two "choking" criteria of Fig. 7-11, and verify by numerical calculation two points on each of the two curves in that figure.

7.7.   For the straight-sidewall contraction shown in Fig. 7-10a, and an upstream Froude number equal to 4, calculate a number of simultaneous values of θ_1 and contraction ratio b_3/b_1. Apply Eqs. (7–6) through (7–10) to the following scheme of calculation:

Choose a value of β_1, hence calculate y_2/y_1 and θ_1; hence v_1/v_2 and Fr_2. Noting that the equations can be applied between zones 2 and 3 in the same way as between zones 1 and 2, calculate β_2 by this trial process: take trial values of y_3/y_2 and calculate β_2 and θ_2 therefrom until a value of θ_2 is obtained equal to the value of θ_1 calculated above. From the values of β_1, β_2, and $\theta_1 = \theta_2$ now obtained determine b_3/b_1 by proving the following result:

$$\frac{b_1 - b_3}{b_3} = \frac{\sin\theta_1 \sin(\beta_1 + \beta_2 - \theta_1)}{\sin(\beta_1 - \theta_1)\sin(\beta_2 - \theta_1)}$$

Plot b_3/b_1 against θ_1, and verify that θ_1 tends to zero when b_3/b_1 reaches the limiting value indicated by Fig. 7-11.

7.8.   Supercritical flow of depth y_1 in a rectangular channel approaches a smooth upward step, height Δz, in the channel bed. Derive the equations governing the two "choking" criteria analogous to those of the width contraction shown in Fig. 7-11, and plot the appropriate two curves of $\Delta z/y_1$ vs. Fr_1.

7.9. Show from Eq. (7–26) that for the banked bend developed in Sec. 7.3 the bed level z_i at the inside wall is equal to

$$z_i = H_0 - y_0 \left(\frac{r_i}{r_o} + \frac{\mathrm{Fr}_0^2}{2} \frac{r_o^2}{r_i^2} \right)$$

where r_i is the radius of the inner wall and Fr_0 is the upstream Froude number. Hence show that if $\mathrm{Fr}_0 < 1$ the flow will be just critical at the inside wall if

$$\frac{3}{2} \mathrm{Fr}_0^{2/3} = \frac{r_i}{r_o} + \frac{\mathrm{Fr}_0^2}{2} \frac{r_o^2}{r_i^2}$$

For the case $\mathrm{Fr}_0 = 0.3$, find the limiting values of r_o/r_i and r_c/b which make the flow critical at the inside wall; r_c is the radius of the channel centerline.

7.10. Show that the momentum Eq. (7–33) may be written as

$$\frac{y_1^2}{2} + \frac{q^2(1-k)}{gy_1} = \frac{y_3^2}{2} + \frac{q^2}{gy_3}$$

where $k = C_D(1-\sigma)/2$, and C_D is the drag coefficient on the pier (based on the upstream velocity and depth) and $\sigma = b_2/b_1$. Hence, setting $r = y_1/y_3$, and using the relation $r^3 = \mathrm{Fr}_3^2/\mathrm{Fr}_1^2$, show that

$$\mathrm{Fr}_3^2 = \frac{r(r^2 - 1)}{2(r + k - 1)}$$

7.11. The piers of a bridge are 2 ft 6 in. thick and are placed at a center distance of 40 ft. A short distance downstream the river is flowing with a velocity of 10 ft/sec at a depth of 8 ft. Find the increase in depth brought about by the bridge piers for the following cases:

Using the result of Prob. 7.10, with C_D assumed equal to (a) 2.0, (b) 2.5.

Using Yarnell's equation (7–34), with K assumed equal to (c) 0.9, (d) 1.05, (e) 1.25.

7.12. Assuming that in Fig. 7-22 the flow at section 2 is critical, and that $E_1 = E_2$, prove Eq. (7–35).

7.13. Assuming that in Fig. 7-22 the flow at section 2 is critical, and that $M_2 = M_3$, prove Eq. (7–36).

7.14. Lateral inflow is taking place into a rectangular channel of length L at a uniform rate per unit length of channel. The channel is closed at the upstream end ($x = 0$) and the bed slope and the resistance are both zero; prove from Eq. (7–40) that

$$\frac{dy}{dx} = \frac{-2qq_x}{gy^2(1 - \mathrm{Fr}^2)}$$

or

$$\frac{2q_x^2 x}{y} dx + \left(gy - \frac{q_x^2 x^2}{y^2} \right) dy = 0$$

where $q_x = q/x = \mathrm{constant}$. Note that this equation is of the form $M dx + N dy = 0$, which is directly integrable if $\partial N/\partial x = \partial M/\partial y$; hence prove that the solution of this equation is

$$\left(\frac{x}{L}\right)^2 = \left(1 + \frac{1}{2\mathrm{Fr}_L{}^2}\right)\frac{y}{y_L} - \frac{1}{2\mathrm{Fr}_L{}^2}\left(\frac{y}{y_L}\right)^3$$

where y_L and Fr_L are the values obtaining at the downstream end of the channel, where $x = L$.

Hence show that if the channel terminates in a critical section at its downstream end, the depth at the upstream end will be equal to $y_L\sqrt{3}$.

7.15. A paved parking lot 200 ft square has a cross slope in one direction only (with a total fall of 6 in.) into a gutter 2 ft wide and of rectangular section running along one side of the lot. The bed of this gutter falls 6 in. in its whole length before ending in a free overfall.

If rain falls on the parking lot at the steady rate of 3 in./hr, find the maximum depth of water lying on the surface of the lot, and the minimum allowable depth of the bed of the gutter below the ground surface if the gutter is not to overflow.

For the calculation of S_f, use Eq. (7–50) for the flow over the ground surface (with $\nu = 1.2 \times 10^{-5}$ ft²/sec) and the Manning equation, with $n = 0.012$, for flow in the gutter. In both cases compute the flow profile by trial from Eq. (7–43); it will be found simplest to proceed from chosen values of x and Δx, and to determine y by inserting trial values into the right-hand member of Eq. (7–43) until the correct value of Δy is obtained.

7.16. Prove Eq. (7–52).

7.17. Show that in flow over a horizontal bed with constant specific energy, and discharge decreasing in the direction of flow, the depth increases downstream in subcritical flow and decreases downstream in supercritical flow. Hence verify the form of the profiles shown in Fig. 7-24.

7.18. Considering the whole length of a channel in which a side-discharge weir has been placed, sketch all the upstream and downstream profiles that can occur in association with the profiles shown in Fig. 7-24, assuming that no other controls are present in the channel. Show that the upstream and downstream slopes required are as given in this table:

Profile shown in:	Upstream slope	Downstream slope
Fig. 7-24a	Mild or Steep	Mild
7-24b	Mild or Steep	Mild or Steep
7-24c	Steep	Mild

7.19. For each of the flow situations shown in Fig. 7-24, devise trial procedures by which Eqs. (7–56) and (7–58) can be used to determine the flow over the side weir. Assume that the discharge in the upstream channel is initially known, that C_1 in Eq. (7–55) is equal to $4.1/\sqrt{2g}$, and that the effective length of the weir is not appreciably reduced by end contractions. (Take trial values of downstream discharge until the upstream discharge calculated from Eqs. (7–56) and (7–58) is equal to the known value.)

A long channel of rectangular section 6 ft wide is laid on a slope of 0.001

and is lined with concrete, $n = 0.014$. Two side weirs are to be installed, of equal length and on opposite sides of the channel. They are to come into operation when the flow reaches 20 cusecs, and are to discharge 5 cusecs when the upstream flow is 30 cusecs. Determine the length of the weirs, and the height of their crests above the channel bed. If the length is now made equal to 5 ft, the crest height remaining the same, what will be the discharge over the weirs for the same upstream flow of 30 cusecs?

7.20. Water is being withdrawn from a channel through uninterrupted longitudinal openings in the bed, as in Fig. 7-25, the outflow velocity head being proportional to the specific energy E. Show that the equation of the flow profile, corresponding to de Marchi's equation (7–58), is

$$x = -\frac{y}{rC_d}\sqrt{1 - \frac{y}{E}} + \text{constant}$$

where r is the ratio of the area of openings to the total bed area, and C_d is the coefficient of discharge for the flow through the openings.

7.21. The situation is as in Prob. 7.20, but this time the openings are in a perforated screen, so that the outflow velocity head is proportional to the depth y. Show that the equation of the flow profile may be expressed in either of the alternative forms

$$x = \frac{E}{rC_d}\left[\frac{1}{4}\sin^{-1}\left(1 - \frac{2y}{E}\right) - \frac{3}{2}\sqrt{\frac{y}{E}\left(1 - \frac{y}{E}\right)}\right] + \text{constant}$$

$$= \frac{E}{rC_d}\left[\frac{1}{2}\cos^{-1}\sqrt{\frac{y}{E}} - \frac{3}{2}\sqrt{\frac{y}{E}\left(1 - \frac{y}{E}\right)}\right] + \text{constant}$$

the two constants of integration differing by $\pi E/8rC_d$, and r and C_d being defined as in Prob. 7.20.

7.22. The invert of a box culvert, 4 ft square in section, has a uniform slope of 1 : 25 in the first 500 ft downstream from the culvert entrance. There is then a change of grade and over the next 5,000 ft of length the invert has a slope of 1 : 2,500; the culvert then discharges into the tailwater pool, whose surface level is 6 ft above the invert level at the downstream end of the culvert. The culvert is made of Class II concrete ($k_s = 0.005$ ft, as in Table 4–1). Find the discharge in the culvert when the headpool surface is level with the soffit at the culvert entrance, assuming that the entry is (a) rounded, (b) square-edged. Use the methods of Sec. 4.2 to determine the head loss coefficient (either Darcy f or Manning n) and verify that the flow is in fact "fully rough."

7.23. In the situation of Prob. 7.22, determine the headpool level at which the flow changes from Case 1 to Case 2 of Fig. 7-18b.

7.24. If the culvert of Probs. 7.22 and 7.23 were laid on a uniform grade throughout, between the same upstream and downstream invert levels, would this make any difference to the magnitude of the headpool level at which the flow changes from Case 1 to Case 2?

7.25. A side-channel spillway has a crest 200 ft long. A collector channel runs across the foot of the spillway on a longitudinal slope of 1 : 20; it is concrete lined ($n = 0.014$), and has a trapezoidal section with a base width of 20 ft and side slopes of 1H : 1V. At the downstream end of the channel there is a transition to a very steep slope; find the depth at the upstream end when the spillway discharge is (a) 10,000 cusecs, (b) 1,000 cusecs, in each case uniformly distributed along the crest. [*Note*: In locating the critical section by trial, it is helpful to obtain first a rough trial value of x from Eq. (7–49), thence obtaining values of B and P for insertion in Eq. (7–46), yielding a second trial value. For plotting the flow profile, use the trial method of Prob. 7.15.]

7.26. A long concrete chute of rectangular section 30 ft wide is laid on a slope of 1 : 25 and carries a maximum discharge of 2,000 cusecs. Manning's $n = 0.014$. Find the maximum angle through which the flow could be deflected by a sharp change in direction like that of Fig. 7-17, and still remain supercritical. If the deflection angle had this maximum value, what would be the width of the downstream channel?

　　　　If, before making the change of direction, the channel were expanded to a width of 40 ft through a suitable transition, what would then be the maximum possible value of the deflection angle if the downstream flow is to remain supercritical?

　　　　In each of the above cases, assume now that the downstream flow need not be supercritical. Prove by numerical computation from Eqs. (7–9) and (7–10) that there is still a maximum allowable deflection angle $\Delta\theta$, and find its value in each case.

7.27. The runoff from a small rural catchment is to be carried beneath a proposed new highway by means of a culvert which will be 150 ft long and laid on a slope of $5°$ to the horizontal. The estimated maximum discharge is 500 cusecs. Determine the required culvert size for each of the following systems:

　　(a) a circular culvert of spun concrete with Blaisdell's hood inlet;
　　(b) as for (a), with a conventional square-edged entry;
　　(c) a box culvert of square section, with a conventional square-edged entry.

Base all the designs on an H/D ratio (Fig. 7-18) of 1.5.

7.28. A wide riverbed is of approximately rectangular section and has a Manning n of 0.028, and a slope of 14 ft per mile. At peak flood flow the discharge is 125 cusecs. The river is spanned by an old masonry bridge whose piers have rounded ends, are 4 ft thick, 16 ft long, and placed at 20-ft center distances.

　　　　Show that during peak flood flow there will be a hydraulic jump downstream of the bridge and estimate how far downstream it will be. Calculate also the backwater Δy produced by the bridge piers.

　　　　There is a proposal to replace the existing bridge with a new one whose piers would be 2 ft 6 in. thick, with lens-shaped ends, at center distances of 40 ft. Calculate the backwater Δy that would be produced by the new bridge.

7.29. In the situation of Prob. 7.19, what is the limiting length of side weir above which the flow cannot be subcritical along the whole length of the weir? Sketch the profile along the weir if the length exceeds this amount.

　　　　For the same channel as in Prob. 7.19, the upstream flow is 40 cusecs and the length of each weir is 15 ft. Determine the flow over the weirs and the form of the profile along the weir.

7.30. For the case of outflow through openings in the channel bed, as in Fig. 7-25, sketch the longitudinal profiles that will occur upstream and downstream of the outflow grille when the channel bed slope is mild for a great distance upstream and downstream. Outline a scheme of computation by which the outflow can be calculated when the upstream discharge is known.

7.31. A long concrete flume ($n = 0.014$) is laid on a slope of 0.001 and has a rectangular section 10 ft wide. As a means of dropping water from this flume to another one, an open grille 10 ft long consisting of longitudinal bars 3 in. wide with 1-in. gaps in between is laid in the bed of the flume. The discharge coefficient for the flow through the grille can be assumed equal to 0.5.

Find the flow through the grille when a discharge of 300 cusecs is admitted at the upstream end of the channel. For what value of upstream discharge will the whole of the discharge pass through the grille?

7.32. Using the results of Prob. C7.1, design a straight-sidewall contraction from a width of 50 ft to 35 ft in a rectangular concrete-lined flume, $n = 0.013$. The contraction is to operate as in Fig. 7-10a at a flow of 1,000 cusecs, the upstream flow being uniform and the upstream bed slope 0.025. What downstream bed slope would be required in order to make the flow uniform immediately downstream of the contraction?

7.33. The situation is the same as in Prob. 7.15, except that the outflow gutter falls 2 ft in its whole length before ending in a free overfall leading into a sump. Verify that there will be a critical-flow section somewhere along the channel provided that the overfall is quite free. If the water level in the sump now begins to rise, how far must it rise before the critical section is drowned out? Compare your result with that indicated by Li's approximate criterion, Eq. (7–51).

Computer Programs

(*Note*: Remember the usual practice of writing programs to accept all possible varieties of input data, with provision for listing this data at the output stage.)

C7.1. For the straight-sidewall contraction shown in Fig. 7-10a, write computer programs which will, for any value of upstream Froude number greater than unity, determine the limiting value of contraction ratio b_3/b_1, and
  (i) produce a complete range of simultaneous values of θ_1, b_3/b_1, and the downstream Froude number Fr_3; or
  (ii) for a given value of b_3/b_1, determine the correct value of θ_1 and Fr_3.

C7.2. Write a computer program to solve the lateral inflow problem of Sec. 7.6, assuming shallow flow over a wide surface with the resistance coefficient given by Eq. (7–50).

C7.3. Write a computer program to solve the lateral inflow problem of Sec. 7.6, assuming that the channel is of finite width and trapezoidal section, and that the flow resistance is governed by the Manning equation.

C7.4. Write computer programs to solve the side weir problem of Fig. 7-24, making the same assumptions as in Prob. 7.19. Write separate programs to deal with each of the three situations of Fig. 7-24, and write a fourth program which, given the discharge and the upstream and downstream slopes, will determine which of these situations is applicable and proceed accordingly.

C7.5. Write a computer program to solve the problem of outflow through openings in a channel bed of mild slope, incorporating the scheme of computation developed in Prob. 7.30.

C7.6. Operate the above programs to solve any of the main list of problems to which they are applicable.

Chapter **8**

Unsteady Flow

8.1 Introduction

Many flow phenomena of great importance to the engineer are unsteady in character, and cannot be reduced to steady flow by changing the viewpoint of the observer as in Sec. 3.5. A complete theory of unsteady flow is therefore required, and will be developed in this chapter. The equations of motion are not soluble in the most general case, but we shall see that explicit solutions are possible in certain cases which are physically very simple but are real enough to be of engineering importance. For the less simple cases, approximations and numerical methods can be developed which yield solutions of satisfactory accuracy.

Applications covered in this chapter include the propagation of tides into rivers and estuaries; changes in flow produced by the operation of artificial controls like sluice gates or hydroelectric machines which the canal may be feeding, or even by the collapse of a dam; and the behavior of oscillatory ocean waves, which must be recognized as a case of unsteady flow.

One further application is of such major importance as to warrant a separate chapter of its own. It is the theory of the movement of flood waves, which will be treated in the next chapter.

8.2 The Equations of Motion

We proceed to obtain equations describing unsteady open channel flow. The terms used are defined in the usual way, and are illustrated in Fig. 8-1.

Now it was shown in Sec. 4.2 that when the flow is steady the gradient, dH/dx, of the total energy line is equal in magnitude and opposite in sign to the "friction slope" $S_f = v^2/C^2R$. Indeed this statement was in a sense taken as the definition of S_f; however in the present context we have to recognize the two independent definitions:

$$S_f = \frac{\tau_0}{\gamma R} = \frac{v^2}{C^2 R} \tag{8-1}$$

285

and
$$\frac{\partial H}{\partial x} = \frac{\partial}{\partial x}\left(h + \frac{v^2}{2g}\right) \tag{8-2}$$

introducing partial derivative signs because the quantities involved may now vary with time as well as with x.

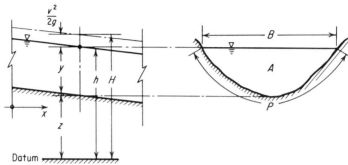

Figure 8-1.   Definition Sketch for the Equations of Motion

To allow for variation with time, we need only reproduce, with appropriate extensions, the argument leading up to Eq. (4–3). The acceleration term vdv/dx in that argument must now be replaced by the more general expression

$$a_x = \frac{dv}{dt} = v\frac{\partial v}{\partial x} + \frac{\partial v}{\partial t}$$

given by Eq. (1–9). The equation of motion therefore becomes

$$-\gamma A\Delta h - \tau_0 P\Delta x = \rho A\Delta x\left(v\frac{\partial v}{\partial x} + \frac{\partial v}{\partial t}\right)$$

i.e.
$$\tau_0 = -\gamma R\left(\frac{\partial h}{\partial x} + \frac{v}{g}\frac{\partial v}{\partial x} + \frac{1}{g}\frac{\partial v}{\partial t}\right)$$

$$= -\gamma R\left(\frac{\partial H}{\partial x} + \frac{1}{g}\frac{\partial v}{\partial t}\right) \tag{8-3}$$

from Eq. (8–2). Substituting from Eq. (8–1), we now have

$$\frac{\partial H}{\partial x} + \frac{1}{g}\frac{\partial v}{\partial t} + \frac{v^2}{C^2 R} = 0 \tag{8-4a}$$

and this equation may be rewritten

$$S_e + S_a + S_f = 0 \tag{8-4b}$$

naming the three terms of Eq. (8–4a) the energy slope, the acceleration slope, and the friction slope respectively.

A more radical restatement of Eq. (8–4a) may be made by using Eq. (8–2), and recalling that the bed slope S_0 is equal to $-\partial z/\partial x$. Since $h = z + y$, we have, from Eq. (8–2)

$$\frac{\partial H}{\partial x} = \frac{\partial z}{\partial x} + \frac{\partial y}{\partial x} + \frac{v}{g}\frac{\partial v}{\partial x}$$

$$= -S_0 + \frac{\partial y}{\partial x} + \frac{v}{g}\frac{\partial v}{\partial x}$$

$$= -\frac{1}{g}\frac{\partial v}{\partial t} - S_f$$

from Eq. (8–4). Hence Eq. (8–4) can be written

$$S_f = S_0 - \frac{\partial y}{\partial x} - \frac{v}{g}\frac{\partial v}{\partial x} - \frac{1}{g}\frac{\partial v}{\partial t} = \frac{v^2}{C^2 R} \qquad (8\text{–}5)$$

Steady uniform flow →|

steady nonuniform flow →|

unsteady non-uniform flow →|

this equation being applicable as indicated. This arrangement shows clearly how nonuniformity and unsteadiness introduce extra terms into the dynamic equation.

Like the steady-flow equations of which they are an extension, Eqs. (8–4) and (8–5) are true only when the pressure distribution is hydrostatic, i.e., when the vertical components of acceleration are negligible.

The equation of continuity for unsteady flow has already been derived in Sec. 1.3 as

$$\frac{\partial Q}{\partial x} + B\frac{\partial y}{\partial t} = 0 \qquad (8\text{–}6)$$

which is a restatement of Eq. (1–3). B is the surface width, as in Fig. 8-1. When the channel is rectangular in section the substitution $Q = Bq$ leads to

$$\frac{\partial q}{\partial x} + \frac{\partial y}{\partial t} = 0 \qquad (8\text{–}7)$$

An alternative form of Eq. (8–6) may be written by expanding the term $\partial Q/\partial x = \partial (Av)/\partial x$, leading to

$$A\frac{\partial v}{\partial x} + v\frac{\partial A}{\partial x} + B\frac{\partial y}{\partial t} = 0 \qquad (8\text{–}8)$$

the three terms of which are known as the prism-storage, wedge-storage, and rate-of-rise terms respectively. The significance of this terminology will become apparent in the treatment of the flood routing problem in Chap. 9.

8.3   The Method of Characteristics

Our problem now, stated in general terms, is to solve the two independent equations (8–4) and (8–5) on the one hand, and (8–6) through (8–8) on the other, for the two unknowns v and y (for a given cross section A is a known function of y). As already noted in Sec. 8.1, the equations are not explicitly soluble except in certain simple cases, but these cases, simple as they are, can be of deep practical interest.

The approach dealt with in this section leads to the so-called *method of characteristics*, a semigraphical method by which explicit solutions, if they exist, are readily obtained, and by which numerical solutions can be worked out in the more general cases where no explicit solutions are possible. There is an extensive literature on the process, going well back into the nineteenth century; the more recent account of it by J. J. Stoker [1,2] is one of the most attractive, and one of the most apposite in the present context. Much of the following material is based on Stoker's treatment.

To introduce the method, we deal first with the simplest possible type of channel: the one having a rectangular section of constant width, and a constant bed slope. The first step is to remove y by the substitution $c^2 = gy$, where c is the speed of a long low wave in water of depth y; accordingly, c becomes our measure of the depth. We note that $d(gy) = d(c^2) = 2c\,dc$, and hence that if (Eq. 8–5) is multiplied by g and rearranged, we obtain

$$2c\frac{\partial c}{\partial x} + v\frac{\partial v}{\partial x} + \frac{\partial v}{\partial t} = g(S_0 - S_f) \tag{8–9}$$

Recalling that $q = vy$, we expand Eq. (8–7) into the following form

$$v\frac{\partial y}{\partial x} + y\frac{\partial v}{\partial x} + \frac{\partial y}{\partial t} = 0 \tag{8–10}$$

whence, multiplying throughout by g, and substituting $c^2 = gy$ as before, we obtain

$$2vc\frac{\partial c}{\partial x} + c^2\frac{\partial v}{\partial x} + 2c\frac{\partial c}{\partial t} = 0 \tag{8–11}$$

or, dividing throughout by c:

$$2v\frac{\partial c}{\partial x} + c\frac{\partial v}{\partial x} + 2\frac{\partial c}{\partial t} = 0 \tag{8–12}$$

By writing first the sum, and then the difference, of Eqs. (8–9) and (8–12), we obtain two further equations:

$$\frac{\partial v}{\partial t} + (v + c)\frac{\partial v}{\partial x} + 2\frac{\partial c}{\partial t} + 2(v + c)\frac{\partial c}{\partial x} = g(S_0 - S_f) \tag{8–13a}$$

$$\frac{\partial v}{\partial t} + (v - c)\frac{\partial v}{\partial x} - 2\frac{\partial c}{\partial t} - 2(v - c)\frac{\partial c}{\partial x} = g(S_0 - S_f) \qquad (8\text{–}14a)$$

So far the argument has consisted mainly of manipulation and rearrangement, producing what appears to be a more orderly layout; but a solution to the equations does not seem to be any closer. However, it is at this point that the argument begins to move directly towards a solution. To understand clearly the following treatment, the reader must recall the basic equations of partial differentiation, namely

$$dy = \frac{\partial y}{\partial x}\,dx + \frac{\partial y}{\partial t}\,dt \qquad (8\text{–}15)$$

or
$$\frac{dy}{dt} = \frac{\partial y}{\partial x}\frac{dx}{dt} + \frac{\partial y}{\partial t} \qquad (8\text{–}16)$$

In these equations y is a variable dependent on the two independent variables x and t, and the equations give the rate of change of y if x and t are simultaneously varied in some prescribed manner, given by dx/dt. If y, x, and t have their usual meanings in open channel flow, we may think of the situation in this way: to an observer walking along the river bank with a speed dx/dt (which he may choose himself) y will appear to vary with time at the rate given by Eq. (8–16). A similar result would of course be true for any other parameter such as v, q, or c.

In the light of this discussion the left-hand sides of Eqs. (8–13a) and (8–14a) now appear as total derivatives, essentially as given by Eq. (8–16). For these equations may be written as

$$(v + c)\frac{\partial(v + 2c)}{\partial x} + \frac{\partial(v + 2c)}{\partial t} = \frac{D_1(v + 2c)}{D_1 t} = g(S_0 - S_f) \quad (8\text{–}13b)$$

and
$$(v - c)\frac{\partial(v - 2c)}{\partial x} + \frac{\partial(v - 2c)}{\partial t} = \frac{D_2(v - 2c)}{D_2 t} = g(S_0 - S_f) \quad (8\text{–}14b)$$

where the total-derivative operators $D_1/d_1 t$ and $D_2/D_2 t$ represent rate of change from the viewpoint of observers moving with velocities $(v + c)$ and $(v - c)$ respectively.

The significance of this result is that the paths of these two imaginary observers can be traced on the x-t plane and a complete solution obtained for any prescribed unsteady-flow situation. Only in the simplest cases does the process lead to explicit solutions, but in the more complex cases numerical methods may be used without great difficulty. We proceed to consider some typical problems.

The Simple-Wave Problem

Here we set out to formulate the simplest possible situation involving unsteady flow. First, we set $S_0 = S_f = 0$, limiting our attention to the case of a horizontal bed without resistance; we then consider the case of flow, initially steady and uniform, which is then disturbed in some way, giving rise to unsteady flow as the disturbance is propagated into the initially steady region.

We have specified that the undisturbed flow is also uniform; this property defines the "simple-wave" problem. It may be thought that the conditions set out above are so special as to be of limited practical interest, but this is not so. There are many practical situations in which the flow changes so quickly that the acceleration terms in Eq. (8–5) are large compared with S_0 and S_f, which may therefore be neglected—to a good first approximation anyway. An example is the release of water from a lock into a navigation canal; in this case the initial surge (which is of most interest to the engineer) can be accurately treated by neglecting slope and resistance, whose effect becomes appreciable only after the wave has traveled some distance. Moreover, quite apart from immediate applications of the simple-wave problem, its study will, as we shall see, disclose principles whose interest and usefulness extend far beyond the simple-wave problem alone.

We now consider Eqs. (8–13b) and (8–14b) in detail. Since $S_0 = S_f = 0$, the total derivatives of $(v \pm 2c)$ in these equations are zero; this means that to observers moving with velocity $(v \pm c)$, the quantities $(v \pm 2c)$ appear to remain constant. The paths of these observers can be traced on the x-t plane, as in Fig. 8-2, giving rise to two families of lines, called *characteristics*. Along

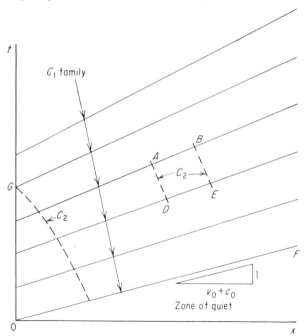

Figure 8-2. Characteristic Curves on the x-t Plane

each member of the first group, which we shall designate the C_1 family, the inverse slope of the line is $(v + c)$, and $(v + 2c)$ is a constant; similarly, along each member of the second group (the C_2 family) the inverse slope is $(v - c)$, and $(v - 2c)$ is a constant.

The two families of curves are therefore contours of $(v + 2c)$ and $(v - 2c)$. We shall see that in the simple-wave problem the members of the C_1 family are also contours of $(v + c)$ and are therefore straight lines, as in Fig. 8-2. To establish this result it is first necessary to prove the following introductory theorem:

THEOREM. If $S_0 = S_f = 0$, and if any one curve of the C_1 or C_2 family of characteristics is a straight line, then so are all other members of the same family.

To prove the theorem we consider the two C_1 lines AB, DE, in Fig. 8-2. DE is a straight line, and we are to prove that AB (which may be any other member of the family) is also straight. Both $(v + c)$ and $(v + 2c)$ are constant along DE, so that their difference c must be constant, and hence v also. It follows that $c_D = c_E$, $v_D = v_E$; also we can write, from the C_2 characteristics AD and BE:

$$v_A - 2c_A = v_D - 2c_D$$

$$v_B - 2c_B = v_E - 2c_E \tag{8-17}$$

whence
$$v_A - 2c_A = v_B - 2c_B \tag{8-18}$$

And since AB is a C_1 characteristic, we have

$$v_A + 2c_A = v_B + 2c_B \tag{8-19}$$

Equations (8-18) and (8-19) can be satisfied only if $v_A = v_B$, $c_A = c_B$, i.e., if AB is a straight line. The theorem is therefore proved for the C_1 family of characteristics; a similar proof can readily be obtained (Prob. 8.1) for the C_2 family.

We now formulate the simple-wave problem in detail. We consider a channel in which the flow is initially uniform, i.e., v and c are constant and equal to v_0 and c_0 respectively. A disturbance is now introduced at the origin of x, at the left-hand end of the channel; this disturbance takes the form of a prescribed variation of v (and/or c) with the time t. We postpone for the present the question of whether $v(t)$ and $c(t)$ may be prescribed independently, or whether the one will be dependent on the other. Also, we assume that the disturbance propagates into the undisturbed region with a velocity c_0 relative to the undisturbed fluid, i.e., with a velocity $(v_0 + c_0)$. This implies that the disturbance sets up a wave front small enough to have a velocity c_0, i.e., we assume for the present that an abrupt wave front of finite height will not form.

It follows that we can draw a straight line OF, of constant inverse slope $(v_0 + c_0)$, dividing the undisturbed flow, or "zone of quiet," from the disturbed region above OF. This line will also be a C_1 characteristic—the first of that family—and since it is straight, so are all the other members of the

family, as in Fig. 8-2. However, the C_2 characteristics are *not* straight lines
(Prob. 8.1). If we could now calculate the values of v and c appropriate to
every C_1 characteristic, we could obtain v and c at every point on the x-t
plane, and we should have the complete solution to the problem.

This calculation is, in fact, easily carried out, given the prescribed values
of $v = v(t)$ and/or $c = c(t)$ along the t axis. Consider any point G on this axis,
of ordinate t; the C_1 characteristic issuing from this point will have an inverse
slope equal to

$$\frac{dx}{dt} = v(t) + c(t) \tag{8-20}$$

We can examine the interdependence of $v(t)$ and $c(t)$ by drawing a C_2 characteristic (shown dotted) from G to OF; whatever the form of this line may be,
it indicates the result

$$v(t) - 2c(t) = v_0 - 2c_0 \tag{8-21}$$

and this equation tells us that $v(t)$ and $c(t)$ are not independent; only one or
the other need be prescribed (indeed only one of them *can* be prescribed), as
a description of the disturbance at the origin of x. From Eqs. (8–20) and (8–21)
it follows that the inverse slope of the C_1 characteristic issuing from G can be
expressed in either of the two forms:

$$\frac{dx}{dt} = \tfrac{3}{2}v(t) - \tfrac{1}{2}v_0 + c_0 \tag{8-22a}$$

$$= 3c(t) + v_0 - 2c_0 \tag{8-22b}$$

and from these equations v and c can readily be obtained at any point in the
x-t plane.

The argument leading up to Eq. (8–22) is undoubtedly circuitous, but the
end result, in the form of Eq. (8–22), presents an extremely simple treatment
of problems which are physically quite real, and indeed of some practical
importance. Consider, for example, uniform flow in a river discharging into
a large lake or estuary. Initially the water level in the estuary is the same as
the water level at the river mouth; then under tidal action the estuary level
begins to fall. It is clearly of some interest to ask how long it will be before
the water level falls by a certain amount at some specified distance upstream
from the mouth. If bed slope and resistance are neglected, Eq. (8–22) gives
the solution very simply, as in the following example:

Example 8.1

Water flows at a uniform depth of 5 ft and velocity of 3 ft per second in a channel
of rectangular section, into a large estuary. The estuary level, initially the same as the
river level, falls at the rate of 1 ft/hr for 3 hr; neglecting bed slope and resistance,
determine how long it takes for the river level to fall by 2 ft at a section 1 mile

upstream from the mouth. At this time, how far upstream will the river level just be starting to fall?

Taking x as positive upstream, we find

$$v_0 = -3 \text{ ft/sec}$$

$$c_0 = \sqrt{5g} = +12.7 \text{ ft/sec}$$

whence
$$v_0 + c_0 = +9.7 \text{ ft/sec} = +6.60 \text{ mph}$$

The accompanying sketch of the x-t plane shows how the solution is arrived at. We are seeking the point H, at which $x = 5{,}280$ ft and the depth $= 3$ ft, i.e., $c = \sqrt{3g} = 9.83$ ft/sec. We consider the characteristic GH, along which c is constant and equal to $\sqrt{3g}$; it originates at the point G, representing a depth of 3 ft at the river mouth. This means that at G, $t = 2$ hr $= 7{,}200$ sec.

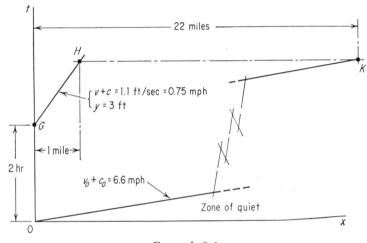

Example 8-1

We now require the inverse slope of GH. From Eq. (8–22b), it is equal to

$$\frac{dx}{dt} = 3c(t) + v_0 - 2c_0$$

$$= 3\sqrt{3g} - 3 - 25.4$$

$$= 1.1 \text{ ft/sec}$$

Hence the time interval between G and H is equal to

$$\frac{5{,}280}{1.1} = 4{,}800 \text{ sec} = 1 \text{ hr } 20 \text{ min}$$

and the total time elapsed is therefore equal to

$$2 \text{ hr} + 1 \text{ hr } 20 \text{ min} = 3 \text{ hr } 20 \text{ min.} \quad \textit{Ans.}$$

And the distance upstream to K, on the boundary of the zone of quiet at this time is equal to

$$3 \text{ hr } 20 \text{ min } \times 6.60 \text{ mph} = 22.0 \text{ miles.} \quad \textit{Ans.}$$

This example brings to light a most important aspect of the theory of characteristics. The numerical working, and the sketch of the x-t plane, suggested this picture of events: After 2 hr, the depth at the river mouth falls to 3 ft (the point G). This depth is then propagated upstream (along the characteristic GH) until it reaches a section one mile from the mouth (the point H). In essence, what we have is a wave motion in which a 3-ft depth is sent, or signaled, along a path represented by GH.

Indeed, the idea that movement along a characteristic represents wave motion across the water surface might have been inferred simply from the fact that $(v + c)$ and $(v - c)$ are the velocities, relative to a stationary observer, of elementary waves moving in opposite directions across the water surface. However, this concept alone would not have sufficed to provide numerical solutions such as in Example 8.1; for this we need to know v, as well as c, to arrive at the resultant wave velocity $(v + c)$, and this can only be done by means of Eq. (8–22).

It may be asked what wave motion is represented by movement along the C_2 characteristics. In the simple-wave problem, a wave motion of this sort is not recognizable because $(v - 2c)$ is constant not only along each C_2 characteristic but also over the whole of the x-t plane; this follows from the constancy of v_0 and c_0 along OF in Fig. 8-2. Now the essence of a wave motion is that some parameter will appear constant to an observer moving with a certain velocity (the wave velocity) and *only* with that velocity. If it appears constant to all observers everywhere (as $(v - 2c)$ does) the wave motion degenerates into one of zero amplitude and can no longer properly be called a wave motion. Nevertheless it must be kept in mind that the simple-wave problem is an exceptional case, and that in general movement along *each* characteristic represents a possible wave motion. This physical significance of the characteristic lines is the most important feature of the whole theory.

So far we have excluded the possibility of an abrupt surge front of finite height, of the kind treated in Sec. 3.5. No reason has been given for this assumption, but we shall see in the next section that the theory affords a simple means of determining the circumstances under which an abrupt surge front will form. The alternative name "bore" for the abrupt surge has been noted in Sec. 3.5; another alternative is the term "shock," which is also applied to the corresponding phenomenon in gas dynamics.

8.4   Positive and Negative Waves; Surge Formation

In the simple-wave problem discussed in the previous section the disturbance introduced at one end of the channel may be positive, i.e., such as to increase the depth, or negative, reducing the depth. An important difference between the two consequent types of wave becomes apparent on considering Eq. (8–22b). If the disturbance is negative, $c(t)$ decreases as t increases, so the inverse slope dx/dt given by Eq. (8–22b) diminishes as we move up to the t axis. It follows that the slopes of the C_1 characteristics increase as t increases, i.e., that the characteristics diverge outwards from the t axis, as shown in Fig. 8-2. From this it follows that negative waves are dispersive, i.e., that sections having a given difference in depth move further apart as the wave moves outwards from its point of origin (Prob. 8.2).

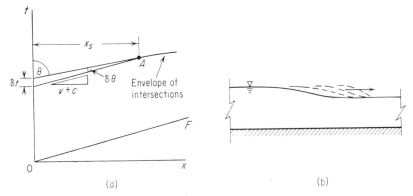

Figure 8-3.   *The Convergence of Characteristics and Steepening of the Wave Front in a Positive Wave*

When the disturbance is positive, on the other hand, the C_1 characteristics converge, as in Fig. 8-3a, and must eventually meet. Such an intersection implies that the depth has two different values in the same place at the same time—an obvious anomaly. What in fact happens is equally obvious: the wave becomes steeper and steeper, as in Fig. 8–3b, until it forms an abrupt steep-fronted wave—the surge, or "bore" treated in Sec. 3.5. While the intersections of neighboring characteristics will form an envelope as in Fig. 8-3a, the surge will actually form at the "first" point A of the envelope—i.e., the point having the least value of t.

The front of the surge will not necessarily be broken and turbulent; it may, like the hydraulic jump, consist of a train of smooth unbroken waves if the depth ratio is small enough. It was pointed out in Sec. 6.7 that the limiting value of upstream Froude number for this "undular" jump ranged from 1.25 to 1.7, depending on the bed roughness. This corresponds to a range of depth ratios from 1.35 to 1.95 approximately. The same limitation applies to the surge, which is simply a moving hydraulic jump. It follows that a surge would certainly not break at the first point of formation, as at A in Fig. 8-3a, for the depth ratio there approaches unity. Breaking would only occur after subsequent development of the surge beyond the point A; tracing this development would be a matter of some difficulty. However, in all the subsequent argument the term surge will be applied to any abrupt change in depth, as indicated by the point A, whether it is undular or broken.

The intersection of any neighboring pair of characteristics can be located by an elementary geometrical argument. With the terms defined as in Fig. 8-3a, the following results are easily obtained:

$$\delta\theta = \frac{\delta t \sin \theta}{x_s/\sin \theta} = \frac{\delta t \sin^2 \theta}{x_s}$$

and

$$\delta\theta = \delta(\tan \theta) \cos^2 \theta$$

whence

$$x_s = \frac{\delta t \tan^2 \theta}{\delta(\tan \theta)}$$

$$= \frac{(v + c)^2}{d(v + c)/dt} \tag{8-23}$$

where $(v + c)$ equals the inverse slope of the characteristic, as given by Eq. (8–22). Substituting Eq. (8–22b) into Eq. (8–23), we obtain

$$x_s = \frac{[3c(t) + v_0 - 2c_0]^2}{3dc(t)/dt} \tag{8-24}$$

From this equation an envelope of intersections can be traced, given a specified disturbance $c = c(t)$ along the t axis (Prob. 8.3). If, as in the case of the release of water from a lock, the disturbance is specified as a variation in $q = q(t)$, corresponding values of c and dc/dt can readily be obtained by the use of Eq. (8–21) and inserted in Eq. (8–24). (See Prob. 8.4.)

Once a surge does develop, there is of course an energy loss across the surge, and characteristics cannot be projected from one side of the surge to the other. But as we shall see in Sec. 8.5, the flow on each side of the surge can be described by a separate system of characteristics.

From the above remarks it should now be clear that the method of Example 8.1 could have been used with confidence only in the case of a negative wave. In the corresponding positive-wave problem, the method is applicable only for values of x less than the one given by Eq. (8–24), as in Prob. 8.3.

We may also note here that in the case of the negative simple wave, Eq. (8–21) can be written in the more general form

$$v - 2c = v_0 - 2c_0 , \text{ a constant} \tag{8-25a}$$

or

$$v = 2\sqrt{gy} + v_0 - 2c_0 \tag{8-25b}$$

applicable to the *whole* of the x-t plane, because a C_2 characteristic can be drawn from any point on the plane to the line OF (Fig. 8-2) bounding the zone of quiet. In many cases it is more convenient to solve problems by the direct use of Eq. (8–25) and the concept of wave motion discussed in Sec.

8.3, rather than by the use of the x-t plane, as in Example 8.1. (See Prob. 8.5).

It need hardly be emphasized that if a surge forms, Eq. (8–25) could be applied only by using different constants on opposite sides of the surge.

The form of the negative-wave profile can readily be deduced by recalling the significance of Eq. (8–22b). This equation gives the speed at which a section of constant depth y moves; since for a given y, this speed is constant, we may replace dx/dt by $x/(t$-$t_1)$, where t_1 is the value of t at $x=0$ (as at the point G in Fig. 8-2). Eq. (8-22b) can then be rewritten as

$$\frac{x}{t-t_1} = 3\sqrt{gy}\,(t_1) + v_0 - 2\sqrt{gy_0} \qquad \textbf{(8–26\,a)}$$

where $y=y(t_1)$ prescribes the initiating disturbance along the t-axis. Since y must be a known function of t_1, Eq. (8-26a) connects x, y, t, and the known constants v_0 and y_0: it therefore defines the wave profile completely in space and time. It will be seen in Sec. 8.5 that this equation takes a particularly simple and useful form in the so-called dam-break problem.

The Effect of Resistance on Surge Formation

Equations (8–23) and (8–24) show that if bed slope and resistance are neglected any positive wave, however gentle, must eventually form a surge with an abrupt wave front. The next question to consider is whether the inclusion of slope and resistance in the working would modify this conclusion in any way. In order to take account of these factors in the most general possible way, it is necessary to resort to the numerical methods which will be detailed in Sec. 8.6. However, it is possible to formulate a simple problem for for which an explicit solution is possible. The corresponding physical situation is this: the initial flow is uniform, as in the simple-wave problem, but uniform in the more usual sense that $S_0 = S_f \neq 0$, as in Chap 4. The positive disturbance which is then set up at one end of the channel is a small one, in which increments of v and c are only small fractions of the original values.

The resulting theory can be appropriately described as "linear," because squares and higher powers of ratios such as $\Delta v/v$ and $\Delta c/c$ can be neglected; one important consequence is that the increments Δv and Δc can be assumed to be of differential order of magnitude, i.e., they can be evaluated by differentiation. A further common consequence—not immediately obvious in this case—is that differential equations which would otherwise be nonlinear and not explicitly soluble become linear and soluble.

This device is extremely common in all branches of mechanics, particularly in those cases where the essence of the problem is some disturbance or perturbation introduced into an otherwise stable, or steady, or uniform, situation. If the disturbance is assumed to be small a linear theory can often be produced, reducing an otherwise intractable problem to manageable size. Such a theory usually yields results which are a fair approximation to the facts even when the disturbance is large, but it is extremely useful also for the general view that it gives of the trend followed by the solution. For example, in the present problem one would intuitively expect flow resistance to make a positive (or negative) wave more dispersive, and to postpone the formation of a surge. The linear theory will show clearly whether this expectation is justified, and willl also provide other results of interest.

We now develop the argument in detail. For an approximate solution, such as this will evidently be, it will be sufficiently accurate to describe the effects of resistance by using the Chézy equation with constant C, and by assuming a wide channel, with $R = y$. Since $S_f = v^2/C^2 R$, we can write

$$S_0 - S_f = S_0\left(1 - \frac{\text{Fr}^2}{\text{Fr}_0{}^2}\right) \tag{8-27}$$

where Fr is the Froude number and the subscript 0 indicates the initial uniform flow condition. We can make further use of the Froude number in Eqs. (8–13) and (8–14) by recalling that $\text{Fr} = v/c$, and recasting the equations in this form

$$\frac{D_1}{D_1 t}[c(\text{Fr} + 2)] = gS_0\left(1 - \frac{\text{Fr}^2}{\text{Fr}_0{}^2}\right) \tag{8-13c}$$

$$\frac{D_2}{D_2 t}[c(\text{Fr} - 2)] = gS_0\left(1 - \frac{\text{Fr}^2}{\text{Fr}_0{}^2}\right) \tag{8-14c}$$

where the operators $D_1/D_1 t$ and $D_2/D_2 t$ have the same meanings as before. A difficulty arises over the sign of the resistance term, for S_f should properly be written as $S_f = v|v|/C^2 R$, to indicate that S_f is not always positive, but has the same sign as v. This precaution is unnecessary in the present case, for a *small* disturbance will not give rise to any reversal of flow, and S_f will always have the same sign as S_0. Equations (8–13c) and (8–14c) are therefore satisfactory in the form given above.

We consider the course of events on the *x-t* plane in Fig. 8-4. Because we are assuming initial conditions to be uniform, the first C_1 characteristic OF will be a straight line, and along this line the right-hand side of Eqs. (8–13) and (8–14) will be zero. Consider now the tracing of the C_2 characteristics upwards from OF; initially we have $d[c(\text{Fr} - 2]/dt = 0$, i.e.,

$$(\text{Fr} - 2)\frac{dc}{dt} + c\frac{d\text{Fr}}{dt} = 0 \tag{8-28}$$

from which it follows that dc/dt and $d\text{Fr}/dt$ are of different sign, or the same sign, according as Fr is greater or less than 2. It will be seen later that this result has important consequences affecting the stability of the motion; meanwhile we note simply that the sign of $d\text{Fr}/dt$ along the C_2 characteristics is not yet determined.

We denote by ΔFr the increase in Fr along a C_2 segment such as DA in Fig. 8-4. The corresponding increase in $(S_0 - S_f)$ is equal to

$$\Delta(S_0 - S_f) = \frac{-2S_0\Delta\text{Fr}}{\text{Fr}_0} \tag{8–29}$$

by differentiation from Eq. (8–27). It is natural (although not perhaps strictly necessary) to take x positive in the direction of wave propagation; if this convention is adopted the signs of S_0 and Fr_0 will depend on whether the positive wave being considered is moving upstream against the initial uniform flow,

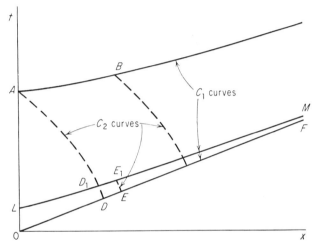

Figure 8-4.   *The Effect of Resistance on a Positive Wave*

or downstream. These two types of wave may be termed "adverse" and "following" waves respectively; in the former case S_0, v, and Fr are all negative, and in the latter they are all positive. (The former type can exist only when $\text{Fr}_0 < 1$.) The result is that the increment of $(S_0 - S_f)$ given by Eq. (8–29) is of opposite sign to ΔFr, whether the wave is adverse or following.

We can now recast Eq. (8–14c) to give, for the average rate of change along a C_2 characteristic over a short time interval starting on OF (e.g., along DA):

$$\frac{d}{dt}[c(\text{Fr} - 2)] = -\frac{gS_0\Delta\text{Fr}}{\text{Fr}_0} \tag{8–30}$$

since from Eq. (8–29) the average value of $(S_0 - S_f)$ over that interval is equal to $-S_0\Delta\text{Fr}/\text{Fr}_0$. Along any C_1 characteristic other than OF the average value of $(S_0 - S_f)$ is equal to $-2S_0\Delta\text{Fr}/\text{Fr}_0$, whence Eq. (8–13c) becomes

$$\frac{d}{dt}[c(\text{Fr} + 2)] = -\frac{2gS_0\Delta\text{Fr}}{\text{Fr}_0} \tag{8–31}$$

Consider the C_1 characteristic LM, very close to OF. Along any C_2 intervals DD_1, EE_1, joining LM and OF, the increment ΔFr will be small; hence from Eq. (8–30) there will be little change in c(Fr − 2) along these intervals. Now, neighboring intervals such as DD_1 and EE_1 have very little proportionate difference in their lengths, because of the slow convergence of LM and OF; it follows that there is little proportionate difference in the corresponding values of ΔFr. The end result is that the change in c(Fr − 2) between D_1 and E_1 is a whole order of magnitude smaller than the change along DD_1 or EE_1, i.e., the rate of change of c(Fr − 2) along LM is negligible compared with $gS_0\Delta$Fr$/$Fr$_0$, so we may write for LM:

$$\frac{d}{dt}[c(\text{Fr} - 2)] = 0 \qquad (8\text{–}32)$$

This equation may now be combined with Eq. (8–31) to explore the variation of c and F along LM, and in particular the convergence of this C_1 characteristic towards its neighbor OF.

Reverting for the moment to Eq. (8–30), we can expand the derivative as in Eq. (8–28):

$$c\frac{d\text{Fr}}{dt} + (\text{Fr} - 2)\frac{dc}{dt} = \frac{gS_0}{\text{Fr}_0}\frac{d\text{Fr}}{dt}\Delta t$$

where Δt is the time increment along an interval such as DD_1. It follows that

$$\frac{d\text{Fr}}{dt} = \frac{2 - \text{Fr}}{c\left(1 + \dfrac{gS_0\Delta t}{c\text{Fr}_0}\right)}\frac{dc}{dt} \qquad (8\text{–}33)$$

and since Δt is small the term $gS_0\Delta t/c\text{Fr}_0$ can be neglected so that Eq. (8–33) reduces to Eq. (8–28); this approximation is at least accurate enough to transform the right hand member of Eq. (8–31)—itself of small magnitude—from $-2gS_0\Delta$Fr$/$Fr$_0$ to $-2gS_0(2 - \text{Fr})\Delta c/c\text{Fr}_0$. Subtracting Eq. (8–32) from Eq. (8–31) we now obtain

$$4\frac{dc}{dt} = \frac{2gS_0(\text{Fr}_0 - 2)\Delta c}{c_0\text{Fr}_0}$$

or
$$\frac{dc}{dt} = M(c - c_0) \qquad (8\text{–}34)$$

where $M = gS_0(\text{Fr}_0 - 2)/2c_0\,\text{Fr}_0$, noting that except where the small difference $(c - c_0)$ occurs explicitly c may be interchanged with c_0, and Fr with Fr$_0$. Integrating Eq. (8–34) and inserting boundary conditions, we obtain

$$c = c_0 + (c_1 - c_0)e^{M(t - \Delta t)} \qquad (8\text{–}35)$$

where Δt is the time interval between OF and LM at $x = 0$, and $c = c_1$ at L. Equation (8–35) indicates how c varies with t along LM, and shows clearly that c decays exponentially from c_1 to c_0 as $(t - \Delta t)$ goes from zero to infinity, *provided* that M is negative, i.e., that $\text{Fr}_0 < 2$. The significance of this result will become apparent shortly.

The geometry of the line LM is easily obtained by using Eq. (8–28), implying that to a first approximation

$$v - 2c = v_0 - 2c_0$$

everywhere along LM. It follows that along this line

$$\frac{dx}{dt} = v + c = 3c + v_0 - 2c_0$$

$$= v_0 + c_0 + 3(c_1 - c_0)e^{M(t-\Delta t)} \qquad (8\text{–}36)$$

Integrating and inserting the condition $x = 0$ when $t = \Delta t$, we obtain:

$$x = (v_0 + c_0)(t - \Delta t) + \frac{3(c_1 - c_0)}{M}[e^{M(t-\Delta t)} - 1] \qquad (8\text{–}37)$$

which is the equation of the C_1 characteristic LM. The form of these equations clarifies the role played by resistance; Eq. (8–36) shows that if M is negative, i.e., $\text{Fr}_0 < 2$, the inverse slope dx/dt diminishes with increasing t, i.e., the line LM is concave upwards, curving away from OF. The effect of resistance is therefore to delay the intersection of neighboring characteristics and the resultant bore formation (it may indeed be postponed indefinitely). On the other hand, if M is positive, $\text{Fr}_0 > 2$, LM is convex upwards and bore formation occurs even earlier than if there were no resistance.

It remains to examine more closely the case where $\text{Fr}_0 < 2$ (M negative) so as to determine the amount by which bore formation is delayed, or the conditions under which it may be postponed indefinitely. The equation of the first C_1 characteristic OF is simply

$$x = (v_0 + c_0)t \qquad (8\text{–}38)$$

and the x interval between the lines LM and OF at any time t is therefore equal to

$$\Delta x = (v_0 + c_0)\Delta t - \frac{3\Delta c_1}{M}[e^{M(t-\Delta t)} - 1] \qquad (8\text{–}39)$$

setting $\Delta c_1 = c_1 - c_0$. The lines intersect if Δx reduces to zero; it reaches a minimum, equal to

$$\Delta x_m = (v_0 + c_0)\Delta t + \frac{3\Delta c_1}{M} \qquad (8\text{–}40)$$

when t tends to infinity. This minimum value cannot reach zero, i.e., LM and OF cannot intersect, unless $\Delta c_1/\Delta t$ exceeds $-M(v_0 + c_0)/3$, i.e.,

$$\frac{\Delta c_1}{\Delta t} \geq \frac{gS_0(2 - \text{Fr}_0)(1 + \text{Fr}_0)}{6\text{Fr}_0} \qquad (8\text{–}41)$$

The physical interpretation of this result is simply that if $Fr_0 < 2$, a bore will still form if the wave is strong enough, i.e., if the initial value of dc/dt at $x = 0$ exceeds the value given by Eq. (8–41). We note that this equation also embodies the criterion $Fr_0 \gtrless 2$; for if $Fr_0 > 2$, the right-hand member is negative (note that the quotient S_0/Fr_0 is positive whether the wave is adverse or following, and that in the former case $|Fr_0| < 1$). It follows that any positive value dc/dt, however small, will exceed the right-hand member of Eq. (8–41); this confirms the previous deduction that if $Fr_0 > 2$ a positive disturbance will always lead to bore formation.

The substitution $gdy = 2cdc$ converts Eq. (8–41) into the form

$$\frac{dy}{dt} \geq \frac{gy_0S_0(2 - Fr_0)(1 + Fr_0)}{3v_0} \tag{8–42}$$

an equation which has been deduced by Lighthill and Whitham [3] from a purely algebraic argument; the technique used by them is commonly employed in the solution of differential equations, and the reader will find it a useful exercise to work through the details (Prob. 8.7).

We consider some further ramifications of the problem, confining our attention to the case $Fr_0 < 2$. The above argument is subject to two restrictions: it relates only to the neighborhood of the wave front, and only to small disturbances. If we remove the first restriction, the argument involves more detail because the term $gS_0\Delta t/cFr_0$ in Eq. (8–33) is not negligible. The result is that bore formation is not so long delayed by resistance, and the critical value of dc/dt is lower than that given by Eq. (8–41). A similar result follows when the critical disturbance is of finite height, and there is a finite angle between the two C_1 characteristics whose convergence is being considered. In this case Eq. (8–32) does not hold true, and $d[c(Fr - 2)]/dt$ is positive along the upper characteristic. The result is that the magnitude of dc/dt is less than the value given by Eq. (8–34), so that c, and $(v + c)$, do not diminish so quickly along the C_1 line as in the case of Eqs. (8–34) through (8–36). The effect is to reduce the curvature of the C_1 line, so that intersection and bore formation is not delayed as long as in the previous argument.

The details of these two arguments are lengthy, and they are left as an exercise for the reader (Probs. 8.8 and 8.9). In fact it is hardly worth while pursuing the argument to these lengths, for the results are still limited by the assumptions that $R = y$ and that the Chézy C is constant. For this reason the detailed tracing of bore formation in a particular case is best done by numerical methods, using the linear theory given above only as an indicator of the outlines of positive wave behavior.

One general feature of the positive wave disclosed by this theory is that if $Fr_0 < 2$, then c diminishes along the C_1 characteristics. Tracing any such characteristic is therefore equivalent to following a section of diminishing depth at a diminishing velocity $(v + c)$.

Another question of interest is the development of the surge, or bore, after the intersection of characteristics. For a steady rate of rise at the left-hand end of the channel, an envelope of intersections would develop somewhat as in Fig. 8-5. Each point on this envelope represents an elementary surge of very small height; this surge will then move forward at the speed $(v + c)$ appropriate to its depth, overtaking and being overtaken by other elementary surges until they are all absorbed into one large surge. As the surge at the very front of the disturbance builds up, its height becomes finite and its speed therefore becomes greater than $v_0 + c_0$; its path will then diverge from OF

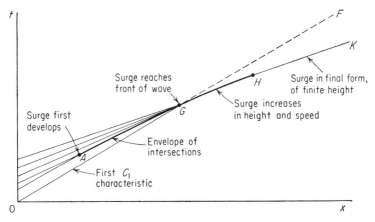

Figure 8-5. The Development of a Surge from a Positive Wave

as shown in Fig. 8-5. Even when resistance is neglected this path is difficult to trace in detail along the curved portion GH, but the slope of the final path HK is readily determined and its position can be obtained approximately (Prob. 8.13). Even at this final stage the surge may be unbroken or "undular" if the depth ratio across the surge is small, as previously discussed. The bore of the River Severn in England is often observed to be of this type.

Equations (8–41) and (8–42) make it clear that the major conditions for tidal bore formation in a river are that the rate of tidal rise be high and that the river velocity be low (Prob. 8.10). The former condition is often produced by the action of a funnel-shaped estuary, which by its convergence towards the river mouth intensifies the tidal rise there. This condition may also be expressed in terms of the rate of discharge increase (Prob. 8.4) in the case of release of water from storage (Probs. 8.11 and 8.12). In these problems the reader can also determine whether surges are more apt to form on waves traveling upstream ("adverse" waves) or downstream ("following" waves).

In connection with Prob. 8.4, it may be pointed out that when the initial condition is that of stationary water on a horizontal bed, the linear theory indicates through Eqs. (8–34), (8–41), and (8–42), that the C_1 characteristics remain straight lines (since $M = 0$) and resistance does not delay their intersection at all. This is to be expected, since the linear theory deals with small disturbances; only for large disturbances would resistance have an appreciable effect on surge formation, and their movement could be traced only by the numerical methods of Sec. 8.6.

The general conclusion from this section is that provided $Fr_0 < 2$, resistance makes positive waves more dispersive, and delays surge formation. The converse also follows immediately from an examination of the equations: that resistance makes negative waves *less* dispersive, and tends to promote surge formation. It is doubtful however whether resistance could ever be large enough to make a surge form in a negative wave, for this would create anomalies in the energy balance across the wave. A negative wave is essentially an invasion of deep water by shallow water; so if we consider movement relative to the wave, water must always pass from the deep-water side of the wave to the shallow-water side. But we have seen in the study of hydraulic jumps and surges in Chap. 3 that the shallow-water side has more energy than the deep-water side; therefore if a negative surge were to form there would be a gain of energy as the water passed through the surge from the deep to the shallow side, and this is impossible. With the positive surge there is of course no such difficulty; deep water is invading shallow water, so that water passes from the shallow side to the deep side of the surge and loses energy in the process. In this context "energy" means the value computed by using velocities taken relative to the surge front.

8.5   The Dam-Break Problem

This problem is just one application of the principles and techniques discussed in the previous section, but it is of sufficient importance to warrant a complete section of its own. The problem deals with the course of events following the sudden insertion or removal of an obstacle in a channel, and is clearly of interest in a number of real situations, ranging from the catastrophic flood following the collapse of a dam, to the operation of sluice gates in an irrigation canal.

Downstream Riverbed Dry

In the simplest form of the problem, bed slope and resistance are assumed negligible, the canal is rectangular in section and is closed by a transverse vertical plate, as in Fig. 8-6, containing still water on its right and having a dry channel bed on its left. The plate is to be suddenly removed and the subsequent motion studied. The best way to approach the problem is to consider first the situation created by moving the plate off to the left at finite speed; if the speed of the plate is increased to a very large value, we then have the case of complete removal of the plate.

Suppose, in the first instance, that the plate is gradually accelerated to a speed w; the path of the plate is then traced on the x-t plane, Fig. 8-6b, by the line $OABC$, the line BC being straight with an inverse slope of $-w$. Assuming for the moment that some water of some unknown depth will remain in contact with the plate, it follows that the water velocity will be the

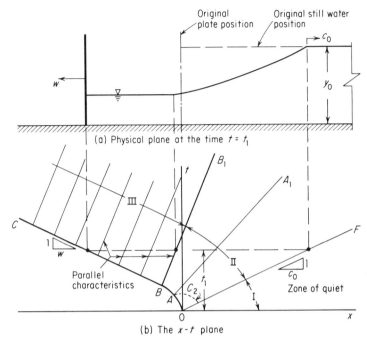

Figure 8-6. Initial Study for the Dam-break Problem

same as the plate velocity. Now we can draw a C_2 characteristic (shown dotted) from any point such as A on $OABC$ to the first C_1 characteristic OF; since from Eq. (8–17)

$$v_A - 2c_A = v_0 - 2c_0 \qquad (8\text{–}43)$$

c_A is readily obtained from v_A (the velocity of the plate). Hence from Eqs. (8–22a) and (8–22b) we deduce the slope of the C_1 characteristic drawn outwards from A:

$$\frac{dx}{dt} = v_A + c_A = \tfrac{3}{2}v_A - \tfrac{1}{2}v_0 + c_0 \qquad (8\text{–}44a)$$

$$= 3c_A + v_0 - 2c_0 \qquad (8\text{–}44b)$$

The interesting general point that emerges is that our initial disturbance does not have to be prescribed along the t axis, as in Sec. 8.3; it may be prescribed along any other line (such as $OABC$) in the x-t plane. Whatever may be the position of this line, we can use the fact that $(v - 2c)$ is constant everywhere in the plane to determine, as from Eq. (8–44), the slope of C_1 characteristics issuing from any point on the line.

With the aid of Eq. (8–44) we can therefore find the values of v and c everywhere on the x-t plane, and the problem is solved. We now examine the main features of the solution. From Eq. (8–44a) it follows that $(v_A + c_A)$, the inverse slope of the C_1 characteristic, decreases as we pass from O to B, because the leftward speed of the plate is increasing and v_A (negative) is decreasing. The C_1 characteristics issuing from OAB therefore diverge, and the ones issuing from BC are parallel, with inverse slopes equal to

$$\left(\frac{dx}{dt}\right)_{BC} = -\tfrac{3}{2}w - \tfrac{1}{2}v_0 + c_0 \tag{8–45}$$

We can now divide the region between Ox and $OABC$ into three zones:

  Zone I—the zone of quiet between OF and Ox, representing undisturbed still water.

  Zone II—the region of diverging characteristics and therefore varying depth and velocity between BB_1 and OF.

  Zone III—the region of parallel characteristics, and therefore constant depth and velocity, between BC and BB_1. All the water represented by this region is moving with the plate.

The situation at any instant can be determined by drawing a line of constant t, as in Fig. 8-6b. The intersections of this line with the boundaries of the various zones are projected upwards to Fig. 8-6a, showing the zone boundaries on the physical plane.

To determine the depth of water in Zone III, adjacent to the plate, we simply use Eq. (8–43) and obtain:

$$c_B = \tfrac{1}{2}v_B - \tfrac{1}{2}v_0 + c_0$$
$$= -\tfrac{1}{2}w + c_0 \tag{8–46}$$

noting that $v_B = -w$ and $v_0 = 0$. It follows that

$$y_B = \frac{c_B{}^2}{g} = \frac{(c_0 - \tfrac{1}{2}w)^2}{g} \tag{8–47}$$

Now c_B cannot be negative, so that w cannot exceed $2c_0$. If $w = 2c_0$, $c_B = y_B = 0$, and the water reaches zero depth at the plate. If the speed of the plate exceeds $2c_0$ it simply loses contact with the water, which then feathers out to an edge of zero depth moving with a speed $2c_0$, as in Fig. 8-7a.

We can now simulate the complete removal of the plate in terms of the problem which has just been treated. We need only postulate that the plate is suddenly moved to the left with a speed equal to or exceeding $2c_0$. On the x-t plane the line OAB shrinks to a point, and the line BC acquires an inverse slope of $-2c_0$. The line OF has of course an inverse slope of $+c_0$, and Zone III shrinks to nothing as BB_1 moves round to coincide with BC, which itself becomes a C_1 characteristic (Prob. 8.14). The field of flow is therefore represented by the fan-shaped region COF in Fig. 8-7b, and the velocity and depth at any point in that region are readily obtained by applying Eq. (8–44b)

to a point A which may be located anywhere in the plane. This gives the slope of the C_1 characteristic through A, which equals x/t since the line passes through the origin, so that $t_1 = 0$ in Eq. (8-26a), which therefore becomes

$$\frac{x}{t} = 3\sqrt{gy} + v_0 - 2\sqrt{gy_0} \qquad (8\text{-}26\,b)$$

and gives a complete description, in space and time, of the flow profile shown

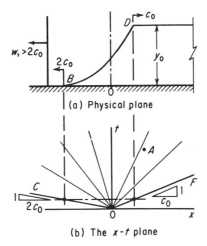

(a) Physical plane

(b) The x-t plane

Figure 8-7.   *The Dam-break Problem with Complete Removal of the Dam*

in Fig. 8-7a. At any instant the water-surface profile is a parabola, tangential to the channel bed. From Eq. (8-26b) the reader can verify (Prob. 8.6) that the leading feather-edge B advances with speed $2c_0$ and the trailing edge D recedes upstream with speed c_0, as shown in Fig. 8-7.

Problem 8.6 also shows that at the original position of the plate (or dam), i.e., at the origin of x, the depth is constant and equal to $4y_0/9$, and the water velocity is constant and equal to $2c_0/3$. This constant rate of outflow is maintained until the negative wave front D reaches the rear wall of the reservoir, is reflected from it, and returns to the origin of x; thereafter the outflow rate gradually diminishes. The existence of the steady-flow section at the origin of x, forming a kind of fixed center to the flow profile, has caused this type of wave to be termed the "centered simple wave."

Downstream Riverbed Submerged

If the downstream riverbed is not initially dry, but contains a layer of still water, the problem is not greatly changed. On the x-t plane, the C_1 characteristics near OF still form a family of straight lines radiating from the origin, as in Fig. 8-7*b*, for there is nothing in the presence of water downstream which changes the basic dynamic relations governing the family of characteristics in Zone II of this figure. The only influence that can be exerted by this downstream water is to limit in some way the *extent* of Zone II, without influencing

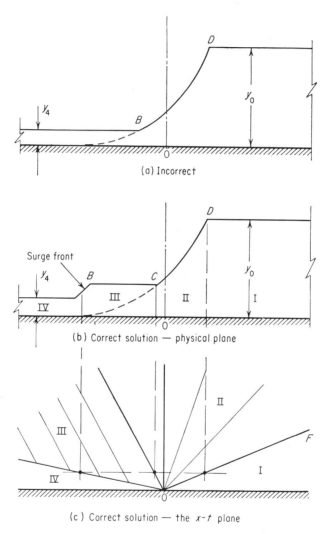

(a) Incorrect

(b) Correct solution — physical plane

(c) Correct solution — the *x-t* plane

Figure 8-8. *The Dam-break Problem,* $0 < y_4 < 0.138 y_0$

the relationships existing *within* this zone, and without of course changing the shape of the negative-wave profile.

The question is, just how does the downstream water limit the extent of Zone II? An apparent possibility is shown in Fig. 8-8*a*, in which the wave profile follows the same curve as in Fig. 8-7a until it terminates on the downstream water surface, at B. This solution, however, involves a discontinuity in the velocity, which is zero immediately downstream of B and has a finite value given by Eq. (8-25b) immediately upstream. Such a discontinuity can only occur via a surge of finite height, as in Fig. 8-8b. The negative-wave profile therefore runs downstream until it reaches a section C having the velocity and depth necessary to form a surge; in fact the velocity-depth relation given by Eq. (8–25b) forms the third equation required to define a surge, as discussed in Sec. 3.5.

Once a surge forms, it outruns the negative-wave profile because the velocity of the surge front B is greater than the wave velocity of the section C (but see Prob. 8.15). There is therefore a region BC, or Zone III, which is of constantly increasing length. The corresponding region on the x-t plane (Fig. 8-8c) contains a family of parallel characteristics, and it is bounded by a line of slope $-c$, where c is the speed of the surge. We could describe the region of still water downstream as Zone IV; the corresponding region on the x-t plane will be filled with a family of parallel characteristics, of slope $+c_4$. It is important to note that the profile CD in Fig. 8-8b is a part of the complete profile in Fig. 8-7a, and that the junction of the profiles BC and CD is therefore a sharp corner at C. In the actual flow situation it is possible that this corner would be rounded off by means of slight local departures from hydrostatic pressure distribution.

Another interesting question arises: will the negative-wave profile CD still be "centered" at O, the original position of the dam? For the case shown in Fig. 8-8, the answer is yes, but this need not necessarily be so. The reason for the existence of a steady flow situation at O in Fig. 8-7 was that the t axis coincided with a characteristic, along which v and c are constant. The same is true in Fig. 8-8, but will remain true only if Zone II includes the t axis. If the characteristic separating Zones II and III were to swing to the right of the t axis (because of increasing downstream depth) then the wave would no longer be centered; C would be on the right of O, as in Fig. 8-9a, and indeed would continue moving away from O to the right. However, the velocity and depth at section O would still remain steady, because this section now lies in the steady-state Zone III.

It is a simple matter to determine the criterion distinguishing the case of Fig. 8-8b from that of Fig. 8-9a. In the limiting condition on the threshold between the two cases, the dividing line between Zones II and III on the x-t plane is the t axis itself, and section C in the physical plane remains stationary at section 0. In the whole region BC, therefore, the velocity and depth are $2c_0/3$ and $4y_0/9$ respectively; it is readily shown (Prob. 8.17) that in this case the downstream depth y_4 is equal to $0.138y_0$. If y_4 is less than this amount, section C moves downstream, and the outflow at O remains steady at $8c_0y_0/27$; if y_4 exceeds this amount section C moves upstream and the outflow at O, although steady, is less than the maximum of $8c_0y_0/27$.

The Effects of Slope and Resistance

In a natural river the downstream water in the previous problem would be
flowing, and slope and resistance would have to be taken into account. A
well-known study of this kind is the one undertaken by Ré [4] during World
War II in connection with the Allied plans for crossing the Rhine. There was
a possibility that the Germans might seek to impede the operation by destroy-
ing dams in the vicinity of Basel; it was therefore important to be able to
predict the flood wave in the Rhine that would result. Ré used numerical and
graphical methods to trace the whole development and progress of the wave,

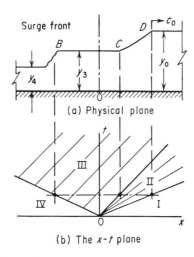

Figure 8-9.   The Dam-break Problem, $0.138y_0 < y_4 < y_0$

and it appears now that his work might have been improved by using the
simple-wave theory described above for the period immediately after the dam
break, when accelerations are very high and numerical methods become
sensitive to error unless very short time intervals are used. Once this initial
period of high acceleration has passed, numerical methods are suitable,
indeed essential if proper allowance is to be made for slope and resistance.

This is a suitable point at which to discuss the physical reality of the simple-
wave solutions for the dam-break problem arrived at in this section. One
obvious source of error lies in the large vertical accelerations arising from the
steep water-surface slopes in the first few seconds of the motion after the dam
break; these accelerations invalidate the assumption of hydrostatic pressure
distribution on which the whole analysis is based, but the overall effect is
unlikely to be serious as the error applies only to the first few seconds. More
serious is the effect of resistance, which may be appreciable when the downstream river bed is dry. Clearly the leading feather edge of Fig. 8-7a will be
strongly impeded by resistance, and one would expect the actual leading edge

of the wave to be much slower and steeper than the theory indicates. Experiments show that this is in fact the case; the results of Schoklitsch [5] and of Dressler [7], plotted in Fig. 8-10 in a dimensionless form due to Keulegan [6], are typical. The effect of resistance on the profile is seen to be quite appreciable.

Theories have been developed taking resistance into effect, but they are beyond the scope of this text and will not be discussed in detail here. The linear theory of Sec. 8.4 is of course inapplicable because the disturbance is by no means a small one. An analysis by Dressler [7], based on a perturbation

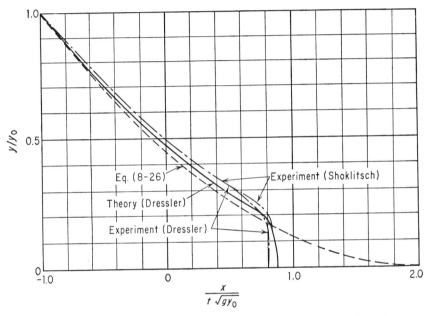

Figure 8-10.   *The Effect of Resistance on the Wave Following a Dam-break*

procedure, indicates a dimensionless wave profile which is plotted in Fig. 8-10 and is seen to be very close to the experimental profiles measured by Schoklitsch and by Dressler.

The case of the dry downstream riverbed is of somewhat limited practical interest, for it is unusual for the bed of a river or canal to be completely dry. When the downstream riverbed contains water initially, it seems likely that this water will reduce the effect of resistance materially by acting as a lubricant for the leading portion of the wave. This is still a matter of conjecture, as the problem has not been systematically investigated, either by experiment or analysis. It remains open as an interesting field to explore; meanwhile it is suggested that when the downstream riverbed contains water the simple-wave theory can be used without taking account of resistance, at least during the initial period of wave formation when the acceleration terms in Eq. (8–5) are high compared with S_f and S_0 (Prob. 8.19). The introduction of the slope and resistance terms can be postponed until the later consideration of the wave movement further down the channel.

These last remarks constitute a safe general rule for all sudden changes in flow such as result from the operation of sluice gates, lock gates, etc.

Sluice-Gate Operation

The operation of sluice gates sets problems which are essentially similar to the dam-break problem, having only the slight additional feature that the water upstream and downstream may be initially in motion. This feature does not introduce any real difficulty, for the negative-wave profiles of Figs. 8-7a and 8-8b can have a uniform velocity superimposed on them without changing their form or their basic dynamic relationships. Equation (8–25) can be applied directly to such a profile. From the viewpoint of the x-t plane, the fact that v_0 is nonzero creates no problems; the arrangement shown in Figs. 8-8b and c could quite well portray the situation after the sudden raising of a sluice gate, with Zones I and IV representing the initial flow upstream and downstream of the gate. An interesting feature of the negative-wave profile is that it can still be centered, from the viewpoint of a stationary observer, because it is still possible to have a C_1 characteristic coinciding with the t axis. However if v_0 is nonzero the center section, although still at the gate section, will have values of v and c which are in general greater than the value $2c_0/3$ which obtains when $v_0 = 0$ (Prob. 8.21).

Figures 8-11a, b, and c show typical situations arising from sluice gate operation. The first two show complete and partial closure of the gate, and the third shows partial opening of the gate. The first case is quite simply dealt with, and apart from the solution of particular numerical cases, a general criterion is easily deduced (Prob. 8.22) which determines whether the downstream face of the gate will be left high and dry, or will have water left in contact with it. If the former, the trailing feather edge on the wave will probably be attenuated rather than steepened by bed resistance.

In both of the situations shown in Figs. 8-11b and c, interesting problems of determinacy arise, in that the number of equations appears at first to exceed by one the number of unknowns. The resolution of these problems is left as an exercise for the reader (Probs. 8.25 and 8.27).

The above remarks cover only the first-wave development immediately following the gate operation. The later movement of the waves over long distances will of course be influenced by slope and resistance; for example the stationary water left downstream of the gate in Fig. 8-11a will eventually move off down the channel slope. In general this movement is best determined by numerical methods.

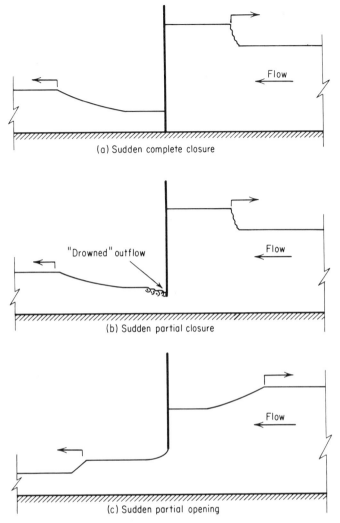

Figure 8-11. *Positive and Negative Wave Formation after Sluice Gate Operation*

8.6 Some Practical Problems

This section will deal with various topics not readily classifiable under other headings. They include means of computation as well as physical problems occurring in practice.

The General Shape of Cross Section

Thus far the method of characteristics has been considered only for channels of uniform rectangular section. The method can be extended to channels of uniform nonrectangular section by a device due to Escoffier [8]. The dynamic equation (8–5) can be applied unaltered to channels of any section; we write it here in rearranged form:

$$\frac{\partial v}{\partial t} + v\frac{\partial v}{\partial x} + g\frac{\partial y}{\partial x} = g(S_0 - S_f) \tag{8–48}$$

but we can no longer, as in Eq. (8–9), substitute $d(gy) = 2c\,dc$. The continuity Eq. (8–8) can be amended as follows, recalling that $dA = B\,dy$:

$$B\frac{\partial y}{\partial t} + A\frac{\partial v}{\partial x} + vB\frac{\partial y}{\partial x} = 0 \tag{8–49}$$

We now multiply Eq. (8–49) by $\sqrt{(g/AB)}$:

$$\sqrt{\frac{gB}{A}}\frac{\partial y}{\partial t} + \sqrt{\frac{gA}{B}}\frac{\partial v}{\partial x} + v\sqrt{\frac{gB}{A}}\frac{\partial y}{\partial x} = 0 \tag{8–50}$$

Adding or subtracting Eqs. (8–48) and (8–50), and recalling that $c = \sqrt{(gA/B)}$ we obtain

$$\frac{\partial v}{\partial t} + (v \pm c)\frac{\partial v}{\partial x} \pm \sqrt{\frac{gB}{A}}\left[\frac{\partial y}{\partial t} + (v \pm c)\frac{\partial y}{\partial x}\right] = g(S_0 - S_f) \tag{8–51}$$

Thus far the argument is closely parallel to that for rectangular channels, but with the difficulty that, in Eq. (8–51), there is no quantity directly corresponding to $2c$ in Eqs. (8–13) and (8–14); there is only a correspondence between the derivatives

$$\sqrt{\frac{gB}{A}}\,dy \qquad \text{and} \qquad d(2c),$$

in which A and B are themselves functions of y. To overcome this difficulty, Escoffier introduced the "stage variable" ω to replace y as a measure of the water level in the channel. It is defined as

$$\omega = \int_0^A c\frac{dA}{A} = \int_0^y \sqrt{\frac{gA}{B}}\frac{B\,dy}{A} = \int_0^y \sqrt{\frac{gB}{A}}\,dy \tag{8–52}$$

From this definition it follows immediately that

$$d\omega = \sqrt{\frac{gB}{A}}\,dy$$

so that Eq. (8–51) can be rewritten

$$\frac{\partial v}{\partial t} + (v \pm c)\frac{\partial v}{\partial x} \pm \left\{\frac{\partial \omega}{\partial t} + (v \pm c)\frac{\partial \omega}{\partial x}\right\} = g(S_0 - S_f) \tag{8–53}$$

and it is possible to write a pair of characteristic equations corresponding exactly to Eqs. (8–13) and (8–14), with ω replacing $2c$. For any given cross section, c and ω can be obtained as functions of depth without difficulty, although the calculations may be lengthy. Once $c(y)$ and $\omega(y)$ are established in graphical or tabular form, simple-wave problems in uniform nonrectangular channels can be solved without difficulty, after developing equations corresponding to Eqs. (8–25) and (8–26). (Probs. 8.29 through 8.31).

Numerical Methods

The major problem here is to take account of slope and resistance by including the term $g(S_0 - S_f)$ in Eqs. (8–5) and (8–48). In the first instance the discussion will be confined to the channel of rectangular section.

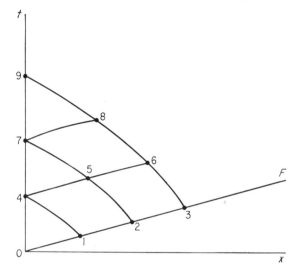

Figure 8-12.   *Scheme for Numerical Computation by Characteristics*

Figure 8-12 indicates a possible scheme of computation, based directly on the use of intersecting C_1 and C_2 characteristics. The particular problem used here by way of an example is that in which initial conditions are uniform and a disturbance $c = c(t)$ is specified along the t axis. Conditions are wholly known along OF, i.e., at 1, 2, 3, ..., so from the point 1 we can draw a C_2 characteristic meeting the t *axis* at 4; the value of Δt along 1–4 is deduced from its slope, and the value of v at 4 is then readily determined from Eq. (8–14). Conditions at 2 and 4 are now wholly known, so characteristics 4–5 (C_1) and 2–5 (C_2) can be drawn; the point 5 is therefore located and $(v + 2c)$, $(v - 2c)$, at that point can be obtained from Eqs. (8–13) and (8–14). Once this is done, it is possible to solve for v and c at 5. Following this general system, a complete network can be built up, establishing next the points 6, 7, 8, 9, in that order.

The process is cumbersome because the location of each meshpoint must be calculated as well as the values of v and c at that point. It is not therefore well adapted either to hand or machine computation. It is mentioned here only in order to draw attention to certain ways in which characteristics govern the course of any system of computation. On Fig. 8-12 the values of v and c at point 5 are clearly determined by the values of v and c over the region 4-0-1-2 on the t axis and OF, and *only* by those values. Outside this region (e.g., at 3) v and c could be given any values we please without influencing the values at 5. Similarly, values of v and c at the point 2 will influence values at the points 5, 7, and through them the points 6, 8, 9, etc., but cannot exert any influence on values at the point 4.

We can generalize these notions by drawing regions termed *ranges of influence* and *domains of dependence* bounded by C_1 and C_2 characteristics, as in Fig. 8-13. Conditions at the point P are determined wholly by conditions

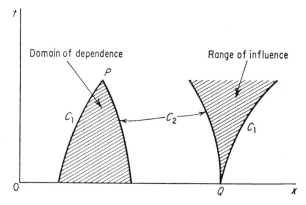

Figure 8-13.   *Range of Influence and Domain of Dependence*

within the shaded region below it, which constitutes its domain of dependence. Similarly, conditions at the point Q can affect conditions only within the fan-shaped range of influence drawn above it. These concepts are, as we shall see, of great importance in considering any scheme of numerical computation covering the x-t plane.

However, one can keep these concepts in mind without tracing out the characteristics themselves, and the systems of computation to be described here do not attempt to do so: they simply use Eqs. (8-9) and (8-12) to calculate v and c at points fixed in advance, and forming part of a regular network on the x-t plane. Two schemes are commonly used—one based on a rectangular array of points, Fig. 8-14a, and the other on a "staggered" array, as in Fig. 8-14b. The latter is usually more convenient, but the former has particular advantages in the region next to an axis, as in Fig. 8-14a. On p. 388, Chap. 9, a method is described which combines the merits of both systems.

We consider the staggered array first. For this case the elements of Eqs. (8–9) and (8–12) in the region LPR can be written in finite-difference form (where M is the mid point of the interval LR):

$$\left.\begin{array}{ll} \dfrac{\partial v}{\partial x} = \dfrac{v_R - v_L}{\Delta x}, & \dfrac{\partial c}{\partial x} = \dfrac{c_R - c_L}{\Delta x} \\[2ex] \dfrac{\partial v}{\partial t} = \dfrac{v_P - v_M}{\Delta t}, & \dfrac{\partial c}{\partial t} = \dfrac{c_P - c_M}{\Delta t} \end{array}\right\} \qquad \text{(8–54)}$$

with S_0 and S_f being given the values obtaining at M, i.e., the mean of the values at L and R. The time derivatives are evaluated along PM, as is clearly

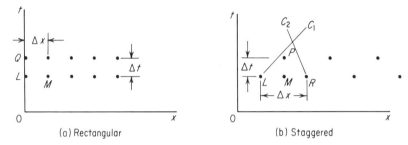

(a) Rectangular   (b) Staggered

Figure 8-14.   *Alternative Schemes for Finite-Difference Computations*

reasonable. It is now a matter of elementary algebra to substitute these values into (Eqs. 8–9) and (8–12) and solve for v_P and c_P:

$$v_P = v_M + \frac{\Delta t}{\Delta x}\left[2c_M(c_L - c_R) + v_M(v_L - v_R) + g\Delta x(S_0 - S_f)_M \right] \quad \text{(8–55)}$$

$$c_P = c_M + \frac{1}{2}\frac{\Delta t}{\Delta x}\left[2v_M(c_L - c_R) + c_M(v_L - v_R) \right] \quad \text{(8–56)}$$

The system is based on the assumption that v and c are known for one value of t, and that the solution is being advanced to a new value, $t + \Delta t$. This situation will be quite common in practice, particularly when a simple-wave solution is adopted for the short period immediately following some sudden change in control, and slope and resistance terms are then introduced at the end of this period (Prob. 8.33). The problem will, of course, occur in many other forms but in all of them it is necessary to keep in mind the questions of influence and dependence, as in Fig. 8-13. In the situation of Fig. 8-14b, this means that Δx and Δt must be chosen so that the triangle formed by drawing a C_1 characteristic through L and a C_2 characteristic through R must enclose the point P (otherwise the values at L and R would not in fact determine the values at P). This means that the ratio $\Delta x/\Delta t$ must always exceed $2(v + c)$.

In the situation of Fig. 8-14a, we assume that v and c are known at L and M, and c only at Q. This problem would arise in the case of a disturbance

$c = c(t)$ prescribed along the t axis. To calculate v_Q we can use Eq. (8–14a). Setting

$$\left.\begin{array}{ll}\dfrac{\partial v}{\partial x} = \dfrac{v_M - v_L}{\Delta x}, & \dfrac{\partial c}{\partial x} = \dfrac{c_M - c_L}{\Delta x} \\[3mm] \dfrac{\partial v}{\partial t} = \dfrac{v_Q - v_L}{\Delta t}, & \dfrac{\partial c}{\partial t} = \dfrac{c_Q - c_L}{\Delta t}\end{array}\right\} \qquad \text{(8–57)}$$

and evaluating S_0 and S_f at L, we obtain by substituting into Eq. (8–14a) and solving for v_Q:

$$v_Q = v_L + \Delta t \left[\frac{1}{\Delta x}(c_L - v_L)(2c_L - 2c_M - v_L + v_M) + g(S_0 - S_f)_L \right]$$
$$+ 2(c_Q - c_L) \quad \text{(8–58)}$$

In this case the ratio $\Delta x/\Delta t$ must exceed $|v - c|$, so that the C_2 characteristic through M passes above Q.

When the channel is rectangular but of variable width, a fairly simple modification is required to Eq. (8–12), and through it to Eqs. (8–56) and (8–58). Details are left as an exercise for the reader (Probs. 8.46 and C8.2). When the channel is nonrectangular there is a further unknown ω_P in the situation of Fig. 8-14b, and a further equation in the form of the geometrical relationship between c and ω. The details of the computation become lengthy, and are probably best handled by a computer program (Probs. 8.34 and C8.3).

In planning a series of calculations, provision should of course be made to terminate the network in the approximate neighborhood of the first C_1 characteristic. Mesh points do not have to fall exactly on this line, but it is convenient if they are either on the line or slightly beyond it, so that they can have values assigned to them appropriate to the zone of quiet.

Further aspects of numerical methods are discussed in connection with flood routing in Chap. 9.

Modification of Surges by Change of Section

Consider a surge approaching a channel junction, as in Fig. 8-15a. After reaching the junction it will continue to travel on through the two branch channels but with diminished intensity because of the increased area of cross section. The region at the junction therefore becomes filled with water of depth intermediate between the two depths in Fig. 8-15a; the result must be that a negative wave is reflected back into the channel from which the surge originally came. The relationships among the various depths can be simply calculated if we assume that the surge is small enough to have a velocity $c = \sqrt{gy_1}$, where y_1 is the original water depth. We can further assume without loss of generality that the water in the two branches, 2 and 3, is initially stationary.

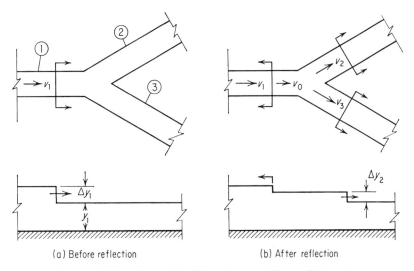

(a) Before reflection   (b) After reflection

Figure 8-15.   *Surge Modification at a Channel Junction*

We denote the heights of the original surge by Δy_1, and of the transmitted surge by Δy_2. The height of the negative reflected wave will therefore be $\Delta y_1 - \Delta y_2$. The velocities behind the three positive surges are v_1, v_2, and v_3; the velocity behind the reflected wave is v_0. We now recall from Eq. (3–17) that the speed of a surge is equal to $c = \Delta q/\Delta y$, where Δq and Δy are the changes in discharge and depth through the surge; since c is also equal to $\sqrt{(gy_1)}$ we can write

$$\Delta y_1 = \frac{v_1 y_1}{c_1} = \frac{v_1 c_1}{g}$$

and similarly

$$\Delta y_2 = \frac{v_2 c_2}{g}$$

$$= \frac{v_3 c_3}{g}$$

$$\Delta y_1 - \Delta y_2 = \frac{c_1}{g}(v_0 - v_1)$$

$$\left.\right\} \qquad \textbf{(8–59)}$$

By continuity $\qquad\qquad A_1 v_0 = A_2 v_2 + A_3 v_3 \qquad\qquad$ **(8–60)**

Solving these equations for Δy_2, we obtain

$$\frac{\Delta y_2}{\Delta y_1} = \frac{2A_1/c_1}{\dfrac{A_1}{c_1} + \dfrac{A_2}{c_2} + \dfrac{A_3}{c_3}} \qquad\qquad \textbf{(8–61)}$$

Maintaining the distinction among c_1, c_2, and c_3 gives the analysis a false

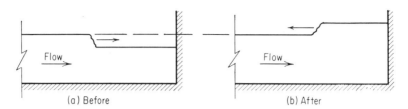

(a) Before   (b) After

Figure 8-16.   *Doubling of Surge Height by Reflection at a Blind End*

appearance of generality, for in fact all these wave velocities have been
assumed equal. It is more realistic to write the result as

$$\frac{\Delta y_2}{\Delta y_1} = \frac{2A_1}{A_1 + A_2 + A_3} \tag{8-62a}$$

$$= \frac{2b_1}{b_1 + b_2 + b_3} \tag{8-62b}$$

to a first approximation, where b is the width of a channel (assumed rectangular). Equation (8–62) can of course be extended to cover any number
of branch channels. If, instead of branches, there is simply a change of section
in the same channel, the surge will increase in height if the change is a contraction, or reduce in height if it is an expansion. When a surge reaches a
blind end, Eq. (8–62) shows that it will double in height, as in Fig. 8-16.

The above analysis relates to fairly abrupt changes of section; a common
example of a gradual change of section that may affect surge movement is the
bed slope on a short artificial channel feeding, say, a hydroelectric scheme.
If the downstream channel outlet is closed abruptly, a positive surge will
develop almost immediately. But as it moves upstream the downstream water
level must keep rising, as in Fig. 8-17, if the surge height is to remain approximately constant. The effect must be to slow down the motion of the surge.

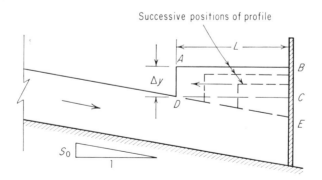

Figure 8-17.   *Positive Surge in a Sloping Channel*

In making an analysis, resistance need not be taken into account, for the water downstream of the surge front has no net motion. It may be assumed that the upstream flow is uniform. Considering first a channel of rectangular section, we note that the continuity relationship $c = \Delta q/\Delta y$ has to be modified, for it implies that the discharge difference Δq has only to fill the rectangle $ABCD$, of area $L\Delta y$. In fact, it also has to fill the triangle DCE, of area $S_0 L^2/2$. The effect is that at time t after the outlet is closed, the total volume of water per unit width that has been delivered by the discharge difference Δq is equal to

$$\Delta q \cdot t = v_0 y_0 t = L\Delta y + \frac{S_0 L^2}{2} \tag{8-63}$$

where the subscript 0 indicates the initial uniform flow conditions. Since the surge velocity c is equal to dL/dt, this equation can be combined with the momentum equation to yield a differential equation in L; but the equation is not explicitly soluble, even if we use the linear form of the momentum equation appropriate to surges of small height:

$$c + v_0 = \sqrt{g y_0}\left(1 + \frac{3\Delta y}{4 y_0}\right) \tag{8-64}$$

which is implied in Eq. (3–20). In general a trial solution is required, and it is convenient to work step-by-step through specified time intervals Δt. Assuming that c, Δy, and L are known at the start of a time interval, choose a trial value of L at the end of the interval. This leads to a value of c, and thence to two values of Δy, via Eq. (8–63) and the momentum equation. If these two values are not equal, a new trial value of L must be taken. The process is well adapted to solution on a high-speed computer (Probs. 8.36, 8.37, C8.4, and C8.5).

It is not quite correct to assume that the water downstream of the surge is always stationary with a horizontal surface. Changes in surface level brought about by the surge movement are not instantaneously propagated over the whole region downstream, and there will in fact be a slight water-surface slope, with a slight downstream movement of the water. This surface slope will itself form the profile of a positive wave moving downstream, so that the maximum water level reached when the surge arrives at the upstream end of the channel will eventually reach the downstream end as the result of this wave motion. But since the designer is interested in the magnitude of the maximum surface level, rather than in how it is reached, it will be sufficient for his purposes to assume, as in the above analysis, that the water surface downstream is horizontal and rising steadily as a whole.

The Forebay Problem

Consider a channel like the one described above, whose function is to deliver water to a hydroelectric scheme. Suppose that the initial situation consists of any steady-state condition, including complete shutdown. There is then a sudden increase in demand from the plant; the upstream source which feeds the channel will not be able to deliver this flow immediately, so some form of storage is required to maintain the supply until the new discharge can be established in the channel. This storage may be supplied by a natural or artificial basin, called a *forebay*, between the end of the channel and the turbine intakes. Determination of the necessary size of this basin constitutes the forebay problem.

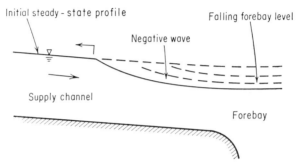

Figure 8-18.   *The Forebay Problem*

The extreme situation considered by the designer is that in which the demand is suddenly increased from zero to the maximum. The design requirement is that the full discharge be established in the channel before the water surface at the intakes falls to a dangerously low level. The complete sequence of events is complex and difficult to analyze, but a useful insight can be attained by noting the similarity between this case and the estuary problem of Example 8.1. As the water level falls at the channel outlet, a negative wave moves up the channel as shown in Fig. 8-18, so that if the channel is rectangular Eq. (8-25b) is *approximately* applicable and the channel outflow velocity is

$$v = 2\sqrt{gy_0} - 2\sqrt{gy} \qquad \qquad (8\text{--}65)$$

being in this case taken positive in the direction opposite to that of wave propagation. The subscript 0 indicates initial conditions. Equation (8-25b) is not exactly applicable because the channel-bed slope makes y_0 and c_0 vary, and the initial characteristic OF on the x-t plane will not be a straight line. The problem is not therefore exactly of the simple-wave type, but it approximates closely enough to this type for Eq. (8-65) to be applicable, taking an average value for y_0.

During the progress of a negative wave up the channel, water is being taken from the channel itself, not from the upstream source. If there were no forebay, the channel would therefore provide the only available storage, whose effectiveness will depend on the discharge capacity of the channel outlet; this capacity depends basically on the properties of the negative wave as disclosed by Eq. (8–65). In a rectangular channel of width b the outflow discharge Q is equal to:

$$Q = 2by\sqrt{g}(\sqrt{y_0} - \sqrt{y})$$ (8–66)

which has a maximum when $y = 4y_0/9$, equal to

$$Q_m = \frac{8}{27} by_0\sqrt{gy_0}$$ (8–67)

This flow is the maximum rate at which a negative wave can withdraw water from the channel, without recourse to forebay storage or to the upstream source. If this rate Q_m exceeds the steady-state channel capacity Q_s, then it is likely that no forebay storage will be needed—provided, of course, that a drawdown of up to $5y_0/9$ at the turbine intakes can be tolerated.

In order to examine this tentative conclusion more critically, we must consider what happens when the negative wave reaches the lake supplying the channel. A positive surge will then run back down the channel; this surge provides the first means by which water flows from the lake into the channel. The discharge rate behind the surge will necessarily be greater than the rate in front of it, established by the negative wave. Accordingly, when the surge reaches the turbine intakes there is temporarily more water than required; further variations in the flow could be traced (with some difficulty) by numerical methods, but the tendency will be towards the steady-state condition and it is unlikely that there will be, at any future instant, a shortage of water at the intakes.

If a negative wave cannot draw the required flow from the channel without lowering the level at the intakes below what is acceptable, forebay storage is required. The forebay must have a large enough plan area for it to supply the flow deficiency until the positive surge arrives, bringing with it water from the upstream source. If the supply channel is long, the required size of forebay may be excessive. For example, if the channel is 10 miles long with a water depth of 10 ft, the time of wave travel to the supply reservoir and back will be nearly 2 hr. To make good a deficiency of even a few hundred cusecs over that period would require a storage capacity of millions of cubic feet.

It remains to be seen how the maximum Q_m compares with the normal delivery of the channel, Q_s. If we assume this to take place at uniform flow with a specific energy of approximately y_0, then its Froude number must be 0.318 (Prob. 8.38) if Q_s is to be equal to Q_m. In order to minimize resistance losses, power canals usually have Froude numbers less than this amount, so in many cases the negative wave will in fact be able to draw the required flow from the channel. If not, forebay storage will be required.

The above discussion has covered general principles in a semiquantitative way, and needs to be supplemented in the case of particular design projects by more detailed studies, taking slope and resistance into account. An

interesting recent account of studies of this kind in French practice is given
by Guelton and others [9]. But however the details may work out, it must be
remembered that the whole sequence of events is dominated, first by the passage of the negative wave upstream to the supply reservoir, and second by the
downstream movement of a positive surge, which is the means of first bringing
water from the supply reservoir to the intakes.

8.7   Oscillatory Waves

Much of the discussion in this chapter has dealt with wave motions defined
in the general sense of the term—i.e., as a moving disturbance without restriction as to form. In this section we consider waves in the more restricted
sense of a moving surface disturbance which is also periodic or oscillatory in
form, Fig. 8-19.

The field of theory and observation opened up by consideration of this
topic is a vast one, including among other things a long history of marine
wave observations, and a substantial body of advanced mathematical theory.
Some of the problems involved have proved intractable up to very recent times;
for instance, as recently as 1956 a survey of theories attempting to explain
the development of waves by wind action resulted in the conclusion that all
current theories contained serious flaws. It is only since that date that the
matter has been cleared up [10].

The treatment given here will only be a bare summary of the results which
are most obviously useful to the engineer dealing with free surface flow. It
may be remarked, however, that even if the engineer's needs are limited, he
will find it helpful (and interesting) to have some general background knowledge of waves and their behavior. The usual source of such knowledge is the
marine observations referred to above, whether in the form of mariner's lore,
or of scientific oceanographic data. Some of the former is of doubtful validity,
such as the belief that every seventh wave is higher than the others, but some
of it is quite soundly based, such as the observation that waves tend to steepen
when moving against a tidal flow. Problem 8.12 shows that this will in fact
happen if the water is shallow enough for the resistance of the sea bed to be
appreciable.

Scientific writers like Cornish [11] and Russell [12] have used this wealth
of marine observation as a basis for interesting general accounts of wave
behavior, keeping mathematical detail to a minimum. Also to be recommended is the admirable work of Rachel Carson [13]. All of these accounts
deal with the sea, and convey something of the sea's universal interest and
appeal; the same feeling is communicated even by parts of Stoker's [2]
sophisticated mathematical treatment of the subject.

The following material will not include rigorous proofs of the basic wave
equations, for these proofs are based on the methods of theoretical hydrodynamics, which are beyond the scope of this text. Complete treatment of the

proofs is given in a number of advanced texts, e.g., those of Stoker [2] and Milne-Thomson [14]. More important, from the engineer's viewpoint, is an appreciation of the way in which the basic equations dictate certain features of the behavior of waves, and it is with this aspect of the question that the following treatment will mainly deal. For a more complete discussion along these lines the reader is referred to the proceedings of a symposium, edited by Johnson [15].

The Basic Equations

The wave motion of a surface disturbance can be described by the equation

$$y = F(x - ct) \tag{8–68}$$

where F indicates any arbitrary function, depending on the type and form of the disturbance. The reason for the form of Eq. (8–68) is that to an observer moving along with velocity c the quantity $(x - ct)$ appears to be constant, and so therefore does y; herein, as previously remarked, lies the essence of wave motion.

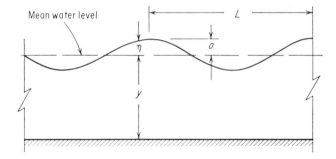

Figure 8-19.   *Definition Sketch for Oscillatory Wave Motion*

For the oscillatory wave of Fig. 8-19 the function F will be periodic in form, and an obvious choice is the sine function. If this is the correct choice, then the appropriate special form of Eq. (8–68) is

$$\eta = a \sin \frac{2\pi}{L} (x - ct) \tag{8–69}$$

where L is the wavelength and η is the height of the surface above the mean depth line, as in Fig. 8-19.

Apart from the wavelength L, the two other important parameters are the wave velocity c and the wave period T. For *any* periodic wave motion, these three are related by the equation

$$L = cT \tag{8–70}$$

and the question arises whether the conditions of our particular problem—liquid surface waves—give rise to a further equation connecting L, c, and T. If such an equation exists, it will show the influence of both gravity and surface tension, for in order to distort a liquid surface as in Fig. 8-19 it is necessary to do work against both gravity (by lifting liquid from below mean water level to above that level) and surface tension (by elongating the free surface). Surface tension has very little effect unless the wavelength is quite small, and it will not be taken account of in the initial statement of results; its effects will be noted later. The effect of viscosity is small: while there is some viscous damping of oscillatory waves, it is a distinctly second-order effect, and for most practical purposes can be neglected.

A consequence of this last statement is that the theory of irrotational flow of inviscid fluids can be used with confidence in deriving the equation of wave motion. Even so, the problem is one of some difficulty except in the simplest case, that of a wave whose amplitude is only a small fraction of its length. For this case it can be shown that

$$c^2 = \frac{gL}{2\pi} \tanh \frac{2\pi y}{L} \qquad (8\text{-}71)$$

where y is the mean depth. Waves described by this equation are known as "Stokesian" waves after the originator of the relevant analysis. Equation (8-71) is one product of a whole complex of results which completely determine the motion of the wave and of the fluid particles. These results show also that the sinusoidal form indicated by Eq. (8-69) is in fact correct, and that the individual water particles move in elliptical orbits whose semiaxes are

$$\frac{a \cosh 2\pi y'/L}{\sinh 2\pi y/L} \quad \text{and} \quad \frac{a \sinh 2\pi y'/L}{\sinh 2\pi y/L} \qquad (8\text{-}72)$$

in the horizontal and vertical directions respectively, where y' is the mean height of a particle above the bed, and y as before is the mean surface level. Figure 8-20 shows typical particle orbits; each particle moves forward as the crest of the wave passes over it, and backward as the trough passes over it.

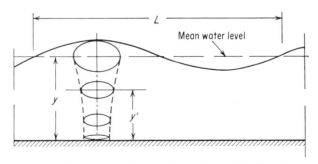

Figure 8-20. Elliptical Particle Orbits in Shallow Water

Clearly the vertical amplitude of the uppermost orbit must be $2a$, the amplitude of the wave, and that of the lowest orbit zero. The horizontal amplitude diminishes with increasing depth below the surface, but does not necessarily reach zero, unless the water is very deep.

It was remarked in Sec. 2.3 that the result $c^2 = gy$ is a special case of Eq. (8–71) when L/y is large. Figure 8-21 shows how c^2/gy approaches its limiting value of unity from below. The behavior of the tanh function shows that when L/y is small, $\tanh 2\pi y/L$ tends to unity and the "deep water" wave equation

$$c^2 = \frac{gL}{2\pi} \tag{8–73}$$

is applicable. The water particles now move in circular orbits (Fig. 8-22); by assuming this to be the case it is possible to verify Eq. (8–73) by an elementary argument (Prob. 8.40).

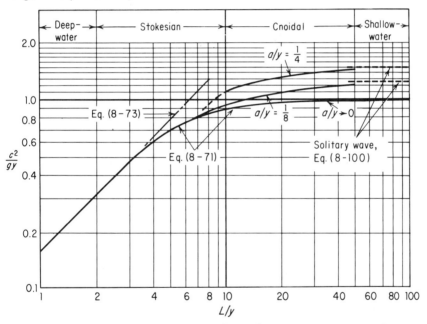

Figure 8-21. Gravity Wave Relationships among c, L, y, and a

The ratio y/L need only exceed about 0.5 for Eq. (8–73) to be true to a very close approximation; for lower values of y/L the wave is said to "feel the bottom" and Eq. (8–71) must be used. When y/L is less than about 0.05 the equation

$$c^2 = gy \tag{8–74}$$

is very nearly true. For this "shallow-water" wave the horizontal amplitude of the water-particle orbits becomes virtually independent of y', since $\cosh 2\pi y'/L$

is close to unity. The horizontal velocity is therefore constant throughout any vertical section, which will accordingly move back and forth as a whole. A further property these waves possess is that of having hydrostatic pressure distribution down each vertical section; this property, so commonly assumed to be present in other forms of open channel flow, is absent from all oscillatory waves except those of this particular type. It follows that the dynamic equation (8–5) can be applied only to this type of oscillatory wave.

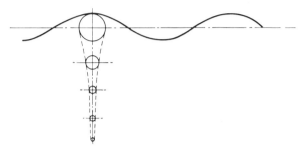

Figure 8-22.   *Circular Particle Orbits in Deep Water*

Suppose that we do in fact apply the dynamic equation (8–5) with $S_0 = S_f = 0$, and the continuity equation (8–10), to a situation in which there is wave action of small amplitude in otherwise still water. The depth is now written as $y + \eta$, y being the mean depth, a constant. Then Eq. (8–5) and (8–10) become

$$\frac{\partial \eta}{\partial x} + \frac{v}{g}\frac{\partial v}{\partial x} + \frac{1}{g}\frac{\partial v}{\partial t} = 0 \tag{8–75}$$

$$v\frac{\partial \eta}{\partial x} + (y + \eta)\frac{\partial v}{\partial x} + \frac{\partial \eta}{\partial t} = 0 \tag{8–76}$$

Now v and η are both small, so that products of these quantities with each other and with their derivatives can be neglected. Removing these terms, we can then eliminate v between Eqs. (8–75) and (8–76) in this way:

$$\frac{\partial^2 v}{\partial x \partial t} = -g\frac{\partial^2 \eta}{\partial x^2} = -\frac{1}{y}\frac{\partial^2 \eta}{\partial t^2}$$

whence
$$\frac{\partial^2 \eta}{\partial t^2} = gy\frac{\partial^2 \eta}{\partial x^2} \tag{8–77}$$

which is well known as the most general possible form of the wave equation. Its solution is

$$\eta = F_1(x - \sqrt{gy}\, t) + F_2(x + \sqrt{gy}\, t) \tag{8–78}$$

which denotes two wave motions traveling in opposite directions with velocity $\sqrt{gy}$, as previously indicated by the theory of characteristics.

For the Stokesian wave theory to apply when y/L is small, it is not sufficient for a/L to be small, since under these conditions a/y could be finite. If this ratio is finite, vertical accelerations will be appreciable and a more general theory is needed to describe the wave. Such a theory has been developed, and the resulting equation for the wave profile, corresponding to Eq. (8–69), involves the elliptic cn-function instead of the sine function. For this reason the appropriate waves are termed "cnoidal." Their profile is no longer symmetrical about the mean depth line; as in Fig. 8-23, the crests become sharper and the troughs become flatter. For this reason, and others, the length a must now be redefined as one-half of the overall wave height H shown in Fig. 8-23,

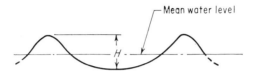

Figure 8-23.   Typical Cnoidal Wave Profile

without the implication that the mean depth is midway between crest and trough levels. It is this definition of a that is used in labeling the upper curves in Fig. 8-21; they show that the speed of the cnoidal wave is greater than that of the small-amplitude Stokesian wave, and in many cases greater even that $\sqrt{gy}$. This behavior is analogous to that of the surge of finite height, Sec. 3.5.

An important feature of such waves is that the particle orbits are no longer closed, as in Fig. 8-20. If the lowest layer of fluid is regarded as stationary then all the fluid particles above it undergo a net translation in the direction of wave motion during the passage of one complete wave. This means that the wave velocity may be defined as the velocity of the wave profile relative to either the lowest fluid layer or the body of the fluid itself; the latter case would involve the definition of a mean fluid velocity relative to the wave profile, and this mean might be (in the notation of Sec. 1.9) either v_m, αv_m, or βv_m, depending on whether the flux of fluid, of energy, or of momentum, was thought to be significant. For large values of a/y the coefficients α and β could be substantially greater than unity.

However, the doubts raised by these alternatives are of limited practical interest in the present context. We shall be concerned here mainly with waves moving through otherwise stationary water, with no effective restraint downstream; in this case it is observed that the wave velocity relative to the stationary surroundings is the velocity of the wave profile relative to the lowest fluid layer, and it is this velocity which is plotted in Fig. 8-21. A consequence of this action is that a net fluid flow, or mass transport, is induced relative to the stationary bed; a familiar example is provided by the approach of ocean waves to a beach, each wave bringing with it a certain inflow of water which runs up the beach before retreating.

The basic Stokesian analysis, leading to Eq. (8-71), is strictly correct only when a/L and a/y are small. It can be extended to take account of finite amplitudes, and the extended theory appears to be applicable for values of L/y less than 10 approximately [6]. For $L/y > 10$, the cnoidal theory is applicable. Both theories indicate substantial departures from the sinusoidal profile, as in Fig. 8-23.

Although appreciable wave amplitudes lead to .appreciable errors in the basic Stokesian theory, these errors can be tolerated in most situations of engineering interest, if only because there is already much uncertainty in the engineer's predictions of the characteristics of naturally occurring waves. The subsequent discussion will therefore be confined to the elaboration of the Stokesian wave theory. We may also note here that the theory is a two-dimensional one—i.e., the waves are assumed to be confined between vertical planes parallel to each other and to the direction of wave motion.

Wave Energy

The kinetic energy in a wave can be obtained by the essentially straightforward process of integration throughout the whole body of liquid. It is equal to

$$\frac{1}{4}a^2\gamma L = \frac{1}{16}H^2\gamma L \qquad (8\text{-}79)$$

over one whole wavelength, per unit width of wave crest. The potential energy is obtained more simply, being equal to the weight of liquid above the mean depth line times the average height through which it has been raised. It happens to have the same value as the kinetic energy; it follows that the total energy possessed by one whole wavelength, per unit width of wave crest is equal to

$$\tfrac{1}{2}a^2\gamma L = \tfrac{1}{8}H^2\gamma L \qquad (8\text{-}80)$$

Figure 8-24.   Wave Groups

Group Velocity

It is well established by observation that waves often travel in groups, as in Fig. 8-24. A simple way of creating a single group of waves is to make a disturbance in still water, whereupon a group of waves will be seen to move radially outwards from the center of the disturbance. In such a case one finds that if attention is fixed on any individual wave, say near the middle of the group, that wave is seen to move up to the front of the group and then disappear. Similarly if attention is fixed on the rear of the group, waves will be seen to originate there before moving forward through the group and disappearing. The conclusion is the curious one that individual waves are moving faster than the group of which they form a part.

The explanation for this behavior can be found without difficulty. Groups like those in Fig. 8-24 correspond to "beats" in the theory of sound, and can therefore be produced by the superposition of two trains of waves having slightly differing wave lengths and velocities. To carry through the analysis, it is convenient to make the substitution $m = 2\pi/L$; Eq. (8–69) then becomes

$$\eta = a \sin m(x - ct)$$

and the displacement η due to the superposition of two wave trains of equal amplitude and slightly differing wavelength is then equal to

$$\eta = a \sin m(x - ct) + a \sin(m + \Delta m)[x - (c + \Delta c)t]$$
$$= 2a \cos \tfrac{1}{2}[x\Delta m - (m\Delta c + c\Delta m)t] \sin m(x - ct) \qquad \textbf{(8–81)}$$

neglecting the product of Δm and Δc. This equation describes an "amplitude modulated" wave train. It may be written as

$$\eta = G \sin m(x - ct)$$

$$\left.\begin{array}{l} \\ \end{array}\right\} \qquad \textbf{(8–82)}$$

where $\qquad G = 2a \cos \tfrac{1}{2}[x\Delta m - (m\Delta c + c\Delta m)t]$

This implies that there is a wave train of velocity c, length $2\pi/m$, and variable amplitude G. The expression for G indicates that all the waves of the "fundamental" train are enclosed in an envelope (Fig. 8-24) which itself forms the outline of a wave profile moving with a velocity

$$\frac{m\Delta c + c\Delta m}{\Delta m}$$

This is the "group velocity," i.e., the velocity of the wave groups. In general Δc and Δm may be taken as vanishingly small, and the group velocity may then be written

$$c_g = c + m \frac{dc}{dm} = c - L \frac{dc}{dL} \qquad \textbf{(8–83)}$$

The above argument holds for any kind of wave. Applying it now to the waves with which we are concerned, i.e., to Eq. (8–71), we find after some manipulation that

$$c_g = \tfrac{1}{2}c\left(1 + \frac{4\pi y/L}{\sinh 4\pi y/L}\right) \qquad \textbf{(8–84)}$$

When $y \to \infty$, the ratio $(4\pi y/L)/\sinh 4\pi y/L$ tends to zero; when $y \to 0$, the same ratio tends to unity. It follows that the group velocity c_g ranges between the values of $\frac{1}{2}c$ for deep water and c for very shallow water, being always less than c and thus verifying the observations described above. The form of Eq. (8–84) is such that the two extremes of deep-water and shallow-water conditions are not approached quite as rapidly with increasing or diminishing y/L as in the case of Eq. (8–71), but these extremes are approximated quite closely (within 2–3 per cent) at the limiting values of $y/L = 0.5$ and 0.05 which were given previously in connection with Eq. (8–71).

Energy Propagation

It is clear that waves not only possess energy, but must also transmit this energy in the direction of motion. The rate of energy transfer across any fixed vertical plane normal to the direction of wave motion is equal to the integral, taken throughout the depth of flow,

$$\int \{p + \tfrac{1}{2}\rho(u^2 + v^2)\}\, u\,dy$$

where u and v are the horizontal and vertical velocities and p is the pressure due to the wave motion—which as we have seen, is not usually hydrostatic in distribution. The known solution provides complete information on the distribution of velocity and pressure; from this, another straightforward integration leads to the mean rate of energy transfer

$$\frac{dE}{dt} = \frac{1}{16} H^2 \gamma c \left(1 + \frac{4\pi y/L}{\sinh 4\pi y/L} \right)$$

$$= \tfrac{1}{8} H^2 \gamma c_g \tag{8–85}$$

a result which has some interesting implications. One's natural expectation might be that the rate of forward energy transfer is equal to $H^2 \gamma c/8$, since $H^2\gamma/8$ is the wave energy per unit length in the direction of motion, and one would expect this energy to be carried forward at the wave velocity c. Equation (8–85) shows, on the other hand, that it is the group velocity c_g rather than the wave velocity c that governs the forward transmission of energy (and this is true whether the waves happen to be traveling in groups or as a single train of constant amplitude).

On account of this result, the group velocity c_g may be described as the velocity of energy transmission, for it is equal to the rate of energy transfer divided by the amount of energy per unit length in the direction of transfer. There is in fact an exact analogy with the ordinary continuity equation, with energy corresponding to fluid volume. It is interesting to note that Eq. (8–85) is derived from a dynamical argument, whereas Eq. (8–84) is kinematical in origin. A discussion of whether, and to what extent, there is a fundamental identity between the two results is beyond the scope of this text.

Wave Resistance and Energy Radiation

When a wave train is set up in otherwise stationary water by a moving disturbance such as a ship, the wave motion gives rise to a special type of "wave resistance." Consider a fixed vertical plane some distance behind the disturbance, and normal to the direction of its motion; the ship is moving at the same speed as the waves, c, and the region between the ship and the fixed plane must therefore be having its energy (per unit width) increased at the rate $H^2 \gamma c/8$. But energy is transmitted forward across this plane by the waves at a rate of only $H^2 \gamma c_g/8$, from Eq. (8–85); the difference, $H^2 \gamma (c - c_g)/8$, must be supplied by the ship. The need for the ship to create energy at this rate constitutes in effect, a "wave resistance" P_w, such that

$$P_w c = H^2 \gamma (c - c_g)/8,$$

i.e.,
$$P_w = \frac{H^2 \gamma (c - c_g)}{8c} \tag{8–86}$$

The same form of energy imbalance underlies the "energy radiation" phenomenon discussed in Sec. 6.7 in connection with the undular hydraulic jump. It is convenient to change the viewpoint of the observer so that the jump becomes a moving positive surge leaving stationary water behind it and to consider what happens to the kinetic energy of the moving water in front of the surge as it absorbed by the surge. If the surge is broken, some of this kinetic energy will be dissipated by turbulence at the face of the surge. If the surge is undular, it becomes a moving disturbance of the kind exemplified by a ship in the previous discussion, and the same conclusion follows: the region between the surge and a stationary transverse plane somewhere behind it is filling with energy faster than energy can be transmitted forward across this plane. The difference must be supplied by some of the kinetic energy of the moving water being absorbed by the surge, so that something analogous to energy dissipation is taking place. Now the energy imbalance arises from a failure to transmit, or "radiate" enough energy forward through a stationary plane; accordingly, if the observer's viewpoint is changed so as to make the surge appear stationary, one can think of energy as being radiated downstream. And since the presence of the energy imbalance is due to the velocity difference $(c - c_g)$, the energy is in effect being radiated at a velocity equal to $(c - c_g)$.

Modification of Waves in Shallow Water

Equation (8–73) is true for deep water, and Eq. (8–71) is true for shallow water. We now consider what happens to a particular train of waves as it moves from deep water into shallow water, say over a shelving beach.

Denoting deep-water conditions by the subscript 0 we obtain from Eqs. (8–71) and (8–73) the equation

$$\frac{c^2}{c_0{}^2} = \frac{L}{L_0} \tanh \frac{2\pi y}{L} \tag{8–87}$$

but this equation alone will not tell us how c and L behave in water of given depth y, even if c_0 and L_0 are known. We need a further equation, and this is provided by considering the passage of waves through a region of varying depth, say near a shoreline as in Fig. 8-25. In particular we consider the wave period T at two typical points A and B of differing depths. The number of wave crests passing each point in unit time is equal to the wave frequency, $1/T$; it follows that if the frequencies differ at A and B, the number of wave

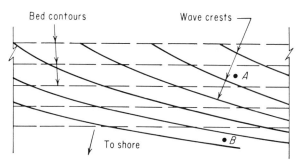

Figure 8-25.   *Waves Approaching a Shelving Beach*

crests in the space between these points must increase or decrease without limit. This is clearly impossible, so we conclude that the wave period is the same at A, B, and everywhere else. It follows from Eq. (8–70) that

$$\frac{c}{c_0} = \frac{L}{L_0} \tag{8–88}$$

and that Eq. (8–87) becomes

$$\frac{c}{c_0} = \tanh \frac{2\pi y}{L} = \tanh \frac{2\pi y}{L_0} \frac{c_0}{c}$$

or

$$\frac{2\pi y}{L_0} = \frac{c}{c_0} \tanh^{-1} \frac{c}{c_0}$$

$$= \frac{c}{c_0} \log_e \sqrt{\frac{c_0 + c}{c_0 - c}} \tag{8–89}$$

from which, given c_0 and L_0, c may be found by trial.

Because $\tanh (2\pi y/L)$ is less than unity, c is always less than c_0 and the wave loses speed as it moves into shallow water. This may give rise to refraction of the wave motion, as in the theory of geometrical optics. When a ray of light passes from air into glass, it is deflected closer to the normal, as in Fig. 8-26a, in accordance with Snell's law

$$\frac{\sin \alpha}{\sin \alpha_0} = \frac{c}{c_0} \qquad (8\text{–}90)$$

the ratio c/c_0 being the refractive index of the glass. Essentially the same thing happens when a surface wave moves into shallower water, but in this case it is convenient to interpret α as the angle between the wave crest and the bed contour, as in Fig. 8-26b. A uniform train of waves moving obliquely towards a uniformly shelving beach will therefore be deflected so that the line of the crests tend towards coincidence with the bed contours, as in Fig. 8-25.

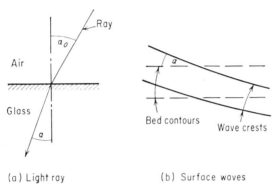

(a) Light ray   (b) Surface waves

Figure 8-26.   Wave Refraction

Wave refraction has some interesting effects in the way of local intensification or dispersion of wave activity. If waves refract so that the normals to the wave crests converge, the wave height will increase because a fixed amount of energy per unit length is being confined within an increasingly narrow passage. A dramatic example of this has been found on the Californian coast, where by an unhappy chance an oil-loading wharf was sited opposite a lens-shaped shoal, Fig. 8-27a. The prevailing wavelength and direction were such that the wharf was near the focal point of the lens formed by the shoal, with the result that wave heights of up to 20 ft have been recorded at the wharf. A more common example is illustrated in Fig. 8-27b, which shows how refraction at the mouth of a bay concentrates wave attack on the headlands flanking the bay—a process well understood and often cited by geomorphologists.

The course of wave refraction over an irregularly varying seabed can be traced by numerical techniques based on Eqs. (8–89) and (8–90). The reader can devise a basic technique for himself without difficulty (Prob. 8.44); a complete discussion of this and other methods will be found in Ref. [15]. Griswold [20] has programmed certain of these methods for the high-speed computer. When the seabed and initial wave motion are quite regular then

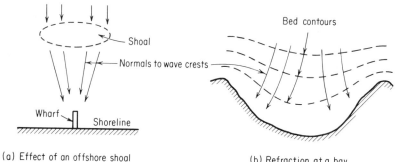

(a) Effect of an offshore shoal   (b) Refraction at a bay

Figure 8-27.   *Examples of Local Wave Intensification by Refraction*

explicit solutions can be obtained, as will be shown in Example 8.2 at the end
of this section.

Standing Waves and Basin Oscillation

When a wave is reflected by a fixed vertical boundary normal to the wave
motion the incident and reflected waves, traveling in opposite directions,
combine in this way:

$$\eta = a \sin \frac{2\pi}{L}(x - ct) + a \sin \frac{2\pi}{L}(x + ct)$$

$$= 2a \sin \frac{2\pi x}{L} \cos \frac{2\pi t}{T} \qquad\qquad \textbf{(8–91)}$$

which is the equation of a *standing wave*, Fig. 8-28. At any instant the $\eta - x$
profile must be as shown in this figure, but as t changes the profile does not
advance; Eq. (8–91) shows that while the amplitude of the $\eta - x$ curve changes
with time, the phase does not. The whole curve rises and falls through the
positions indicated by broken lines, the sections of zero vertical motion being

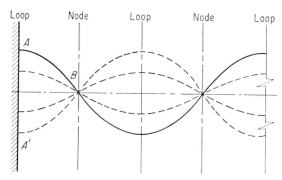

Figure 8-28.   *Standing Wave System formed by Reflection at a Vertical Wall*

called *nodes*, and those of maximum vertical motion being called *loops*. Clearly the horizontal motion will be zero at the loops and a maximum at the nodes, through which the volume of water enclosed between the extreme positions of the profile (e.g., within ABA') must flow twice in every complete cycle. We note also that the period and wavelength of the standing wave are the same as those of the incident wave.

The motion of the individual water particles is obtained by compounding the orbital motions due to the two opposing wave trains which make up the standing wave. Since we know that the motion at the loops is purely vertical and at the nodes purely horizontal, we can deduce immediately that the former is due to the cancellation of the two horizontal movements and the compounding of the two vertical movements arising from the two component wave trains; similarly the latter arises from the compounding and cancellation of horizontal and vertical movements respectively. This conclusion can of course be verified by considering the orbital motions in detail (Prob. 8.41); in the present discussion we take it as established and deduce from Eq. (8–72) that the half-amplitudes of the horizontal motion at the nodes, and the vertical motion at the loops, are

$$\frac{2a \cosh 2\pi y'/L}{\sinh 2\pi y/L} \quad \text{and} \quad \frac{2a \sinh 2\pi y'/L}{\sinh 2\pi y/L} \tag{8–92}$$

respectively, y' being again defined as the mean height of a particle above the bed. Care must be taken to read these results in conjunction with Eq. (8–91), according to which the half-amplitude of the standing wave is $2a$, not a.

Standing waves may completely fill enclosed or semienclosed basins, as in Fig. 8-29. Any mode of oscillation is possible provided only that closed ends coincide with loops and open ends with nodes. Such oscillations are the exact analogue of organ-pipe vibrations in the theory of sound, and can be sustained at high amplitude with very little supply of energy from outside, provided that

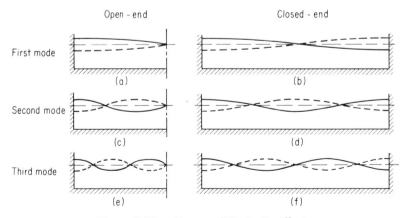

Figure 8-29.   *Forms of Basin Oscillation*

this energy, or "excitation" is supplied at one of the natural frequencies at
which the basin oscillates. The problem is essentially one of resonance. In
Figs. 8-29a, c, and e, the first, second, and third modes of oscillation are
shown for semienclosed basins, the basin length L_b being equal to

$$\frac{1}{4}, \frac{3}{4}, \quad \text{and} \quad \frac{5}{4}$$

times the wavelength in the three cases. When the basin is closed at both ends
the first three modes of oscillation are as shown in Figs. 8-29b, d, f. Here the
basin length L_b is equal to

$$\frac{1}{2}, 1, \quad \text{and} \quad \frac{3}{2}$$

times the wavelength.

In practice, the open-ended basin is approximated by an open bay, which
may be set in resonance by waves or tides arriving at the mouth of the bay.
One of the most dramatic examples of this action is provided by the Bay of
Fundy in the northeastern corner of the North American continent between
New Brunswick and Nova Scotia. This bay is rectangular in plan view and
about 100 miles long. It is noted for the extremely high tides which occur at
the head of the bay, although the tides near the mouth are of normal size.
The explanation is provided by Fig. 8-29a, i.e., this is a case of an open-ended
basin oscillating in the first mode, excited by tidal action at the mouth. It is
easily verified (Prob. 8.42) that this behavior is consistent with a mean depth
of approximately 70 ft.

Examples of closed-end basins are provided by lakes or harbors. In the
former case the oscillation can be set off by differences in barometric pressure
or by "wind set-up"—i.e., the action of wind in piling up water on the lee
shore. When the wind drops, the water rocks backwards and forwards in the
basin, usually in the first mode, the motion often persisting for long periods.
In the case of harbors, the motion can be excited by the arrival of waves at a
comparatively narrow harbor entrance. In both cases the term "seiche" is
often used, either as a noun or a verb, to describe the motion, but in the case
of harbors the term "harbor surging" is more common.

In the case of harbor surging considerable interest centers on the amount
of horizontal motion at the nodes, for this motion can have serious effects on
ship moorings in the vicinity. The amplitude is obtained from Eq. (8–92) and
the maximum velocity by combining amplitude and frequency, recalling that
the motion must be simple harmonic, arising as it does from a sinusoidal
wave motion. When the ratio y/L is small enough for the shallow water theory
to hold, the appropriate special form of Eq. (8–92) can be obtained from
Eq. (8–5), setting $S_0 = S_f = 0$:

$$g \frac{\partial y}{\partial x} + v \frac{\partial v}{\partial x} + \frac{\partial v}{\partial t} = 0 \qquad \qquad (8\text{–}93)$$

Now at the nodes $\partial y/\partial t = 0$, hence from Eq. (8–7), $\partial q/\partial x = 0$, i.e.,

$$v\frac{\partial y}{\partial x} + y\frac{\partial v}{\partial x} = 0$$

or

$$v\frac{\partial v}{\partial x} = -\frac{v^2}{y}\frac{\partial y}{\partial x} = -\mathrm{Fr}^2 g\frac{\partial y}{\partial x} \qquad (8\text{–}94)$$

where v is the horizontal velocity at the node. Now in most cases of practical interest the velocities at the nodes are only a few feet per second, and depths are at least 20 ft. It follows that Fr^2 is small, i.e., that the term $v\,\partial v/\partial x$ is only a small fraction of $g\,\partial y/\partial x$. To a first approximation we can therefore drop the former term from Eq. (8–93), leaving only

$$g\frac{\partial y}{\partial x} + \frac{\partial v}{\partial t} = 0 \qquad (8\text{–}95)$$

Now the simple harmonic variation in $\partial y/\partial x$ at the nodes can readily be obtained from the geometry of the profile; hence the horizontal motion at the nodes can be determined completely as a simple harmonic motion. Details are left as an exercise for the reader (Prob. 8.43).

An example of the computations necessary to determine basin oscillation in a particular case will be given in Example 8.3 at the end of this section.

Wave Breaking in Shallow Water

The possibility of the breaking of positive waves has already been discussed in Sec. 8.4. It was remarked there that the intersection of characteristics on the x-t plane marked the beginning of surge development, but did not necessarily indicate breaking of the wave, for in the first instance the surge might only be a weak one of the undular type.

In that discussion, the intersection of characteristics could be interpreted as the overtaking of the lower portion of a wave profile by a higher part having a higher wave velocity. Qualitatively, the same mechanism should govern the movement of an oscillatory wave up a shelving beach, although the wave eventually becomes so high that it is governed by cnoidal wave theory rather than by the shallow-water theory of Sec. 8.4. The higher parts of the wave have higher wave velocities, with the result that the wave eventually curls over and breaks. (In this case, the horizontal length scale is much shorter than in the waves of Sec. 8.4, so a discontinuity in depth is always a sharp discontinuity, never a wave front of moderate length as in the case of the undular surge.) The criterion for the breaking point is based on a detailed consideration of the wave profile [15] but reduces to the simple form

$$\frac{H}{y} = 0.78 \qquad (8\text{–}96)$$

where H is the wave height as defined in Fig. 8-23, and y as before is the mean depth.

As the waves move into very shallow water the wave profile of Fig. 8-23 develops and becomes more and more pronounced until eventually the wave crests have the appearance of isolated ridges separated by wide areas of almost undisturbed water. Although the waves are now distinctly non-Stokesian in character, the tracing of the variation of H as they move into shallow water can be done by the theory given above and as worked out in Example 8.2. This method gives a good approximation because one of its most important elements depends on Eq. (8–85) for the transmission of energy, and this equation is true independent of the wave form.

The Effect of Surface Tension

It can be shown [14] that the effect of surface tension is to change Eq. (8–71) to the following:

$$c^2 = \left(\frac{gL}{2\pi} + \frac{2\pi\sigma}{\rho L}\right)\tanh\frac{2\pi y}{L} \qquad (8\text{–}97)$$

which for deep-water waves reduces to

$$c^2 = \frac{gL}{2\pi} + \frac{2\pi\sigma}{\rho L} \qquad (8\text{–}98)$$

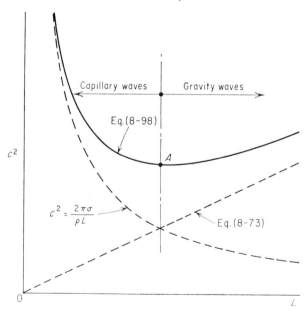

Figure 8-30.   *The Relative Influence of Surface Tension and Gravity on Wave Relationships*

where σ is the surface tension. Figure 8-30 shows a graph of c^2 against L for Eq. (8–98), the two broken lines showing separately the gravity and surface tension components. It is a matter of elementary calculus to show that c^2 has a minimum value equal to $2\sqrt{\sigma g/\rho}$ when $L = 2\pi\sqrt{\sigma/\gamma}$, represented by the point A on Fig. 8-30. This point can be used as an approximate dividing line between waves in which surface tension is the dominant influence—capillary waves, or ripples—and those in which gravity is the dominant influence. For water at 68°F, the values of c and L at A are 0.76 ft/sec and 0.056 ft respectively, so it is only in very small waves that capillary influence is appreciable. Nevertheless it is of some interest, particularly in the operation of hydraulic models, to see that this influence is not always negligible.

The Solitary Wave

It is well established by observation, in particular by the classic measurements of J. Scott Russell [16], that a single disturbance in a liquid may give rise to a single wave, the so-called solitary wave, Fig. 8-31. A solution can be found

$$\eta = 2a \operatorname{sech}^2 \sqrt{\frac{3a}{2y^3}}\,(x - ct) \qquad (8\text{–}99)$$

according to which the wave does not change in form as it moves forward.

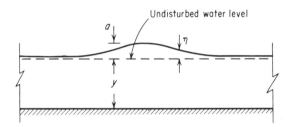

Figure 8-31.   The Solitary Wave

The wave velocity is given by

$$c = \sqrt{g(y + a)} \qquad (8\text{–}100)$$

The solitary wave is the special form of cnoidal wave for which $L/y \to \infty$ and its velocity c is plotted accordingly on Fig. 8-21. The above equations take no account of boundary resistance, whose effect will presumably be to make the solitary wave subside, i.e., become longer and lower, as it moves downstream. Keulegan [17] has examined this question and found that by a suitable application of boundary-layer theory the rate of subsidence can be calculated. The theoretical results corresponded closely with the observations of Scott Russell made more than a century before.

Stability and Roll Waves

Mechanical oscillations of finite amplitude may in general arise in two
different ways—through an initial disturbance of finite size, or through a
small initial disturbance in a situation having some inherent instability. The
oscillatory waves discussed in this section are in the second category, the
usual source of instability being wind shear applied to the free liquid surface;
this creates a condition in which any slight disturbance will magnify until it
grows into a progressive wave train. As mentioned previously, the mechanism
of wave growth in these circumstances was not clearly understood until
recent years [10].

Instability arising from wind shear can occur independently of any movement which the water may already possess. A further type of instability can
arise in a long channel from the slope and resistance of the channel and the
characteristics of the flow; the details have already been worked out in Sec.
8.4, in the argument leading up to Eqs. (8–41) and (8–42). It was shown there
that if the Froude number Fr exceeds 2, any positive disturbance, however
small, will become more intense until a surge forms. In this situation therefore
we should expect surges to be constantly forming; empirical evidence for this
supposition is provided by the phenomenon of *roll waves*, illustrated in Fig.
8-32. These waves are commonly observed to occur on steep slopes where

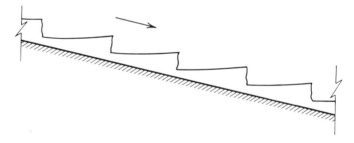

Figure 8-32. Roll Waves

Fr > 2, usually in artificial channels. They consist of a series of surges at
regular intervals, separated by regions in which the depth reaches a minimum
just downstream of the surge front.

The analysis of the roll waves themselves is a matter of some difficulty, and
is outside the scope of this text. The matter has been substantially elucidated
by the work of Dressler [18], to which the reader is referred for details. One
further problem which can be dealt with here is the generalization of the
criterion Fr ≥ 2, which it will be remembered was derived for the case of a
wide channel with constant Chézy C. Solutions have been presented by a
number of investigators; the most compact appears to be that of Escoffier [8],
which can be expressed in terms similar to those of the argument leading up

to Eq. (8–41) and (8–42). The right-hand member of Eqs. (8–13c) and (8–14c) is written as

$$gS_0\left(1 - \frac{v^2}{v_0^2}\right)$$

where v_0 is the uniform-flow velocity appropriate to the particular depth and is therefore an indirect measure of the depth. Here the subscript 0 has quite a different significance from the one attached to Fr in Eqs. (8–13c) and (8–14c), where it indicated the original undisturbed condition. Suppose now that small variations Δv, Δv_0, $\Delta \omega$ occur in an otherwise uniform flow, i.e., initially $v = v_0$; then by differentiation

$$\Delta\left[gS_0\left(1 - \frac{v^2}{v_0^2}\right)\right] = \frac{2gS_0}{v}(\Delta v_0 - \Delta v) \qquad (8\text{–}101)$$

We refer again to Fig. 8-4, and consider the behavior of the increments Δv and $\Delta \omega$ along the characteristic LM close to the first characteristic OF. In the original argument, Eqs. (8–31) and (8–32) were found to be true along LM; it follows from Eqs. (8–53) and (8–101) that the general form of these equations must be

$$d(\Delta v + \Delta \omega) = \frac{2gS_0}{v}(\Delta v_0 - \Delta v)\, dt \qquad (8\text{–}102)$$

and

$$d(\Delta v - \Delta \omega) = 0 \qquad (8\text{–}103)$$

reintroducing the stage variable ω. From Eq. (8–103) it follows that Δv can be replaced by $\Delta \omega$ in Eq. (8–102); performing this operation we obtain

$$d(\Delta \omega) = \frac{gS_0}{v}\left(\frac{\Delta v_0}{\Delta \omega} - 1\right)\Delta \omega\, dt \qquad (8\text{–}104)$$

along LM. The conclusion is that if $\Delta \omega$ is positive (i.e., for a positive wave), $\Delta \omega$ will increase without limit, and the motion will be unstable, if

$$\frac{dv_0}{d\omega} > 1 \qquad (8\text{–}105)$$

and this is the general form of the criterion Fr > 2. The same result has been obtained by a number of other investigators using various forms of stability analysis, but a further and more general significance can also be read into this result. It arises in connection with the flood-routing problem, and is discussed in Chap. 9.

The above argument is applicable to either laminar or turbulent flow, for if the friction slope is taken as proportional to v^m, the only effect on the analysis is to replace the factor 2 by m on the right-hand side of Eqs. (8–101) and (8–102) and to introduce the factor $m/2$ on the right-hand side of Eq. (8–104). Clearly this does not alter the conclusion expressed by Eq. (8–105). In fact field and laboratory observations appear to indicate two distinct classes of roll waves: those occurring in turbulent flow in fairly large channels, at high values of the Reynolds number Re; and those occurring at low values of Re, in flows involving thin sheets of liquid running down smooth steep surfaces. The report of an experimental investigation by Mayer [19] on the latter type prompted interesting discussion from a number of sources on the two types of wave; these discussions, and the original paper, also provide a comprehensive bibliography on the subject.

Example 8.2

An ocean beach has a regular bed slope of 1 in 50, continuing seawards for some miles. Waves in the sea are moving along and towards the coast; about 3 miles out from the shoreline they are observed to have a height of 2.5 ft and a length of 250 ft, and the wave crests are seen to be making an angle of 60° with the shoreline. Find the distance from the shoreline at which the waves will begin to break, and the angle that the wave crests make with the shoreline at this point.

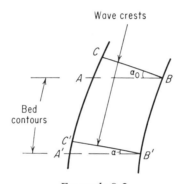

Example 8-2

Before making any calculations, we examine the geometry of the wave crests and the trajectories drawn orthogonally—i.e., at right angles to the crests—for this governs the energy transmission of the waves. In the diagram, let CB and $C'B'$ be corresponding segments of wave crests drawn in deep water and shallow water respectively. AB and $A'B'$ are drawn parallel to the bed contours, and $CAC'A'$ and BB' are orthogonals which we can think of as sidewalls within which energy is being carried forward at a certain fixed rate. From Eq. (8–85) the energy transmission per unit crest length varies as H^2c_g, whence it follows that

$$H \propto \frac{1}{\sqrt{c_g b}}$$

where b is the crest length intercepted between orthogonals, e.g., CB and $C'B'$. Now since all orthogonals are similar and similarly situated, $AB = A'B'$, whence

$$\frac{b}{b_0} = \frac{C'B'}{CB} = \frac{\cos \alpha}{\cos \alpha_0}$$

Also, $c/c_0 = \sin \alpha / \sin \alpha_0$, from Eq. (8–90); setting $r = c_g/c$, we can now write, still denoting deep-water conditions by the subscript 0,

$$\frac{H}{H_0} = \sqrt{\frac{c_{g0}b_0}{c_g b}} = \sqrt{\frac{r_0 \sin \alpha_0 \cos \alpha_0}{r \sin \alpha \cos \alpha}}$$

$$= \sqrt{\frac{\sin 2\alpha_0}{2r \sin 2\alpha}}$$

since $r_0 = \frac{1}{2}$. Also, from Eqs. (8–84) and (8–88)

$$r = \frac{c_g}{c} = \frac{1}{2}\left[1 + \frac{4\pi y c_0/L_0 c}{\sinh(4\pi y c_0/L_0 c)}\right]$$

The following system of tabulation is based on taking a series of values of c/c_0, from which L is obtained via Eq. (8–88), y via Eq. (8–89), and H from the equations developed above. The angle α also emerges in the tabulation, via Eq. (8–90). Where required, the appropriate equation number is noted at the head of the column. The tabulation proceeds by taking trial values of c/c_0 until a value of $H/y = 0.78$ is reached—see Eq. (8–96). The last line is obtained by interpolation.

c/c_0	$2\pi y c_0/L_0 c$ Eq.(8–89)	y, ft	r, Eq. (8–84)	$\sin \alpha$, Eq. (8–90)	α	H/H_0	H, ft	H/y
0.5	0.546	10.88						
0.4	0.424	6.75						
0.3	0.310	3.70	0.969	0.260	15°04′	0.943	2.36	0.64
0.2	0.204	1.62	0.987	0.173	9°58′	1.134	2.84	1.75
0.28	0.287	3.20	0.974	0.242	14°02′	0.973	2.43	0.76
0.278	0.285	3.15	0.975	0.240	13°54′	0.977	2.44	0.78

The first two lines have not been completed because the values of y obtained were obviously too large. The distance from the shoreline at which the waves will begin to break is equal to

$$50 \times 3.15 = 160 \text{ ft (approx.)}$$

and the angle the wave crests make with the shoreline is

$$14° \text{ (approx.)}$$

Ans.

The precision of the method does not warrant expressing distances to the nearest foot, or angles to the nearest minute.

It is noteworthy that when the wave first reaches shallow water the wave height actually decreases because of the increase in the group velocity c_g; only when very shallow water is reached does the effect of convergence begin to overcome the effect of increasing c_g, so that H/H_0 begins to rise towards unity again.

Example 8.3

A yacht harbor approximates in shape to a rectangular recess 300 ft long set in, and at right angles to, a straight shoreline. The water depth in the harbor is 30 ft; determine the period and the deepwater length of waves that would cause the basin to oscillate in the first mode. If the basin does oscillate in this mode with a vertical amplitude of 5 ft, find the amplitude and maximum velocity of the horizontal motion at the entrance to the harbor.

For the basin to oscillate in the first mode, the wavelength must be $4 \times 300 = 1{,}200$ ft. The ratio y/L then equals $30/1{,}200$ or $1/40$, which is low enough to make the waves of the shallow-water type, with

$$c = \sqrt{30g} = 31 \text{ ft/sec}$$

Hence the wave period $T = 1{,}200/31 = 38.6$ sec. To determine the deep-water wavelength L_0, we obtain a T-L_0 relation by eliminating c between Eqs. (8–70) and (8–73); it is

$$T^2 = \frac{2\pi L_0}{g}$$

whence

$$L_0 = \frac{g}{2\pi} (38.6)^2 = 7{,}600 \text{ ft} \left.\begin{array}{c} \\ \\ \end{array}\right\} \quad Ans.$$

and

$$T = 38.6 \text{ sec}$$

It may be remarked that this is an unusually long wave. For the second part of the problem, we can use the results of Prob. 8.43 because the wave is of the shallow water type. The harbor entrance is at a node, and the amplitude of the horizontal motion will be

$$\frac{LH}{2\pi y} = \frac{1{,}200 \times 5}{60\pi} = 31.8 \text{ ft} \quad Ans.$$

and the maximum horizontal velocity will be

$$\frac{LH}{2Ty} = \frac{1{,}200 \times 5}{77.2 \times 30} = 2.6 \text{ ft/sec} \quad Ans.$$

Note that it is the shallow-water wave length which is inserted in these expressions for amplitude and velocity.

References

1. J. J. Stoker. "The Formation of Breakers and Bores," *Communications on Pure and Applied Mathematics, New York University*, vol. 1, p. 1 (January 1948).
2. J. J. Stoker. *Water Waves* (New York: John Wiley & Sons, Inc., 1957).
3. M. J. Lighthill and G. B. Whitham. "On Kinematic Waves: I—Flood Movement in Long Rivers." *Proc. Roy. Soc.* (London) vol. 229, no. 1178 (May 10, 1955), p. 281.
4. R. Ré. "Etude du lacher instantané d'une retenue d'eau dans un canal par la méthode graphique," *La Houille Blanche*, vol. 1, no. 3 (May 1946).
5. A. Schoklitsch. "Über Dambruchwellen," *Sitzungberichte der K. Akademie der Wissenschaften*, Vienna, vol. 126 (1917), p. 1489.
6. G. H. Keulegan. "Wave Motion," in H. Rouse (ed.), *Engineering Hydraulics* (New York: John Wiley & Sons, Inc., 1950), Chap. 11.

7. R. F. Dressler. "Comparison of Theories and Experiments for the Hydraulic Dam-Break Wave," *International Association of Hydrology, Assemblée générale de Rome*, vol. III (1954), p. 319.

8. F. F. Escoffier and M. B. Boyd. "Stability Aspects of Flow in Open Channels," *Proc. A.S.C.E.*, vol. 88, no. HY6 (November 1962), p. 145.

9. M. Guelton, P. Weingaertner, and P. Sevin. "Fonctionnement en éclusées du canal industriel de Basse-Durance," *La Houille Blanche*, vol. 16, no. 5 (October 1961), p. 597. This paper is one of a symposium of five papers on various aspects of unsteady flow under the general title of "Intumescences." The symposium provides an interesting view of many aspects of French practice in this field.

10. M. J. Lighthill. "Physical Interpretation of the Mathematical Theory of Wave Generation by Wind," *J. Fluid Mech.*, vol. 14, part 3 (November 1962), p. 385. This paper gives a physical discussion of the mathematical theories of J. W. Miles and O. M. Phillips, to which complete references are given.

11. V. Cornish. *Ocean Waves* (London: Cambridge University Press, 1934).

12. R. C. H. Russell and D. H. MacMillan. *Waves and Tides* (London: Hutchinson & Co., 1952).

13. Rachel Carson. *The Sea Around Us* (New York: New American Library, 1956).

14. L. M. Milne-Thomson. *Theoretical Hydrodynamics*, 4th ed., (New York: The Macmillan Company, 1958).

15. J. W. Johnson (ed.). *Proceedings, 1st Conference on Coastal Engineering, Long Beach, California*, The Engineering Foundation, 1951.

16. J. Scott Russell. "Report on Waves," *British Association Reports*, 1944.

17. G. H. Keulegan. "Gradual Damping of Solitary Waves," *J. Res. Nat. Bur. Standards*, vol. 40, no. 6 (June 1948), p. 499.

18. R. F. Dressler. "Mathematical Solution of the Problem of Roll-Waves in Inclined Open Channels," *Communications on Pure and Applied Mathematics, New York University*, vol. 2 (1949), p. 149.

19. P. G. Mayer. "Roll Waves and Slug Flows in Inclined Open Channels," with discussions by F. F. Escoffier, R. H. Taylor, J. F. Kennedy, T. Ishihara, Y. Iwagaki, and Y. Iwasa; *Trans. Am. Soc. Civil Engrs.*, vol. 126 (1961), p. 505.

20. G. M. Griswold. "Numerical Calculation of Wave Refraction," *J. Geophys. Res.*, vol. 68, no. 6 (March 1963), p. 1715.

Problems

8.1. Prove that if $S_0 = S_f = 0$, and if any one member of the C_2 family of characteristics is a straight line, then so is every other member of the same family. Prove also that in the simple wave problem the C_2 characteristics are *not* straight lines.

8.2. Examine the dispersion of the negative wave in Example 8.1 by completing the following table, in which the distances listed in each horizontal line are those occurring at the same instant.

	Depth = 3 ft	4 ft	5 ft (wave front)
Distance upstream (miles)	1.00		22.0
	2.00		

8.3.   The initial situation is the same as in Example 8.1, but in this case the estuary
        level *rises* at the rate of 1 ft/hr for 3 hr, and then remains steady. Calculate,
        and plot on the *x-t* plane, the envelope of intersections of the C_1 character-
        istics, and hence determine when, where, and at what depth a surge will first
        develop. At this instant, what will be the depth midway between the surge and
        the river mouth?

8.4.   In a channel having negligible bed slope and resistance, the flow is initially
        uniform and the discharge at one end of the channel is then varied in a
        specified way with time, $q = q(t)$. Using the relation $q = vy = vc^2/g$, show that
        at this end of the channel

$$g \frac{dq}{dt} = 2c(v + c) \frac{dc}{dt}$$

        A navigation canal of rectangular section contains stationary water at a
        depth of 10 ft. Water is now released into the canal from a lock, and q
        rises at a uniform rate to a maximum value of 50 cusecs/ft after a period of
        5 min. Neglecting bed slope and resistance, plot the envelope of intersections
        of C_1 characteristics, and determine when, where, and at what depth a surge
        will first develop.

8.5.   With the initial situation as in Example 8.1, the estuary level falls at the rate of
        1.5 ft/hr for 2 hr. By algebraic methods rather than by reference to the *x-t*
        plane, determine when the depth will reach 3 ft at a section 2 miles upstream,
        and find what the velocity will be at that section at that instant.

8.6.   From Eq. (8–26b) for a negative-wave profile, determine the wave velocities
        of the sections where $y = 0$, and $y = y_0$; does the first of these have a physical
        meaning? Show also that there is a section where the depth remains constant
        at all times.

8.7.   The technique used by Lighthill and Whitham [3] to prove Eq. (8–42) was
        based on the assumption that y and v could be expanded in series

$$y = y_0 + \tau y_1 + \tau^2 y_2 + \cdots$$

$$v = v_0 + \tau v_1 + \tau^2 v_2 + \cdots$$

        where $\tau = t - x/(v_0 + c_0)$ and the coefficients y_1, v_1, etc., are functions of
        t alone. To follow the method through, first show that y_1 is the value of
        $\partial y/\partial t$ at the wave front, where $\tau = 0$. Then substitute the above equations and
        their derivatives into Eqs. (8–5) and (8–10), assuming $R = y$ and that the
        Chézy C is constant. Set $\tau = 0$ so that the resulting equations describe
        conditions at the wave front; it will be found that y_2, v_2, and the higher
        series coefficients all drop out. It will also be found that v_1 and dv_1/dt are
        associated in the same way in each of the two resulting equations, so that they
        can be eliminated together. Performing this elimination, show that the result
        is the equation

$$\frac{dy_1}{dt} = \frac{3y_1^2}{2y_0(1 + \text{Fr}_0)} - \frac{gS_0 y_1(2 - \text{Fr}_0)}{2v_0}$$

        and hence that if y_1 is initially greater than the right-hand member of Eq.
        (8–42), y_1 will continually increase as the wave advances until a bore forms.

8.8.   Consider a C_1 characteristic, such as AB in Fig. 8-4, arising from a small positive wave and separated from the first characteristic OF by a substantial time interval Δt. Assuming that this time interval has the same value along any C_2 characteristic, derive the equation of the line AB corresponding to Eq. (8–37). Condense the working by setting $1/k = 1 + gS_0\Delta t/c\mathrm{Fr}_0$, and $2r = (1 - k)^2(2 - \mathrm{Fr})^2$. Obtain the x interval for a given t between this and a neighboring characteristic by differentiating x with respect to the interval Δt; hence determine the conditions under which the two characteristics will intersect.

8.9.   Formulate and solve the problem corresponding to Prob. 8.8 for the case where Δt is small enough to make $gS_0\Delta t/c\mathrm{Fr}_0$ small, but the relative wave magnitude $\Delta c/c_0$ is not small.

8.10. Rearrange Eq. (8–42) so that the right hand member is a function of g, v_0, Fr_0, and the Chézy C.

       A wide river has a depth of 4 ft and a slope of 0.0002 near its mouth. Manning's n is 0.018. Initially flow is uniform in the river, and at the mouth sea level matches river level. The tide then begins to rise; calculate the initial rate of rise, in feet per hour, needed to make a bore form. If the rate of rise is in fact 50 percent greater than this critical amount, determine when and where the first elementary surge will form. Assume that the Chézy C remains constant at the value appropriate to the initial depth of 4 ft. (Care must be taken over the sense-sign convention.)

8.11. The problem concerns the release of water at a controlled rate into a river in which the initial flow is uniform. Assuming that the equation derived in Prob. 8.4 is applicable over short time intervals even when resistance is taken into account, derive from Eq. (8–41) an equation for the limiting value of the initial rate of discharge increase dq/dt.

       In a wide river, flow is initially uniform at a depth of 4 ft. The bed slope is 0.0002, and the Chézy C can be assumed constant at 110. Surplus water from a storage reservoir is now released into the river, so that the discharge per unit width q rises at a uniform rate till it has increased by 50 cusecs per foot after a period of 10 min; thereafter it falls again. Determine (a) whether a surge will form at the wave front which moves downstream; (b) if so, when and where it will first appear; and (c) if not, what rate of increase in q would be necessary to create a surge.

8.12. For the same situation as in Prob. 8.11, answer the same questions about the wave front which moves upstream. Hence determine whether, for a given initial dq/dt, a surge is more apt to form on an adverse or a following wave.

8.13. In the situation of Prob. 8.4, trace the development of the elementary surges beyond the envelope of first intersections of C_1 characteristics, by plotting to scale a graphical display corresponding to Fig. 8-5. Calculate the speed of the surge which finally develops, and determine the approximate position of the line HK by assuming that the curve GH is a parabola, whose inverse slope increases linearly with x. If q at the lock begins to fall immediately after reaching its maximum value of 50 cusecs/ft, estimate qualitatively the effect that resistance would have on the height and speed of the surge which finally develops.

8.14. From Eq. (8–45) show that if the speed of withdrawal of the plate in the dam-break problem exceeds $2c_0$, then the lines BB_1 and BC on Fig. 8-6b coincide.

8.15. Reconcile the statement that B is outrunning C in Fig. 8-8b with the result obtained in Sec. 3.5 that small positive disturbances on the deep-water side of a surge must catch the surge up and be absorbed into it. Does not this suggest that C should overtake B? If you are unable to resolve the question in general algebraic terms, work through the next problem and decide what light is thrown on the matter by the results of this problem.

8.16. In a canal of rectangular section and horizontal bed, still water is contained in a lock at a depth of 10 ft. Outside the lock there is still water at a depth of 1 ft. The lock gates are now suddenly opened; find the speed and height of the resulting surge, and the rate at which the region of constant depth behind it (BC in Fig. 8-8b) is lengthening.

8.17. Show that if, in the dam-break problem of Figs. 8-8 and 8-9, the velocity and depth in the region BC are $2c_0/3$ and $4y_0/9$, i.e., if the point C remains stationary at the original dam position, then the undisturbed downstream depth must be $0.138y_0$, where the subscript 0 relates to undisturbed upstream conditions.

8.18. The canal of Prob. 8.16 is 10 ft wide. Find the outflow rate following sudden opening of the lock gates if the initial depth of still water inside the lock is 10 ft, and outside the lock is (a) zero; (b) 1 ft; (c) 3 ft.

8.19. For the case of the dam-break problem in which the downstream water level is low enough for the wave to remain centered at O, the original dam or gate position, use Eqs. (8–25) and (8–26) to determine the values at O (in terms of t, c_0, and y_0) of the three acceleration-slope terms of Eq. (8–5).

Slope and resistance are to be taken as negligible, and the simple-wave theory directly applicable, if the sum of the unsigned magnitudes of these three acceleration terms exceeds $20(S_0 - S_f)$, where S_f is evaluated at O. For a rectangular channel in which $y_0 = 12$ ft, $R = y$, Chézy $C = 100$, $S_0 = 0.001$, determine the initial time period during which this criterion is satisfied.

8.20. In the situation of Probs. 8.16 and 8.18, find the speed and height of the surge resulting from sudden opening of the lock gates for the following values of initial downstream depth: 1.5, 2.0, 2.5 ft.

8.21. If the subscript 0 denotes initial conditions upstream of a sluice gate, and the sluice gate is then suddenly raised clear of the water, show that if the resulting negative wave is centered the central values of v and c will be of magnitude $(2c_0 + v_0)/3$, where v_0 is the (unsigned) magnitude of the initial upstream velocity. (Care is needed with signs.) Why must the flow always be critical at the center section?

8.22. A sluice gate is placed in a channel of rectangular section, and the initial upstream and downstream depths are 10 ft and 2 ft respectively. If the sluice gate is suddenly closed completely, find the height and speed of the resultant upstream surge, and the speed with which the front of the resultant negative wave moves downstream. Will the gate be left high and dry on the downstream side? If not, what will be the depth of water next to the gate? If so, what initial values of upstream and downstream depth would be needed (for the same discharge) so that after sudden gate closure the trailing edge of the negative wave will be stationary at the gate? Derive in general terms the criterion which must be met for this latter condition to obtain.

8.23. Given the initial situation of the last problem, the sluice gate is suddenly raised clear of the water. Calculate all the elements of the resultant flow situation, and determine whether the resultant negative wave will be centered. If not, will the discharge at the gate be steady?

8.24. For a sluice gate mounted in a channel of rectangular section, determine the limiting initial ratio of upstream to downstream depth at which sudden removal of the sluice gate will cause the boundary C between the negative wave CD and the uniform flow region BC to remain stationary at the gate section (cf. Prob. 8.17).

8.25. For the case of sudden partial sluice-gate closure shown in Fig. 8-11b, where the initial upstream and downstream flow conditions are given, as well as the new gate opening (which determines the outflow depth immediately downstream of the gate), write the available equations in general form, and establish that the number of equations is equal to the number of unknowns.

8.26. Given the initial situation of Prob. 8.22, the gate is suddenly closed far enough to make the depth immediately downstream equal to 1.5 ft. Calculate all the elements of the new flow situation, including the new discharge through the gate.

8.27. Carry out the same general study as in Prob. 8.25 for the case of sudden partial gate opening shown in Fig. 8-11c.

8.28. Given the initial situation of Prob. 8.22, the gate is suddenly opened far enough to make the depth immediately downstream equal to 2.5 ft. Calculate all the elements of the new flow situation, including the new discharge through the gate.

8.29. For the case of a simple wave in a uniform nonrectangular channel, show that the equation corresponding to Eq. (8–26) is:

$$\frac{x}{t} = c + \omega + v_0 - \omega_0$$

8.30. For a channel of trapezoidal section, examine the functions $c = c(y)$ and $\omega = \omega(y)$ and decide whether they can be expressed as functions of $y' = my/b$ alone, where b is the base width of the channel and its side slopes are mH : 1V. Using any suitable means (including a computer program), calculate and list values of c and ω for values of y up to 10 ft in a trapezoidal channel of base width 20 ft and side slopes $1\frac{1}{2}$H : 1V.

8.31. A long irrigation channel having the trapezoidal section of Prob. 8.30 carries a flow of 1,000 cusecs at a uniform depth of 8 ft. If the inflow to the channel at the upstream end is suddenly cut off by a sluice gate, determine the depth of stationary water just downstream of the gate, and the speed with which the leading edge and trailing edge of the resulting negative wave move downstream. Neglect slope and resistance. (*Note*: for this and the next few problems, it is convenient to take x positive downstream.)

8.32. The same as Prob. 8.31, except that the channel is of rectangular section 20 ft wide.

8.33. In the situation of Prob. 8.32, the bed slope of the channel is 0.001. It is
       required to determine the course of events at a section 5 miles downstream of
       the channel inlet, until the discharge at that section falls to 50 cusecs. Carry
       out a complete numerical solution, taking slope and resistance into account,
       from the instant which is (a) zero, (b) 30 sec, (c) 60 sec after the moment of
       gate closure. In cases (b) and (c) use the simple-wave solution in the initial
       period before slope and resistance are taken into account.

8.34. Rewrite Eqs. (8–48) and (8–50) in terms of v, c, and the stage variable ω.
       Deduce equations corresponding to Eqs. (8–55) and (8–56) for the staggered-
       net system of Fig. 8-14b, without attempting to solve explicitly for ω_P;
       solve only for y_P and c_P, leaving ω_P wherever it occurs in the right-hand
       members of the resulting equations. Devise a trial-and-error system by which
       values of v_P, c_P, and ω_P may be obtained from these equations and from
       tabulated data giving c and ω as functions of y.

8.35. Apply the method developed in Prob. 8.34 to solve the situation of Prob. 8.33
       for the channel section of Prob. 8.31.

8.36. Derive the equation corresponding to Eq. (8–63) for a channel of trapezoidal
       section.

8.37. A channel of rectangular section 30 ft wide has a Manning n of 0.014, a bed
       slope of 0.0002, and a length of one mile. It delivers 1,000 cusecs to the
       headworks of a hydroelectric scheme, and under normal operation the flow
       in the channel is uniform. Find the maximum height of surge reached at the
       headworks on sudden complete shutdown from normal flow.

8.38. Flow takes place with a specific energy y_0 and a discharge per unit width of
       $8y_0\sqrt{(gy_0)}/27$, as given by Eq. (8–67). Show that the Froude number is 0.318.

8.39. The hydroelectric scheme of Prob. 8.37 is shut down, and the supply reservoir
       is at such a level as would deliver normal discharge through the channel under
       uniform flow conditions. The delivery channel is open at its upstream end and
       filled with stationary water at reservoir level. Full load is now suddenly
       thrown on to the power station; assuming that the initial depth in the channel
       is equal to average depth, and neglecting slope and resistance, determine
       whether the negative wave invading the channel can withdraw the required
       flow from the channel, and if so, how far the water level at the head works will
       be drawn down below normal operating level.

8.40. Imagine that the deep-water wave train of Fig. 8-22 has been brought to rest
       by superimposing a velocity equal and opposite to the wave velocity. Taking
       for granted that the orbits of the water particles are circular, verify Eq. (8–73)
       by applying the Bernoulli equation to the situation.

8.41. Determine the phase relationship between incident and reflected waves at a
       section which is a node of the resultant standing wave. Hence verify that the
       orbital motions resulting from the two component waves are so related that
       the vertical components cancel each other and the horizontal motions are
       added together. Deduce the corresponding result for the loops of the standing
       wave.

8.42. An open-ended basin 100 miles long is subject to the action of tides having a
       period of $12\frac{1}{2}$ hr. Find what the mean depth of water must be to make the
       basin oscillate in the first mode, Fig. 8-29a.

8.43. Show from Eq. (8–95) that at the node of a standing wave which is long enough for the shallow-water theory to hold, the half-amplitude of the horizontal motion is $gHT^2/4\pi L$, and the maximum horizontal velocity is $gHT/2L$, where H is the amplitude of the standing wave. Prove also that these results can be written as $LH/4\pi y$ and $LH/2Ty$, and show that the last result verifies the previous assumption that Fr is small, provided that H/y is small. Finally, verify that the result $LH/4\pi y$ is a special case of the general result given by Eq. (8–92).

8.44. Repeat Example 8.2 by the following numerical method: given the line of a wave crest (starting in deep water) calculate c from y for a number of points along the crest, and lay off normals from the crest representing the distance traveled by the wave in some standard time interval. Hence draw a new crest line, and repeat the process. Compare the results with those obtained in Example 8.2.

8.45. A progressive wave train of low amplitude forms at the interface between two fluids, of density ρ_0 and ρ_1 for the lower and upper fluids respectively. By the argument of Prob. 8.40, derive the equation corresponding to Eq. (8–73) and show that it differs from Eq. (8–73) only in that g is replaced by $g(\rho_0 - \rho_1)/\rho_0$. Neglect shear stress and surface tension at the interface, and assume that the thickness of each fluid layer is much greater than the wavelength.

8.46. Show that the effect on the equations of motion of variable width b in a rectangular channel is to multiply Eq. (8–12) by b and add the term $vc\,db/dx$ to the left-hand side, there being no change in Eq. (8–9). Hence determine the corresponding changes to be made in Eqs. (8–56) and (8–58).

8.47. The northern side of an artificial harbor is bounded by a wall running east and west, as shown in the figure. The highest storm waves come from a

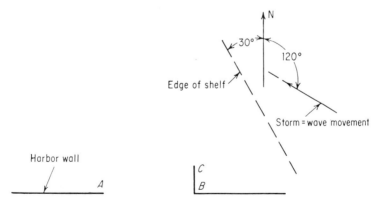

Problem 8-47

direction 30° South of East, as shown, and are refracted towards the harbor entrance AB by the edge of an offshore shelf, running 30° west of north. The shelf is 100 ft below mean sea level, and beyond the shelf is deep water. Consideration is being given to protecting the harbor entrance from these waves by building a wall BC as shown; find the period of the shortest waves that can enter the harbor if the angle between AC and AB is (a) zero; (b) $10°$; (c) $20°$. What must this angle be if *all* waves of this type are to be kept out of the harbor?

8.48. The situation is similar to that of Example 8.2, except that the bed slope is 1 in 200, and in deep water the wavelength is 600 ft, the wave height 3 ft, and the wave crests make an angle of $45°$ with the shoreline. Find the distance from the shoreline at which the waves will begin to break, and the angle that the wave crests make with the shoreline at this point. Also, compute and draw to scale one complete wave crest from deep water to the breaking point, and determine the variation of wave height along this crest.

8.49. For the situation of Example 8.3, determine the periods and the deep-water lengths of waves that would cause the basin to oscillate in the second and third modes. In each case locate the nodes and for a vertical amplitude of 5 ft find the amplitude and maximum velocity of the horizontal motion at the nodes. (*Note*: It cannot in this case be assumed that the shallow-water theory will apply within the basin.)

Computer Programs

C8.1. Using Eqs. (8–55) through (8–58), write a computer program which will determine the propagation of a tidal rise upstream from the mouth of a river, assuming it to be wide rectangular in section and that the Chézy C is constant. Operate the program for the case of Prob. 8.3, assuming that the initial flow in the river is uniform and that its slope is 0.0003.

C8.2. Using the results of Prob. 8.46, write a program which will solve the same problem as C8.1, except that the width of the river decreases linearly upstream from the mouth until a constant width is reached. Apply it to the case specified in C8.1, but with the width of the river 1,000 ft at the mouth, decreasing linearly to 400 ft at one mile upstream, remaining constant thereafter. The bed slope is uniform throughout the reach of variable width, and the initial depth at the mouth is 5 ft, the same as the initial uniform depth in the 400-ft-wide reach of the river. Variations in depth along the initial backwater curve in the variable-width reach may be neglected.

C8.3. Write a computer program which will apply the system of Prob. 8.34 to a channel of trapezoidal section, for the case of a negative wave moving along the channel. Use the program to rework Prob. 8.31.

C8.4. Write a computer program which will trace the propagation of a positive surge in a sloping channel of rectangular section following a complete stoppage of flow downstream. Use the program to rework Prob. 8.37.

C8.5. Write the same program as in C8.4, except that the channel section is trapezoidal. Apply the program to a situation which is the same as that of Prob. 8.37, except that the channel has a base width of 30 ft, side slopes of 2H : 1V, $n = 0.023$, bed slope 0.0002, and a normal flow of 1,500 cusecs.

C8.6. Use the program of C8.1, modified as may be necessary, to rework Prob. 8.33.

Chapter 9

Flood Routing

9.1 Introduction

We consider in this chapter one of the most important of all the unsteady flow phenomena the engineer has to deal with. It is the movement of a flood wave down a channel, usually that of a natural river, and the associated problem is the tracing of this movement and any related changes in the form and height of the wave. It is essential that the engineer should possess theoretical means of determining the behavior of a flood wave in a channel of specified form and slope, for he must be able to predict the effect on flood propagation of the changes that he must often make in a natural river in the interests of channel improvement and flood control. For instance, if a certain reach of a river channel is to be confined within stopbanks or levees in order to protect the surrounding country from floods, will this confinement of the flood wave result in higher flood levels further downstream? If so, how much higher will they be? The engineer must be able to provide, at least approximately, answers to these questions.

The process of tracing by calculation the course of a flood wave is known as flood routing, and in large measure the problem is an application of the principles of unsteady flow developed in the previous chapter. However, a flood wave does have its own special features, and may require its own special techniques for exploring these features. For instance, the rise and fall of a flood occurs much more slowly than many of the flow changes discussed in Chap. 8, and some of the acceleration terms in the equations of motion may therefore be much reduced in importance. Or again, one may expect a flood wave to become longer and lower as it moves downstream as in Fig. 9-1; this "subsidence" of the flood wave is evidently of prime interest to the engineer. He may therefore prefer, instead of a general solution, an approach which concentrates on those aspects of the solution that are particularly relevant to the subsidence problem.

Now subsidence may be controlled by the resistance and acceleration terms in the dynamic equations of motion, or by a much more simple mechanism: the "pondage" or storage in lakes through which the flood passes, or by irregularities in a natural river channel having the effect of a chain of small lakes. This is an example of a general principle which can be conveniently expressed in electrical terms: a pulse fed into a system can be attenuated either by a capacitance or by a resistance. The distinction drawn between these two mechanisms does not raise any questions of basic principle,

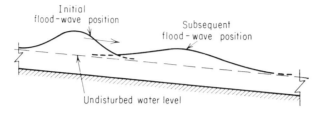

Figure 9-1.   *Attenuation, or Subsidence, of a Flood Wave as it Moves Downstream*

for a completely general treatment, such as that of Sec. 8.6, will describe both of them adequately. However if one mechanism alone dominates a certain form of flood movement, a theory taking account of this mechanism alone may be very much simpler than the general theory, and well worth using on this account. This is certainly the case when the flood movement is dominated by storage effects, and the first specific topic treated in this chapter will therefore be the special methods suitable for dealing with situations of this kind.

9.2   Storage Routing

We shall use this term to describe the analysis of flood movement in which storage effects are dominant. The simplest example is *level-pool routing*, in which we consider the movement of a flood through a lake. The main element in the calculation is the balancing of inflow, outflow, and volume of water in the lake; this is simply a matter of using the continuity equation. A secondary, but none the less important, role is played by the dynamic equation governing the outflow conditions; through this equation the outflow rate is uniquely determined by the lake level and therefore by the volume of water stored in the lake.

We define V as the volume of water in storage at any instant, with the outflow and inflow rates denoted by O and I respectively. Over a time interval Δt, we can write:

Inflow volume $-$ outflow volume $=$ increase in storage

i.e.,
$$\tfrac{1}{2}(I_1 + I_2)\Delta t - \tfrac{1}{2}(O_1 + O_2)\Delta t = V_2 - V_1 \qquad \textbf{(9–1)}$$

where the subscripts 1 and 2 indicate the start and end of the time interval respectively. Now if Eq. (9–1) represents one step in a numerical process, we can assume that we know I_1 and I_2 (because the whole inflow hydrograph is known) and also the values O_1 and V_1 obtaining at the start of the time interval. The object is to determine O_2 and V_2, then to proceed to the next interval, and so on. Collecting the unknowns of Eq. (9–1) on one side of the equation, we obtain:

$$V_2 + \tfrac{1}{2}O_2\Delta t = V_1 - \tfrac{1}{2}O_1\Delta t + \tfrac{1}{2}(I_1 + I_2)\Delta t$$

Dividing throughout by Δt and introducing the parameter

$$N = \frac{V}{\Delta t} + \frac{O}{2} \qquad\qquad (9\text{–}2)$$

we have $\qquad\qquad N_2 = N_1 + \tfrac{1}{2}(I_1 + I_2) - O_1 \qquad\qquad (9\text{–}3)$

The convenience of arranging the equation in this way will become apparent in the tabulation of Example 9.1. Meanwhile it can be pointed out that the further information required to solve the problem, i.e., a V-O relationship, can readily be put in the form of an N-O relationship since N is a function of V and O. No difficulty is created by the dependence of N on the choice of time interval Δt: in fact we shall see that Δt can be changed midway through the computation without difficulty. The details are best shown by an example.

Example 9.1

A lake having steep banks and a surface area of 500 acres discharges into a steep channel which is approximately rectangular in section, with a width of 25 ft. Initially conditions are steady with a flow of 1,000 cusecs passing through the lake; then a fresh comes down the river feeding the lake, giving rise to the following inflow hydrograph:

Time from start (hours)	0	3	6	9	12	15	18	21
Inflow (cusecs)	1,000	1,200	1,600	2,100	2,630	2,950	3,050	3,000

Time from start	24	27	30	33	36	39	42	45	48
Inflow	2,840	2,600	2,300	2,000	1,700	1,430	1,200	1,050	1,000

Calculate and plot the outflow hydrograph for the 48-hr period.

The first step is to obtain N and V as functions of the outflow O, and this is done in the following tabulation, using the fact that critical flow exists at the lake outlet. The following conversion is used

$$1 \text{ acre-in.} = \frac{4840 \times 9}{3600 \times 12} \text{ cusec-hr}$$

$$= 1 \text{ cusec-hr, very nearly}$$

in calculating the lake volume V, which is measured upwards from a plane 5 ft above the channel-bed level at the lake outlet. However, any other convenient reference level could have been used.

EXAMPLE 9.1 Preliminary Tabulation

Lake level H, ft	Channel depth $y_c=\tfrac{2}{3}H$	$v_c=\sqrt{gy_c}$ ft/sec	Outflow O $=25v_cy_c$ cusecs	Lake volume V, cusec-hr	$V/\Delta t$ (thousands of cusecs) $\Delta t=3hr$	$\Delta t=6hr$	$N=\dfrac{V}{\Delta t}+\dfrac{O}{2}$ $\Delta t=3hr$	$\Delta t=6hr$
5.0	3.33	10.35	862	0	0	0	431	431
6.0	4.00	11.35	1135	6,000	2	1	2,567	1,567
7.0	4.67	12.25	1430	12,000	4	2	4,715	2,715
8.0	5.33	13.10	1745	18,000	6	3	6,873	3,873
9.0	6.00	13.90	2084	24,000	8	4	9,042	5,042
10.0	6.67	14.63	2440	30,000	10	5	11,220	6,220

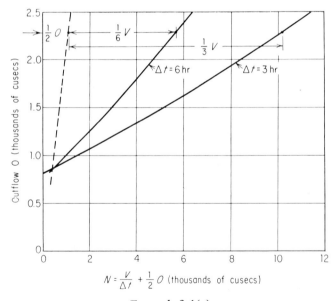

Example 9-1(*a*).

The two N-O relationships are plotted in the accompanying graph, from which values are read off for use in the main tabulation. This is given below.

EXAMPLE 9.1 Main Tabulation

Time t hr	Δt hr	Inflow I, cusecs	$\bar{I}=\frac{1}{2}(I_1+I_2)$	N cusecs	ΔN $=\bar{I}-O_1$	Outflow O, cusecs
0		1,000		1,500 ←		1,000
	3		1,100 —	↓ → 100 ←		
3		1,200		1,600 ←		→ 1,015
	3		1,400		385	
6		1,600		1,985		1,060
	3		1,850		790	
9		2,100		2,775		1,165
	3		2,365		1200	
12		2,630		3,975		1,330
	3		2,790		1460	
15		2,950		5,435		1,530
	3		3,000		1470	
18		3,050		6,905		1,755
	3		3,025		1270	
21		3,000		8,175		1,950
	3		2,920		970	
24		2,840		9,145		2,100
	3		2,720		620	
27		2,600		9,765		2,200
	3		2,450		250	
30		2,300		10,015		2,240
30		2,300		5,570		2,240
	6		2,000		−240	
36		1,700		5,330		2,170
	6		1,450		−720	
42		1,200		4,610		1,960
	6		1,100		−860	
48		1,000		3,750		1,710

The essence of the calculation is in the last three columns, and it proceeds as indicated by the arrows in the first three lines. From the known initial value of O (1,000 cusecs) the corresponding value of N (1,500 cusecs) is found from the N-O graph. Then from Eq. (9–3) the increase in N is equal to

$$\Delta N = \bar{I} - O_1$$

$$= 1,100 - 1,000 = 100 \text{ cusecs}$$

Therefore the next value of N is 1,600 cusecs, whence from the N-O graph the outflow O = 1,015 cusecs. This completes one step, and the process is now repeated.

The method of changing the time interval is illustrated at $t = 30$ hr. A line is drawn and the parameters for $t = 30$ hr are repeated below the line, including a new value of N appropriate to the new value of Δt. These values below the line then form the starting point for a fresh tabulation.

The inflow and outflow hydrographs are drawn on the accompanying figure. It is seen that considerable attenuation, or subsidence, has taken place, as well as a substantial time lag between the inflow and outflow peaks. There is another feature of interest: that the outflow peak occurs where the two hydrographs intersect. The reader will be able to prove for himself why this must happen.

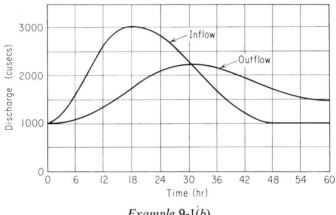

Example 9-1(*b*)

The differential equation of which Eq. (9–1) is the finite-difference form is a very simple one, and it would be explicitly soluble if the form of the inflow hydrograph were correspondingly simple, and if the *V-O* relationship were linear. These conditions can in fact be met without departing too far from physical reality, as Example 9.1 has shown. In this exercise the *V-O* curve is a fair approximation to a straight line, and the rising and falling limbs of the inflow hydrograph can be approximated separately by sine curves. It may therefore be worthwhile to examine Eq. (9–1) from this point of view. Expressed in differential form, the equation becomes

$$I - O = \frac{dV}{dt} \tag{9–4}$$

and if

$$O = kV \tag{9–5}$$

where k is a constant, then

$$\frac{dO}{dt} + kO = kI \tag{9–6}$$

where I is a known function of t. This equation is directly integrable after being multiplied by an integrating factor e^{kt}. The result, after integration, is

$$O e^{kt} = \int k I e^{kt} \, dt \qquad (9\text{–}7)$$

and further developments will be dependent on the form of $I = I(t)$. If for instance we assume that

$$I = I_0 + A_0(1 - \cos at) \qquad (9\text{–}8)$$

as in Fig. 9-2, where I_0 is the initial inflow, $a = 2\pi/T_0$, and T_0 is the duration

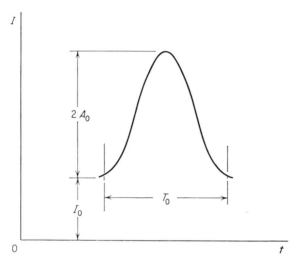

Figure 9-2.   Sinusoidal Approximation to the Inflow Hydrograph

of the inflow hydrograph (assumed symmetrical), then Eq. (9–7) becomes (Prob. 9.3)

$$O = I_0 + A_0 - \frac{A_0 k (a \sin at + k \cos at)}{k^2 + a^2} - \frac{A_0 a^2 e^{-kt}}{k^2 + a^2} \qquad (9\text{–}9)$$

where the last term derives from a constant of integration and is chosen so as to make $O = I_0$ when $t = 0$. Differentiation of Eq. (9–9) shows that O has a maximum value when

$$k \sin at + ae^{-kt} = a \cos at \qquad (9\text{–}10)$$

This maximum can occur only when $t > T_0/2$ (Prob. 9.3). This result verifies the conclusion of Example 9.1 that the crest of the outflow hydrograph will be lower and later than the crest of the inflow hydrograph.

Because the inflow hydrograph will not normally be symmetrical, the coefficient a will have to be changed at the crest of the hydrograph (Prob. 9.4). The reader will be able to judge for himself whether the work involved is less than in the numerical process of Example 9.1, and whether the approximations used have seriously affected the result.

River Routing

The question now arises whether the storage-routing technique, obviously successful when applied to a level pool, can also be applied to a river reach. Clearly the continuity Eq. (9–1) will still be true, but the storage V will no longer be uniquely determined by the outflow. We may expect the inflow to a certain river reach to be related to the upstream depth, and the outflow to the downstream depth; it follows, as in Fig. 9-3, that the storage in the reach

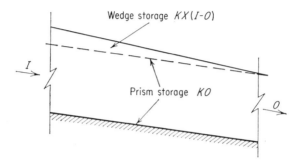

Figure 9-3. Prism and Wedge Storage

will depend on the inflow as well as the outflow. In this connection it is usual to subdivide storage, in the terms introduced with Eq. (8–8), into prism storage (dependent on outflow alone) and wedge storage, dependent on the difference $(I - O)$. The terms are illustrated in Fig. 9-3. If we assume a linear relationship analogous to Eq. (9–5), we obtain

$$V = K[O + X(I - O)] \qquad (9\text{–}11)$$

an equation which is the basis of the *Muskingum method* of storage routing, so-called because it was first developed (by the U. S. Corps of Engineers) in connection with flood-control schemes in the Muskingum River Basin, Ohio. Whether or not we assume a special form of relationship like Eq. (9–11), we can assemble enough data on the river characteristics to be able to plot curves such as those in Fig. 9-4, in which a $V\text{-}O$ curve is drawn for each value of I, or $(I - O)$. Similarly, an $N\text{-}O$ curve can be drawn for each value of I, or $(I - O)$.

Given such a set of curves for a particular river reach, the tabulation of Example 9.1 can proceed with no more difficulty than in the level-pool case. On each line of the table we determine O from N by means of the particular $N\text{-}O$ curve drawn for the particular value of I appropriate to that line. The fact that we need a separate $N\text{-}O$ curve for each line of the table does not add

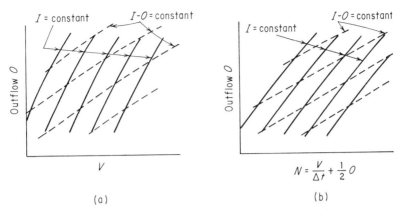

Figure 9-4.   V-O and N-O Curves for a River Reach

any real difficulty (Prob. 9.5). Moreover, an explicit solution based on linearizing assumptions, as in Eqs. (9–7) through (9–10), is only slightly more complex than in the level-pool case. The extension of Eq. (9–5) becomes

$$V = \frac{O}{k_1} + \frac{I}{k_2} \tag{9-12}$$

which is an alternative form of Eq. (9–11), with

$$\frac{1}{k_1} = K(1 - X); \qquad \frac{1}{k_2} = KX \tag{9-13}$$

Accordingly Eq. (9–4) becomes

$$\frac{dO}{dt} + k_1 O = k_1 I - \frac{k_1}{k_2} \frac{dI}{dt} \tag{9-14}$$

which is essentially no more complex than Eq. (9–6), for the right-hand member is still a known function of t alone. And if I is assumed to be a sine function the solution corresponding to Eq. (9–7), i.e.,

$$O e^{k_1 t} = \int \left(k_1 I - \frac{k_1}{k_2} \frac{dI}{dt} \right) e^{k_1 t} \, dt \tag{9-15}$$

is still explicitly integrable (Prob. 9.6). A comprehensive range of solutions of this kind has been developed by Yevdjevich [5].

While explicit solutions of this type have received little attention, the numerical method of Example 9.1, modified as in Fig. 9-4, has been very widely used in engineering practice. There is an extensive literature on the subject, a thorough survey of which is given by Chow [1], and much attention has been given to speeding the computations by such devices as special slide rules and analog computers, the latter being often of substantial size. A comprehensive discussion of these methods is given by Gilcrest [2]. While many variants of the basic method exist, there has been a tendency of recent years to apply the term *Muskingum method* to any finite-increment system based on Eq. (9–1), although this term strictly applies only to systems which assume a linear *V-I-O* relation, as in Eq. (9–11).

Despite the popularity of the basic method, it cannot be claimed that it is logically complete, for the equations of motion do not strictly justify the belief that the storage V is determined by I and O alone. To clarify the matter we examine the terms that make up the friction slope S_f in Eq. (8–5), first substituting S_f into the resistance equation

$$Q = CA\sqrt{R\left(S_0 - \frac{\partial y}{\partial x} - \frac{v}{g}\frac{\partial v}{\partial x} - \frac{1}{g}\frac{\partial v}{\partial t}\right)} \qquad (9\text{–}16)$$

using the Chézy equation as a matter of convenience, without implying that C is a constant. Now if the last three slope terms are small compared with S_0, the discharge can be computed as in uniform flow and it is dependent on the depth alone (even if C varies). Indeed this assumption was implied in the argument leading up to Eq. (9–11), but it may not always be justified. It is however very close to the truth for natural floods in steep rivers, whose slopes are of the order of 10 ft per mile or more. Typical values for each of the slope terms in Eq. (9–16) taken from an actual river in steep alluvial country are:

$$S_0, \quad \frac{\partial y}{\partial x}, \quad \frac{v}{g}\frac{\partial v}{\partial x}, \quad \frac{1}{g}\frac{\partial v}{\partial t},$$

$$\text{ft/mile} \quad 26, \quad \tfrac{1}{2}, \quad \tfrac{1}{8}-\tfrac{1}{4}, \quad \tfrac{1}{20}$$

These figures relate to a very fast-rising flood in which the flow increased from 10,000 to 150,000 cusecs and decreased again to 10,000 within 24 hr. Even in this case, where the acceleration terms were comparatively large, the last three slope terms are so small that only S_0 need be retained in Eq. (9–16). However when the bed slope is very flat the $\partial y/\partial x$ term may well be of the same order as S_0 but in this case the Froude number will be very low, so that the third term will be negligible. It can be shown that the third and fourth terms are of the same order of magnitude, but detailed discussion of this question must be postponed until Sec. 9.4. Meanwhile we note that on steep slopes only S_0 need be retained in Eq. (9–16), and on very flat slopes only $(S_0 - \partial y/\partial x)$ need be retained. In the former case the flood wave has certain very simple properties, as we shall see in the next section, that make the numerical methods of this section unnecessary, so we shall assume here that in applying these methods we are dealing with the latter case, i.e., rivers with flat slopes in which $S_f = S_0 - \partial y/\partial x$, approximately.

Under this assumption we can readily work out the elements of the curves in Fig. 9–4 by the following trial process: Given a pair of values of I and O, calculate the upstream and downstream depths, assuming $S_f = S_0$ in Eq. (9–16). Then from these depths a value of $\partial y/\partial x$ is obtained and inserted in

Eq. (9–16), which is then used to obtain corrected values of upstream and downstream depth. If this correction makes a substantial change in $\partial y/\partial x$, a further correction will be necessary, but this is not usually the case. When final values of upstream and downstream depths are obtained, the volume V is readily calculated (Prob. 9.5).

A practical difficulty which emerges in the working of this problem is that the numerical method is not very precise near the crest of the hydrograph, so it is difficult to be sure whether the flood wave is or is not subsiding as it moves downstream. These difficulties may have led to the view, which has been expressed, that flood waves in uniform channels do not subside at all, subsidence occurring only when reservoir-type storage is available to absorb some of the peak flow, as in Example 9.1. However it does not follow that subsidence is negligible in uniform channels, but to resolve the question it is necessary to resort to methods more precise than the numerical methods of this section.

9.3   The Movement of a Flood Wave in a River Channel

Before dealing with specific problems like subsidence, it is desirable to develop certain notions concerning the movement of the flood wave as a whole. One basic notion is that of the wave speed, which we shall now explore.

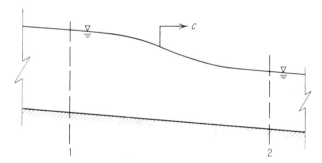

Figure 9-5.   The Monoclinal Flood Wave

The Speed of a Flood Wave

The most elementary form of this problem relates to the so-called "monoclinal" flood wave, shown in Fig. 9-5. This is simply a "step" increase in discharge; the problem is to find the speed c with which this step moves downstream.

Dynamically, the monoclinal flood wave is very different from the simple surge dealt with in Sec. 3.5. It may extend over a great length of channel, and it is substantially influenced by slope and resistance. But from the viewpoint of the continuity equation, the two phenomena are basically similar. As in the case of the simple surge, we can imagine the monoclinal flood wave brought to rest by the superimposition of a velocity equal and opposite to the wave velocity c. The discharge through the now stationary wave is constant and equal to $(v - c)A$, where v is the velocity relative to the bank; we may therefore write

$$Q - cA = \text{constant over the wave profile}$$

or $$dQ - c\,dA = 0$$

where Q is the discharge measured relative to the bank. It follows that the wave velocity is equal to

$$c = \frac{dQ}{dA} = \frac{1}{B}\frac{dQ}{dy} = v + A\frac{dv}{dA} \qquad \text{(9–17)}$$

which is an alternative form of Eq. (3–17).

The argument leading to Eq. (9–17) is a standard and accepted form, but it is subject to two important limitations. First, it has been assumed that the motion, like the surge movement of Sec. 3.5, may be reduced to steady flow by letting the observer move with a suitable velocity, and this implies that the wave-front profile is permanent in form without dispersion or subsidence. Doubts are thrown on this assumption by the arguments of Sec. 8.4, which showed that a positive wave tends to become steeper as it moves downstream. A further limitation is implied in the term dQ/dy, which is a total derivative and has a unique meaning only if Q is a function of y alone.

Further doubts about Eq. (9–17) are raised by its appearance of being in conflict with the dynamical arguments of Chap. 8, according to which the wave velocity, measured relative to the bank, is equal to

$$c = v \pm \sqrt{gy} \qquad \text{(9–18)}$$

Note that in Eqs. (9–17) and (9–18) we use c to denote the net wave velocity relative to the bank. This definition will be used throughout this chapter, except in Sec. 9.7, in which c will be defined as $\sqrt{gy}$, the speed of a small wave relative to the water, as in Chap. 8.

There are therefore a number of doubts about whether Eq. (9–17) properly gives the speed of a flood wave consisting of a step increase in discharge; there may be even more doubts concerning its application to the crest of the more normal shape of flood wave, consisting of a hump or intumescence as in Fig. 9-1. Despite all these reservations, extensive measurements of flood-crest speed, ranging from the classic observations of Seddon [3] in 1900 until recent times, have yielded values close to those obtained from Eq. (9–17) by assuming that dQ/dy has a unique meaning derived from the uniform flow condition, i.e., that $S_f = S_0$. For a constant Chézy C and a wide rectangular section, Eq. (9–17) then leads to

$$c = \frac{3}{2} v \qquad (9\text{–}19)$$

and to slightly different values of the ratio c/v for different types of section and resistance equations (Prob. 9.7). The ratio is always greater than unity, because dv/dA is always positive.

The Kinematic Wave

The uncertainties discussed above can be resolved only be recognizing that there is a form of wave movement different from that discussed by means of the theory of characteristics in Chap. 8. The matter has been cleared up by the theoretical studies of Lighthill and Whitham [4]. These investigators defined a "kinematic" wave as one in which Q is a function of y alone; as we have seen, this implies that $S_f = S_0$ and that the other three slope terms in Eq. (9–16) are negligible. We must first establish that there can be a true wave motion when Q is a function of y alone; to do this, consider the equation of continuity

$$\frac{\partial Q}{\partial x} + B \frac{\partial y}{\partial t} = 0 \qquad (9\text{–}20)$$

which can be rewritten

$$\frac{dQ}{dy} \frac{\partial y}{\partial x} + B \frac{\partial y}{\partial t} = 0$$

or

$$\frac{dy}{dt} = \frac{dx}{dt} \frac{\partial y}{\partial x} + \frac{\partial y}{\partial t} = 0 \qquad (9\text{–}21)$$

where

$$\frac{dx}{dt} = c = \frac{1}{B} \frac{dQ}{dy} \qquad (9\text{–}22)$$

It follows from Eq. (9–21) that to an observer moving with a velocity c given by Eq. (9–22), y and therefore Q will appear to be constant. As we have seen in Chap. 8, this situation is the essence of a true wave motion, so the existence of a kinematic wave is proved. The form of the proof shows also why the term "kinematic" was chosen to describe this type of wave motion: it is because the wave property follows principally from the equation of continuity, whereas the waves discussed in Chap. 8 arose from the continuity and the dynamic equations; we shall therefore refer to them as "dynamic" waves.

The speed of the kinematic wave, given by Eq. (9–22), is the same as that of the monoclinal wave, given by Eq. (9–17). It is noteworthy that only *one* value of speed is indicated by Eq. (9–22), whereas in the case of the dynamic waves of Sec. 8.3, there are two possible wave speeds, indicated by the two characteristic directions on the x-t plane. Physically, this means that a "dynamic" disturbance—i.e., one in which all four slope terms of Eq. (9–16) are significant, will propagate in both the upstream and downstream directions. A kinematic disturbance, on the other hand, will propagate only in the downstream direction.

Evidently both types of wave movement—kinematic and dynamic—may be present in any natural flood wave. The bed slope S_0 is usually by far the most important slope term, even if the other three terms are not negligible; the main bulk of the flood therefore moves substantially as a kinematic wave, although as we shall see in the next section its character may be somewhat modified by the other slope terms of Eq. (9–16). In particular, the speed of the main flood wave may be expected to approximate to that of the kinematic wave, given by Eq. (9–22), and this result was in fact proved in Lighthill and Whitham's study, by methods that are beyond the scope of this text. But unless the other slope terms are absolutely negligible (which they seldom are) they will produce dynamic wave fronts also, moving at speeds $v \pm \sqrt{gy}$ in front of and behind the main body of the flood wave, as shown in Fig. 9-6.

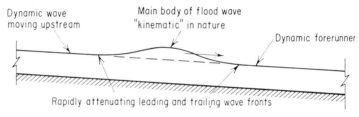

Figure 9-6. Movement of the Natural Flood Wave, after M. J. Lighthill and G. B. Whitham [4]

We can verify the position of these wave fronts relative to the main flood wave by considering the relative size of the kinematic and dynamic wave speeds. In a wide rectangular channel with constant Chézy C we have seen that Eq. (9–17) gives

$$c = \frac{3}{2} v \qquad (9\text{–}19)$$

for the speed of the kinematic wave. This is less than the speed of the leading dynamic wave, $v + \sqrt{gy}$, provided that $\frac{1}{2}v < \sqrt{gy}$, i.e., that Fr < 2, a condition which will be fulfilled in all natural rivers except mountain torrents. If it is fulfilled, the leading dynamic wave will be ahead of the main (kinematic) wave, as in Fig. 9-6. The other dynamic wave will be behind the main wave, since $(v - \sqrt{gy})$ must always be less than dQ/dA.

We have now established that the leading dynamic wave acts as a forerunner of the main wave; the question now arises whether it will bring about any appreciable rise of the water level before the arrival of the main wave. The question has in fact been answered in Sec. 8.4; if Fr < 2, a positive wave of this sort will attenuate as it moves downstream unless the initial rate of rise exceeds the value given in Eqs. (8–41) or (8–42). It is most unlikely that this critical value will be exceeded in a natural flood; Eq. (8–42) indicates, for common values of flow parameters, a rate of rise of one foot every few minutes. Only a catastrophic natural event, or the operation of man-made controls, would produce a rate of rise of this magnitude.

In the normal natural flood, therefore, the dynamic forerunner will attenuate rapidly unless $Fr > 2$, as given by the argument of Sec. 8.4. It is interesting to note from the above argument that this is also the condition that the kinematic wave should overtake the dynamic forerunner; clearly such an event would lead to the steepening of the dynamic wave front, verifying the result of Sec. 8.4. This particular physical interpretation gives a deeper significance to the criterion $Fr \gtrless 2$, for it suggests an actual mechanism by which the wave front steepens when $Fr \geq 2$.

The identity of these two criteria is not a mere numerical coincidence, since it holds for any shape of channel cross section and any form of resistance equation. Escoffier's analysis, reproduced in Sec. 8–5, has shown that the general form of the criterion $Fr \gtrless 2$ in the argument of Sec. 8.4 is

$$\frac{dv_0}{d\omega} \gtrless 1 \qquad (9\text{–}23)$$

where v_0 is the uniform flow velocity appropriate to a given depth, and ω is the stage variable, defined by the equation

$$d\omega = c_d \frac{dA}{A} \qquad (9\text{–}24)$$

where $c_d = \sqrt{(gA/B)}$, the dynamic wave velocity relative to the water. Now from Eq. (9–17), the kinematic wave velocity will be greater or less than the dynamic wave velocity according as

$$A \frac{dv_0}{dA} \gtrless c_d \qquad (9\text{–}25)$$

where v has been replaced by v_0 because the existence of the kinematic wave postulates a uniform flow relationship between v and A. It follows from Eq. (9–24) that Eqs. (9–23) and (9–25) are identical; therefore the kinematic wave criterion is identically the same as the one derived in Secs. 8.4 and 8.5, whatever the shape of cross section or form of resistance equation.

This discussion of the dynamic forerunner may appear to have little significance for the engineer, since in most cases the forerunner will attenuate down to a scarcely perceptible feather edge, making little contribution to the progress of the flood wave as a whole. However, as we shall see in Sec. 9.7, it does not follow that the forerunner has no significance in flood-routing computations.

The Monoclinal Wave as a "Kinematic Shock"

We consider now the movement of the main bulk of the flood wave, which approximates to a kinematic wave. If it is exactly kinematic, i.e., if $S_f = S_0$, it follows from Eq. (9–17) and from the form of the usual resistance equations that the kinematic wave speed c increases with y, and hence with Q. Any kinematic wave profile must therefore advance as in Fig. 9-7, with the front face becoming continually steeper. We note that since c is a function of y alone, a horizontal chord such as BB' in Fig. 9-7 will remain the same length; the wave therefore does not lengthen, or disperse, so it does not subside and the crest A maintains the same maximum depth. This result accords well with the common observation that flood waves in steep rivers of uniform section do not subside as they move downstream.

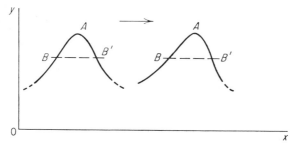

Figure 9-7. *Distortion of the Kinematic Wave Profile with Down-stream Movement*

The steepening of the wave front shown in Fig. 9-7 may suggest that ultimately a surge will form. However, as the wave front becomes steeper the secondary slope terms of Eq. (9–16) come into play, and as will be shown in the next section the effect of these terms is to introduce dispersion and attenuation, not present in the kinematic wave proper; we may expect therefore that they will retard the steepening of the wave front, thus delaying, or perhaps preventing altogether, the formation of a surge. If the steepening process is eventually arrested, short of surge formation, the result must be the smooth, steady-state, step profile shown in Fig. 9-5 and described as the "monoclinal wave." For this wave front Lighthill and Whitham [4] proposed the alternative name "kinematic shock," since it derives ultimately from the steepening of a kinematic wave front. We now consider some further properties of this form of wave front.

From the argument leading up to Eq. (9–17) we can write, taking the viewpoint of an observer traveling with the wave:

$$A(c - v) = Q_r, \text{ a constant} \tag{9–26}$$

Q_r being termed the "overrun." For the wave velocity c we can obtain a finite-difference form, as an alternative to the differential form of Eq. (9–17) Writing the continuity equation (9–26) between sections 1 and 2 in Fig. 9-5, we have

$$A_1(c - v_1) = A_2(c - v_2) = Q_r$$

whence

$$c = \frac{Q_1 - Q_2}{A_1 - A_2} \qquad (9\text{–}27)$$

and

$$Q_r = \frac{A_2 Q_1 - A_1 Q_2}{A_1 - A_2} \qquad (9\text{–}28)$$

These results can be represented geometrically as in Fig. 9-8, the points 1 and 2 representing sections 1 and 2 of Fig. 9-5. We note that the finite difference form Eq. (9–27) corresponds exactly to Eq. (3–17).

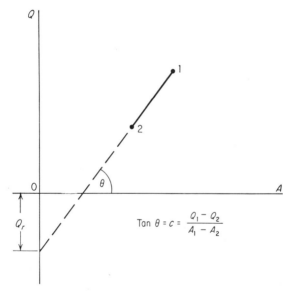

$$\text{Tan } \theta = c = \frac{Q_1 - Q_2}{A_1 - A_2}$$

Figure 9-8.   *Features of the Monoclinal Flood Wave*

The profile of the wave front can be determined [4] by combining Eq. (9–26) with the dynamic equation, which in this case is more conveniently written as an equation for the velocity:

$$v = C\sqrt{R\left(S_0 - \frac{\partial y}{\partial x} - \frac{v}{g}\frac{\partial v}{\partial x} - \frac{1}{g}\frac{\partial v}{\partial t}\right)} \qquad (9\text{–}29)$$

In the special case where the channel is wide rectangular ($R = y$) and the Chézy C is constant, the profile can be determined explicitly by eliminating v between Eqs. (9–26) and (9–29). The argument is a straightforward but fairly lengthy exercise in elementary calculus, and details are left to the reader

(Prob. 9.8). The result is that to an observer moving with the wave front, the wave profile appears to have the following differential equation

$$\frac{dy}{dx} = -S_0 \frac{(y - y_2)(y_1 - y)(y - Y)}{y^3 - y_{cr}^{\ 3}} \tag{9-30}$$

which can be integrated to

$$e^{S_0 x / y_2} = \frac{(y_1 - y)^{a_1} e^{y/y_2}}{(y - y_2)^{a_2}(y - Y)^G} \tag{9-31}$$

where

$$
\left.
\begin{aligned}
&a_1 = \frac{y_1^{\ 3} - y_{cr}^{\ 3}}{y_2(y_1 - y_2)(y_1 - Y)}, \quad a_2 = \frac{y_2^{\ 3} - y_{cr}^{\ 3}}{y_2(y_1 - y_2)(y_2 - Y)}, \\
&G = \frac{y_{cr}^{\ 3} - Y^3}{y_2(y_2 - Y)(y_1 - Y)}, \quad Y = \frac{y_1 y_2}{(\sqrt{y_1} + \sqrt{y_2})^2}, \quad y_{cr}^{\ 3} = \frac{Q_r^2}{B^2 g}
\end{aligned}
\right\} \tag{9-32}
$$

Now the preceding argument gives us the means of determining not only the established wave profile, but also the conditions under which a stable wave profile will form. When $y = y_{cr}$, dy/dx is infinite, and a surge must form. This situation will be avoided if $y_1 > y_2 > y_{cr}$, in which case $dy/dx = 0$ at $y = y_1$ and $y = y_2$, so that the curve is asymptotic at each end to the lines $y = y_1$ and $y = y_2$ (Fig. 9-9a). The length of the profile is then infinite, but a

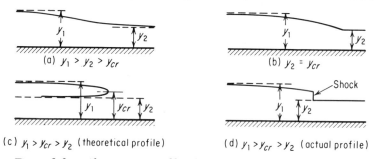

(a) $y_1 > y_2 > y_{cr}$

(b) $y_2 = y_{cr}$

(c) $y_1 > y_{cr} > y_2$ (theoretical profile)

(d) $y_1 > y_{cr} > y_2$ (actual profile)

Figure 9-9.   Alternative Profiles for the Monoclinal Flood Wave

suitable practical measure of its length is the distance between the sections at which $y = y_1 - \varepsilon(y_1 - y_2)$ and $y = y_2 + \varepsilon(y_1 - y_2)$, where ε is small, say 0.05. This distance is approximately equal to (Prob. 9.9)

$$\frac{y_2}{S_0}(a_1 + a_2)\log_e \frac{1}{\varepsilon} \tag{9-33}$$

If we take $\varepsilon = 0.05$, and note that $(a_1 + a_2)$ is in the neighborhood of 6–10 (Prob. 9.9), then for normal values of channel parameters the length given by Eq. (9–33) is of the order of tens of miles at least.

Now if y_{cr} is increased until it becomes equal to y_2, Eq. (9–30) shows that dy/dx is finite at the downstream end of the profile, where $y = y_2$ (Fig. 9-9b).

Surge formation is now imminent; if y_{cr} is further increased, the profile develops the physically impossible form shown in Fig. 9-9c and this anomaly can only be removed by fitting in an abrupt surge front, as in Fig. 9-9d.

This argument proves that under suitable conditions, i.e., $y_1 > y_2 > y_{cr}$, a stable monoclinal wave can exist, and will maintain a steady form as it moves downstream. The substantial length indicated by Eq. (9–33) suggests that this wave will occur only in long rivers; we can throw some further light on this question by considering the development of the wave (thinking of it as a kinematic shock) from an initial kinematic form like that of Fig. 9-7. The problem is the basic one of shock formation, already discussed in Sec. 8.4 in connection with Fig. 8-3a. The result is given in Eq. (8–23); we modify this equation according to the new definition of c adopted in this chapter, and write

$$x_s = \frac{c^2}{\partial c/\partial t} = -\frac{c}{\partial c/\partial x} \tag{9–34}$$

or

$$t_s = \frac{c}{\partial c/\partial t} = -\frac{1}{\partial c/\partial x} \tag{9–35}$$

where x_s is the distance traveled, and t_s is the time elapsed, before a shock develops. The point on the profile at which a shock ultimately develops will be that for which t_s is a minimum. The partial derivatives have their usual meanings, and the relation

$$\frac{dc}{dt} = \frac{\partial c}{\partial t} + c\frac{\partial c}{\partial x} = 0 \tag{9–36}$$

implied in Eqs. (9–34) and (9–35), follows from the basic kinematic-wave property.

Now in natural floods c will be of the order of 10 ft/sec and the derivative $\partial c/\partial t$ will hardly exceed 1 ft/sec/hr, even in a very fast-rising flood. It follows that x_s will be of the order of at least tens, and probably hundreds, of miles. This figure assumes that the wave remains kinematic; in fact the attenuation introduced by the slope terms other than S_0 will make the distance x_s longer still. The conclusion is that the monoclinal wave, or kinematic shock, will occur only in very long rivers and may therefore be comparatively rare in practice.

Indeed much of the material treated in this section may be of limited direct interest to the river engineer, who is mainly concerned to trace the progress and subsidence (if any) of the flood crest; the kinematic shock, on the other hand, occurs well down on the leading flank of the flood wave. Nevertheless it is important to dispose of the matters of principle treated in this section before moving on to the consideration of crest subsidence. If these matters of principle were neglected, the questions posed at the beginning of this section would remain unanswered and would continue to raise doubts about the validity of the subsequent discussion.

9.4 The Subsidence of the Flood Crest

In the previous section a result was quoted [4] to the effect that the main bulk of a natural flood wave moves substantially as a kinematic wave. The analysis leading to that result was an approximate one, strictly applicable only to a wave of small height; it does not exclude the possibility that slope terms other than S_0 may modify somewhat the kinematic character of the wave. In particular, they may possibly make the crest speed differ from the kinematic value given by Eq. (9–17), and they may introduce appreciable dispersion and subsidence. We first consider the effect of the absolute magnitude of S_0 on the relative magnitude of the four slope terms in Eq. (9–29). For a kinematic wave with a wave velocity c, it follows that for any point on the wave profile

$$\frac{dy}{dt} = \frac{\partial y}{\partial t} + c\frac{\partial y}{\partial x} = 0 \qquad (9\text{–}37)$$

We now consider the magnitude of $\partial y/\partial x$ in a wide rectangular channel for a given inflow hydrograph $q = q(t)$. From Eq. (9–37),

$$-\frac{\partial y}{\partial x} = \frac{1}{c}\frac{\partial y}{\partial t} = \frac{1}{c^2}\frac{\partial q}{\partial t} \propto \frac{1}{(yS_0)}\frac{\partial q}{\partial t} \propto \frac{1}{(qS_0)^{2/3}}\frac{\partial q}{\partial t} \qquad (9\text{–}38)$$

eliminating y by the relation $q \propto y^{3/2}S_0^{1/2}$. Then

$$\frac{\partial y/\partial x}{S_0} \propto S_0^{-5/3} \times \begin{pmatrix}\text{terms characteristic of the}\\ \text{inflow\ \ hydrograph\ \ alone}\end{pmatrix} \qquad (9\text{–}39)$$

and must become small as S_0 increases. Also

$$\frac{v}{g}\frac{\partial v/\partial x}{\partial y/\partial x} = \frac{\partial(v^2/2g)/\partial x}{\partial y/\partial x} = O(\mathrm{Fr}^2) \qquad (9\text{–}40)$$

where Fr is the Froude number and the notation $O(\ )$ means "is of the order of magnitude of". Further,

$$\frac{1}{g}\frac{\partial v}{\partial t} = O\!\left(\frac{c}{g}\frac{\partial v}{\partial x}\right) = O\!\left(\frac{v}{g}\frac{\partial v}{\partial x}\right) \qquad (9\text{–}41)$$

so the two remaining slope terms in Eq. (9–29) are of the same order of magnitude, and are of no higher order than $\partial y/\partial x$, unless $\mathrm{Fr}^2 \gg 1$, which is true only in mountain torrents. It follows from Eqs. (9–39) through (9–41) that flood waves will be kinematic in rivers that are sufficiently steep, but short of the torrential. In rivers of sufficiently gentle slope, $\mathrm{Fr}^2 \ll 1$, even during floods, and whereas $(\partial y/\partial x)/S_0$ will be appreciable, the last two terms will be negligible. Conceivably there could be intermediate values of slope on which all four slope terms would be appreciable, but for the present we shall consider only the case of very gentle slopes. Before doing so, it is convenient to deduce an alternative proof of the nonsubsidence of the kinematic wave, by considering the crest A in Fig. 9-7. At this point

$$\frac{\partial y}{\partial x} = 0, \quad \text{and} \quad \frac{\partial Q}{\partial x} = 0 \tag{9–42}$$

since Q is a function of y alone. Hence from Eq. (9–20)

$$\frac{\partial y}{\partial t} = 0, \quad \text{and} \quad \frac{\partial Q}{\partial t} = 0 \tag{9–43}$$

It follows that Eq. (9–37) must be true at the crest A, which does not therefore subside. This form of argument can readily be adapted to the problem of detecting subsidence in other forms of wave.

Flood Waves on Gentle Slopes

Under this heading we consider the case in which Eq. (9–16) becomes, for the wide rectangular section and constant Chézy C:

$$Q = BCy\sqrt{y\left(S_0 - \frac{\partial y}{\partial x}\right)} \tag{9–44}$$

Consider the instantaneous wave profile shown in Fig. 9-10. At the point A, $\partial y/\partial x = 0$, but because Q depends on $\partial y/\partial x$ as well as on y, the point of maximum Q will not be at A, but at a point B somewhat downstream of A. We prove this by differentiating Eq. (9–44); we then have

$$\frac{\partial Q}{\partial x} = \tfrac{3}{2}BC\frac{\partial y}{\partial x}\sqrt{y\left(S_0 - \frac{\partial y}{\partial x}\right)} - \tfrac{1}{2}BCy\frac{\partial^2 y}{\partial x^2}\sqrt{\frac{y}{S_0 - \dfrac{\partial y}{\partial x}}} \tag{9–45}$$

$$= 0 \quad \text{when} \quad 3\frac{\partial y}{\partial x}\left(S_0 - \frac{\partial y}{\partial x}\right) = y\frac{\partial^2 y}{\partial x^2} \tag{9–46}$$

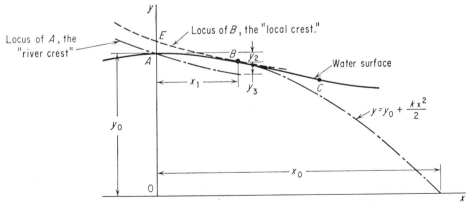

Figure 9-10. *Profile of the Non-Kinematic Wave*

Because $\partial^2 y/\partial x^2$ is negative, the solution of this equation near A will have $\partial y/\partial x$ negative—i.e., B is downstream of A.

Definition of the Wave Crest

This matter is not so simple as in the kinematic wave, where the point A in Fig. 9-7 was clearly to be defined as the wave crest. In the present case $\partial Q/\partial x = 0$ at B, so that from Eq. (9–20) $\partial y/\partial t = 0$ at B, and the depth at B is the greatest that will occur at this section. It follows that B, not A, should be defined as the wave crest. This definition agrees both with common experience and the requirements of the river engineer, who is concerned not with the maximum depth existing in the river at any instant, but with the maximum level reached at any particular section.

The true flood peak is therefore at B, which we shall term the "local crest"; the point A will be termed the "river crest," because the depth at A is the greatest in the entire river at that instant. It may be asked how the depth at B can be the greatest that will occur at that section, when greater depths exist upstream at that instant; the answer is that the upstream depths (e.g., at A) will have subsided to less than the maximum depth at B by the time the corresponding points on the profile have reached B. In other words, the river crest A must subside in its downstream motion if it is to slide under the point B.

The Existence of Subsidence

A more formal proof of the existence of subsidence can be deduced in this way. Let the subscripts 0 and 1 indicate conditions at A and B respectively, with $\partial y_1/\partial x$ indicating $\partial y/\partial x$ at B, and so on. Thus

$$\frac{dy_1}{dx} = \frac{\partial y_1}{\partial x} + \frac{dt}{dx}\frac{\partial y_1}{\partial t}$$

in which $dx/dt = c_1$, the wave speed of the point B, as yet unknown. Because $\partial y_1/\partial t = 0$, it follows that

$$\frac{dy_1}{dx} = \frac{\partial y_1}{\partial x} \qquad\qquad (9\text{–}47)$$

which expresses an important conclusion. The partial derivative $\partial y_1/\partial x$ is the slope of the wave profile at the point B in Fig. 9-10; the total derivative dy_1/dx is the slope of the locus of the point B on successive wave profiles. Equation (9–47) indicates that because the two slopes are equal, the locus of B (shown as a broken line in Fig. 9-10) is tangential to the wave profile at B, and is therefore an envelope of the entire family of instantaneous water-surface profiles.

Because $\partial y_1/\partial x$ is negative, so is dy_1/dx, and the existence of subsidence is proved. Equation (9–47) also validates the traditional engineering practice of using the high-water marks left by a receding flood to obtain values of depth and slope from which the discharge may be calculated by a resistance formula (the "slope-area" method). These water marks trace out the envelope of B, which by Eq. (9–47) has the same slope as the water surface, and may therefore be used via Eq. (9–44) to calculate the discharge.

The Amount of the Subsidence

Granted that subsidence exists, it remains to determine its magnitude. If we are to use Eq. (9–47) as the basis for this determination, we must make some assumption about the shape of the water surface profile between A and B. If it is assumed simply that the curve AB is a parabola with vertex at A and axis coincident with the y axis (Fig. 9-10), the argument will then apply only to those waves that are long enough for the point B to occur well within that part of the profile that is convex upwards. This is probably true of all natural floods in rivers. However in other cases, for example storm-water drains, flood waves may be so sharply peaked that B occurs well down the flank of the wave, perhaps even on that part of the profile which is concave upwards.

Confining our attention to natural flood waves, and assuming that AB is a parabola, we can write

$$\left.\begin{aligned} \frac{\partial^2 y}{\partial x^2} &= k, \text{ a function of time alone} \\[2mm] \frac{\partial y}{\partial x} &= kx \\[2mm] y &= y_0 + \frac{kx^2}{2} \end{aligned}\right\} \qquad (9\text{–}48)$$

This parabola, if extended to the right and left, meets the x axis at a distance x_0 from the origin. The "wave slope" is introduced

$$S_w = \frac{y_0}{x_0} \qquad (9\text{–}49)$$

as a convenient measure of the steepness of the flood wave. From Eq. (9–48), we can write the results

$$kx_0{}^2 = -2y_0; \quad ky_0 = -\frac{2y_0{}^2}{x_0{}^2} = -2S_w{}^2; \quad \frac{ky_0}{S_0{}^2} = -\frac{2S_w{}^2}{S_0{}^2} = -\frac{2}{r^2} \quad (9\text{–}50)$$

where $r = S_0/S_w$, a ratio that will occur frequently in the following results. In natural floods r is invariably much greater than unity, and usually greater than 10; the term $1/r^2$, which frequently arises in this argument, can therefore be treated as small.

The analysis is therefore based on two special assumptions: that the profile AB is of constant curvature, and that r^2 is large. Both assumptions are justified if the flood wave is long and slow-rising; such a wave can conveniently be termed a "mild" wave, and appears to occur in virtually all natural floods.

The distance $AB = x_1$ is now calculated. Substituting Eq. (9–48) into Eq. (9–46), and eliminating y_0 by means of Eq. (9–50), we obtain the quadratic

$$7x_1{}^2 + 3rx_1x_0 - x_0{}^2 = 0 \tag{9–51}$$

which has a positive root given by

$$\frac{x_1}{x_0} = \frac{-3r + \sqrt{9r^2 + 28}}{14} \tag{9–52}$$

Because r^2 is large the square root can be approximated by

$$3r\left(1 + \frac{14}{9r^2} - \frac{98}{81r^4}\right)$$

(see the Appendix to Chap. 3). The solution Eq. (9–52) then becomes

$$\frac{x_1}{x_0} = \frac{1}{3r}\left(1 - \frac{7}{9r^2}\right)$$

or

$$x_1 = \frac{y_0}{3S_0}\left(1 - \frac{7}{9r^2}\right) \tag{9–53a}$$

from Eq. (9–49). To a first approximation the term $-7/9r^2$ may be dropped, leaving simply

$$x_1 = \frac{y_0}{3S_0} \tag{9–53b}$$

We can now write down an expression for dy_1/dx, the slope of the locus of B, i.e., the rate of crest subsidence with distance. From Eqs. (9–47), (9–48), and (9–53b), the expression is

$$\frac{dy_1}{dx} = \frac{\partial y_1}{\partial x} = \frac{ky_0}{3S_0} \tag{9–54}$$

but its direct usefulness is limited by the fact that it is difficult to obtain k from flood records. These records usually give the variation of stage at a given section, so that quantities like k, characteristic of the instantaneous profile, can be determined only by comparing at-a-section records taken at neighboring sections. This cannot be done accurately unless recording stations are set at close intervals along the river, and it would be better if Eq. (9–54) could be recast in terms of quantities available from at-a-section records.

This recasting can in fact be done. First, we consider the relative magnitudes of y_1 and y_0; from Eqs. (9–48), (9–50), and (9–53b) we have

$$y_1 = y_0 + \frac{kx_1{}^2}{2}$$

$$= y_0 + \frac{ky_0{}^2}{18S_0{}^2}$$

i.e.,
$$\frac{y_1}{y_0} = 1 - \frac{1}{9r^2} \tag{9-55}$$

so that the proportional difference between y_1 and y_0 is of the order of the small second-order quantity $1/r^2$, which can usually be neglected. It would be reasonable therefore to assume that the total derivatives of y_1 and y_0 differ only by second-order terms. This can in fact be proved; it means among other things that the loci of A and B on Fig. 9-10 are approximately parallel. It follows that y_3 in this figure is equal to the interval AE; from the properties of the parabola AE is also equal to y_2. Hence $y_2 = y_3$; i.e.,

$$x_1^2 \frac{\partial^2 y}{\partial x^2} = t_1^2 \frac{\partial^2 y}{\partial t^2}$$

in which t_1 is the time required for the river crest to move from A to B and $\partial^2 y/\partial t^2$ is the second derivative (assumed constant) of the y-t curve at B. Then, because $x_1/t_1 = c_0$, the wave speed of A,

$$k c_0^2 = \frac{\partial^2 y}{\partial t^2} \tag{9-56}$$

and this equation establishes the desired connection between the instantaneous profile and the stage record taken at a section. Equation (9–54) now becomes

$$\frac{dy_1}{dx} = \frac{y_0}{3 S_0 c_0^2} \frac{\partial^2 y}{\partial t^2} \tag{9-57}$$

and all the elements of this equation can be obtained readily from flood records at a single river section. [In view of Eq. (9–55), y_0 and y_1 can be regarded as interchangeable in Eq. (9–57).] Finally, we can assess the order of magnitude of dy_1/dx by noting that

$$\frac{dy_1/dx}{S_0} = \frac{k y_0}{3 S_0^2} = -\frac{2}{3r^2} \tag{9-58}$$

from Eqs. (9–50) and (9–53b). Therefore the space rate of subsidence is itself a small fraction of the bed slope. However, it need not be negligible; whereas bed slopes are usually of the order of feet per mile, subsidence rates are appreciable if they are of the order of tenths of a foot, or less, per mile.

We have not yet discussed whether the existence of subsidence will make the wave speed differ appreciably from the kinematic wave speed $3v/2$. This matter will be dealt with in the next section.

It is noteworthy that the above argument can be applied to a wave trough as well as to a wave crest, as shown in Fig. 9-11. The trough behaves in every way as a mirror image of the crest, so that $\partial y/\partial t = 0$ at a point B_2, somewhat downstream of the point A_2 where $\partial y/\partial x = 0$, and the locus of B_2 slopes upwards in the downstream direction (Prob. 9.10).

The above arguments can be extended to fit the cases where the surface width B is related to the depth y by the equation

$$\frac{B}{B_s} = \left(\frac{y}{y_s}\right)^i \tag{2-37}$$

and the resistance equation takes the general form

$$v = C_0 R^j S_f^{1/2} \tag{9-59}$$

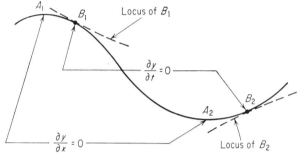

Figure 9-11.   *Attenuation of Crest and Trough in a Non-Kinematic Wave*

The details have been worked out by Henderson [6] and can be reproduced by the reader as an exercise (Prob. 9.11). Equation (9–53b) becomes

$$x_1 = \frac{y_0}{2pS_0} \tag{9-60}$$

and Eq. (9–56) remains unaltered. Equation (9–57) then becomes

$$\frac{dy_1}{dx} = \frac{y_0}{2pS_0 c_0^2} \frac{\partial^2 y}{\partial t^2} \tag{9-61}$$

where

$$p = i + j + 1 \tag{9-62}$$

A further possible extension is to the case where all four slope terms are included in Eq. (9–16). Analysis is possible if it is assumed that

1. The Froude number Fr is small enough for Fr^4 and higher powers to be neglected;

2. The last two slope terms in Eq. (9–16) are small enough to be evaluated with sufficient accuracy from the resistance equation by including only the first two slope terms; and

3. Over the crest region, y is close to y_0, the ratio $(\partial y / \partial x)/S_0$ is small, and Fr remains approximately constant.

Item 2 above involves an approximation that is quite admissible and in fact is a consequence of item 1. If Fr is small, the extra terms now being included are second-order terms and great precision is not needed in computing them.

The details are left as an exercise for the reader (Prob. 9.12). Equation (9–53a) becomes

$$x_1 = \frac{y_0}{3S_0} \frac{1 - \text{Fr}^2/4 - 7/9r^2}{1 + 4\text{Fr}^2/3r^2} \tag{9–63}$$

whence, dropping terms in $1/r^2$, Eq. (9–57) becomes

$$\frac{dy_1}{dx} = \frac{y_0}{3S_0 c_0{}^2} \frac{\partial^2 y}{\partial t^2} \left(1 - \frac{\text{Fr}^2}{4}\right) \tag{9–64}$$

from which it appears that Fr is "small" in the sense used above if $\text{Fr}^2 < 0.5$ approximately, which will be true in practically all natural floods. Equation (9–64) now becomes the general form which should be used instead of Eq. (9–57); finally, the extension embodied in Eq. (9–61) can be incorporated in Eq. (9–64) by writing it as

$$\frac{dy_1}{dx} = \frac{y_0}{2pS_0 c_0{}^2} \frac{\partial^2 y}{\partial t^2} \left(1 - \frac{\text{Fr}^2}{4}\right) \tag{9–65}$$

which is the most general possible form of the subsidence equation.

One last remark must be made about the material in this section. The argument has culminated in Eq. (9-65), which gives the slope of the crest envelope; there still remains the question of possibly integrating this equation to give y as a function of x for the complete profile of the crest envelope. This matter will be investigated in Sec. 9.6.

9.5   The Speed of a Subsiding Flood Wave

When a flood wave is subsiding, as described in the previous section, the wave velocity may depart from the kinematic wave velocity, for two reasons:

1. The discharge is no longer a function of depth alone, so that a unique meaning cannot be assigned to dQ/dy in Eq. (9–22).

2. Even if Eq. (9–22) gave the speed with which a section of constant Q or y moves downstream, this would not necessarily apply to the wave crest, at which neither Q nor y remains constant.

However, we can determine the velocity of the crest by the same type of argument as the one that led up to Eq. (9–22), except that in this case we are tracing the path of a point having constant $\partial y/\partial x$ or constant $\partial y/\partial t$ rather than constant Q or y. It is convenient to trace first the path of the river crest A (Fig. 9-10); at this point $\partial y/\partial x$ remains constant and equal to zero. Setting $y_x = \partial y/\partial x$, we can write

$$\frac{dy_x}{dt} = \frac{\partial y_x}{\partial t} + \frac{dx}{dt} \frac{\partial y_x}{\partial x} \tag{9–66}$$

which equals zero if

$$\frac{dx}{dt} = -\frac{\partial y_x/\partial t}{\partial y_x/\partial x} = -\frac{\partial^2 y/\partial x \partial t}{\partial^2 y/\partial x^2} = \frac{\partial^2 Q/\partial x^2}{B\partial^2 y/\partial x^2}$$

thus,

$$c_0 = \left[\frac{\partial^2 Q/\partial x^2}{B\partial^2 y/\partial x^2}\right]_0 \tag{9-67}$$

Differentiating Eq. (9–45) with respect to x and inserting the conditions $y = y_0$, $\partial y/\partial x = 0$ yields

$$\frac{\partial^2 Q}{\partial x^2} = \frac{3BCk\sqrt{y_0 S_0}}{2}\left(1 - \frac{ky_0}{6S_0^2}\right) \tag{9-68}$$

Writing

$$v_0 = C\sqrt{y_0 S_0} \tag{9-69}$$

the velocity at A, the substitution of Eqs. (9–50) and (9–68) into Eq. (9–67) yields

$$c_0 = \tfrac{3}{2}v_0\left(1 + \frac{1}{3r^2}\right) \tag{9-70}$$

a result which differs only slightly from the kinematic wave speed given by Eq. (9–22). The wave velocity c_1 of the local crest B is obtained from this argument:

$$c_0 - c_1 = -\frac{dx_1}{dt.} \tag{9-71a}$$

$$= -\frac{1}{3S_0}\frac{dy_0}{dt} \tag{9-71b}$$

from Eq. (9–53b). Now because $\partial y_0/\partial x = 0$,

$$\frac{dy_0}{dt} = \frac{\partial y_0}{\partial t}$$

and

$$\frac{dy_0}{dx} = \frac{1}{c_0}\frac{\partial y_0}{\partial t}$$

Thus

$$\frac{dy_0}{dt} = c_0\frac{dy_0}{dx} \approx c_0\frac{dy_1}{dx} = c_0\frac{\partial y_1}{\partial x} = c_0 kx_1 \tag{9-72}$$

assuming again that the loci of A and B on Fig. 9-10 are parallel. Equation (9–71) now becomes

$$c_0 - c_1 = \frac{c_0 kx_1}{3S_0} = \frac{-c_0 ky_0}{9S_0^2}$$

whence

$$\frac{c_1}{c_0} = 1 - \frac{2}{9r^2} \tag{9-73}$$

so that the wave velocities at A and B differ only in second-order terms from each other and from the kinematic wave value given by Eq. (9–19).

As with the subsidence results of the previous section, the results of this section can be extended to take account of more general forms of cross section and of resistance equation. If Eqs. (2–37) and (9–59) are assumed true, then Eq. (9–70) becomes (Prob. 9.13)

$$c_0 = \frac{pv_0}{i+1}\left(1 + \frac{1}{2pr^2}\right) \tag{9-74}$$

(cf. Problem 9.7). Similarly Eq. (9–73) becomes

$$\frac{c_1}{c_0} = 1 - \frac{1}{2p^2r^2} \tag{9-75}$$

Taking account of the last two slope terms in Eq. (9–16) generalizes Eqs. (9–74) and (9–75) further to

$$c_0 = \frac{pv_0}{i+1}\left(1 + \frac{1}{2pr^2} - \frac{\text{Fr}^2}{2pr^2}\right) \tag{9-76}$$

and

$$\frac{c_1}{c_0} = 1 - \frac{1}{2p^2r^2} + \frac{3\text{Fr}^2}{2p^2r^2} \tag{9-77}$$

The details of these final modifications do not obscure the general conclusion, that c_0 and c_1 differ only in second-order terms from each other and from the kinematic wave speed. In view of the small rates of subsidence detected by the theory of Sec. 9.4, we may conclude from that section and this one that if the wave is "mild," i.e., slow-rising, as natural flood waves are, then substantial rates of subsidence, and substantial departures from the kinematic wave velocity, can be produced only by storage effects such as in the level-pool routing of Sec. 9.2. It remains to be seen whether these storage effects can also be simulated by channel irregularities having the effect of a chain of small lakes.

9.6 Channel Irregularities and the Diffusion Analogy

From the results of Sec. 9.4, a useful and interesting line of argument can be developed which throws further light on the process of flood-wave subsidence. First, we note that the results of Sec. 9.5 have justified, for the gentle-slope case of Eq. (9–44), the approximate result

$$c = \tfrac{3}{2}v = \tfrac{3}{2}C\sqrt{y\left(S_0 - \frac{\partial y}{\partial x}\right)} \tag{9-78}$$

Substituting this in Eq. (9–45), we obtain

$$\frac{\partial Q}{\partial x} = Bc\frac{\partial y}{\partial x} - \frac{Bcy}{3\left(S_0 - \dfrac{\partial y}{\partial x}\right)}\frac{\partial^2 y}{\partial x^2} \tag{9–79}$$

Eliminating $\partial Q/\partial x$ between Eqs. (9–20) and (9–79) leads to the result

$$\frac{dy}{dt} = \frac{\partial y}{\partial t} + c\frac{\partial y}{\partial x} = K_1\frac{\partial^2 y}{\partial x^2} \tag{9–80}$$

where
$$K_1 = \frac{cy}{3\left(S_0 - \dfrac{\partial y}{\partial x}\right)} \quad \text{or} \quad \frac{cy}{3S_0}\text{(approx.)} \tag{9–81}$$

The essential feature of the argument now lies in comparing Eqs. (9–37) and (9–80). The former is the standard wave equation common to all branches of mechanics and applicable in the present context to kinematic waves; it indicates that y appears constant to an observer moving with velocity c. Equation (9–80) also is a standard form of wave equation, containing an extra "diffusion" term $K_1\,\partial^2 y/\partial x^2$, whose effect is to make y decrease in the view of an observer moving with velocity c. In this diffusion term is summed up the effect of the second slope term in Eq. (9–16), which modifies the kinematic character of the flood wave and makes it subside. Indeed Eq. (9–80) conveys the same information as does Eq. (9–54) in Sec. 9.4; one equation can be converted into the other by noting from Eq. (9–72) that near the crest of the flood wave $dy/dt = c\,dy/dx$ approximately. The term K_1 may be called a *diffusion coefficient*.

So far the argument has dealt with a uniform channel, as in Secs. 9.4 and 9.5. It is also clear from the treatment of level-pool routing in Sec. 9.2 that channel irregularities, having the effect of a chain of small lakes, will also contribute to the subsidence of the wave. In fact any form of off-channel storage, such as seepage to the surrounding country, will have the same effect. Now it was suggested by Hayami [7] that these storage effects could be indicated by a second diffusion coefficient K_2 added to the first one, the combined coefficient $K = K_1 + K_2$ replacing K_1 in Eq. (9–80). There is no strict theoretical basis for this proposal, but when applied in practice it gives satisfactory results even when K and c are assumed constant, which can only be approximately true.

The great advantage of this "diffusion analogy" method is that if K and c are assumed constant it can make use of a well-known wave equation, i.e., Eq. (9–80), which has known explicit solutions. Because of this the movement of the entire wave profile, however irregular it may be, can be traced with confidence. On the other hand, the method of Sec. 9.4 can only deal with the crest region, and then only by assuming its profile to be of a regular geometric shape. The major disadvantage of the diffusion analogy is that there is no known way of relating the second coefficient K_2 to the geometry of the channel irregularities or to other properties influencing the off-channel storage, and usually the combined coefficient K must be determined by observing the progress of a known flood. In this way K_2 is found to be of the order of 10^5–10^6 ft²/sec in many natural channels. The need for this empirical determination of K_2 makes the method less useful when applied to a "design" channel, which exists only on the drawing board.

We consider now the solution of Eq. (9–80). The form of the solution is very much dependent on the form of the initial flood wave introduced at the upstream end of the reach; if, for instance, this initial wave is a sine curve then Eq. (9–80) can be solved explicitly, the solution indicating a progressive sine wave moving at the kinematic wave velocity with an exponentially decaying amplitude—the classical solution to the diffusion problem. However the extension of this solution to any arbitrary wave form is difficult, requiring the technique of Fourier analysis; the most practically useful initial form is simply that of a stepwise increase in depth, as indicated by Fig. 9-12a. Given this situation at the upstream end ($x = 0$), Hayami obtains the following solution of Eq. (9–80), for constant K and c.

$$\frac{\Delta y_2}{\Delta y_1} = 1 - \frac{2}{\sqrt{\pi}} \int_0^X \exp\left[\frac{cx}{2K} - Z^2 - \left(\frac{cx}{4KZ}\right)^2\right] dZ \qquad (9\text{--}82)$$

where $X = x/2\sqrt{(Kt)}$, Δy_1 is the initial rise in water level given by Fig. 9-12a, and Δy_2 is the rise occurring at a section 2, x units downstream from section 1, after a time t. The effect of a rise of limited duration can then be simulated by combining the upward step of Fig. 9-12a with a later downward step of the same amount, Fig. 9-12b, giving the final result of Fig. 9-12c. A hydrograph of any given shape can then be approximated by a number of such unit rises, as in Fig. 9-12d. The integration of Eq. (9–82) must of course be done numerically, and is well suited to a high-speed computer.

For the unit rise of Fig. 9-12c, Eq. (9–82) is directly applicable when $t = t_1 < t_0$; when $t = t_1 > t_0$ it is readily deduced from Eq. (9–82) that

$$\frac{\Delta y_2}{\Delta y_1} = \frac{2}{\sqrt{\pi}} \int_{X_1}^{X_2} \exp\left[\frac{cx}{2K} - Z^2 - \left(\frac{cx}{4KZ}\right)^2\right] dZ \qquad (9\text{--}83)$$

where $X_1 = x/2\sqrt{(Kt_1)}$, $X_2 = x/2\sqrt{(Kt_2)}$, and $t_1 - t_2 = t_0$, the duration of the rise, as in Fig. 9-12c. When $t = t_1 = t_0$, Eqs. (9–82) and (9–83) give the same result. Hayami tested the approximation shown in Fig. 9-12d by considering the case of an upstream flood profile in the form of a sine wave with an 8-hr period. At a section 14 kilometers downstream, the flood profile computed by the explicit solution of Eq. (9–80) was virtually indistinguishable from the profile computed by using Eq. (9–83) and the approximation of Fig. 9-12d,

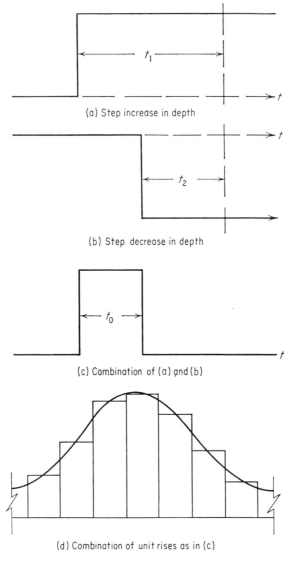

(a) Step increase in depth

(b) Step decrease in depth

(c) Combination of (a) and (b)

(d) Combination of unit rises as in (c)

Figure 9-12.   *Method of Approximating to the Flood Profile in the Diffusion Analogy, after S. Hayami* [7]

with strips of width $t_0 = 1$ hr. It appears therefore that this approximation will be quite satisfactory if the strip width t_0 does not exceed, say, one-tenth of the duration of appreciable flood rise.

As the reader will find in working through Probs. 9.19 and 9.20, the diffusion analogy method overestimates the speed of the forerunner resulting from an abrupt flood rise which approximates to the unit rise of Fig. 9-12c. The reason

is not difficult to see. Basic to the method is the assumption that the center of
the flood wave is moving with an average kinematic wave velocity c and that
the front and rear of the flood wave are diffusing outwards from its center
with velocities that will be quite high if the flood wave is a steep one. The
predicted forerunner speed will therefore be high. Now the complete theory
of Sec. 8.5 shows that this speed is limited to the value $v_0 + \sqrt{gy_0}$, or that a
surge will form, but the diffusion analogy cannot predict this limitation, for
the basic reason that it does not take into account the last two slope terms of
Eq. (9–16). The difficulty is not a serious one, for it does not materially influence the crest region, to which most practical interest attaches. However
the existence of this difficulty lends more interest to the application, in the
next section, of the complete theory of Secs. 8.3 through 8.5.

9.7 The Method of Characteristics

In Sec. 8.6 a computational system was described for using the complete
equations of motion, taking account of all slope and resistance terms.
Although the calculations were not directly based on the geometry of the
characteristic lines on the x-t plane, they still had to be carried out with an
eye to the general behavior of these lines. For this reason the system will still
be referred to (following the general practice) as the method of characteristics.

The system was based on Eqs. (8–54) through (8–56), applicable to a staggered array of points on the x-t plane, Fig. 8-14b, and Eqs. (8–57) and (8–58),
applicable to a rectangular array, as in Fig. 8-14a. The applications discussed
in Chap. 8 generally related to the fast-rising type of disturbance, for which
the method of characteristics is particularly suitable. The question arises
whether it is also suitable for application to natural flood waves, and the
answer is affirmative.

Stoker [8] has demonstrated the application of the method to floods in the
Ohio and Mississippi rivers, and has investigated the situation arising at the
junction of the two rivers on the arrival of a flood coming down the Ohio.
This latter problem leads to some numerical complexities, and for complete
details the reader is referred to Stoker's description [8]. However, the principle of the method is no different from that already described in Chap. 8.
The importance of this particular application is that no other method of flood
routing appears able to deal with the problem, and in particular to predict
the backwater effect upstream of the junction in the unflooded branch.
(Clearly this involves an *upstream* wave motion, which can be predicted only
by the complete theory of dynamic, as opposed to kinematic, waves.) Cases
like this one serve as a useful reminder that while other flood-routing methods
have their uses, the method of characteristics is still the most general and
therefore the most powerful.

It is helpful to consider the computational process in more detail than was done in Sec. 8.6. The problem is set up by specifying a certain y-t relationship at the upstream end of the reach, which we take as the origin of x. Initially conditions are known (and usually uniform) everywhere downstream of this section—i.e., v and y are known along the x axis. We can therefore apply Eqs. (8–54) through (8–58), which are based on an advance from known values at one instant t to unknown values at a later instant $(t + \Delta t)$. In applying these equations it will be convenient to reintroduce the definition of c as the dynamic wave speed $\sqrt{gy}$, used in Chap. 8.

Considering now the arrangement of points on the x-t plane, either a rectangular or staggered array may be used, or a combination of both. Figure 9-13 shows a convenient arrangement which combines the merits of both systems. Indicating rows by letters and columns by numbers as shown, the calculation proceeds as follows:

Assuming that v and c are known at the points A_1, A_3, A_5, A_7, ..., etc., the values at A_2 are obtained by linear interpolation. The points B_1, A_1, A_2, are then used as a rectangular array, the value of v at B_1 being thus obtained from Eq. (8–58). The values of v and c at B_2, B_4, B_6, ..., etc., are now obtained by using the staggered-array Eqs. (8–54) through (8–56). Then D_1, B_1, B_2, are used as another rectangular array by which v at D_1 is obtained from Eq. (8–58), and values of v and c obtained at D_3, D_5, D_7, ..., etc., from Eqs. (8–54) through (8–56). As before, it is important that the ratio $\Delta x/\Delta t$ should be large enough for a point such as D_5 to lie within the triangle formed by a C_1 characteristic through B_4 and a C_2 characteristic through B_6. Also, B_1 must lie below the C_2 characteristic drawn through A_2. These conditions will be met provided simply that

$$\frac{\Delta x}{\Delta t} > 2(v + c) \qquad \text{(9–84)}$$

at all points on the x-t plane.

This scheme of calculation begins at the x axis and proceeds upwards. The first characteristic OF plays no direct part in the calculation, but merely indicates the region (between OF and the t axis) within which the calculation is to proceed. If the calculations are extended into the region to the right of OF, no great harm is done, although some effort has been wasted. Within any computing interval which straddles this line, interpolation is less accurate because of the discontinuity in surface slope that exists within the interval. The obvious remedy is to choose $\Delta x/\Delta t$ so that mesh points fall on or close to OF. However a similar difficulty arises on a less well-defined line OG to the left of OF. It has been noted in Sec. 9.3 that the dynamic forerunner traced by OF attenuates very quickly, the first substantial flood rise being brought about by the main bulk of the flood moving more slowly as a kinematic wave, its front being traced by a line such as the curve OG. In the wedge-shaped region FOG, therefore, the flood rise will be insignificant, and along OG the water surface slope will change quickly, making interpolation less accurate within

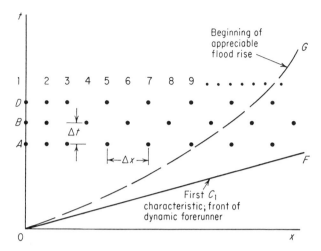

Figure 9-13. *Scheme of Computation for Flood Routing by Characteristics*

any intervals that straddle this line. In fact the whole region *FOG* is subject to inaccuracies arising from the small size of the departures from the undisturbed condition; these inaccuracies led Stoker [8] to use explicit algebraic methods to determine the behavior of v and c in this region. These methods are based on series expansions very similar to those of Prob. 8.7, and the process is carried a step further to the evaluation of the second-order coefficients v_2, y_2 in those equations. The result is an expression for the wave profile, valid near the dynamic wave front [i.e., for small values of the variable $\tau = t - x/(v_0 + c_0)$].

The complete development is lengthy when expressed in general terms, but the first coefficient of the expansion for y in Prob. 8.7 will be given here. This coefficient, y_1, is the rate of rise $\partial y/\partial t$ at the dynamic wave front; if its initial value $y_1(0)$ is less than the right-hand member of Eq. (8–42), which will be denoted as T, then the solution of the equation is as follows

$$y_1(t) = \frac{T y_1(0)e^{-bt}}{T - y_1(0)(1 - e^{-bt})} \tag{9–85}$$

where $b = gS_0(1 - \tfrac{1}{2}\mathrm{Fr}_0)/v_0$, the subscript 0 indicating as before the undisturbed condition just ahead of the wave front. Equation (9–85) gives the rate of rise $y_1(t)$ at any time t, and clearly indicates exponential decay of the wave front.

In a natural flood the initial value $y_1(0)$ will usually be a very small fraction of T, which as we have seen is of the order of one foot every few minutes; Eq. (9–85) can therefore be approximated by the relation

$$y_1(t) = y_1(0)e^{-bt} \tag{9–86}$$

and for low slopes of the order 0.0001, bt will approximate to unity when $t = 1$ hr, indicating that $y_1(t)$ reduces by a factor e every hour. This is quite a fast rate of attenuation, and for higher slopes the rate would be faster still.

Given the initial value $y_1(0)$, then $\partial y/\partial t$ at the wave front at any instant can be obtained from Eq. (9–86), and the slope of the wave profile obtained by setting $\partial y/\partial t = -(v_0 + c_0) \, \partial y/\partial x$. The second-order coefficient y_2 would be required in order to determine the curvature of the wave profile further back from the wave front and hence to locate, at least approximately, the line OG where the rise becomes appreciable. The coefficient y_2 may be determined by continuing the process of Prob. 8.7, and for some further details the reader is referred to Stoker [8].

A difficulty in applying these results is that initial values like $y_1(0)$ are seldom known with any certainty; however, the difficulties of the algebraic treatment of the region FOG can of course be avoided simply by retaining numerical methods and improving the accuracy by using a finer network on Fig. 9-13. If the work is being done on a high-speed computer, the use of a finer net is the simplest way of dealing with the problem.

As to the choice of interval size, Stoker found that in routing natural floods down the Ohio, it was satisfactory to use $\Delta x = 10$ miles, $\Delta t = 0.3$ hr, for a rectangular net. In a staggered net the same value of Δx would be retained, and Δt would be halved. As a general test of the accuracy of the numerical method, Stoker applied it to the case where the *initial* condition was the monoclinal wave, or kinematic shock, for which an exact solution is given by Eq. (9–31). The actual flood rise was from a depth of 20 ft to a depth of 40 ft in an idealized model of the Ohio. It is known that the monoclinal wave does not alter in form, but propagates unchanged with a kinematic wave velocity given by Eq. (9–27); it therefore makes a useful test of the accuracy of the numerical method. Using $\Delta x = 5$ miles, $\Delta t = 0.08$ hr, Stoker found that after 12 hr the numerical method gave values of v and c which were nowhere in error by more than 0.8 percent.

A related problem also treated was that of a flood rise from 20 to 40 ft at the upstream end ($x = 0$), the depth increasing linearly with time over a 4-hr period. This extremely fast rate of rise made it necessary to use small intervals of $\Delta x = 1$ mile and $\Delta t = 0.048$ hr in the first hour or two after the start of the rise. The solution to this problem is plotted on Fig. 9-14 in the form of depth contours on the x-t plane. Here the kinematic wave speed is 5 mph, and the forerunner speed 19.7 mph. It is seen that the inverse slopes of the y contours drift steadily away from the dynamic towards the kinematic wave speed, tending ultimately to the monoclinal wave of steady form. The figure also shows clearly how the divergence of these two speeds, or x-t slopes, gives rise to a wedge-shaped region of practically undisturbed flow, corresponding to the zone FOG in Fig. 9-13.

On the basis of these examples it would appear that natural floods can be routed with sufficient accuracy by choosing $\Delta x = 5$–10 miles; Δt must then be fixed in conformity with Eq. (9–84). Fast-rising floods produced by artificial control may require values of Δx as low as 1 mile.

In Stoker's use of the method, average values of width, depth, etc., were assumed applicable to the natural waterways dealt with. No significance was attributed to irregularities as such, and no effect was recognized which could be ascribed to their action. Therefore the logical basis of the diffusion analogy, and its compatibility with the method of characteristics, are still open questions.

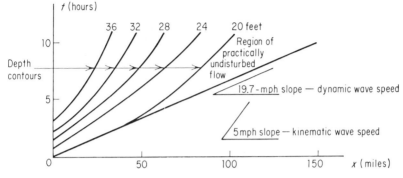

Figure 9-14.   *Routing of a 20-40 ft Flood Rise by Characteristics,*
after J. J. Stoker [8]

Mention should also be made here of the successful application of the method of characteristics by Ackers and Harrison [9] to flood waves in circular storm-water drains running part full. The waves dealt with were mainly quite fast-rising, with appreciable acceleration terms—i.e., all four slope terms in Eq. (9–16) would be significant. Such an application is particularly well suited to the method of characteristics, since it exploits the full generality and power of the method.

9.8   Rating Curves and Expressions for Discharge

It is not difficult to obtain a continuous record of the variation of river water level, or stage, with time. Instruments are set up on many rivers for that purpose; they commonly consist of a float gage and a clockwork-driven pen-and-chart recording mechanism. However the stage-time records so obtained must be converted into discharge-time records if they are to serve fully the many needs of the river engineer, and to this end it is usual to build up, through a number of discharge measurements at various stages, a systematic stage-discharge relationship which can be used to convert stage records into discharge records. A graphical display of such a relationship is termed a rating curve; an example is shown in Fig. 9-15.

Unfortunately there cannot be a unique stage-discharge relationship unless the flow is uniform. We have already seen that during the progress of a flood the influence of slope terms other than S_0 in Eq. (9–16) means that the discharge is not a function of depth alone. For a given depth y, the discharge will be greater on the rising stage of a flood than on the falling stage, so that the course of a single flood is traced on the stage-discharge plane by a closed loop like that shown in Fig. 9-15. This is the well-known *loop-rating curve*, and it

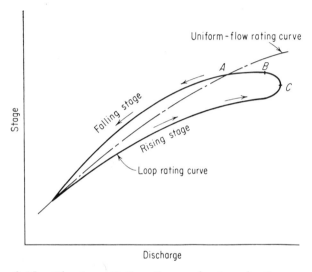

Figure 9-15. *The Loop-Rating Curve, showing the Progress of a Typical Flood Wave*

is desirable to be able to construct such a curve by theoretical methods if flood records are to be properly interpreted. For this purpose the basic requirement is a discharge equation whose terms can be derived readily from the flood record itself, taken at a single river section.

Consider first the case in which S_0 and $\partial y/\partial x$ are the only significant slope terms in Eq. (9–16), i.e., that

$$Q = BCy\sqrt{y\left(S_0 - \frac{\partial y}{\partial x}\right)} \tag{9–44}$$

If Q_0 is the discharge as given by the "normal" rating curve, i.e., the one based on uniform flow, then

$$Q_0 = BCy\sqrt{yS_0} \tag{9–87}$$

and

$$\frac{Q}{Q_0} = \sqrt{1 - \frac{1}{S_0}\frac{\partial y}{\partial x}} \tag{9–88}$$

The problem now is to replace $\partial y/\partial x$ by some alternative quantity, deducible from at-a-section flood records. If subsidence were to be neglected, then Eq. (9–37) would be true, and we could write

$$\frac{\partial y}{\partial x} = -\frac{1}{c}\frac{\partial y}{\partial t} \tag{9–89}$$

so that Eq. (9–88) becomes

$$\frac{Q}{Q_0} = \sqrt{1 + \frac{1}{S_0 c}\frac{\partial y}{\partial t}} \tag{9–90}$$

which is well known as the "Jones formula." The logical basis of the formula is not strictly correct, because as we have seen in Sec. 9.4, the influence of slope terms other than S_0 brings about subsidence. And, of course, it is these very terms which create the need for a loop-rating curve.

We can readily take account of subsidence by using Eq. (9–58) to deduce the result

$$\frac{\partial y}{\partial x} = -\frac{1}{c}\frac{\partial y}{\partial t} + \frac{dy}{dx}$$

$$= -\frac{1}{c}\frac{\partial y}{\partial t} - \frac{2S_0}{3r^2} \tag{9–91}$$

whence Eq. (9–88) becomes

$$\frac{Q}{Q_0} = \sqrt{1 + \frac{1}{S_0 c}\frac{\partial y}{\partial t} + \frac{2}{3r^2}} \tag{9–92}$$

so that the modification required to the Jones formula is slight. The total derivative dy/dx in Eq. (9–91) applies in the neighborhood of the crest region; it can be shown [6] that down on the flanks of the wave subsidence effects are negligible and the correction term $2/3r^2$ in Eq. (9–92) is not required. For the purpose of this argument, the division between flanks and crest may be set rather arbitrarily at the sections $x = \pm 2y_0/S_0$.

If the last two slope terms in Eq. (9–16) are taken into account, and if we make the same assumptions as those following Eq. (9–62) in Sec. 9.4, an expression for discharge follows readily from the equation for S_f in Prob. 9.12. From this equation and from Eq. (9–91), we obtain

$$\frac{Q}{Q_0} = \sqrt{1 + \frac{2}{3r^2} + \frac{5Fr^2}{6r^2} + \frac{1}{S_0 c}\frac{\partial y}{\partial t}\left(1 - \frac{Fr^2}{4}\right)} \tag{9–93}$$

applicable over the crest of the wave; it can be shown [6] that on the flanks the equation

$$\frac{Q}{Q_0} = \sqrt{1 + \frac{Fr^2}{2r^2} + \frac{1}{S_0 c}\frac{\partial y}{\partial t}\left(1 - \frac{Fr^2}{4}\right)} \tag{9–94}$$

is applicable.

Finally, some features of the geometry of the loop-rating curve may be pointed out. The points A and B on Fig. 9-15 correspond to points A and B

on Fig. 9-10, as is readily verified by considering the properties of those points. The point C, not previously referred to, is the point at which $\partial Q / \partial t = 0$. From Eq. (9–44) it is readily shown (Prob. 9.14) that the distance AC is equal to $2y_0/3S_0$, i.e., twice AB. It follows that an observer on the river bank watching the passage of a flood sees first the passage of the point of maximum discharge, then the point of maximum stage, and then (on the falling stage) the river crest A, where the flow is momentarily uniform.

9.9   Lateral Inflow and the Runoff Problem

In all the material dealt with so far in this chapter, it has been assumed that there is no inflow or outflow at the sides of the channel, although a reference was made in Sec. 9.6 to the possibility that losses to the surrounding country might contribute to the diffusion effect. In fact, Appleby [10] has presented a form of the diffusion analogy in which allowance can be made for lateral inflow, and has developed an electrical analogue computer to solve the resulting equations.

Lateral inflow is often negligible in the movement of a flood down a major river channel, but it is an essential element of what may be called the "runoff problem," involving the process by which the surface runoff from rainfall flows overland into small rivulets, then into larger channels and finally into a major river channel. Traditionally, this problem has been regarded as part of hydrology rather than of hydraulics, and the influence of fluid dynamics on the runoff process has been somewhat neglected. For instance, the Unit Hydrograph theory, very commonly used by hydrologists, is based on an assumption of linearity, in the following form. Suppose that a certain unit depth of "rainfall excess," i.e., rainfall not absorbed into the ground and therefore available as surface runoff, is deposited on a catchment over a given period of time, and a certain outflow hydrograph results. The theory then postulates that n times that unit depth of rainfall excess would give rise to an outflow hydrograph with ordinates n times as great as those of the original one.

It is by no means certain that the dynamic equations of motion would justify this assumption of linearity. But there has traditionally (and unfortunately) been a tendency to neglect the dynamics of fluid flow when dealing with the problems of hydrology; indeed the term "hydrologic" is often applied to the storage-routing technique of Sec. 9.2, in which the dynamic equation is of secondary importance. A notable exception to this tendency has been the development of a Japanese school of engineers who have attacked the runoff problem by kinematic wave theory rather than by the traditional hydrological methods. An account of this work is given by Ishihara [11].

Because of the steep slopes usually present in the upper catchment regions where the runoff problem exists, S_0 will be the only significant slope term and the discharge will be a function of depth alone. The problem is therefore an ideal one for kinematic wave theory, by which some interesting results can be simply obtained. Henderson and Wooding [12], by treating an elementary case, have contrasted the results of this dynamical approach with those of the Unit Hydrograph theory.

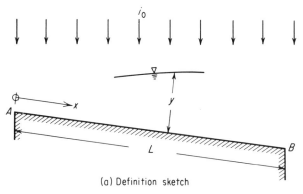

(a) Definition sketch

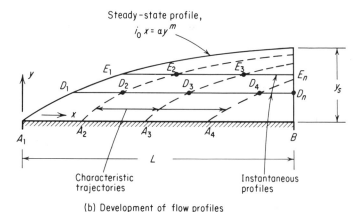

(b) Development of flow profiles

Figure 9-16.   *The Runoff Problem with Lateral Inflow*

We consider the case shown in Fig. 9-16, in which rain falls with an intensity i_0 on a steep impermeable surface of length L, of unit width normal to the plane of the paper. This simulates the case in which i_0 is the rainfall excess remaining after the absorption of some rainfall into the ground; i_0 has the dimensions of velocity, (e.g., inches per hour).

The discharge q will be related to the depth y by an equation of the form

$$q = \alpha y^m \tag{9–95}$$

which covers both laminar and turbulent flow, depending on the magnitudes of α and m. When a steady state is reached, the flow profile will be given by

$$q = i_0 x = \alpha y^m \qquad (9\text{–}96)$$

We are however concerned with the build-up of flow towards this steady state. The kinematic wave property arises from the continuity equation, which in view of the lateral inflow i_0 takes the form

$$\frac{\partial q}{\partial x} + \frac{\partial y}{\partial t} = i_0 \qquad (9\text{–}97)$$

Reverting to the use of c to denote the kinematic wave velocity, we can write

$$c = \frac{dq}{dy} = m\alpha y^{m-1} \qquad (9\text{–}98)$$

whence
$$\frac{\partial q}{\partial x} = \frac{dq}{dy}\frac{\partial y}{\partial x} = c\frac{\partial y}{\partial x}$$

and Eq. (9–97) becomes

$$c\frac{\partial y}{\partial x} + \frac{\partial y}{\partial t} = \frac{dy}{dt} = i_0 \qquad (9\text{–}99)$$

which is essentially a wave equation like Eq. (9–21), except that in this case y does not remain constant in the eyes of an observer moving with velocity c, but appears to increase with time at the rate i_0. To such an observer, therefore, the following relations will appear to be true:

(i) $dy/dt = i_0$

whence $y = i_0 t \qquad (9\text{–}100)$

there being no constant of integration since the surface is initially dry, i.e., $y = 0$ at $t = 0$ for all x.

(ii) $dq/dt = ci_0$

[from Eqs. (9–98) and (9–99)]   and $dq/dx = i_0$

because for the observer, $dx/dt = c$. It follows that in the observer's view

$$q = \alpha y^m = i_0 x + \text{constant} \qquad (9\text{–}101)$$

Now the kinematic wave theory, like the dynamic wave theory, can be discussed in terms of a method of characteristics, i.e., the tracing of the wave motion on the x-t plane. In the kinematic case there will only be one set of characteristics on this plane, corresponding to a single direction of wave motion. In the present discussion a characteristic will be a path traced on the x-t plane by the motion of our observer moving with velocity c, but it is of more interest in this case to follow the motion of this observer by the y-x

curve that he traces out on the physical plane. Such a curve may be described as a "characteristic trajectory," or simply a "trajectory."

Now a comparison of Eqs. (9–96) and (9–101) shows that all these trajectories are curves identical with the steady state profile but displaced from it by an amount depending on the starting position of the observer at $t = 0$. Figure 9-16b shows a number of these trajectories drawn in broken lines, one for each of a number of imaginary observers. All observers would start walking at the same instant, $t = 0$, from different initial positions A_1, A_2, A_3 ..., etc.

Our main interest lies in tracing successive positions of the water-surface profile, and the trajectories are only a means to this end; we obtain the water surface profile by joining up points on the trajectories having the same value of t. Now Eq. (9–100) appears true to all observers, so points D_1, D_2, D_3 etc., having the same value of t, will also have the same value of y. The water-surface profile is therefore parallel to $A_1 B$ in the region to the right of the steady-state profile; $A_1 D_1 D_n$ and $A_1 E_1 E_n$ are successive positions of the profile. In view of Eq. (9–100), the plateau $D_1 D_n$ rises through the position $E_1 E_n$, etc., with constant velocity i_0 until the steady state is reached after a time

$$t_s = y_s/i_0 = L/\alpha y_s^{m-1} = (L/\alpha i_0^{m-1})^{1/m} \qquad (9\text{–}102)$$

where y_s is the maximum (steady state) depth at the section B. The time t_s corresponds to the "time to equilibrium," which is a feature of all standard hydrological methods of estimating runoff. In all these methods t_s is assumed to be a property of the catchment and independent of the rainfall intensity, but Eq. (9–102) shows clearly the dependence of t_s on i_0.

There is another important respect in which results of the foregoing dynamical argument differ from those predicted by standard hydrological methods. From the assumptions made in the Unit Hydrograph theory it follows that for a given depth of rainfall excess h_0 the runoff hydrograph becomes steadily longer and lower with increasing rainfall duration t_0, behaving in fact like an attenuating wave. The foregoing argument, on the other hand, shows that if $t_0 < t_s$, then the peak runoff discharge is independent of t_0, being equal to

$$q_m = \alpha h_0^m \qquad (9\text{–}103)$$

because from Eq. (9–100) the maximum outflow depth is simply equal to $h_0 = i_0 t_0$, whatever the individual values of i_0 and t_0 may be. Indeed this result can be generalized further to cover the case where i_0 varies with time over the duration of the rainfall. In this case Eq. (9–100) simply becomes

$$y = \int i_0 \, dt \qquad (9\text{–}104)$$

and the maximum outflow depth is still equal to the total depth of precipitation h_0.

When $t_0 > t_s$ the situation is different, but this case is of lesser practical importance. In catchments other than the very smallest, the worst floods arise from rainfall durations t_0 less than the time to equilibrium t_s.

Further comparisons with hydrological methods can be developed out of the foregoing analysis. For details the reader can consult the original paper [12] or work them out as an exercise (Probs. 9.15 and 9.16). Comparisons based on the elementary model of Fig. 9-16 should not however be pushed too far, for it does not follow that this situation is a realistic model of a natural catchment. Also, the theory excludes the possibility of a critical outflow section at B, and any consequent drawdown in the steady-state curve. This drawdown curve would, however, make an appreciable difference only in a small catchment (Prob. 9.25); for natural catchments it would make little difference to the runoff profile. This conclusion is borne out by Ishihara's [11] application of the kinematic wave method to natural catchments.

References

1. Ven Te Chow. "Flood Routing," in his *Open-Channel Hydraulics* (New York: McGraw-Hill Book Company, Inc., 1959), Chap. 20.

2. B. R. Gilcrest. "Flood Routing," in H. Rouse (ed.), *Engineering Hydraulics* (New York: John Wiley & Sons, Inc., 1950), Chap. 10.

3. J. A. Seddon. "River Hydraulics," *Trans. Am. Soc. Civil Engrs.*, vol. 43 (1900), p. 179.

4. M. J. Lighthill and G. B. Whitham. "On Kinematic Waves: I—Flood Movement in Long Rivers," *Proc. Roy. Soc.* (London), (A), vol. 229, no. 1178 (May 1955), p. 281.

5. V. M. Yevdjevich. "Analytical Integration of the Differential Equation for Water Storage," *J. Res. Nat. Bur. Standards*, vol. 63B, no. 1 (July-September 1959), p. 43.

6. F. M. Henderson. "Flood Waves in Prismatic Channels," *Proc. Am. Soc. Civil Engrs.*, vol. 89, no. HY4 (July 1963), p. 39; with discussion by J. I. Collins, A. J. M. Harrison, and D. L. Brakensiek, vol. 90, no. HY1 (January 1964), p. 329; and closure by F. M. Henderson, vol. 90, no. HY4, part 1 (July 1964), p. 241.

7. S. Hayami. "On the Propagation of Flood Waves," Bulletin No. 1, Disaster Prevention Research Institute, Kyoto University, Japan (December 1951).

8. J. J. Stoker. *Water Waves* (New York: John Wiley & Sons, Inc., 1957), Chap. 11.

9. P. Ackers and A. J. M. Harrison. "The Attenuation of Flood-Waves in Part-Full Pipes," *Proc. I.C.E.* (London), vol. 28 (July 1964), p. 361.

10. F. V. Appleby. "Runoff Dynamics; A Heat Conduction Analogue of Storage Flow in Channel Networks," *Assemblée Générale de Rome*, Int. Assoc. Sci. Hydrol., vol. 3 (1954), p. 338.

11. Y. Ishihara. "Hydraulic Mechanism of Runoff," *Proceedings of a Conference on Hydraulics and Fluid Mechanics*, Perth, Australia, December 1962 (New York: Pergamon Press, 1963).

12. F. M. Henderson and R. A. Wooding. "Overland Flow and Interflow from Limited Rainfall of Finite Duration," *J. Geophys. Res.*, vol. 69, no. 8 (April 15, 1964), p. 1531.

Problems

9.1.   A lake having steep banks and a surface area of 1,500 acres discharges into
a steep channel which is approximately rectangular in section, with a width of
200 ft. Initially conditions are steady with a flow of 6,000 cusecs passing
through the lake: then a fresh comes down the river feeding the lake, giving
rise to the following inflow hydrograph:

Time from start (days)	0	0.5	1.0	1.5	2.0	2.5	3.0
Inflow (cusecs)	6,000	6,800	8,200	10,300	12,300	12,100	10,400

Time from start (days)	3.5	4.0	4.5	5.0	5.5	6.0
Inflow (cusecs)	8,800	7,700	6,900	6,300	6,000	6,000

Compute the outflow hydrograph, and plot it on the same graph with the
inflow hydrograph.

9.2.   A steep-sided lake of area 500 acres discharges into a trapezoidal channel
of base width 30 ft, side slopes $1\frac{1}{2}$H : 1V, bed slope 0.0005, Manning's
$n = 0.025$. After an initially steady flow of 2,000 cusecs through the lake, the
following inflow hydrograph occurs:

Time from start (hours)	0	3	6	9	12	15
Inflow (cusecs)	2,000	2,280	2,770	3,500	4,280	4,140

Time from start (hours)	18	21	24	27	30	33
Inflow (cusecs)	3,540	2,980	2,600	2,320	2,100	2,000

Compute and plot the outflow hydrograph together with the inflow hydrograph.

9.3.   Prove Eqs. (9–9) and (9–10). Examine the behavior of the terms of Eq.
(9–10) and hence show that it can be satisfied only when $t = 0$, or $t = T_0/2$.

9.4.   Recalculate Prob. 9.1 by using Eq. (9–7), approximating to the V-O curve
by a straight line through the origin and fitting a separate sine curve to each
limb of the inflow hydrograph, i.e., do not assume that the hydrograph is
symmetrical. (This will require two separate integrations, the two resulting
O-t equations being fitted together at the crest of the inflow hydrograph by
the choice of suitable constants.) Locate the crest of the outflow hydrograph
and determine its height. All results are to be obtained by algebraic methods.
Compare the results so obtained with those determined by numerical methods
in Prob. 9.1.

9.5.   Over a 10-mile reach of a certain river the section is approximately rectangular and 1,000 ft wide. The Manning $n = 0.027$ and it can be assumed that
$R = y$; the slope is 0.5 ft per mile. Flow is initially steady at 10,000 cusecs;
then a flood wave enters the upstream end of the reach. The inflow hydrograph is approximated by the equation

$$I = 10,000 + 20,000 \left(1 - \cos\frac{\pi t}{24}\right)$$

where I is measured in cusecs and t in hours; the duration of the flood at
the upstream end is 48 hours. Prepare data for, and plot, a series of N-O

curves as in Fig. 9-4b, and using these curves route the flood through the 10-mile reach. Use 4-hr intervals for the numerical process, and use the corresponding values of I for the plotting of the N-O curves.

9.6.   In the situation of Prob. 9.5, choose the values of K and X which will make the V-I-O relationship fit most closely to Eq. (9–11). Using these values and the I-t equation of that problem, determine the outflow hydrograph by integration of Eq. (9–15), and compare it with the one obtained by numerical methods in Prob. 9.5.

9.7.   For a wide ($B = P$) channel cross section described by Eq. (2–37)

$$\frac{B}{B_s} = \left(\frac{y}{y_s}\right)^i$$

and for a resistance equation in which $v \propto R^j$, show that the ratio of the kinematic wave velocity to the stream velocity is equal to

$$\frac{c}{v} = \frac{i+j+1}{i+1}$$

and list the values of this ratio for wide rectangular, triangular, and parabolic channels, for both the Chézy and Manning equations.

9.8.   Assuming a wide rectangular channel ($R = y$) of width B, and a constant Chézy C, deduce Eq. (9–30) from Eqs. (9–26) and (9–29), and hence obtain Eq. (9–31) by integration. Note that to an observer moving with velocity c, v will appear to be constant and therefore

$$\frac{dv}{dt} = \frac{\partial v}{\partial t} + c\frac{\partial v}{\partial x} = 0$$

This relation may be used to simplify Eq. (9–29).

9.9.   Assuming that ε is very small, deduce Eq. (9–33) from Eq. (9–31). Making the further assumption that y_{cr} is negligibly small compared with y_1 or y_2, express the sum $(a_1 + a_2)$ as a function of the relative wave strength $(y_1 - y_2)/y_2$, and prepare a graph showing the variation of $(a_1 + a_2)$ over the range $0 \le (y_1 - y_2)/y_2 \le 5$. Verify by differentiation the existence of the minimum shown by this graph.

9.10.   Prove by an examination of Eqs. (9–45) and (9–46) that if the wave trough profile A_2B_2 in Fig. 9-11 is obtained by inverting the crest profile A_1B_1, then B_2 is a local depth minimum corresponding to the local crest B_1, and that the slopes of the loci of B_1 and B_2 are equal in magnitude and opposite in sign. In applying Eq. (9–57) to the trough, how must the term y_0 be interpreted? As the depth at the trough? If not, why not?

Initially flow is uniform at a depth of 5 ft in a wide riverbed. There is then a series of releases of water from storage behind a dam, giving rise to a depth-time curve at the upstream end approximated by the equation

$$y = 10 - 5\cos\frac{\pi t}{12}$$

where t is the time in hours. Assuming that the Chézy C is constant and equal to 80, and that the curvature over the crest and trough is approximately constant, find the difference between crest and trough depths at a section 30 miles downstream.

9.11. Prove Eqs. (9–60) and (9–61) from Eqs. (2–37) and (9–59).

9.12. Including all four slope terms in Eq. (9–16), and adopting the approximations outlined just before Eq. (9–63), prove that the friction slope S_f over the crest region becomes

$$S_f = S_0 \left(1 + \frac{\text{Fr}^2}{r^2}\right) - \frac{\partial y}{\partial x}\left(1 - \frac{\text{Fr}^2}{4}\right)$$

and hence that Eqs. (9–63) and (9–64) hold true.

9.13. Assuming Eqs. (2–37) and (9–59) to hold true, show that the wave speed of the river crest A (Fig. 9-10) is given by Eq. (9–74). Also, prove Eq. (9–75) from Eqs. (9–60) and (9–71a).

9.14. Prove from Eq. (9–44) that $\partial Q/\partial t = 0$ at a section which is $2y_0/3S_0$ downstream of the river crest A, neglecting terms in $1/r^2$ and assuming that the required section falls within the parabolic crest region.

9.15. Consider, in the situation of Fig. 9-16, the subsidence of the runoff profile after the cessation of rainfall. Prove that the characteristic trajectories are lines parallel to the surface AB, and that at a time t after the cessation of

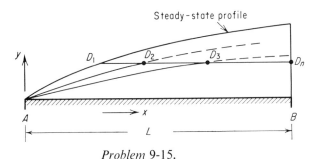

Problem 9-15.

rainfall at least part of the water surface profile consists of a curve having the equation

$$x = \alpha y^m / i_0 + m \alpha y^{m-1} t$$

this curve making up the whole of the profile if the initial condition was steady state, i.e., arising from a rainfall duration $t_0 \geq t_s$. If $t_0 < t_s$, the profile at the cessation of rainfall being say AD_1D_n in the figure, show that the above curve makes up only the left hand part of the profile, successive positions being AD_1D_n, AD_2D_n ..., etc. until the profile reaches D_n, the whole profile thereafter being made up of the above curve.

Using these results, sketch the outflow hydrograph, i.e., the q-t curve at the section B, and give the equations of the various parts of the hydrograph.

9.16. Consider, in the situation of Fig. 9-16, the whole outflow hydrograph resulting from a rainfall excess of depth h_0 and limited duration t_0. Show that if $t_0 = t_s$, the time to reach a steady state, then the hydrograph has a sharp peak, but that if $t_0 > t_s$ or $t_0 < t_s$ the hydrograph is topped by a plateau because the maximum outflow lasts for a finite duration, just before (if $t_0 > t_s$) or just after (if $t_0 < t_s$) rainfall ceases. Sketch a number of hydrographs having the same value of rainfall depth h_0 and various values of t_0.

9.17. A river flows through hilly country and then out on to a plain whose general level is not much above that of the river bed, so that flooding is frequent. It is proposed to control this flooding by stopbanking the river so as to confine it within a uniform channel having the same width, 1,500 ft, as the riverbed where it first reaches the plain. The maximum flood stage recorded at this point gave a depth of 25 ft in the river; four hours after reaching this peak, the water level had fallen by one foot. The bed slope is 2 ft per mile, Manning's $n = 0.030$, and the channel can be taken as wide rectangular. Assuming that the stopbank scheme is carried out, estimate the reduction in depth that would have taken place after the flood crest described above had reached a point 100 miles downstream.

9.18. In the situation of Prob. 9.17, it has been suggested that a large ponding area might be set aside and confined within stopbanks at the junction of the plains and the hilly country, where land is cheap. You are asked to provide a quick approximate answer to these questions: (a) How large would this area have to be so as to lower the flood crest of Prob. 9.17 by 5 ft, assuming that the whole flood had a duration of 48 hr, and that the outflow channel was 1,500 ft wide as before; (b) would the flood wave flowing out of this ponding area subside in the downstream channel at a rate materially different from that of the original flood wave? Use Eqs. (9–5), (9–9), and (9–10) to answer these questions.

9.19. Water is flowing at a depth of 10 ft in a wide rectangular channel whose slope is 0.5 ft per mile, with Manning's $n = 0.030$. There is then a controlled release of water from a hydroelectric plant which suddenly raises the depth to 20 ft for 3 hr, and then lowers it to 10 ft again. Assuming a diffusion coefficient of 2×10^5 ft^2/sec, calculate and plot the stage-time curve at a section 20 miles downstream. Assume, as a constant value, a suitable average value of wave velocity.

9.20. Water is flowing at a depth of 10 ft in a wide rectangular channel whose slope is 2 ft per mile, with Manning's $n = 0.027$. A controlled release of water is then made from a hydroelectric plant, suddenly raising the depth to 20 ft for 3 hr; this release is made in order to obtain data for estimating the diffusion coefficient K. At a section 20 miles downstream, the depth is measured when the water is first released, and at 1-hr intervals thereafter. The measured depths are: 10.0, 10.0, 13.8, 16.8, 17.0, 14.4, 12.5, 11.5, 10.9, 10.5, 10.3, 10.2, 10.1, and 10.1 ft. Assuming a suitable average value of wave velocity, determine the value of K by a trial process, and verify that it is consistent with the entire shape of the stage-time curve at the downstream section.

9.21. In the river reach of Prob. 9.20, a flood wave gives rise to a depth-time curve at the upstream end approximated by the equation

$$y \text{ (ft)} = 20 - 10 \cos \frac{\pi t}{24}$$

where t is the time in hours from the start and the duration of the whole flood is 48 hr. Use the value of K obtained in Prob. 9.20 to determine the depth-time curve at the downstream end of the reach.

9.22. For the normal type of flood wave which rises to a crest and then falls again, sketch the arrangement of y contours on the x-t plane corresponding to that of Fig. 9-14, taking care to trace clearly the motion of the flood crest and to show how the contours indicate the presence of crest subsidence.

9.23. In a river of wide rectangular section, bed slope of 1 ft per mile, and Manning's $n = 0.030$, the initial depth is 10 ft and a flood wave then enters the upstream end of the channel, giving rise to the depth-time curve

$$y \text{ (ft)} = 20 - 10 \cos \frac{\pi t}{24}$$

where t is the time in hours and the flood duration is 48 hr. Use the method of characteristics to trace the motion of the flood wave for a distance of 100 miles downstream. Check the magnitude of any subsidence detected against the result given by Eq. (9–65).

9.24. For the river channel of Prob. 9.23, compute and plot a normal (i.e., uniform-flow) stage-discharge curve. Also plot a loop-rating curve and determine the maximum discharge, (a) for the inflow hydrograph given in Prob. 9.23, (b) for a hydrograph having the same amount of rise and the same sinusoidal form, but one half the period.

9.25. In the situation of Prob. 7.15, determine the steady-state profiles over the paved surface and in the channel, adopting the assumptions of the runoff problem in Sec. 9.9, i.e., taking $S_0 = S_f$ throughout and neglecting the influence of the critical section downstream. Compare these profiles with those calculated in Prob. 7.15.

9.26. In the situation of Prob. 7.15 and under the assumptions of Prob. 9.25, calculate the time from the start of the rainfall at which the flow from the paved surface to the drainage channel reaches a maximum, and the further time required for the outflow from the drainage channel to reach a maximum. Determine also the complete Q-t relationship at the channel outflow, from the start of the rainfall until a maximum discharge is reached.

9.27. Suppose that in the situation of Prob. 9.25 the rain has lasted long enough for the channel outflow to reach its maximum. The rain then ceases abruptly; determine the subsequent Q-t relationships, both at the channel outflow and at the edge of the pavement where it joins the channel.

9.28. Water discharges from a lake under a sluice gate into a wide rectangular channel having a slope of 0.005. At the gate the channel bed is 10 ft below lake level, and a few feet downstream from the gate the depth reaches the value of 2 ft, at which it remains uniform for a great distance downstream. The gate is now suddenly lifted clear of the water; discuss the subsequent motion and calculate the shape and speed of any stable profile which forms at the resulting wave front. Determine whether a surge will form, and if so what its height and speed will be.

Computer Programs

(*Note*: Remember the usual practice of writing programs to accept all possible varieties of input data, with provision for listing this data at the output stage.)

C9.1. Write programs to handle the level-pool routing problem, with the following alternative methods of dealing with the outflow conditions:

(a) Values of N and O are first computed and listed in the machine store, with access to these values either by a table-look-up instruction or a programmed equivalent.

(b) Values of N and O are computed as required during the course of the main calculation, corresponding to the main tabulation in Example 9.1, separate programs being written for the following outflow conditions:

(i) Steep channel of rectangular section.

(ii) Mild-slope channel of rectangular section.

(iii) Channel of rectangular section with provision for classifying the slope by the methods of Chap. 4 (e.g., Example 4.1).

(iv) As for (i), (ii), and (iii) but with trapezoidal section.

(v) The actual N-O relationship for any of the above channels is approximated by a relation of the form $N \propto O^m$, which is then used to calculate N-O values during the main tabulation.

C9.2. Write a program to apply the storage routing method to flow in a river, using the Muskingum equation (9–11) to relate V, O, and I.

C9.3. Write a program which will route a unit rise, as in Fig. 9-12c, through a channel reach by Hayami's diffusion analogy. Then extend the program so as to cover the case where a given inflow hydrograph is approximated by a series of unit rises, as in Fig. 9-12d.

C9.4. Write a program to carry out the trial process of Prob. 9.20 by which a diffusion coefficient is estimated from the observed progress of a known flood. Apply the program to the case of Prob. 9.20.

C9.5. Write a program which will route a flood, specified at the inflow end by a depth-time curve, through a channel reach by the method of characteristics, using the network of Fig. 9-13.

C9.6. Operate the above programs to solve any of the main list of problems to which they are applicable.

Chapter 10

Sediment Transport

10.1 Introduction

It is common knowledge that water flowing in natural or artificial channels often has the ability to scour sand, gravel, or even large boulders from the bed or banks and sweep them downstream. This phenomenon is usually termed sediment transport, whatever the size of the transported material, although in everyday speech the term "sediment" is usually reserved for fine materials like silt or sand.

In many situations the phenomenon is of great economic importance. As previously mentioned in Sec. 7.5 in connection with bridge piers, the cost of a flood control scheme is critically dependent on the estimated maximum flood level; this level in its turn may be seriously affected by the scour and subsequent downstream deposition of sediment, either temporarily during the course of a single flood, or as part of a more permanent long-term process. A more indirect influence on flood levels may be exerted by the attack of a river on its banks, creating sharp and irregular curves which increase the flow resistance of the channel and thereby raise the flood level for a given flood discharge. The deposition of sediment may also reduce the storage capacity and therefore the value of reservoirs being used for some form of water supply; similar deposition in harbors may require costly dredging or other measures for the continuous removal of banks and bars. Rivers themselves are often used for navigation and they too may require costly operations to maintain a clear navigable waterway. Artificial channels, too, present their problems; irrigation canals carrying silt-laden water from a river may have their action seriously impaired by the scour or deposition of sediment. Many poorly designed canals have silted up and become inoperable through failure to carry the sediment load admitted at the canal headworks.

Sediment transport with its attendant problems governs, therefore, a great many situations that are of major importance to civilized man. Indeed it is a major geological influence in the shaping of landforms, and the examples listed above are only short-term aspects of the long-term geological process. In dealing with these examples the engineer is seeking to control this process (at least to a limited extent), and the task is formidable not only for its size but also for its complexity. Many features of sediment transport are still imperfectly understood, but progress continues to be made on the general problem by many investigators. In this chapter some of the basic existing knowledge of the problem will be described, together with an outline treatment of those areas of the subject in which present knowledge is uncertain. Discussion of these latter areas by investigators tends to range far and wide, and in many cases it will be necessary to refer the reader to the original papers for details.

10.2   Modes of Sediment Motion and Bed Formation

In the following treatment it will normally be assumed that the granular bed material whose motion is being discussed is incoherent, i.e., that there are no adhesive forces between particles, as there would be in clayey sands or soils. While incoherent sands and gravels account for most of the practical problems met in the field of sediment transport, cohesive soils are also encountered and cannot be completely neglected. Systematic knowledge of their behavior is difficult to compile, but reference will be made in a later section to some preliminary work that has been done in this direction.

The basic mechanism responsible for sediment motion is the drag force exerted by the fluid flow on individual grains. The cumulative effect of all such drag forces is a retarding shear stress exerted by the bed on the flow, discussed in Chap. 4. The converse view of this stress is of one exerted in the direction of motion by the flow on the bed, and its magnitude will clearly be one of the important determinants of the sediment motion.

In Sec. 10.1 a reference was made to a canal's ability to "carry a sediment load." It is so natural to think of the sediment as a burden to be borne by the flowing water that the word "load" has been well established for many years as the term describing the sediment being carried forward. A distinction is commonly made between "bed load," in which the grains roll along the bed with occasional jumps up into the main stream, and "suspended load," in which the material is maintained in suspension up in the mainstream by the turbulence of the flowing water. The distinction is clear enough when the two materials involved are different—e.g., when a silt-laden river is flowing over a bed of coarse gravel, and the gravel is moving as bed load. However, when both loads are of the same material (for example, a river "flowing in its own silt") the distinction becomes rather an arbitrary one.

Bed Formation

In moving sediment along the bed the channel flow can distort the bed into a variety of different shapes. The following classification of these shapes is essentially as given by Simons and Richardson [1], and with some minor qualifications is the one now generally accepted.

At low velocities the bed does not move at all; we consider now the sequence of events as the velocity is steadily increased. First, the "threshold" of movement is reached and the bed begins to move. If the bed is of fine material such as sand, then on further increase of the velocity the bed develops *ripples* of the sawtooth section shown in Fig. 10-1a; they can be seen in the sand on any

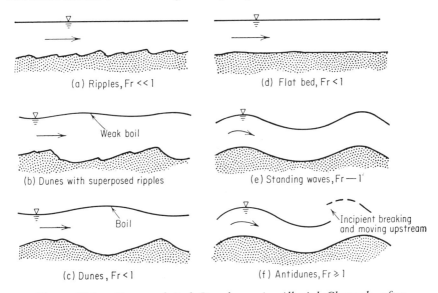

Figure 10-1.   *Forms of Bed Roughness in Alluvial Channels, after D. B. Simons and E. V. Richardson* [1]

beach, or on wind-blown desert sand. At higher velocities larger periodic irregularities appear on the bed, Fig. 10-1b; these are known as *dunes*, and when they first begin to appear they carry the ripples superimposed, as in Fig. 10-1b. At higher velocities still, the ripples disappear and only the dune-pattern remains, Fig. 10-1c. Dunes are both larger and more rounded than ripples. Neither dunes nor ripples have crests which extend across the full width of the stream; both formations tend to occur in the form of "short-crested" waves appearing as staggered arrays when seen from above, ripples having a more regular echelon-type formation than dunes. Both dunes and ripples migrate slowly downstream through the scour of material from the upstream face and its subsequent deposition on the downstream face.

While the distinction between dunes and ripples is quite clear in terms of the above sequence of events, the two phenomena are essentially similar in nature and maintained by the same mechanism of upstream scour and downstream deposition. Empirically, the distinction between them is clearly made by the observation that ripples may occur superimposed on dunes. A further difference between the two is that while no limit has yet been found to the grain size of the sediment in which dunes may form, ripples are not found in sand coarser than 2 mm approximately, or—in more general terms—having a fall velocity greater than 8 mm per sec approximately [18]. The significance of fall velocity as a measure of grain size will be discussed more fully later. Despite the clearness of the observed distinctions between ripples and dunes, the approximate theories which have been put forward to explain their origin have so far been unable to differentiate between them.

Both dunes and ripples are observed in wind-blown sand, or in the open channel flow of water. In the latter case, they occur when the Froude number is low; in order to deal with higher Froude numbers, we resume our consideration of the sequence of events on gradually increasing velocity. The next step beyond the formation of dunes, Fig. 10-1c, is that the dunes in their turn are erased by the flow, leaving a flat bed as in Fig. 10-1d. Further increase in the velocity, bringing the Froude number to unity and beyond, leads to the formation of sand waves, Fig. 10-1e, which occur in association with, and in phase with, the surface waves shown in the figure. As the Froude number approaches unity the surface waves become so steep that they break, as in Fig. 10-1f; at the same time there is a gradual movement upstream of the whole wave system. The sand waves are then called *antidunes*; since they depend for their existence on an interaction between the bed and the water surface they are found only in open channel flow and not in wind-blown sand.

No completely satisfactory theories have yet been put forward which account for the existence of ripples, dunes, and antidunes, but some elementary general statements can be made here about their formation and action. First, the downstream migration of ripples and dunes can be accounted for by recognizing that for low values of the Froude number, or in the complete absence of a free fluid surface (as in wind-blown sand), the velocity will be a maximum at the dune crests and a minimum at the troughs. It is reasonable to suppose that the volumetric rate of sediment transport per unit width, q_s, will increase with the velocity and will therefore be a maximum at the crests and a minimum at the troughs. Now it is possible to write a continuity equation for the sediment flow rate, analogous to Eq. (8–7) for fluid flow. The equation is

$$\frac{\partial q_s}{\partial x} + \beta \frac{\partial z}{\partial t} = 0 \qquad\qquad \textbf{(10–1)}$$

where z, as usual, is defined as the height of the bed above datum, and β is the ratio of grain volume to total volume in the bed. It follows from Eq. (10–1) that on the upstream face of a dune, AB in Fig. 10-2a, $\partial q_s/\partial x$ will be positive and $\partial z/\partial t$ negative, i.e., that the face AB is being eroded away. Similarly, $\partial z/\partial t$ will be positive along the downstream face BC, which is therefore building up, or aggrading; it follows that the dune will move downstream as indicated in the figure.

This argument has been put into quantitative form by Exner [3], at the expense of some rather severe approximations. It is assumed first that q_s is directly proportional to the velocity v, so that Eq. (10–1) may be written

$$K \frac{\partial v}{\partial x} + \frac{\partial z}{\partial t} = 0 \qquad (10\text{–}2)$$

where K is a constant that includes β. It is then assumed that v in this equation is the average velocity over a complete vertical section of the flow, although

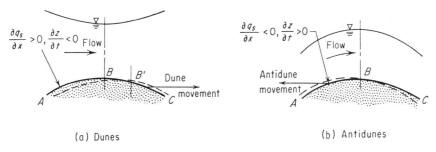

(a) Dunes   (b) Antidunes

Figure 10-2.   *Wave Motion of Dunes and Antidunes*

strictly speaking it should be the local velocity near the dunes, which may be very different from the average (cf. the situation of Fig. 2-13). From the fluid flow continuity equation we then have

$$vy = v(h - z) = q, \quad \text{a constant} \qquad (10\text{–}3)$$

where h is the height of the water surface above datum and may be assumed constant if the Froude number Fr is small.

It follows that

$$(h - z) \frac{\partial v}{\partial x} - v \frac{\partial z}{\partial x} = 0 \qquad (10\text{–}4)$$

and the elimination of $\partial v / \partial x$ between Eqs. (10–2) and (10–4) leads to

$$\frac{\partial z}{\partial t} + \frac{Kq}{(h - z)^2} \frac{\partial z}{\partial x} = 0 \qquad (10\text{–}5)$$

which is readily seen to be the equation of a wave traveling with velocity $Kq/(h - z)^2$. This expression for velocity shows that in subcritical flow the highest points on the dune or ripple have the highest wave velocity—suggesting that its front face will steepen until it reaches a limiting slope determined by the angle of response of the material. The argument therefore provides a simple explanation of the sawtooth form of the sand ripple, although it does not of course explain the origin of ripples or dunes.

If there is a free surface and the Froude number Fr is small, the surface will not show any waviness reflecting that of the bed. As Fr increases the surface develops a wave form which is out of phase with the bed wave, in accordance with the properties of subcritical flow. This effect accentuates the velocity difference between the troughs and crests of the bed wave. When sand waves or antidunes form, however, as in Fig. 10-2b, the picture of events is reversed. In this case, the flow is moving into the supercritical range and the surface wave is in phase with the bed wave. In fact, referring back to Chap. 2, the discussion centered on Fig. 2-3 shows that the water depth is greater, and the velocity is less, over the sand-wave crests than over the troughs. The argument previously adduced from Eq. (10–1) would show, therefore, that the upstream face AB in Fig. 10-2b is aggrading and the downstream face BC is degrading. The wave form is therefore moving upstream.

None of these elementary remarks show why dunes or antidunes should form and grow in the first instance. Kennedy [4] has shown by a theoretical argument that for low values of Fr an initial small wavy disturbance will grow into dunes of finite amplitude, provided that variations in the sediment discharge rate q_s are assumed to lag behind corresponding variations in the velocity at the bed. This implies that the maximum q_s would occur not at the crest B in Fig. 10-2a but at a point B' somewhat downstream. Physically, this assumption appears reasonable, for one would not expect q_s to respond immediately to changes in velocity; it would require some time to accommodate itself to these changes. The same investigator showed that antidunes will grow from small initial disturbances when Fr is near unity, whether or not q_s lags behind the velocity.

Details of Kennedy's theory are beyond the scope of this text, but it may be pointed out here that for both dunes and antidunes the theory turns on an interaction between bed waves and surface (gravity) waves. This gravity-wave influence means, among other things, that the theory predicts for both dunes and antidunes a minimum wavelength L_m equal to

$$L_m = \frac{2\pi v_0^2}{g} \tag{10–6a}$$

where v_0 is the mean velocity of the stream. The similarity of Eq. (10–6a) to Eq. (8–73) is evident. Reynolds [2] extended Kennedy's work and qualified it in some respects. In particular he pointed out that Eq. (10–6a) should be amended to the more general form

$$L_m = \frac{2\pi v_0^2}{g} \tanh \frac{2\pi y}{L_m} \tag{10–6b}$$

These theories are remarkably complete but do not predict the occurrence of ripples, whose presence in wind-blown sand (and indeed at great depths on the ocean bed) cannot be due to the influence of any gravity wave at a free surface. Further, the observed wavelengths of ripples are often very much less than the values indicated by either form of Eq. (10–6). However, the theories open promising lines of attack on the problem of dune and antidune formation; in particular they suggest investigation of the interesting problem of the lag between q_s and the velocity.

In the previous discussion the nomenclature proposed by Simons and Richardson [1] has been used, but it is not universally used in this exact form. Kennedy, for instance, prefers to apply the term " antidune " to all configurations in which the surface wave and the bed wave are in phase—i.e., those of Figs. 10-1e and *f*, and not to reserve the term for those bed waves which are traveling upstream. However, the nomenclature of Fig. 10-1 is probably the most commonly used, and will be employed generally in this chapter.

10.3   The Threshold of Movement

The term *threshold* has been introduced in the previous section, with its obvious connotation of a limiting condition marking the boundary between one state of affairs and another. Like many threshold conditions, the threshold of sediment motion cannot be defined with absolute precision; as the velocity of flow over an initially stationary bed of sediment is steadily increased there is no point at which sediment movement suddenly becomes general. There is first a condition in which a grain is detached from the bed every few seconds, and the movement could possibly be ascribed to the unstable initial position of each particular grain. Then as the velocity is increased grain movement gradually becomes more frequent until it is general over the whole bed.

The difficulty of detecting the threshold condition is, however, more apparent than real. If a number of observers are asked to watch the whole process and to nominate the point at which bed movement has become generally established, and is no longer just the accidental dislodging of an occasional grain, their decisions are remarkably consistent, leading to values of the threshold velocity which vary by only a few percent. In the subsequent discussion, therefore, the threshold condition can be regarded as a sound and well-founded concept.

The Experimental Approach

The problem now is to determine the criterion which must be satisfied at the threshold condition, and the first approach described will be through dimensional analysis and experiment. The relevant parameters are:

τ_0, the shear stress at the bed	d, the grain diameter
ρ_s, the grain density	g, the acceleration of gravity
ρ_f, the fluid density	μ, the fluid viscosity

Here, in accordance with commonly accepted practice, the shear stress τ_0 is taken as the important factor measuring the power of the flow to dislodge the sediment. The velocity v is important only in so far as it helps to determine τ_0. Dimensional analysis yields $6 - 3 = 3$ dimensionless numbers:

$$\frac{\tau_0}{\rho g d}, \quad \frac{\rho_s}{\rho_f} \quad \text{and} \quad \frac{d\sqrt{\tau_0 \rho}}{\mu}$$

The density ρ in the first and third numbers need not yet be identified as ρ_s or ρ_f. Although velocity is not included it is dimensionally convenient to have a parameter with the dimensions of velocity. We therefore introduce the "shear velocity"

$$v^* = \sqrt{\frac{\tau_0}{\rho_f}} \qquad\qquad (10\text{–}7)$$

which is not a real velocity but is related to the real fluid velocity which would give rise to a shear stress τ_0. For instance, it can be shown (Prob. (10.1) from the Darcy equation for flow in a circular pipe that

$$\frac{v^*}{v} = \sqrt{\frac{f}{8}} \qquad\qquad (10\text{–}8)$$

where f is the Darcy coefficient.

Eliminating τ_0 between Eq. (10–7) and the dimensionless numbers, and setting $\rho = \rho_f$ in the first and third, the three numbers become:

$\dfrac{v^{*2}}{gd}$, a kind of Froude number;

$\dfrac{\rho_s}{\rho_f}$, as before, which can be written s_s, equal to the sediment specific gravity if the fluid is water; and

$\dfrac{v^* d}{v}$, a kind of Reynolds number, known in fact as the "particle Reynolds number," and indicated by Re*.

When experimental results are plotted, one would expect a series of curves on the v^{*2}/gd – Re* plane, one for each value of s_s. In fact the classic experimental work of Shields [5], and of other investigators, show a more simple result: v^{*2}/gd and s_s turn out to be related in such a way that when results are plotted on the plane of $v^{*2}/gd(s_s - 1)$ and $v^* d/v$, they fall on a single line, shown in Fig. 10-3. The term $v^{*2}/gd(s_s - 1) = \tau_0/\gamma d(s_s - 1)$ is called the "entrainment function," and will be indicated by F_s. The curve in Fig. 10-3 shows the classic pattern of variation of a dimensionless coefficient with Reynolds number (cf. f – Re curves for pipe flow and C_D – Re curves for submerged bodies). There is a laminar flow region in which the coefficient varies inversely as Re*, a transition region, and a region of fully developed turbulence in which the coefficient is substantially constant. A further point

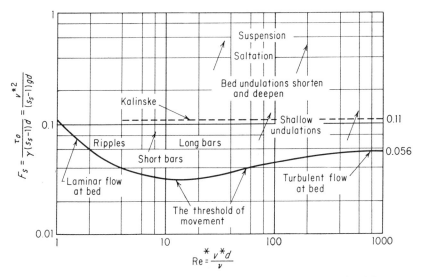

Figure 10-3.   *The Entrainment Function, after A. Shields* [5]

of interest lies in the annotations in the region above the curve, representing a moving bed. These are based mainly on the work of Shields, whose experiments covered developed sediment motion as well as the threshold of motion. His results indicated, as in Fig. 10-3, that bed formation is determined by position on the $F_s - Re^*$ plane. This conclusion was a rather tentative one, but has not been seriously contested since the time of Shields' experiments, at least for low values of the Froude number. For high values of Fr, the conclusion must be qualified in the light of the sequence of events shown in Fig. 10-1.

An Analytical Approach

C. M. White [6] has attempted to analyze the threshold condition by considering the equilibrium of a single typical grain. It is assumed that the bed shear τ_0 is an average compounded from the drag forces on a number of grains which are more exposed or prominent than others in the bed, the drag force on the other, protected, grains being neglected. The number of these exposed grains per unit area of the bed will clearly vary as $1/d^2$, and is assumed equal to r/d^2, where r is a " packing factor," which will clearly be less than unity. This is another way of saying that there is, corresponding to each prominent grain, an " area of influence" equal to d^2/r, as in Fig. 10-4.

Now if P_g is the drag force on each prominent grain, the average shear stress τ_0 is equal to

$$\frac{P_g}{\text{Area of influence}} = \frac{P_g r}{d^2}$$

whence
$$P_g = \frac{\tau_0 d^2}{r}$$
(10–9)

and the threshold condition is described simply by equating this force to the resistance force, equal to the effective weight of the grain (assumed spherical) times the effective coefficient of friction (the tangent of the angle of repose ϕ).

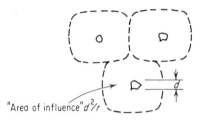

"Area of influence" d^2/r

Figure 10-4.   Distribution of Prominent Grains, after C. M. White [5]

It follows from Eq. (10–9) that

$$\frac{\tau_0 d^2}{r} = \frac{\pi d^3}{6} \gamma(s_s - 1) \tan \phi$$

i.e., that
$$\tau_0 = \frac{\pi}{6} r \gamma d(s_s - 1) \tan \phi$$
(10–10)

This equation is strictly true only if the force on each grain acts through the center of gravity of the grain. Probably the force acts somewhat above this point, because some of the lower part of the grain is protected from the direct action of the fluid flow. The critical shear stress $\tau_0 = \tau_c$ would therefore be somewhat less than the value given by Eq. (10–10). The argument is over-simplified in other respects too: it takes no account of the effect of turbulent velocity fluctuations and of uplift due to the nearness of the boundary and the strong velocity gradient, which tends to move any submerged body towards the faster-moving fluid layers. Nevertheless these imperfections should affect only the numerical constant in Eq. (10–10) (which must be found by experiment anyway) and not the form of the equation; in this respect it is interesting to note that Eq. (10–10) contains Shields' entrainment function, for it can be rewritten

$$F_s = \frac{\tau_0}{\gamma d(s_s - 1)} = \frac{r\pi \tan \phi}{6}$$
(10–11)

providing a useful verification of the form of Shields' results. In fact Shields did adduce a physical argument justifying the form of the entrainment function, but only White's argument has been given here because it is a great deal simpler than Shields'. The whole problem is a good example of how physical reasoning can, even though it is approximate, supplement the results of dimensional analysis, making them more compact by showing how certain dimensionless numbers can be combined.

White suggested that in laminar flow the resultant force P_g, having a greater surface drag component than in turbulent flow, would act at a higher level. If so, the effect would be to reduce the critical value of τ_0, explaining the dip in Shields' curve in Fig. 10-3. In turbulent flow one might expect P_g to act at a certain fixed level, making F_s a constant, as Shields' results showed. Taking this constant as 0.056, it follows from Eq. (10–11) that

$$r = \frac{6 \times 0.056}{\pi \tan \phi}$$

$$= 0.15 \text{ approximately} \tag{10–12}$$

taking $\phi = 35°$ as a typical value. This value of r is not far from those observed by White in an actual count of exposed grains on a sand surface. This makes quite a close check in view of the approximations involved in the analysis.

Applications

The curve in Fig. 10-3 forms a satisfactory basis for river and canal work where it is required to prevent bed movement or to keep it to a minimum. The application becomes particularly simple when the flow is governed by the right-hand limb of the curve, where F_s is a constant. If we assume that the fluid is water and that $s_s = 2.65$, which is true in a very large proportion of real situations, it can be readily shown (Prob. 10.2) that on the right-hand limb of the curve, to the right of $Re^* = 400$, the grain size d exceeds $\frac{1}{4}$ in. approximately, i.e., the bed material is coarse alluvium. The equation

$$\frac{\tau_0}{d\gamma(s_s - 1)} = 0.056 \tag{10–13}$$

is then applicable; it can be converted to more practical terms by recalling that

$$\tau_0 = \gamma RS \tag{4–4}$$

whence
$$\frac{RS}{d(s_s - 1)} = 0.056$$

which becomes, on inserting the value $s_s = 2.65$

$$d = 11 \ RS \tag{10–14}$$

giving in simple form the minimum size of stone that will remain at rest in a channel bed of given R and S. In applying this equation it must be remembered that in the course of time river and canal beds in coarse alluvium become "armored," i.e., the smaller stones are flushed out of the surface lining of coarser stones. As in the choice of a value of Manning's n from Eq. (4–22), it may be assumed for design purposes that the D-75 size is typical of the stone size in this surface layer. The actual size will be somewhat larger than the D-75 size, so this assumption is a conservative one in design calculations aimed at protecting the bed and banks from scour. The details can be illustrated by an example.

Example 10.1

An irrigation canal is to carry 100 cusecs through country consisting of a coarse alluvial gravel with a D-75 size of 2 in. The canal is to be laid along a line having a slope of 0.01 and it may be assumed that the banks will be grassed and protected from scour. Find the minimum width for the canal.

From Eq. (10–14), the required value of R is equal to

$$\frac{2}{12 \times 11 \times 0.01} = 1.52 \text{ ft}$$

and this is a maximum allowable value. Now from Eq. (4–22) the Manning n is equal to

$$0.031 \times (\tfrac{1}{6})^{1/6} = 0.023$$

whence

$$v = \frac{1.49 \times (1.52)^{2/3} \times \sqrt{0.01}}{0.023}$$

$$= 8.55 \text{ ft/sec}$$

Assuming to a first approximation that $R = y$, it follows that

$$q = 1.52 \times 8.55 = 13 \text{ cusecs/ft width}$$

Hence the required width is equal to

$$\frac{100}{13} = 7.7 \text{ ft}$$

and since this is a minimum, a conservative value would be 10 ft.   *Ans.*

For grain sizes finer than $\tfrac{1}{4}$ in., Eq. (10–14) is no longer available but the Shields' curve of Fig. 10-3 still gives a unique d-RS relationship for given values of s_s and v. Taking $s_s = 2.65$ and $v = 1.2 \times 10^{-5}$ ft²/sec as for Eq. (10–14), it is possible to solve for d, as in Prob. 10.2, and also for RS, at every point on the Shields' curve. Carrying out the numerical solution at a number of points leads to a complete d-RS relationship which is plotted in Fig. 10-5. Details are left as an exercise for the reader (Prob. 10.3). From this curve problems corresponding to Example 10.1 can be solved for grain sizes less than $\tfrac{1}{4}$ in. (Prob. 10.4). The appropriate value of Manning's n will be quite low, for Strickler's equation (4–22) gives $n = 0.016$ for $d = \tfrac{1}{4}$ in., and smaller values of d will make n lower still. Higher values of n could only be caused by bed ripples or dunes due to established bed movement.

The U. S. Bureau of Reclamation [7] has recommended safe values of the shear stress τ_0, which is often referred to in this context as "tractive force," the corresponding design process applied to rivers and canals being called the "tractive force method." For coarse noncohesive materials the recommended value of τ_0 in pounds per square foot is equal to 0.4 times the diameter of the

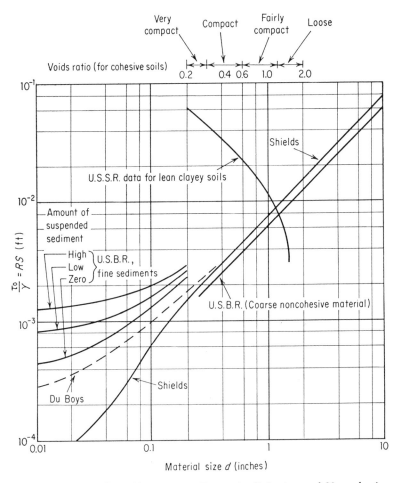

Figure 10-5.   *Allowable Tractive Forces in Cohesive and Noncohesive Bed Material*

D-75 size in inches. The corresponding value of RS for water ($=\tau_0/\gamma = \tau_0/62.5$) is plotted in Fig. 10-5 and it is seen to be quite close to the Shields' curve. Similar curves are plotted showing the U.S.B.R. recommended values for finer sediments; these have higher values of RS than the Shields' curve, for they allow for some degree of sediment movement.

Reference can be made here to the little that is known about permissible shear stresses, or tractive forces, on cohesive soils. Ven Te Chow [8] has interpreted some early Russian data on permissible velocities and expressed them in terms of permissible tractive force, which is stated to depend on the clay content of the soil and on the voids ratio, and covers a substantial range of values —from 0.02 to 0.8 lb/ft^2. The highest of these values is the force needed to move 2-in. gravel, so it appears that the resistance of a compact clayey soil to erosion can be considerable. The appropriate values of RS for lean clayey soils are plotted in Fig. 10-5 as a function of the voids ratio. For heavier clayey soils and sandy clays values of τ_0 can be up to twice those indicated. A more detailed investigation by Smerdon and Beasley [9] of a limited range of soil types and void ratios, mainly of the loosely compacted type, indicated that other soil properties such as plasticity index may be significant. Clearly, much work still remains to be done on the question.

Bank Stability

So far consideration has been given only to the stability of sediment on a substantially horizontal bed, the only disturbing force being that due to the fluid shear stress. But on a channel bank having a substantial side slope, a further disturbing force is provided by the weight of the particles resolved down the side slope. Consider, for instance, the trapezoidal channel section

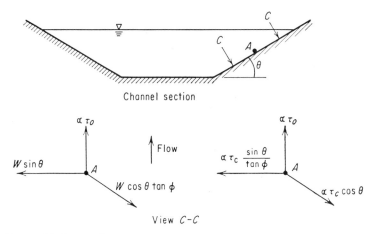

Channel section

View *C-C*

Figure 10-6.   *Definition Sketch, Stability of Grains on Channel Side Slopes*

shown in Fig. 10-6. The view normal to one of the banks shows the forces acting on an individual grain A of weight W. There is a force proportional to τ_0 urging it in the direction of flow, its own weight resolved down the slope— i.e., $W \sin \theta$, and these two are balanced by a resistance force whose limiting value is $W \cos \theta \tan \phi$, where ϕ as before is the angle of repose of the bed material. To compare forces expressed in terms of τ_0 with those expressed in terms of W, we recall that the critical shear stress τ_c is proportional to W $\tan \phi$ (as in C. M. White's argument) and hence that the limiting resistance force in the diagram is proportional to $\tau_c \cos \theta$. Also, the $W \sin \theta$ force will be proportional to $\tau_c \sin \theta / \tan \phi$. Hence we can draw a second diagram in which all the forces are proportional to τ_0 or τ_c; it follows that

$$\tau_c^2 \cos^2 \theta = \tau_0^2 + \tau_c^2 \frac{\sin^2 \theta}{\tan^2 \phi}$$

whence

$$\left(\frac{\tau_0}{\tau_c}\right)^2 = \cos^2 \theta - \frac{\sin^2 \theta}{\tan^2 \phi}$$

or

$$\frac{\tau_0}{\tau_c} = \cos \theta \sqrt{1 - \frac{\tan^2 \theta}{\tan^2 \phi}} \qquad (10\text{–}15)$$

a formula presented originally by the U. S. Bureau of Reclamation [10]. It can be expressed in the simpler form

$$\frac{\tau_0}{\tau_c} = \sqrt{1 - \frac{\sin^2 \theta}{\sin^2 \phi}} \qquad (10\text{–}16)$$

since

$$\cos^2 \theta - \frac{\sin^2 \theta}{\tan^2 \phi} = 1 - \sin^2 \theta \left(1 + \frac{1}{\tan^2 \phi}\right) = 1 - \frac{\sin^2 \theta}{\sin^2 \phi}$$

Equations (10–15) and (10–16) now give the means of designing channel side slopes. They show clearly that the shear stress τ_0 needed to move a grain is less than the critical shear stress τ_c because both shear and side slope are combining their influence to dislodge the grain. But to apply the method properly it is necessary to know what the shear stress τ_0 on the channel side

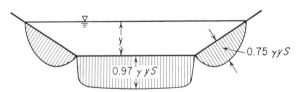

Figure 10-7. *Typical Shear Stress Distribution in a Trapezoidal Channel, after U. S. Bureau of Reclamation*

slopes will in fact be, and it is no longer satisfactory to use an average value of τ_0 given by Eq. (4–4), because the shear stress on the side slopes will be materially less than the average. The U. S. Bureau of Reclamation has made comprehensive tests of shear-stress distribution on trapezoidal channel sections, and a typical result is shown in Fig. 10-7. The coefficients 0.750 and

0.970 shown in this figure vary somewhat with side slope angle θ and the ratio b/y; complete details are given by Ven Te Chow [8], but for practical design purposes it may be conservatively assumed that on the bed

$$\tau_0 = \gamma y S \qquad \qquad \textbf{(10–17)}$$

and that on the side slopes

$$\tau_0 = 0.75\, \gamma y S \qquad \qquad \textbf{(10–18)}$$

In order to apply Eqs. (10–15) and (10–16), information is required on the angle of repose ϕ. Extensive tests by the U. S. Bureau of Reclamation have shown that ϕ depends on the size of the stones and on whether they are angular or rounded. The results of these tests are plotted in Fig. 10-8.

The application of Eqs. (10–14) through (10–18) to the design of a trapezoidal channel in coarse alluvium is best shown by an example.

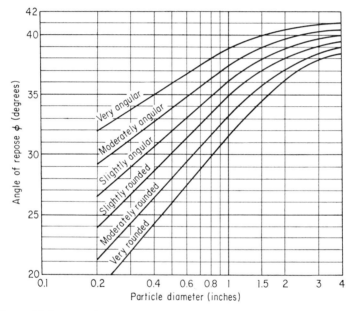

Figure 10-8.   *Angles of Repose of Noncohesive Material, after U. S. Bureau of Reclamation*

Example 10.2

A channel which is to carry 2,000 cusecs is to be excavated on a slope of 0.001 through country made up of coarse alluvium having a D-75 size of 2 in. The stones can be described as "moderately rounded." Assuming that the channel is to be unlined and of trapezoidal section, determine suitable values for the base width and side slopes.

From Fig. 10-8, $\phi = 37°$, whence $\cot \phi = 1.33$. Any convenient value would do for the side slope θ, provided it is materially less than ϕ. A slope of $1\frac{1}{2}$H : 1V, i.e. cot $\theta = 1.5$, might be too close to ϕ for comfort, so we try a slope of $1\frac{3}{4}$H : 1V, i.e. cot $\theta = 1.75$; then cos $\theta = 0.868$. Then from Eq. (10–15),

$$\frac{\tau_0}{\tau_c} = 0.868 \sqrt{1 - \left(\frac{1.33}{1.75}\right)^2}$$

$$= 0.565$$

on the threshold of movement. Therefore the design criterion to be used is that

$$\frac{\tau_0}{\tau_c} \leq 0.565$$

where τ_0 is the actual shear stress on the bank, equal to 0.75 $\gamma y S$, and τ_c is the critical shear stress required to move stones of this size on a flat bed. We can now use a modified form of Eq. (10–14) by recalling that this equation was obtained by, in effect, equating $\tau_0 = \gamma R S$ to $\tau_c = \gamma d/11$. In the present case τ_c remains the same, τ_0 becomes 0.75 $\gamma y S$, so the above inequality becomes

$$\frac{0.75 \; y S}{d/11} \leq 0.565$$

i.e., $yS \leq 0.069d$, or 0.0115 ft. Hence finally, the depth $y \leq 0.0115/0.001$, or 11.5 ft, less 20 percent safety factor, $= 9.0$ ft.

We must now choose the base width b so that the channel will deliver 2,000 cusecs at a depth ≤ 9 ft. From Strickler's equation (4–22)

$$n = 0.031 \left(\frac{1}{6}\right)^{1/6} = 0.023$$

whence

$$\frac{1.49 S^{1/2}}{n} = \frac{1.49\sqrt{0.001}}{0.023} = 2.04$$

We then choose a number of trial values of b, and assuming a depth of 9 ft, proceed by the following tabulation until a discharge of 2,000 cusecs is obtained. [Note that $A = 9(b + 15.75)$ and $P = b + 36.3$.]

b, ft	A, ft^2	P, ft	$R = A/P$	$R^{2/3}$	v, ft/sec	$Q = vA$, cusecs
5	187	41.3	4.52	2.74	5.6	1,048
10	232	46.3	5.01	2.93	5.98	1,390
15	276	51.3	5.37	3.07	6.28	1,730
20	322	56.3	5.72	3.20	6.55	2,110
18.5	308	54.8	5.62	3.17	6.48	2,000

The final result is 18.5 ft, and since this is a minimum, a conservative value would be 20 ft, with side slopes of $1\frac{3}{4}$H : 1V. *Ans.*

The 20 percent reduction in the depth from 11.2 to 9.0 ft is the only explicit safety factor in the calculation, but another one is implied in the choice of the D-75 size as the typical bed material instead of the D-85 or D-90 size with which the bed actually becomes armored.

The General Problem

This section will be concluded with a discussion of the way in which Examples 10.1 and 10.2 fit into the general context of the problem of designing canals in erodible material. We are here considering only the flow of substantially clear water over a bed which is not to be permitted to scour; the alternative problem, in which scour is permitted because some sediment load is introduced at the canal headworks, will be deferred to a later section. The known elements of the problem are the discharge, the bed material, and often the bed slope, although sometimes the designer has some freedom to choose the slope by varying the course of the canal.

If a very gentle bed slope can be used, then it may be possible to use a trapezoidal section, as in Example 10.2. This section is the most vulnerable to scour because of its exposed side slopes. If the bed slope has to be fairly steep, it may be necessary to use a channel with grassed (or otherwise protected) banks, as in Example 10.1. The limiting design criterion then hinges on the possibility of bed scour, and the requirement is to make the canal wide enough to keep the depth and hence the tractive force down to a safe figure. Such canals are often wide enough to make $R = y$, so that the design problem is very simple, as in Example 10.1. If the bed slope becomes greater still, the required width may become excessive, and a better solution may be to reduce the bed slope and absorb the extra fall by means of drop structures, as in Prob. 6.14 of Chap. 6. In the present case, the details can be worked out in Prob. 10.8.

Most of the numerical work in this section has been based on Eq. (10–14), which is true only for a particular value of the specific gravity, 2.65. For other values of s_s this equation must be changed and any subsequent numerical work modified accordingly (Prob. 10.9).

The problem of determining a suitable channel width in view of the tractive-force criterion for the threshold of movement applies to the control of natural rivers as well as to the design of artificial canals. However, this particular application is more conveniently dealt with in a later section, as a special case of a general treatment involving the established movement of sediment as well as the threshold of sediment motion.

10.4   The Suspended Load

After the threshold of motion has been passed, and sediment movement is well established, some of the sediment will be carried in the form of suspended load; in the case of strong flows and fine sediment the amount of suspended load may be substantial. This material is maintained in suspension by the action of turbulence, and in this section the exact form of the mechanism of suspension will be considered.

The discussion introducing Sec. 1.8 dealt with the momentum transfer mechanism by which the exchange of fluid between adjacent fluid layers gives rise to an effective shear stress on the interface between the layers. In turbulent flow, the rate of this exchange is measured by the "eddy viscosity" η which, like the dynamic viscosity μ, may be defined as a rate of mass exchange per unit area between adjacent layers which are separated by a (small) unit distance. This definition enables the basic shear-resistance equation

$$\tau = (\mu + \eta) \frac{dv}{dy} \qquad (10\text{--}19)$$

to be deduced immediately from Eq. (1–14).

Now the random interchange of material between adjacent layers not only brings about momentum transfer as described above; it also accomplishes the mixing of any fluid property which may have different magnitudes in the two layers, such as temperature, colour, salt concentration, or—in the present case—concentration of sediment. An elementary model to consider is that of two buckets, each containing a volume V of, say, salt solution, the concentrations by volume in the two buckets being c_1 and c_2 ($c_1 \neq c_2$). Now if a volume V_1 is taken out of each bucket and transferred to the other one, the new concentrations will be, assuming that $c_1 > c_2$:

$$c_1 - \frac{V_1}{V}(c_1 - c_2) \qquad \text{and} \qquad c_2 + \frac{V_1}{V}(c_1 - c_2) \qquad (10\text{--}20)$$

i.e., an amount of concentration equal to $V_1(c_1 - c_2)/V$ has been transferred from the high concentration bucket to the other. If the process continues, there will be a steady movement of salt from one bucket to another at a rate proportional to the rate of volumetric transfer and the concentration difference, just as the shear stress in the previous case is proportional to the rate of mass transfer and the velocity difference. Just as Eq. (10–19) follows from Eq. (1–14), we can deduce from Eq. (10–20) that in a turbulent flowing fluid of eddy viscosity η carrying sediment of volume concentration c, the volumetric rate of sediment transfer across the streamlines, per unit area, in the direction of positive y is equal to

$$-\frac{\eta}{\rho}\frac{dc}{dy} \quad \text{or} \quad -\varepsilon\frac{dc}{dy} \qquad (10\text{--}21)$$

introducing $\varepsilon = \eta/\rho$, the kinematic eddy viscosity. The need to divide by ρ arises out of the shift of interest from a mass-transfer to a volume-transfer situation.

Now Eq. (10–21) gives the upward rate of sediment movement, and this is balanced by the downward volumetric rate of sediment transfer per unit area movement due to gravity. The volumetric rate per unit area must be equal to cw, where w is the velocity of free fall of the grains in the fluid. It follows that

$$cw + \varepsilon \frac{dc}{dy} = 0 \tag{10–22}$$

and if this equation can be integrated the c-y distribution can be obtained. To perform the integration some knowledge of ε is required and this must come from experiment. If ε is assumed constant, Eq. (10–22) can be integrated immediately, yielding

$$\frac{c}{c_a} = e^{-w(y-a)/\varepsilon} \tag{10–23}$$

where c_a is the concentration at a height a above the bed. Since there is no way of knowing c_a initially, Eq. (10–23) cannot give absolute values of the concentration; it can only give their relationship to the unknown c_a. What this means is that the *total* sediment load cannot be predicted by the above theory; to do this it is necessary to know something of the mechanism of entrainment *at the bed*. From such knowledge it might be possible to predict the concentration at some point a short distance above the bed, which could serve as the reference value c_a for insertion in Eq. (10–23).

A more refined integration of Eq. (10–22) can be made by allowing for the variation in ε; for this purpose it is convenient to review briefly the results of the classical work of Prandtl and von Kàrmàn on turbulent shear flow. We introduce first Prandtl's concept of a "mixing length" l, which is a measure of the length scale of the turbulence. It may be regarded as an indicator of eddy size, or (more usefully) of the average distance traveled by small parcels of fluid in the random movements (superposed on the mean velocity) that make up the turbulent motion. In this latter sense the mixing length is roughly analogous to the "mean free path" in the kinetic theory of gases.

Clearly the mixing length will be an important determinant of the process of momentum exchange between adjacent fluid layers, and therefore of the eddy viscosities η and ε. In fact a combination of theory and experiment has shown that l may be defined by the equation

$$\varepsilon = l^2 \frac{dv}{dy} \tag{10–24}$$

whence from Eq. (10–19) it follows that

$$\tau = \eta \frac{dv}{dy}$$

$$= \rho l^2 \left(\frac{dv}{dy} \right)^2 \tag{10–25}$$

dropping the viscosity μ, which is normally a small fraction of η. In the close neighborhood of a solid boundary it follows from dimensional considerations that the mixing length is directly proportional to the distance y from the boundary,

$$l = \kappa y \tag{10-26a}$$

where κ is von Kàrmàn's universal constant, equal to 0.4 for all homogeneous fluids. In the same neighborhood we can assume

$$\tau = \tau_0 = \rho \kappa^2 y^2 \left(\frac{dv}{dy}\right)^2 \tag{10-27a}$$

or

$$\frac{dv}{dy} = \frac{v^*}{\kappa y} \tag{10-27b}$$

from which follows the well-known law of logarithmic velocity distribution

$$v = \frac{v^*}{\kappa} \log_e y + \text{constant} \tag{10-28}$$

Now it is found by experiment that this equation describes the velocity distribution very accurately even at substantial distances from the wall, where τ differs materially from τ_0. Since in fact $\tau = \tau_0(1 - y/y_1)$, where y_1 is the total depth of flow, we can reconcile Eq. (10-25) with this observation by assuming that at a distance from the solid boundary

$$l = \kappa y \sqrt{1 - \frac{y}{y_1}} \tag{10-26b}$$

and it then follows from Eqs. (10-24) and (10-27b) that

$$\varepsilon = \kappa v^* y \left(1 - \frac{y}{y_1}\right) \tag{10-29}$$

The constant in Eq. (10-28) depends on Re in the case of a smooth boundary or the roughness size k_s (Sec. 4.2) of a rough boundary. For the purpose of the present argument, it is unnecessary to explore the behavior of this constant any further than this.

The substitution of Eq. (10-29) into Eq. (10-22) gives an equation which can be integrated to yield

$$\frac{c}{c_a} = \left[\frac{a(y_1 - y)}{y(y_1 - a)}\right]^{w/\kappa v^*} \tag{10-30}$$

where a and c_a are defined as for Eq. (10-23). Equation (10-30) suffers from the same limitation as Eq. (10-23), namely that it can predict only the relative and not the absolute concentration. Figure 10-9, based on the work of Vanoni and taken from an A.S.C.E. report [11], shows curves drawn from Eq. (10-30) and compared with experimental results. The agreement is good, but it is an agreement in form only, for the values of $w/\kappa v^*$ have been chosen so as to achieve the best fit instead of being determined independently.

A difficulty in determining all the members of $w/\kappa v^*$ independently is that in a sediment-laden flow κ may not have its clear-water value of 0.4. Up to this point it has been assumed that the parameters l, ε, and κ will have the same values for the sediment motion as for the fluid motion, but this is not necessarily true. If any small parcel of fluid is to pull the sediment along with it, in steady or turbulent motion, there must be some relative movement, or slip,

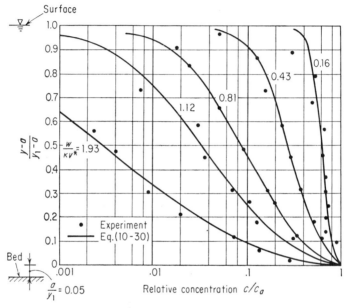

Figure 10-9. *Distribution of Suspended Sediment, after A.S.C.E. Task Committee* [11]

if the fluid is to apply the necessary force to the sediment. This implies that the sediment lags behind the fluid and therefore that its effective mixing length l will be less than that of the fluid. Similarly, ε and κ will be lower, and it is usual to distinguish between the momentum-transfer coefficient ε_m and the sediment-transfer coefficient ε_s, equal to $\beta\varepsilon_m$. Since $\varepsilon \propto \kappa$, the factor β will be equal to the ratio of κ values as well as of ε values.

The above discussion would suggest that

$$\varepsilon_s < \varepsilon_m, \quad \text{i.e.,} \quad \beta < 1 \tag{10–31}$$

A further inference to be drawn is that the sediment, while lagging behind the fluid, also exerts a retarding force on the fluid. The conclusion suggested is that the presence of suspended sediment damps out the irregular oscillations which characterize turbulent motion and thereby reduces l, ε, and κ. This conclusion may be expressed thus

$$\varepsilon_m < (\varepsilon_m)_0 \tag{10–32}$$

where $(\varepsilon_m)_0$ is the clear-water value of ε_m. From Eq. (10–25) it follows that the fluid resistance of a sediment-laden stream would then be less than that of clear-water flow having the same depth, velocity, and boundary roughness.

In field and laboratory observations it has proved very much easier to verify Eq. (10–32) than Eq. (10–31). Figure 10-10, for example, shows typical

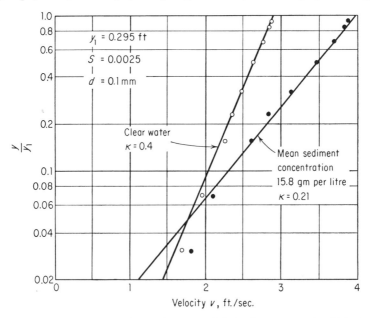

Figure 10-10. *The Effect of Suspended Sediment on the Velocity Profile, after A.S.C.E. Task Committee* [11]

velocity profiles comparing clear-water and sediment-laden flow for the same depth, slope, and bed surface formation. In the latter case the velocity is much greater than in the former, indicating from Eq. (10–28) that κ must be smaller; since $\varepsilon \propto \kappa$, it follows that Eq. (10–32) is verified. The experiment whose results are displayed in Fig. 10-10 was one of a series conducted by Vanoni and Nomicos [12], who used the following technique. Water was run through a flume having a sand bed until a stable condition developed, in which sand was being carried by the water in suspension and the bed had assumed a certain form. In this condition velocity profiles and sediment concentrations were measured. Then the flume was drained and the bed sprayed with a plastic which fixed the bed shape and prevented further erosion; finally clear water was run through the flume at the same depth as before. This technique ensured that the clear-water and sediment-laden flows were similar in every way except in the presence or absence of suspended sediment, and the results showed that sediment-laden water in every case had a lower value of κ and therefore of ε_m.

The effect of lowering κ is to reduce the flow resistance, but the above experiments also showed that this reduction in resistance is small compared with the increase in resistance that can be brought about by bed formations; a dune bed, for instance, can have a Darcy coefficient f many times greater than that of a flat bed. Variations of this sort can completely mask the smaller variations attributable to the suspended sediment itself, and can incidentally provide an explanation for field observations to the effect that sediment-laden flows experience greater resistance than clear-water flows. It is most probable that this increased resistance is due to dune formations which would not exist during clear-water flow.

The careful experimental technique of Vanoni and Nomicos appears to have separated the effects of bed formation from those of suspended sediment; in particular, values of κ close to 0.40 were measured for all clear-water flows, whatever the bed formation. However, in other experiments cited by Laursen [11], pronounced bed roughness produced values of κ as low as 0.30 in clear-water flow; he suggested accordingly that the presence of suspended sediment and the lowering of κ may not be cause and effect, but may both be effects of the bed formation. There may therefore be scope for further work on the question.

More complications have been introduced by the experiments of Elata and Ippen [13], which showed that the decrease in κ occurred even in suspensions of particles having the same density as the fluid. This result is not predicted by the approximate argument, originally due to Vanoni, which introduces Eq. (10–32); according to this argument the lag between particles and fluid occurs only if the particles are denser than the fluid. Another surprising result of the same experiments was that although κ is reduced, the intensity of the turbulence (as measured by the fluctuations in velocity) is actually increased by the presence of suspended particles. A theoretical analysis has been advanced by Hino [14] as an explanation of these results, but it is beyond the scope of this text.

Attempts to verify Eq. (10–31) by experiment introduce still more difficulties. The usual approach is by examining the behavior of the exponent $w/\kappa v^*$ (usually written as z) in Eq. (10–30). This exponent is usually observed to increase with mean sediment concentration. The value of κ in this term is the one appropriate to the sediment, and should therefore be written as $\kappa_s = \beta\kappa_m$. As we have seen, κ_m itself is reduced by the presence of sediment, so any observed increase in z and consequent reduction in κ_s is not necessarily evidence that β is less than unity. Allowance must also be made for the decrease in the fall velocity w with increase in sediment concentration; in the end result it appears that if there is a change in β due to suspended load, it is small compared with the relative change observed in κ.

Some river observations have even indicated that β is greater than unity; Einstein and Ning Chien [11] have shown, however, that this increase in β is only apparent, and may be accounted for by developing a more refined theory than the one on which Eq. (10–30) is based. Among other refinements, the theory distinguishes between the upward and downward velocities with which sediment is transferred across any horizontal plane. The theory is outlined in Ref. [11]; the details are beyond the scope of the present discussion.

Effect of Sediment Motion on Flow Resistance

In the preceding paragraphs it has been pointed out that suspended sediment reduces flow resistance but that the effect is small compared with the influence of bed formations like dunes which are induced by the sediment motion. It might be expected, therefore, that systematic data would be available relating bed formation to standard resistance coefficients such as the Manning n. Unfortunately this is far from being the case; the data so far collected, while showing certain systematic trends, display so much scatter and so many inconsistencies that they cannot yet be recommended as giving accurate and reliable guidance to the river or canal engineer.

The following discussion will outline some of the particular aspects of the problem. Logically the first step is to attempt the separation of the bed form resistance (arising from dunes, etc.) from the intrinsic resistance due to the roughness of the sand or sediment itself. The latter can be estimated fairly closely from existing knowledge, as in Table 4–2. The former can then be derived from observations and displayed in generalized form. A study of this sort was undertaken by Einstein and Barbarossa [15]; the bed shear stress τ_0 was assumed to be made up of two components τ_0' and τ_0'', due to intrinsic roughness and form (e.g., dune) roughness respectively. The total cross-sectional flow area A was assumed to be divisible in a corresponding way into two components A' and A'', so that

$$\tau_0' = \gamma A'S/P = \gamma R'S, \qquad \text{and} \qquad \tau_0'' = \gamma A''S/P = \gamma R''S$$

whence $\qquad \tau_0 = \tau_0' + \tau_0'' = \gamma RS = \gamma(R' + R'')S \qquad$ (10–33)

the hydraulic mean radius R being similarly divisible into components R' and R''. It is now assumed that the sediment transport, and the bed formation, will be a function of Shields' entrainment function F_s with the shear stress taken equal to τ_0' alone. This assumption is an essential feature of Einstein's theory of sediment motion, to be discussed in the next section; in this context it means that the resistance coefficient due to bed form resistance will be a function of $\tau_0'/\gamma d(s_s - 1)$. This coefficient will also be a function of $v/v*''$ (cf. Eq. (10–8), where $v*'' = (\tau_0''/\rho) = \sqrt{(gR''S)}$). Noting that $\tau_0'/\gamma = R'S$, we can write finally

$$\frac{v}{v*''} = \frac{v}{\sqrt{gR''S}} = f\left[\frac{d(s_s - 1)}{R'S}\right] \qquad \text{(10–34)}$$

and this relationship can be verified by observation. The total shear stress $\tau_0 = \gamma RS$ is obtained directly from observations, and $\tau_0' = \gamma R'S$ is calculated from standard data on resistance coefficients. Hence R'' is obtained by subtraction, and the ratio $v/v^{*''}$ can then be calculated.

This process was worked out by Einstein and Barbarossa [15] for a number of natural rivers, using the D-65 size of sand to estimate the grain roughness and the D-35 size to compute the entrainment function. This latter assumption is in accord with Einstein's method for applying his theory of sediment motion. A plot of the two variables in Eq. (10–34) is shown in Fig. 10-11, the

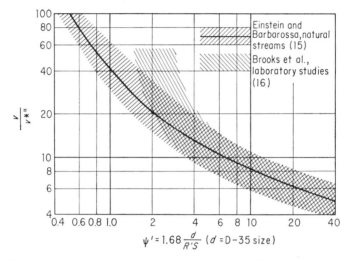

$$\psi' = 1.68 \frac{d}{R'S} \quad (d = \text{D-35 size})$$

Figure 10-11. *Bar Resistance as a Function of the Sediment Transport Rate*

solid line showing the mean relationship and the shaded band on each side of the line the approximate range covered by the observations. The assumed value of s_s was 2.68. The degree of scatter is not excessive by the standards of an investigator making an exploratory study, although it may appear considerable to a practicing engineer seeking a reasonably exact estimate of the resistance coefficient. When the results of laboratory studies [16] are plotted on the same graph they cover a somewhat different region with even more pronounced scatter, as indicated by the further shaded region in Fig. 10-11. The only reason that could be advanced [16] for this difference was that the laboratory studies were conducted with fairly uniform sand and the natural river channels would naturally have nonuniform sand. It was argued by Einstein and Ning Chien [16] that dune formation is accompanied by a sorting of the sediment into finer material on the dune crests and coarser material in the troughs, and in fact that on this account dunes are more likely to be stable in nonuniform than in uniform sand. If so, dunes will be less persistent at higher velocities in uniform sand, and their earlier disappearance will lead to lower resistance and higher values of $v/v^{*''}$, as Fig. 10-11 indicates.

Further tests would be needed to verify this explanation, and indeed further refinements appear to be necessary if the method is to become a reasonably precise guide to the practicing engineer. In considering the relative importance of dune resistance or "bar resistance," and grain resistance, it is natural to consider the sequence of events as the velocity is increased (raising the resistance) and dunes give way to a flat bed (lowering the resistance). The existence of these opposing influences prompts an enquiry whether in some cases there might be a net reduction in resistance as the velocity increases; the work of Brooks [16] gives an affirmative answer. Figure 10-12a is a plot of friction slope S_f against velocity for a fixed depth, width, and bed material.

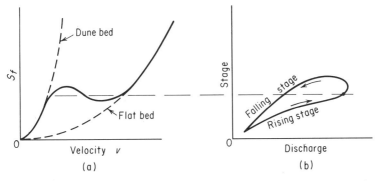

Figure 10-12. *The Effect of Changing Bed Formation on the Velocity-Slope Relationship and on the Loop-Rating Curve, after N. H. Brooks* [16]

As the velocity increases the dunes reduce in size and finally disappear: the S_f-v curve therefore crosses over to the line representing flat-bed conditions. In the process the line will certainly pass through an inflection and may pass through a maximum and minimum as shown; the experimental work of Brooks has shown this to be a possibility. An interesting corollary is that for a given slope and depth there may be as many as three possible velocities, each with its corresponding bed formation.

Changes in bed formation have been cited as an explanation for the loop-rating curve discussed in Sec. 9.8. Figure 10-12b shows the type of rating curve that might be developed as a consequence of the relationship shown in Fig. 10-12a; i.e., with a flat bed on the rising stage and dune bed on the falling stage. Loop-rating curves have been observed [17] in which the rising stage discharge was up to twice as much as the corresponding falling-stage figure. A discrepancy as large as this could hardly be explained by the methods of Sec. 9.8. The flat bed on the rising stage may simply be due to the high velocities and sediment loads appropriate to that condition, or to the fact that dunes have not yet had time to form. The latter explanation is favored by Simons et al. [18], but the question must depend on how fast-rising the flood is; if the time scale is of the order of weeks, the time-lag explanation must be less plausible. However, it may well be applicable to floods which rise in days rather than weeks, in view of Blench's observation [15] that a sand bed in a small flume may take as long as 80 hr to accommodate itself to changes in regimen.

Bagnold's "Penetration" Concept

The earlier discussion of the transfer coefficients ε_s and ε_m has centered on the fact that the sediment follows, but lags behind, the fluid in the random and irregular oscillations that make up turbulent motion. The discussion, although admitting the existence of this lag, nevertheless assumed that the sediment is substantially under the control of the fluid, and responds, although belatedly, to changes in the fluid motion. This assumption must be close to the truth when the sediment is fine and easily suspended, but one can also readily visualize sediment so coarse or dense that given some initial impulse its inertia will make it cut right through the fluid, substantially uninfluenced by any local velocity variations which it meets on its way.

A convenient measure of the sediment's freedom from the fluid influence is provided by Bagnold's definition of the "penetration" of a grain in a fluid [19]. It is assumed that the grain is projected through the stationary fluid with some initial velocity v_0. The distance traveled by the grain before the fluid drag reduces its velocity to $v_0/2$ is defined as the *penetration* of the grain in the fluid. The concept is clearly analogous to that of the "half-life" of a radioactive substance. If the Reynolds number Re of the relative particle motion (based on the particle diameter) is greater than 1000, the drag coefficient C_D may be assumed constant throughout the motion. In this case the penetration distance is independent of v_0, as the following calculation shows. If the grains are moderately angular we can take $C_D = 1.0$, and the retarding force is then $A_1 d^2 \cdot \frac{1}{2}\rho_f v^2$ acting on a mass of $A_2 d^3 \rho_s$, where A_1 and A_2 are constants dependent on the geometry of the particle. If the particle is assumed approximately spherical, $A_1 = \pi/4$ and $A_2 = \pi/6$. Setting $\rho_s/\rho_f = s_s$, the retardation is then equal to $3v^2/4s_s d$, i.e.,

$$\frac{3v^2}{4s_s d} + v\frac{dv}{dx} = 0 \qquad (10\text{--}35)$$

where x is the direction of motion. The integral of this equation is

$$\frac{v}{v_0} = e^{-3x/4s_s d} \qquad (10\text{--}36)$$

from which it follows that $v = v_0/2$ when x is equal to the penetration distance

$$x_p = \frac{4s_s d}{3}\log_e 2 \qquad (10\text{--}37)$$

whence for $s_s = 2.65$,

$$\frac{x_p}{d} = 2.5 \quad \text{(approx.)} \tag{10-38}$$

and for $s_s = 2,400$

$$\frac{x_p}{d} = 2,200 \quad \text{(approx.)} \tag{10-39}$$

the former value applying to grain-water systems and the latter to grain-air systems. If $\text{Re} < 1000$, C_D is no longer independent of Re and x_p is no longer independent of v_0. It is left as an exercise for the reader to show (Problem 10.19) that $x_p/s_s d$ is now a function of Re_0, the Reynolds number of the initial particle motion. When Re_0 is low enough for Stokes' Law to hold, $x_p/s_s d$ is simply equal to $\text{Re}_0/36$; this implies values of x_p somewhat larger than those given by Eq. (10-37), but of the same general order of magnitude.

The penetration concept, while it may not play an explicit part in any theoretical development, is clearly useful for the general picture it conveys of the interaction between grains and fluid. The definition fits any situation in which a velocity difference between grain and fluid is suddenly introduced, e.g., through a turbulent fluctuation in fluid velocity. In this situation the penetration of the particle becomes an inverse measure of the success that the fluid will have in dragging the particle along with it. Similarly when fluid is deflected around a moving body such as an automobile; in this case insects of low penetration are deflected with the air, while dense compact insects of high penetration keep going until they strike the body of the automobile.

The penetration length x_p may usefully be compared with the turbulent mixing length l. If x_p/l is small, then turbulent eddy movements will have ample time (and distance) to bring the grains under their control, and the suspended sediment theory previously discussed will be applicable. But if x_p/l is large, turbulent movements will not last long enough, or move far enough, to influence the sediment motion materially.

The mixing length is of the order of $y_1/20$, where y_1 is the total depth as previously defined. In grain-water systems, where x_p is only a few grain diameters, x_p/l will normally be small unless the grains are very coarse—i.e., unless d is of the order of $y_1/100$ or more. But in grain-air systems like wind-blown sand, x_p will be measured in feet for all but the very finest sands. The depth y_1 will approximate to the boundary-layer thickness, of the order of tens, possibly hundreds, of feet. It follows that x_p and l will be of the same order of magnitude, with x_p possibly larger than l. It is unlikely therefore that sand will be continuously suspended by turbulence in the sense of Eq. (10-22); sand grains become airborne by being projected upwards from the bed almost vertically, and follow a trajectory almost undisturbed by fluid turbulence before returning to the bed. This phenomenon is known as *saltation*; the exact mechanism of the initial upward projection is not fully understood, but it is most likely due to a sudden violent change of pressure near the grain, or to the landing of another grain nearby, or to a mixture of these two causes.

As mentioned previously, this discussion of the penetration concept is of general rather than specific usefulness. Nevertheless it will play some part in the subsequent treatment of theories of sediment motion.

Pressure Distribution

It is of some further interest to inquire into the pressure distribution in flow containing suspended sediment. Consider an element $ABCD$ (Fig. 10-13a) of an unbounded mass of fluid; within this element a number of solid particles

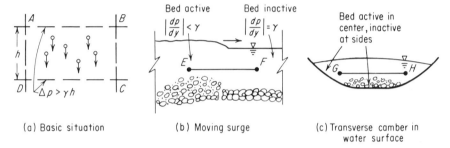

(a) Basic situation   (b) Moving surge   (c) Transverse camber in water surface

Figure 10-13.   *The Effect of Sediment Suspension on Pressure Distribution*

are falling at their terminal velocity. The combined drag force of the fluid on the particles brings about an equal and opposite reaction of the particles on the fluid, which must be balanced by an extra pressure difference Δp between CD and AB, added to the hydrostatic pressure γh. In effect, the total pressure difference between CD and AB must supply a thrust equal to the combined weight of particles and fluid.

If the particles are not moving with their terminal velocity, or if some of them are moving upwards, the pressure difference Δp will be less than in the previous case, and may even be less than γh. In any particular case Δp is easily calculated (provided that the particle velocities are known) from the drag forces and the fluid weight, excluding the particle weights (Prob. 10.11).

In a suspended load of high-penetration particles the upward velocities will be systematically higher than the downward velocities, as may be seen by considering a typical saltation path. Because of fluid resistance, the vertical component of velocity with which the particle returns to the bed must clearly be less than the initial upward velocity. As the reader can verify in Prob. 10.11, this will mean that the pressure gradient in the fluid can be materially less than the hydrostatic value.

The above conclusion may explain certain rather curious effects observed during floods in rivers having coarse gravel beds. One such effect is described in the following account [20]:

On a certain river it was noticed during several floods that at a critical value of flood stage the bed boulders of about 18 inches diameter began moving with a thundering roar and an almost instantaneous increase in flood height—about 2.2 feet in one minute. This gave the effect of a surge moving downstream at a slower rate than the previous river flow. Behind the surge the flow was more turbulent than in front; in fact it had some of the irregular character of a breaking wave. The turbulence and rock thumping lasted until the river level began to drop, generally two to three hours later.

The phenomenon is illustrated in Fig. 10-13b, and the apparent anomalies in the situation may possibly be resolvable by invoking the previous argument and deducing from it that the pressures at E and F, say, may be equal although their depths of submergence differ. A similar variation in surface level may arise, as in Fig. 10-13c, over the cross section of the river. Higher velocities in the center may activate the bed there while at the sides the bed is still inactive. The pressure along the horizontal line GH could then be constant, keeping the system in equilibrium, even though the water level at the center is higher than at the sides. Indeed accounts are commonly heard of the existence of this cambered surface effect during floods in gravel-bed rivers, although it is difficult to find a properly documented account.

The above discussion of Fig. 10-13a, and the argument of Prob. 10.11, are strictly applicable only when the fluid is assumed to be infinite in extent and not bounded by any container. When the fluid is contained within a vessel the situation is complicated by the back flow along the container walls of the fluid originally entrained by the particles and carried along with them; this back flow introduces a drag against the cylinder walls which can only be balanced by an increase in the pressure difference Δp in Fig. 10-13a. It has been shown that in the case of a sphere moving at low Reynolds numbers within and along the axis of a vertical cylinder, the difference between actual and hydrostatic pressure gradients is exactly twice as great as is indicated by the result of Prob. 10.11, however large the containing cylinder may be. An account of this work is given by Pliskin and Brenner [21]. It yet remains to be seen how this rather special result might apply to suspensions of more than one particle in turbulent flow; further work is needed to clear the matter up.

10.5 Bed-Load Formulas and Entrainment at the Bed

Introduction

In the previous section it was pointed out in a discussion of Eqs. (10–23) and (10–30) that a suspended-load theory cannot give absolute values of sediment concentration, or of total sediment load, and that for this purpose it was necessary to know something of the mechanism of entrainment at the bed. Following up this concept, it is convenient to postulate a horizontal plane, close enough to the bed for vertical turbulent movements to be suppressed by the nearness of the bed. There is therefore no turbulent transfer term,

$\varepsilon_s dc/dy$, to balance the sediment-fall term cw in Eq. (10–22); upward movement of sediment across this plane has to be supplied by some mechanism which projects some of the bed load upward into the main body of the flow. If this mechanism were understood, this plane might be taken as the base-level $y = a$ in Eq. (10–30), from which the total suspended load could then be calculated.

There would still remain the problem of determining the bed load, consisting of particles which move by rolling or sliding along the bed. This is a most difficult extension of the "threshold" problem, and the difficulty arises from the statistical nature of the problem. In order to amplify this remark, we first reconsider the threshold problem.

Now attempts to analyze this problem which are based on a simple statical analysis, e.g., White's [6], have tacitly assumed that the fluid's transporting power is uniform over the whole bed, and that the grains on the bed have a certain uniform degree of stability. Carried to its logical conclusion, this assumption would suggest that the threshold of motion is a sharply defined condition marking the boundary between a stationary bed and a generally moving bed. Observations show that this is not true, and that the threshold condition is a somewhat ill-defined one, through and beyond which the rate of transport increases steadily with increasing shear stress at the bed.

The conclusion is that either the transporting power of the fluid or the stability of the grains varies over the surface of the bed; probably both vary in this way. It is reasonable therefore to assume that grains are dislodged not by the exertion of a steady fluid force but by the action of isolated fluid disturbances which in line with normal practice we can describe as eddies. (One can readily conceive of an eddy creating a momentary pressure fluctuation strong enough to dislodge a grain.) Now if there are a number of eddies of varying strength distributed at random through the flow at bed level, and a number of grains of varying stability, then movement will occur when a strong enough eddy meets an unstable enough grain. As the strength of the flow increases, the number of such encounters will increase and the rate of bed-load transport will rise.

The mechanism is essentially statistical in nature, and it would be difficult therefore to predict the relationship between bed-load transport and the strength of the flow. It is questionable even whether the bed shear stress τ_0 will continue, as at the threshold condition, to be a satisfactory measure of the strength of the flow; it has been shown by Thompson [22] that the strength of the eddies will depend not merely on the strength of the flow but also on the size of the grains immediately upstream, from which the eddies are shed as a turbulent wake. However, this question does not pose a real difficulty, since *for a given granular bed* the shear stress τ_0 will be a satisfactory measure of the strength of the flow; we may therefore expect that the volumetric rate of bed-load transport per unit width, q_s, will be related in some way to the bed shear stress τ_0.

It is reasonable to assume that the suspended load also will be related in some way to τ_0, which may be expected to control the mechanism of entrainment at the bed, whatever it may be. Surprisingly, there has been little speculation as to the exact nature of this mechanism, but experimental observations indicate certain ways in which suspended load develops. One is by gradual intensification of the bed load into a thick layer of moving grains, from whose upper surface grains are easily picked up by the turbulence of the overlying flow. Another is "saltation," already discussed at the end of the last section; it occurs mainly with high penetration systems like sand in air, or coarse grains in water. It is noteworthy that saltation can be going on vigorously even when the bed load is quite small, in fact when it consists only of the "surface creep" of a few grains dislodged by the landing of saltating particles. Figure 10-18 shows a typical saltation trajectory.

Laursen [27] draws attention to a third possibility: that particles moving as bed load may be launched upwards into the suspended load either by their movement up the inclined upstream face of a dune, or by skipping upwards from an impact with a grain or group of grains projecting upwards from the bed. So far, few attempts have been made to deduce a sediment transport equation from a direct consideration of any of these mechanisms.

Empirical Formulas

Many such formulas have been developed, some purely empirical, others having a background of semirational, semidimensional argument. They usually involve the difference, or the ratio, between the actual bed shear stress τ_0 and the critical shear stress τ_c at which movement begins. In some cases the shear velocity v^* is used as the measure of τ_0, and the ratio v^*/w then becomes a convenient measure of the balance of flow strength, represented by v^*, against the particle resistance to motion, represented by its fall velocity w.

The formulas are usually known as bed-load formulas, but some of them appear able to cover the suspended-load range as well. Indeed the transition from bed load alone to bed load plus suspended load is no more clearly defined than the initial threshold of motion, and it is possible that one continuous function may well give the total transport rate q_s, whether or not some of the material is suspended load. Some discussion of this point has occurred in connection with the development of the following formulas, but none of it has so far been conclusive. The preceding discussion of Bagnold's penetration concept has indicated that the ratio l/x_p might be a good indicator of the fluid's power to keep material in suspension, whereas the immediately foregoing remarks indicate that the ratio v^*/w would be preferable as an indicator of the fluid's power to move the grains in the first instance. However, tentative notions such as this one have not yet been invoked in order to differentiate between bed-load and suspended-load formulas, and the precision of the

empirical data so far available would hardly make such distinctions worthwhile. Indeed, as mentioned previously, there has been little attempt to relate
any of these formulas to a fully detailed consideration of the mechanism of
sediment movement.

Some of the better-known empirical formulas will now be described. The
oldest of these is the classical Du Boys equation

$$q_s = C_s \tau_0 (\tau_0 - \tau_c) \tag{10-40}$$

whose coefficient C_s is by no means constant. Straub found by examining
the work of many investigators that for grain-water systems C_s is equal to
$0.173/d^{3/4}$, where d is the grain size in millimeters; Eq. (10–40) can therefore be rewritten as

$$q_s = \frac{0.173 \tau_0 (\tau_0 - \tau_c)}{d^{3/4}} \tag{10-41}$$

where q_s and τ are in fps (foot-pound-second) units. This behavior of C_s has
been verified only in the range $\frac{1}{8}$ mm $< d < 4$ mm.

The Du Boys formula preceded Shields' work by many years, and it was
based on empirical values of τ_c that differ somewhat from Shields' values.
These values were retained by Straub in his evaluation of C_s; corresponding
values of RS are plotted against d in Fig. 10-5, assuming that the fluid is water
and $s_s = 2.65$. All other bed-load formulas to be given in this section are
based on the values of τ_c arrived at by Shields and plotted in Figs. 10-3 and
10-5.

Shields [5], besides determining his well-known threshold relationship, also
put forward a bed-load formula

$$\frac{q_s s_s}{q S} = 10 \frac{(\tau_0 - \tau_c)}{\gamma(s_s - 1)d} \tag{10-42}$$

where q, S, and γ are defined in the normal way as water discharge per unit
width, bed slope, and water specific weight. The equation is dimensionally
homogeneous, so that any consistent system of units can be used. It fits fairly
well to a wide variety of experimental data, the range of the scatter being the
equivalent of a factor of 10. Although this amount of scatter may appear
extreme, it is by no means out of the ordinary in this type of work.

Perhaps the best-known formula is that of Einstein [23], which is based
partly on physical reasoning and partly on dimensional considerations. In the
derivation of the formula it was assumed that each grain, after being dislodged
from the bed, traveled an average distance L before coming to rest again, and
that L was proportional to d. Then if we consider an area of the bed having
unit width and length L in the direction of flow (the shaded region in Fig.
10-14), every grain which is dislodged within that region will cross the end AB
before coming to rest again. Then the number of grains being dislodged every

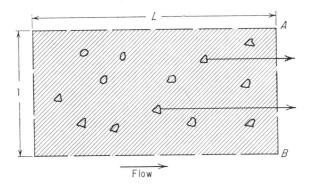

Figure 10-14.   *Definition Sketch for Analysis by H. A. Einstein* [23]

second within the shaded area will also equal the number of grains crossing AB every second. If p_s is the probability that any given grain will be dislodged within a given second, and $A_1 d^2$ is the bed area occupied by each grain, then the number of grains dislodged every second will be equal to

$$\frac{L p_s}{A_1 d^2}$$

Hence if $A_2 d^3$ is the volume of each grain, then the total volume of grains crossing AB every second is equal to

$$q_s = \frac{L p_s A_2 d^3}{A_1 d^2}$$

$$= \frac{A_2}{A_1} \lambda p_s d^2 \qquad (10\text{–}43)$$

where $L = \lambda d$. Now the coefficients A_1 and A_2 are dependent on the geometry of the particle; λ can be assumed, at least tentatively, to be a constant. Since the distance L has something in common with the penetration distance x_p, this last assumption is a plausible one in view of the x_p-d relations obtained in the last section. This leaves the all-important p_s to be evaluated. Einstein postulated that it would be a function of the ratio between the lift force which the flow can exert on a grain, and the submerged weight of the grain itself. The former of these is proportional to $\rho_f v^2 d^2$, where v is some characteristic fluid velocity near the grain, and the latter is equal to $\rho_f(s_s - 1)g A_2 d^3$. Now $\rho_f v^2$ will clearly be proportional to τ_0, so that the ratio lift force : submerged weight becomes proportional to

$$\frac{\tau_0}{A_2 \gamma (s_s - 1) d}$$

which is proportional to the Shields' entrainment function F_s. But F_s is dimensionless, and p_s has the dimensions of $(\text{time})^{-1}$; it follows that p_s must be equal to

$$p_s = \frac{1}{t} f(F_s)$$

where t is some time interval characteristic of the process of entrainment. Einstein argued that a suitable time interval would be the time taken for a grain to move a distance equal to its own diameter when moving at its fall velocity w; the above equation then becomes

$$\frac{p_s d}{w} = f(F_s)$$

whence from Eq. (10–43), incorporating the constants A_1, A_2, and λ in the functional sign, we obtain finally

$$\Phi = \frac{q_s}{wd} = f(F_s) = f\left(\frac{1}{\Psi}\right) \tag{10–44}$$

following the now well-established practice of writing $F_s = 1/\Psi$ in the present context. The term Φ is known as the bed-load function; in the form originally presented by Einstein, w was given in its complete form as

$$w = G\sqrt{gd(s_s - 1)} \tag{10–45}$$

where

$$G = \sqrt{\frac{2}{3} + \frac{36v^2}{gd^3(s_s - 1)}} - \sqrt{\frac{36v^2}{gd^3(s_s - 1)}} \tag{10–46}$$

and approximates closely to $\frac{2}{3}$ when $d \geq 1/16$ in. for grains of specific gravity 2.6 in water.

The empirical Φ-$1/\Psi$ relation is plotted in Fig. 10-15, in a form due to Brown [24], who also deduced the equation

$$\Phi = 40\left(\frac{1}{\Psi}\right)^3 \tag{10–47}$$

for the upper straight-line portion of the curve. At low values of Φ, and hence of q_s, the curve swings away from this straight line to the asymptote

$$\frac{1}{\Psi} = 0.056$$

which corresponds, as one would expect, to the threshold condition of the Shields' data. Empirical data relating to uniform sand and gravel sizes fit moderately well to the curve in Fig. 10-15, but as with the Shields' function the degree of scatter is approximately equivalent to a factor of 10. For non-uniform, or graded, sands a modified method has been worked out by Einstein [25] and will be described later.

By following a rather different line of reasoning, Kalinske [26] arrived at

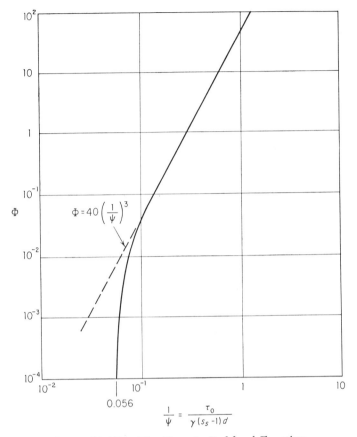

Figure 10-15.   *The Einstein Bed-load Function*

a relationship very similar to Einstein's. In this argument, r is defined as the proportion of the bed area covered by moving particles at any instant. The number of moving particles per unit bed area is then equal to $r/A_1 d^2$, and the volumetric rate of sediment discharge is equal to

$$q_s = \frac{A_2 d^3}{A_1 d^2} rv_s = \frac{A_2}{A_1} rv_s d$$

where v_s is the average particle velocity and A_1, A_2, are defined as in Einstein's argument. Full details of Kalinske's argument will not be given here; in the end result, it amounted to the contention that the particle velocity v_s would be proportional to the shear velocity v^*, and that the coverage factor r would be a function of the ratio τ_0/τ_c, which is directly proportional to the entrainment function $1/\Psi$. It follows that

$$\frac{q_s}{v^* d} \propto r = f\left(\frac{1}{\Psi}\right) \tag{10-48}$$

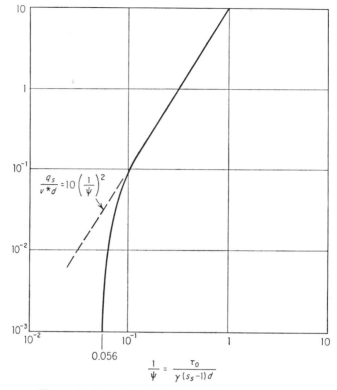

Figure 10-16.   *The Kalinske Bed-load Function*

and comparison of this equation with Eq. (10–44) suggests that the choice between v^* and w for the left-hand member is a rather arbitrary one.

The line plotted in Fig. 10-16 has been fitted to a number of empirical data, which show about the same degree of scatter as for the Einstein function. The upper straight line portion of the curve has the equation

$$\frac{q_s}{v^*d} = 10\left(\frac{1}{\Psi}\right)^2 \tag{10–49}$$

and as in Fig. 10-15 the lower part curves away to the asymptote $1/\Psi = 0.056$. Some of the data originally plotted by Kalinske related to nonuniform sands, the median diameter being used as the characteristic sand size. An alternative method proposed by Kalinske for nonuniform sand mixtures consists of calculating the transport rate q_s for each size fraction in turn and summing all the resulting values. The transport rate q_s for each fraction is assumed to be proportional to the coverage factor r_i for that fraction, and the total coverage r must first be distributed among the size fractions in proportion to the ratio p_i/d_i, where p_i is the proportion by weight of the fraction of size d_i.

A peculiarity of Kalinske's approach is that the critical shear stress τ_c is not taken from Shields' results, but is deduced by an argument similar to that of White [6]. Certain assumptions about the size of the constants involved lead to the result that the entrainment function is constant and equal to 0.11 approximately, i.e., twice as much as Shields' turbulent-flow value; it is plotted accordingly in Fig. 10-3, as a broken line.

The formula proposed by Laursen [27] makes explicit use of the ratio v^*/w. The presentation of the formula is backed by some interesting discussion of the mechanism of entrainment, including Laursen's concepts of particle "launching" already referred to, although not all of the postulates put forward are finally invoked in the development of the formula. The final result was:

$$\bar{c} = \Sigma p \left(\frac{d}{y_0}\right)^{7/6} \left(\frac{\tau_0'}{\tau_c} - 1\right) f\left(\frac{v^*}{w}\right) \tag{10–50}$$

in which $\bar{c}$ is the mean concentration of suspended load in percentage by weight, equal to $265q_s/q$ for a sediment of specific gravity 2.65; p is the proportion by weight of the fraction of size d which is present in the bed

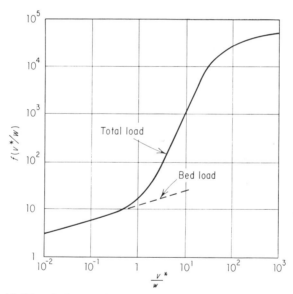

Figure 10-17. *Sediment-load Function, after E. M. Laursen* [27]

material; y_0 is the water depth; τ_c, v^*, and w are as normally defined; and τ_0' is the shear stress due to grain resistance alone, as in the argument preceding Eq. (10–33). From Strickler's equation (4–23) it follows that, for water,

$$\tau_0' = \frac{v^2 d^{1/3}}{30y_0^{1/3}} \tag{10–51}$$

where in this case d is defined as the mean diameter of the total sediment sample in feet, and all other quantities are in fps units (the equation is not, of course, dimensionally homogeneous). The function $f(v^*/w)$ was obtained from experimental measurements and is plotted in Fig. 10-17.

Yalin [28] has derived an expression for total sediment load by considering in detail the mechanism of "saltation" previously referred to, which is then assumed to account for all of the sediment motion. A typical path of a saltating grain is shown in Fig. 10-18; its initial movement is vertically upwards

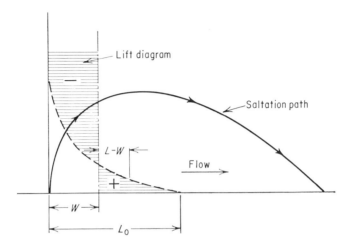

Figure 10-18.   *Lift and Saltation, after M. S. Yalin [28]*

from the bed, due to an uplift force L_0 which is assumed to be a function of the shear velocity v^*. The lift force L is then assumed to reduce exponentially to zero with distance from the bed, as indicated in the figure. The net lift force $L - W$, where W is the particle weight, will therefore be indicated by the shaded region in the figure, and from an assumed knowledge of L as a function of y, and of the distribution of fluid velocity which carries the particle forward, the particle path may be worked out. The equation finally obtained was

$$\frac{q_s(s_s - 1)}{v^* d s_s} = 0.635 r \left[1 - \frac{\log_e(1 + ar)}{ar} \right] \tag{10–52}$$

where $r = (\tau_0 - \tau_c)/\tau_c$

and $a = 2.45 s_s^{-0.4} \sqrt{\tau_c / \gamma s_s d}$

Only the constant 0.635 had to be determined empirically.

Yalin's argument is not of the statistical type, for the uplift force L is not assumed to vary with time. The increase of sediment discharge with increasing τ_0 is due therefore to the increased range of each saltation path, not to any increase in the number of grains in motion. In this respect the analysis may be unrealistic, but the result nevertheless fits some well-known experimental data with remarkable closeness. On a graph of Eq. (10–52) the data which have elsewhere been compared with Einstein's and Kalinske's formulas can be plotted, showing a scatter equivalent to a factor of only 1.5 approximately for coarse grain sizes, 1.72 mm $< d <$ 28.6 mm. This compares with a factor of about 10 for Einstein's and Kalinske's formulas. For $d = 0.7$ mm the scatter was equivalent to a factor of 2; it remains to be seen how the results will appear for smaller values of d. Since, as we have seen, saltation is appropriate to coarse particles of high penetration, it may be that the formula will be less successful when checked against data for finer sands.

All the methods described so far, except Laursen's and Kalinske's, have assumed that the grain size is uniform. Einstein [25] has also presented a method, based essentially on his bed-load formula [23], for calculating the sediment discharge of a nonuniform sand mixture. The method is lengthy and complex, and for full details the reader must be referred to the original paper [25]. In its essence the method consists of applying the empirical form of the bed-load Eq. (10–44), to each size fraction in turn, making some allowance for the way in which the finer grains may be shielded from the flow by the coarser grains. The D-35 grain size is used in the transport equation, and the D-65 size to determine the shear stress τ_0' due to grain resistance alone. It is this shear stress, and not the total due to grain plus bar resistance, which is assumed to be responsible for sediment movement, and is therefore to be inserted in the function Ψ of Eq. (10–44). The bar resistance is obtained from Fig. 10-11, so that when the transport rate is finally determined it may be related to the flow rate (determined from the total resistance) and not merely to the grain resistance.

A further modification of the method has been produced by Colby and Hembree [29]. It has the advantage that certain critical operations in the process are based on measured rather than calculated local values of velocity and sediment concentration.

The most sophisticated approach so far made to the general problem is that of Bagnold [30,31], who considered a number of aspects both of bed formation and of sediment movement. A prominent feature of his work is the consideration of the behavior of a thick suspension near the bed, in particular the development of a normal, or "dispersive" pressure between adjacent layers of the moving suspension, due to the impacts of particles in one layer on those of the other. The reaction arising from any such impact has a normal as well as a tangential component, and the summation of these normal components constitutes the normal pressure, which then provides the mechanism feeding material upwards from the bed.

The situation considered by Bagnold is that in which the whole bed is "live" and moving as a thick grain-fluid mixture. This is in contrast to the work of other investigators, whose postulates about the bed movement have generally implied a number of isolated grains moving over an essentially fixed bed.

Despite this difference, the two variables deduced by Bagnold were the Φ and Ψ introduced by Einstein [23]. Details of the method developed by Bagnold for calculating Φ from Ψ are beyond the scope of this text; ingenious though the method is, it is still dependent on empirically determined constants.

The live-bed condition investigated by Bagnold undoubtedly exists in natural channels even when the bed is of coarse gravel, although very coarse material will become live only in high floods. There is evidence that this general movement of coarse gravel may persist down to some feet below the surface, in that bridge piers in river beds of this kind have been moved bodily downstream by floods, even when the piers have still remained deeply embedded in the gravel of the bed.

Application and Verification of Bed-Load Formulas

Various methods have now been described for calculating the rate of sediment transport. These methods have grown up at different times and in different ways; accordingly the terms used are not always defined in the same way. This is particularly true of the shear stress terms τ_0 and τ_c, and indeed of the grain size d. These differences in definition arise mainly when dealing with sands or fine gravel below say $\frac{1}{4}$ in. in diameter, for in these cases the distinction between grain resistance and bar resistance is important. For the coarser grain sizes the distinction is of less importance.

In Table 10–1 the various definitions are listed. In all the formulas except the Du Boys and Kalinske's, τ_c is defined to be the limiting value as given by

TABLE 10–1 Definitions of Terms in Sediment Load Formulas

Formula	τ_0	τ_c (or RS)	d
Du Boys	Total resistance $=\gamma RS$	Fig. 10-5 (Broken line)	Mean
Shields	Total resistance $=\gamma RS$	Shields (Fig. 10-3 or Fig. 10-5, Full line)	Mean
Einstein-Brown (Fig. 10-15)	Total resistance $=\gamma RS$	Shields	D-35
Kalinske	Total resistance $=\gamma RS$	Fig. 10-3 (Broken line)	Fraction by fraction
Laursen	Grain resistance (Eq. (10-51))	Shields	Fraction by fraction
Yalin	Total resistance $=\gamma RS$	Shields	Mean
Einstein Ref. [25]	Grain resistance (Fig. 10-11)	Shields	D-65 for τ_0' D-35 in Eq. (10-44)

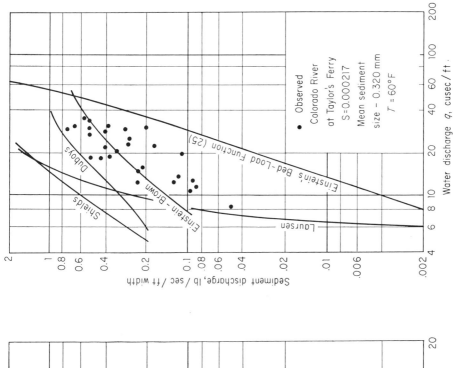

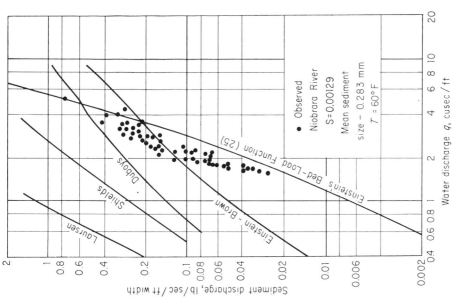

Figure 10-19.   *Sediment Rating Curves and Observed Values in Natural Streams, after V. A. Vanoni et al.* [32]

Shields' experiments, Fig. 10-3; for the Du Boys formula the coefficient C_s has been evaluated in terms of τ_c as given by the broken line in Fig. 10-5, and for Kalinske's formula τ_c is given by the broken line in Fig. 10-3. In Einstein's and Laursen's formulas, τ_0 is defined as being due to grain resistance alone; in the other cases τ_0 is the total shear stress. Yalin's formula has so far been compared only with experimental results for the coarser grain sizes, so the question of distinguishing between grain and bar resistance has not yet risen.

In the right-hand column, the mean size for d means the gravimetric mean obtained by summing the various size fractions by weight, thus

$$\text{Mean } d = \Sigma p d \tag{10–53}$$

where p is defined, as in Eq. (10–50), as the proportion by weight of the fraction of size d.

In Prob. 10.12 the reader can apply the formulas to some particular cases.

Because of practical difficulties, very few systematic field measurements have been made of rates of sediment transport. In a number of these cases Vanoni and others [32] have compared the field results with an exhaustive series of results calculated from many of the established sediment formulas. The results of two of these studies are shown in Fig. 10-19; it is seen that the amount of scatter in the field measurements is quite formidable, and that the differences between the various formulas are equally extreme. Colby [33], in a discussion of the problems in making sediment discharge computations, draws attention to the unsatisfactory nature of total shear as a measure of sediment discharge, and to the difficulty of isolating and determining the grain-resistance shear; he points out that in many sand-bed streams the sediment discharge correlates just as well with the mean velocity as it does with the shear. In view of all these uncertainties, it is clear that the theory is still very far from being able to make accurate predictions of the rate of sediment transport.

10.6   The Stable Channel

Introduction

It is clear that the bed-load formulas discussed in the preceding section will provide, at least approximately, a basis for the design of engineering works on rivers and canals with movable beds. Certain points of immediate practical importance emerge from a rearrangement and inspection of a formula such as Einstein's. First, the equation for bed-shear stress

$$\tau_0 = \gamma R S \tag{4–4}$$

already used to rearrange the Shields criterion, Eq. (10–13), can equally well be substituted into Brown's form of the Einstein relationship, Eq. (10–47), leading to the result

$$\frac{q_s}{Gd\sqrt{gd(s_s-1)}} = 40\left[\frac{RS}{d(s_s-1)}\right]^3 \tag{10–54}$$

where in this case τ_0 is the total shear stress and R accordingly is the total hydraulic mean radius. For sediment of a given size and specific gravity we can write

$$q_s \propto \frac{R^3 S^3}{d^{1\frac{1}{2}}} \tag{10–55}$$

and if it is further assumed that the channel is wide and the Chézy C is constant the water discharge per unit width, q, is given by

$$q \propto R^{1\frac{1}{2}} S^{1/2} \tag{10–56}$$

Consider now the question of how q_s and q will be related in a specified channel having specified bed material. Algebraically, this implies the elimination of R between Eqs. (10–55) and (10–56), so that only q, q_s, and parameters characteristic of the channel alone are retained. The result is

$$q_s \propto \frac{q^2 S^2}{d^{1\frac{1}{2}}} \tag{10–57a}$$

or

$$\frac{q_s}{q} \propto \frac{q S^2}{d^{1\frac{1}{2}}} \tag{10–57b}$$

The form of Eq. (10–57) displays an important feature of the Einstein formula (and other similar formulas), namely, that the sediment concentration increases with the water discharge, or alternatively that q_s varies as q^2. In other formulas the power of q is not always equal to 2, but is always substantially greater than unity.

Although Eq. (10–57) is based on some rather severe approximations, including the implicit assumption that the bed formation does not change as q_s and q vary, records of q_s and q in natural rivers show a moderately close approximation to the $q_s \propto q^2$ relationship. We now consider a few simple situations in which Eq. (10–57) gives some specifically useful information.

1. Where the width of a river increases, q will decrease and so will q_s/q. Sediment will therefore be deposited until the bed slope is increased to a value sufficient to maintain q_s/q constant over the whole river length, as it must be if the channel is to remain stable. Local increases of bed slope at local enlargements of width are in fact found to be a feature of natural rivers.

2. Equation (10–57) also indicates how, if one distributary of a river begins to abstract flow from another one, the process may accelerate rapidly until the waters of the second distributary are completely diverted into the first. An important example of this process of capture is provided by the lower reaches of the Mississippi River; at about 300 miles from the mouth there is a branch leading into the Old River-Atchafalaya River channel, which offers a distinctly shorter path to the sea.

For many decades this secondary channel has been capturing flow from the main channel, and the process has, to a certain limited extent, been encouraged by engineers as a means of diverting flood waters.

The prime cause of the phenomenon is the existence of favorable gradients in the branch channel, but the mechanism which keeps the process going is that indicated by Eq. (10–57). Once the flow in the main channel diminishes by a material amount, the value of q_s/q also diminishes and deposition begins to take place. The capacity of the main channel is thereby reduced, leading to more deposition, with more and more flow being taken by the branch channel; by 1956 it was taking 23 percent of the total flow. Experience indicates that when this proportion reaches 40 percent the capture will proceed so rapidly as to be irreversible; the end result would be the isolation of the port of New Orleans, with other consequences equally disastrous to the economy of the region. However, the process and its dangers have been recognized for decades and preventive operations have been set in train, mainly in the form of control structures at the junction between the two channels [34].

3. Another important deduction from Eq. (10–57) concerns the distribution of a river's annual sediment load over the year. Since $q_s \propto q^2$, a flood with a flow ten times the average can carry one hundred times the sediment load of the average flow. It follows that flood flows must be responsible for a high proportion of a river's annual sediment load; in fact in rivers in coarse alluvium it is likely that the whole of the annual load is brought down by no more than a dozen floods and freshes, with an aggregate duration of only a few weeks.

Now if the flood peaks are smoothed out by, for example, the building of a dam, the flow coming through the dam will then have an aggregate transporting capacity substantially less than it had before. The flow coming through a dam is normally clear water, so no transporting power appears to be needed; however, if there is a tributary entering the river at some point downstream of the dam, this tributary will continue to deliver sediment to a main stream which is now less capable of transporting it. Hence there will be deposition downstream of the confluence until the slope has built up sufficiently to enable the river to handle the sediment load.

The Channel Stability Problem

The bed-load formulas introduced in the preceding section are set up to deal with this problem: given a suitable description of the transported sediment and of the flow conditions, calculate the rate of sediment discharge. However, the above examples make it clear that this problem is only one element of a wider question facing the engineer. This is the general problem of channel stability, which may be expressed in this form: how will a river or canal adjust its channel so as to accommodate itself to the water flow *and* sediment flow which are fed into it? The wider scope of this general problem is well exemplified by the part played by the longitudinal bed slope. In the bed-load formulas slope is treated as an independent variable, whereas we have seen in the examples quoted that a river may adjust its slope in response to changes in the externally imposed conditions; in this sense slope is a dependent variable.

It is also clear from the first example discussed that width and slope are mutually dependent; in fact if the discharge Q and the concentration q_s/q are held constant, then it follows from Eq. (10–57) that the ratio S^2/B must be constant, where B is the average width of the channel. A river may therefore adjust its slope and width simultaneously, but in order to follow up this question it would be necessary to consider the mechanism by which the width might be changed; this brings up the question of bank stability. Clearly the banks of a river are more sensitive to erosion than the bed, and this factor imposes a lower limit on the width-to-depth ratio at which a stable channel can be maintained.

In order to visualize the interdependence of all the variables involved, it is convenient to adopt the viewpoint of the engineer engaged in, say, irrigation work, who has to design a canal to take a certain discharge carrying a specified concentration of a certain sediment. The line of his argument has been clearly set out by Laursen [35]. The governing relationships must be a resistance equation, a sediment transport equation, and the limitation imposed by the bank stability, or as Laursen terms it, "bank competence." There are altogether six pertinent variables:

$$Q, d, q_s/q, S, P, \text{ and } R$$

where d as before is the sediment size and P and R have their usual definitions. They have been chosen as being characteristic of the lateral and vertical channel dimensions, respectively.

The first three variables are known, and form the original specification of the problem. The remaining three are unknowns and would be determined absolutely if the three governing relationships were absolutely determinate. However, this is not so; the bank competence criterion does not completely determine the width-to-depth ratio, but merely sets a lower limit to it. If the ratio is given this limiting value, then S, P and R are completely determined and the slope is a dependent variable, but the ratio may be given larger values provided that the slope is increased correspondingly. It follows that the slope associated with the limiting width-to-depth ratio is a minimum, and that the designer may accept any larger value of the slope which is imposed by the terrain, provided he adjusts the width-to-depth ratio accordingly. In this case the slope becomes, in effect, an independent variable.

It is questionable to what extent a natural river would reproduce in its behavior the conclusions of our hypothetical canal designer. If the width of a channel is too narrow for equilibrium, it is easy to visualize the width being increased by erosion of the banks; it is less easy to visualize how a river would reduce the width of a channel that was too wide. Some further remarks are made on this matter in the next section; for the present the discussion will be confined to the design of artificial canals.

The simplest case to begin with, indeed the only one capable of explicit solution, is the special case where $q_s = 0$, i.e., the water is clear. The limiting design condition will therefore be the threshold of motion, and the problem

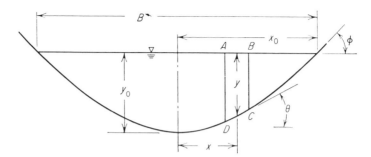

Figure 10-20.   *Definition Sketch for Stable-Channel Analysis, after*
E. W. Lane et al. [10]

will be particularly relevant to channels in coarse alluvium, since for fine
sands there is bound to be some sediment motion unless the cross section is
made extremely large. The limiting cross-sectional shape is that for which the
bed material is just on the point of motion at every part of the boundary;
such a section has been developed at the U. S. Bureau of Reclamation using
methods suggested by E. W. Lane [10,36,37,38].

The method depends on the use of Eq. (10–15) to determine a condition of
limiting equilibrium at each point on the cross-sectional profile, shown in
Fig. 10-20. It is assumed that the bed shear τ_0 varies directly as the local depth
y, i.e., that the shear force on the surface element DC is due only to the weight
of the prism whose section is $ABCD$ resolved down the longitudinal slope
of the channel. This is an approximation, but it at least conforms to the
requirement that the shear force over the whole bed is due to the whole
cross section of water resolved down to the slope. Moreover, it was found in
the U.S.B.R. studies that more precise assumptions yielded very similar
results.

The shear stress τ_0 is therefore given by the equation

$$\frac{\tau_0}{\tau_c} = \frac{y}{y_0} \cos \theta \qquad (10\text{–}58)$$

where τ_c is the maximum shear stress at the center of the bed, equal to Shields'
critical value if the whole bed is on the point of motion. Combination of Eq.
(10–58) with Eq. (10–15) then leads to a y-θ relation which is readily integrated
(Prob. 10.13) to give the result

$$\frac{y}{y_0} = \cos \frac{x \tan \phi}{y_0} \qquad (10\text{–}59)$$

so that the required profile is a sine curve. A further integration shows that the
area, A, of the cross section is equal to

$$A = \frac{2y_0^2}{\tan \phi} \qquad (10\text{–}60)$$

The $y - \theta$ relationship which leads to Eq. (10–59) does not uniquely define a complete channel cross section; the equation would be equally well satisfied by inserting a section of constant depth between the two curved banks (type A in Fig. 10-21). The complete section of Fig. 10-20 is described as type B, and is in effect the narrowest possible section consistent with bank stability.

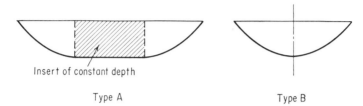

Insert of constant depth

Type A   Type B

Figure 10-21.   *Alternative Solutions in Lane's Stable-Channel Theory*

The theory can be extended so as to relate the type B cross section parameters to the discharge and slope [39]. It can be shown (Prob. 10.13) that the wetted perimeter P of the type B section is equal to

$$P = \frac{2y_0 E}{\sin \phi} \tag{10–61}$$

where E is the complete elliptic integral of the second kind, as defined in Prob. 10.13. In this context it is a function of ϕ alone. It follows that the hydraulic mean radius R is equal to

$$R = \frac{A}{P} = \frac{y_0 \cos \phi}{E} \tag{10–62}$$

For a typical value of ϕ, equal to $35°$, the following key values are readily computed from Eqs. (10–59) and (10–61) through (10–62):

$$\left.\begin{array}{ll} \text{Surface width } B & = 4.49 y_0 \\ \text{Area } A & = 2.86 y_0{}^2 \\ \text{Perimeter } P & = 4.99 y_0 \\ \text{Hydraulic Radius } R & = 0.572 y_0 \end{array}\right\} \tag{10–63}$$

We now consider the matters of flow resistance, discharge, and slope. First, Eq. (10–14) is introduced and modified by recognizing that the maximum shear stress is equal to $\gamma y_0 S$, not $\gamma R S$. Equation (10–14) should therefore be rewritten

$$d = 11 y_0 S \tag{10–64a}$$

or, from Eq. (10–63),

$$d = \frac{11 R S}{0.572} = 19 R S \tag{10–64b}$$

These equations can now be combined with Strickler's formula for the Manning n

$$n = 0.031d^{1/6} \qquad (4\text{-}22)$$

whence
$$v = \frac{1.49R^{2/3}S^{1/2}}{0.031d^{1/6}}$$

It is now possible to substitute from Eq. (10–64) into the above equation, yielding, for wide channels ($d = 11RS$)

$$v = 32R^{1/2}S^{1/3} \qquad \textbf{(10–65a)}$$

and for the type B section ($d = 19RS$)

$$v = 30R^{1/2}S^{1/3} \qquad \textbf{(10–65b)}$$

The above argument has incorporated all three determining factors—resistance, sediment transport, and bank competence. The type B section is the limiting condition dictated by the last criterion; for this section we would therefore expect to find explicit solutions for S, P, and R. From Eqs. (10–63), (10–64b), and (10–65b) we can write

$$Q = 30PR^{1\frac{1}{2}}S^{1/3}; \qquad R = \frac{d}{19S}; \qquad P = 8.75R$$

The elimination of P and R from these equations yields

$$S = 0.44d^{1.15}Q^{-0.46} \qquad \textbf{(10–66)}$$

which gives the limiting (minimum) value of slope at which the type B channel section would just be stable. At lower values of slope it would certainly be stable, and at higher values of slope the wider type A channel of less scouring capacity is required. The size of the type B channel section is readily obtained from Eqs. (10–63), (10–64a), and (10–66). The surface width B becomes

$$B = 4.49y_0 = \frac{4.49d}{11S} = 0.93d^{-0.15}Q^{0.46} \qquad \textbf{(10–67)}$$

Similarly for the wetted perimeter P:

$$P = 1.03d^{-0.15}Q^{0.46} \qquad \textbf{(10–68)}$$

From these equations the designer could, given Q and d, determine the limiting value of slope and the dimensions of the section. In practice, he would not use the results in this form, if only because of the difficulty of excavating the canal section to the exact profile of Fig. 10-20. A trapezoidal channel near the threshold condition might eventually have its banks shaped by the flow to approximately the type B section, but it is not necessary that the designer should be able to predict the resulting shape of the banks.

The main value of the argument leading to Eqs. (10–66) through (10–68) has been in furnishing an example which clarifies the previous statements about the interdependence of the variables in the stable-channel problem, and in particular which shows the part played in the theory by the limiting value of slope. This is of some indirect value to the canal designer; it is also helpful in considering the logical basis of similar empirical equations which, as we shall see, have been advanced to describe the flow in canals having a live bed of fine sediment. Consider, for example, the two alternative forms of Eq. (10–65); there is very little difference in their coefficients although one relates to a very wide channel and the other to the narrow type B channel. However, when we consider equations for B and P the difference between the wide and the narrow channel becomes much more pronounced. From Eq. (10–65a) it follows that

$$Q = 32PR^{1\frac{1}{2}}S^{1/3}$$

Eliminating R between this equation and Eq. (10–14) we obtain

$$P = B = \frac{1.14QS^{7/6}}{d^{3/2}} \qquad (10\text{–}69)$$

which is true for a very wide channel (indeed it could have been used in Example 10.1), and is markedly different from Eqs. (10–67) and (10–68).

Now we have seen that the change from the narrow type B channel to a very wide channel must also be accompanied by a change in slope; the above argument therefore has shown that while the form of Eq. (10–65) changes little with wide variations in slope, Eqs. (10–67) and (10–68) are true only for a particular limiting value of slope and may be described as "slope-dependent." For wide channels of varying slope they must be replaced by Eq. (10–69).

The preceding argument could, strictly speaking, apply to any sediment size, but in practice a theory that postulates a bed on the threshold of motion is most likely to be applicable to coarse gravel, greater than $\frac{1}{4}$ in. in diameter. Indeed for smaller sizes the coefficients of Eqs. (10–14) and (4–22) would have to be changed, for Strickler's equation is based on data taken from coarse-gravel riverbeds. A further point to note is that, as in Chap. 4, the value of d should be taken as the D-75 size, with which the bed tends to become armored.

The "Live-Bed" Channel and the Regime Theory

In the canal design problem described above the objective has been the comparatively simple one of preventing bed movement; the designer can always err on the side of safety by making the channel wider and so reducing the tractive force. But when the canal is excavated in fine bed material, and/or a sediment load is admitted at the canal headworks, the designer's problem becomes the highly critical one of choosing a canal design which will pass the required sediment load without undue deposition or scour; he has little margin for error on either side.

It has already been pointed out that existing bed-load formulas have a number of uncertainties even when the effect of side slopes is neglected; introducing the question of side slopes or bank competence makes the problem one degree more difficult, and indeed puts it beyond the scope of present knowledge. In the circumstances the most immediately profitable approach would seem to be the empirical one of recording the leading parameters of canals which are known to have operated successfully. This empirical approach has been vigorously developed in India, particularly in connection with irrigation systems, and has given rise to what is termed the "regime" theory. A river or canal is said to be "in regime" when it has adjusted its slope and section to an equilibrium condition; this implies, among other things, that its bed, although in movement, is stable because the rate of transport equals the rate of sediment supply. This "live-bed" situation would normally occur when the bed is of fine sand, the "fixed-bed" situation just described normally occurring in coarse alluvium.

The regime theory is not a theory in the strict sense of the term, for it does not incorporate physical explanations for its findings. The essence of the system lies in the development of convenient and simple empirical equations from field data collected from rivers and from successfully operating artificial canals. From the end of the nineteenth century it was observed by many engineers, notably Kennedy and Lindley, that there was a correlation between velocity and maximum depth y in regime channels, and various relationships of the form

$$v = ky^n$$

were proposed. Kennedy's formula, in which $k = 0.84$ and $n = 0.64$, was put forward as early as 1895. Unfortunately the value of n appeared to vary from one canal to another between the limits 0.52 and 0.73; this difficulty was removed by Lacey, who in a series of papers [40,41,42,43] presented a complete system of equations. The first of these was based on the observation that if v was correlated with R instead of y the exponent of R was close to 0.5 for all results noted up to that time. The coefficient of $R^{1/2}$ varied from one canal system to another, appearing to depend on the type of silt in the canal. Accordingly his first equation was

$$v = 1.17\sqrt{fR} \qquad \textbf{(10–70)}$$

where f is a "silt factor," at first tentatively related to the silt size d by the equation

$$f = 8\sqrt{d} \qquad \textbf{(10–71)}$$

where d is in inches; since the first presentation of the theory more detailed attention has been given to the definition of f. Two more equations were presented, both independent of f:

$$v = 16R^{2/3}S^{1/3} \qquad \textbf{(10–72)}$$

and
$$P = 2.67\sqrt{Q} \qquad \textbf{(10–73)}$$

The first of these corresponds to the fixed-bed equation (10–65), being based on the elimination of a sediment term (the silt factor in this case) between a sediment transport equation and a resistance equation. The second corresponds to Eq. (10–68), and the great similarity in form between these two equations suggests that Eq. (10–73) is, like Eq. (10–68), applicable only to a limiting channel section at a limiting (minimum) value of slope.

After making the substitution $v = Q/PR$ it is possible to eliminate P and R from Eqs. (10–70) through (10–73), with the result

$$S = \frac{f^{5/3}}{1750Q^{1/6}} \tag{10–74}$$

which corresponds to Eq. (10–66) but is markedly different in form. This equation is, like the others, based on field observations of canals in good condition, and its inference is clear: that a given discharge Q with a given silt factor will adjust not only the section parameters P and R, but also the slope itself. The three Eqs. (10–70), (10–72), and (10–73) are therefore dependent as a group on there being a certain value of slope. The correspondence noted between the Lacey equations and the fixed-bed equations suggest that this slope dependence is concentrated in Eq. (10–73). This conclusion is borne out by observations on natural rivers, in which Eqs. (10–70) and (10–72) are more commonly found to be true than Eq. (10–73); the reason may be that the rivers have not yet adjusted their slope to the value given by Eq. (10–74).

Equations (10–70) through (10–74) are dimensionally inhomogeneous. The constants given are correct if d is measured in inches and all other quantities in foot-second units. Equations (10–70) and (10–72) are based, not on measured values of v, but on values inferred from a resistance equation. The directly measured quantities in these equations are R and S, and the observations indicated that

$$R^{1/2}S = \text{constant} \tag{10–75}$$

as one proceeds downstream through any canal system, although the discharge and channel dimensions vary widely as the water is distributed. In fact the Einstein-Brown bed-load formula verifies this result; it follows from Eqs. (10–55) and (10–56) that if d and q_s/q are kept constant (as they would be through a single canal system) then $R^{3/2}S^{5/2}$ will be a constant; if Eq. (10–56) is replaced by the Manning equation, then $R^{4/3}S^{5/2}$ will be constant. Each result indicates that $R^{1/2}S$ will be approximately constant, and the same result follows from other forms of resistance equation [39]. For a wide fixed-bed channel we have seen that the shear stress, and therefore RS, must remain constant; the contrast of this condition with Eq. (10–75) draws a clear and convenient distinction between fixed-bed and live-bed systems.

Since the first appearance of Lacey's equations, much more development work has been done in this field, particularly by Blench [44]. He presents three equations based on two silt factors—a bed factor f_b and a side factor f_s. They are defined in this way:

$$f_b = \frac{v^2}{y}; \quad f_s = \frac{v^3}{b}$$

where y is the mean depth and b is the mean width. From these definitions it follows that

$$b = \sqrt{\frac{f_b Q}{f_s}} \tag{10–76}$$

$$y = \sqrt[3]{\frac{f_s Q}{f_b{}^2}} \tag{10–77}$$

A further equation gives the slope:

$$S = \frac{f_b{}^{5/6} f_s{}^{1/12} y^{1/4}}{3.63 g Q^{1/6}(1 + c/2330)} \tag{10–78}$$

where c is the sediment-load concentration in parts per million by weight (ppm). Thus Blench's system, like Lacey's, contains three independent equations. The recommended values of side factors f_s are 0.1, 0.2, and 0.3 for bank material of very slight, medium, and high cohesiveness respectively. The recommended value of bed factor is

$$f_b = 9.6\sqrt{d}\,(1 + 0.012c) \tag{10–79}$$

for subcritical flow, where d, as in Eq. (10–71), is the median bed size in inches and c, as in Eq. (10–78), is the concentration in ppm. This equation is not greatly different from Eq. (10–71), due to Lacey.

The importance of bank competence in the matter of channel stability has already been emphasized; a further development of this question is the relative strength of bed and banks, when they happen to be of differing materials. The Indian canals which supplied much of the data for the Lacey and Blench equations had sand beds and slightly cohesive to cohesive banks, which suggests a correspondingly limited degree of applicability, at least for the Lacey equations; Blench's equations do allow for varying degrees of bank cohesion, although the form of Eqs. (10–76) and (10–77) shows that b and y are very sensitive to errors in bed factors which must be estimated rather arbitrarily. These points suggest that the system should be extended to incorporate field data covering a wider variety of bed and bank conditions. A further limitation of the Indian data is that they come from canals whose sediment charge is usually controlled by sediment exclusion and/or ejection structures so that it is, generally, less than 500 ppm. This, too, suggests the need for a more comprehensive range of field data.

Simons and Albertson [45] have pointed out these deficiencies and have to a large extent remedied them. They made a collection of field data from Indian and North American sources, and in this collection distinguished five types of canal:

TABLE 10–2 Types of Canal Bed and Banks (after Simons and Albertson)

1. Sand bed and banks
2. Sand bed and cohesive banks
3. Cohesive bed and banks
4. Coarse noncohesive material
5. Same as for 2, but with heavy sediment loads, 2,000–8,000 ppm

To each of these types a separate equation could be fitted. The equations are given with a slight rounding off of the exponents and coefficients in the original forms:

$$P = K_1 Q^{1/2} \tag{10–80}$$

$$b = 0.9P \tag{10–81}$$

$$b = 0.92B - 2.0 \tag{10–82}$$

$$R = K_2 Q^{0.36} \tag{10–83}$$

$$y = 1.21R \text{ for } R \le 7 \text{ ft} \tag{10–84a}$$

$$y = 2 + 0.93R \text{ for } R \ge 7 \text{ ft} \tag{10–84b}$$

$$v = K_3 (R^2 S)^m \tag{10–85}$$

$$\frac{C^2}{g} = \frac{v^2}{gyS} = K_4 \left(\frac{vb}{v}\right)^{0.37} \tag{10–86}$$

where P, R, S, and Q have their usual meanings, C is the Chézy coefficient, b and B are mean and surface widths respectively and y is the depth from bed to water surface. It follows that the area $A = by = PR$. The coefficients K_i and the exponent m are given in the following table:

TABLE 10–3 Coefficients and Exponents for Eqs. (10–80), (10–83), (10–85) and (10–86)

Coefficient	Channel type (Table 10-2)				
	1	2	3	4	5
K_1	3.5	2.6	2.2	1.75	1.7
K_2	0.52	0.44	0.37	0.23	0.34
K_3	13.9	16.0	—	17.9	16.0
K_4	0.33	0.54	0.87	—	—
m	0.33	0.33	—	0.29	0.29

The order in which the above equations are given follows Vanoni's explanation of the method [32]. P, R, b, B, and y are obtained from these equations and a channel section is chosen (usually trapezoidal) so as to fit these values. For instance, the relative size of b and B determines the channel side slopes, subject to the limitation that the slope must be less than the angle of repose. The final step is the determination of the slope from either Eq. (10–85) or Eq. (10–86), the second of which is a modified form of Eq. (10–78), due to Blench; some judgment is required finally in choosing between these alternative values of slope.

Type 4 channels in coarse noncohesive material need not be treated by means of Eqs. (10–80) through (10–86), because the method already described in Sec. 10.3 is adequate. It can also be shown that Eq. (10–65), which incorporates a more general statement of the method of Sec. 10.3, fits the type 4 field data as closely as does Eq. (10–85).

The methods of this section are best illustrated by an example. The data are taken from Simons and Albertson [45].

Example 10.3

The channel parameters in the following table are taken from field observations made on these canals:

> No. 1—Fort Laramie I, west of Torrington, Wyo.
> No. 2—Fort Laramie II, south of Lyman, Nebr.
> No. 3—Fort Morgan I, west of Fort Morgan, Colo.

All parameters are in foot-second units except for the median bed-material size, which is in inches. Assume an average water temperature of 74°F ($v = 1.0 \times 10^{-5}$ ft²/sec). All bank material is cohesive ("highly cohesive" for the purposes of Blench's method), and the bed is sandy in all cases.

No.	Q	A	$S \times 10^3$	R	P	y	B	b	Median bed size, in.	Bed concentration c, ppm
1	1031	603	0.058	6.70	90.0	8.29	80.0	72.8	0.0100	87
2	445	232	0.063	4.66	49.6	6.01	44.0	38.5	0.0037	266
3	146	108	0.135	2.83	38.0	3.51	34.0	30.6	0.0125	227

The problem is to determine, for canal No. 3, the channel cross section and slope, by each of the three methods described—the Lacey, Blench, and Simons-Albertson systems—and to compare the calculated with the actual values. Only the discharge and the sediment size and concentration are to be assumed known in the first instance. It is left as an exercise for the reader to make the same calculation for canals 1 and 2 (Prob. 10.14). First, the Lacey method:

From Eq. (10–73) $P = 2.67\sqrt{146} = 32.3$ ft

From Eq. (10–71) $f = 8\sqrt{0.0125} = 0.895$

From Eq. (10–74) $$S = \frac{(0.895)^{5/3}}{1750(146)^{1/6}} = 0.000208$$

From Eqs. (10–70) and (10–73) a further equation can be obtained, relating R to f and Q. From these two equations

$$Q = 1.17PR\sqrt{fR} = 1.17 \times 2.67R\sqrt{fRQ}$$

whence $$R = 0.47f^{-1/3}Q^{1/3}$$

$$= 0.47\sqrt[3]{146/0.895} = 2.57 \text{ ft}$$

Second, the Blench method:

The side factor $f_s = 0.3$

From Eq. (10–79) $f_b = 9.6\sqrt{0.0125}\,(1 + 0.012 \times 227)$

$$= 4.0$$

From Eq. (10–76) $b = \sqrt{4.0 \times 146/0.3} = 44 \text{ ft}$

From Eq. (10–77) $y = \sqrt[3]{0.3 \times 146/16} = 1.40 \text{ ft}$

From Eq. (10–78) $$S = \frac{4^{5/6}(0.3)^{1/12} \times 10^{-5/4}}{3.63g \times 146^{1/6} \times 1.098}$$

$$= 0.00055$$

Third, the Simons-Albertson method:

The channel is of type 2, with sandy bed and cohesive banks.

From Eq. (10–80) and Table 10–3 $P = 2.6\sqrt{146} = 31.4 \text{ ft}$

From Eqs. (10–81) and (10–82) $b = 28.3 \text{ ft}$ and $B = 33.0 \text{ ft}$

From Eq. (10–83) and Table 10–3 $R = 0.44(146)^{0.36} = 2.65 \text{ ft}$

From Eq. (10–84a) $y = 1.21 \times 2.65 = 3.21 \text{ ft}$

   From these results $A = 83$ sq ft or 91 sq ft;
       take a mean of 87 sq ft

whence $v = 146/87 = 1.68 \text{ ft/sec}$

   From Eq. (10–85) $v = 16(R^2S)^{0.33}$

whence $S = 0.000165$

Alternatively, from Eq. (10–86),

$$\frac{1.68^2}{3.21gS} = 0.54\left(\frac{1.68 \times 28.3}{10^{-5}}\right)^{0.37}$$

whence $S = 0.000171$

The mean of these two slope values, which are remarkably close together, gives $S = 0.000168$.

Results are summarized in the following table:

	Q	A	v	$S \times 10^3$	R	P	y	B	b
Lacey	146	83.0	1.76	0.208	2.57	32.3	—	—	—
Blench	146	61.6	2.37	0.55	—	—	1.40	—	44.0
Simons-Albertson	146	87.0	1.68	0.168	2.65	31.4	3.21	33.0	28.3
Actual	146	108	1.35	0.135	2.83	38.0	3.51	34.0	30.6

The Simons-Albertson method gives values closest to those of the actual canal. This is not surprising, because this canal supplied some of the data from which Simons and Albertson derived their equations. Lacey's equations, too, give good results but Blench's equations overestimate the width and the slope and underestimate the depth, at least in this particular case. It is arguable whether this result implies a serious flaw in Blench's approach; it has been shown that the fixed-bed channel may have any slope above a certain value, provided it is wide enough, and there seems no reason why this should not be true of live-bed channels as well. Accordingly, Blench's results could possibly represent an alternative solution, but the designer would probably be safer with the Simons-Albertson results, which have been shown to fit a comprehensive range of field data.

Because of the approximate nature of the Simons-Albertson equations, there are certain minor geometrical incompatibilities in the results given by these equations. One of them has already been seen in the working: that $PR = 83$ and $by = 91$ although these two should be equal. Also, the equations give $P = 31.4$ and $B = 33.0$, although P should be greater than B. None of these anomalies are serious, for the errors introduced are no larger than 10 percent, which is quite a modest margin of error in this type of work. It is a matter of elementary geometry to show that within this margin of error the Simons-Albertson results indicate a trapezoidal channel of base width 20 ft and side slopes 2H : 1V. The last figure is chosen so as to be well below the angle of repose; values of this angle are given by Simons and Albertson [45] and are shown in Fig. 10-22. This figure acts as a supplement to the values previously given in Fig. 10-8 for coarse gravel only.

The preceding bare description of regime methods does something less than justice to the discussion, at times heated, which has centered on these methods since the publication of Lacey's first paper in 1930. For a critical review of the controversy, the reader is referred to Leliavsky [52]. The same author gives a comprehensive bibliography of more recent developments in the method. The main criticisms of the method are:

1. That it applies to a limited range of conditions occurring in India, and is inapplicable elsewhere.

2. That the silt factor f is ill defined in that it must be determined by other factors besides the size d, notably the sediment concentration.

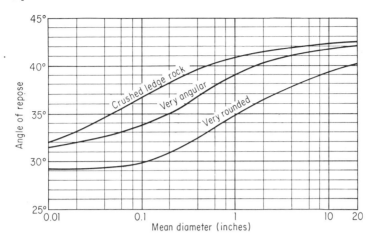

Figure 10-22. *Angles of Repose of Noncohesive Material, after D. B. Simons and M. L. Albertson* [45]

3. It is not clear to what extent the channel slopes in the Indian data were self-adjusted; on this factor depends the ability of Eq. (10–74) to determine the slope correctly as a dependent variable.

Of these criticisms, the first is met by Simons and Albertson's extension of the method, and the second by subsequent work, much of which is quoted by Leliavsky [52]. In general this work is aimed at determining the influence of the sediment concentration on the silt factor, and it is typified by Eq. (10–79), due to Blench. A further development was the definition of two distinct silt factors, at least one of which was shown by Ning Chien [54], using Einstein's bed-load function, to be almost independent of the sediment concentration.

The third criticism is less easily answered, and indeed the slope question introduces many uncertainties. Even if equations like Eqs. (10–74) and (10–78) give correct values of a minimum slope to which the channel will automatically adjust itself, it may still be possible to use greater slopes with a wider channel. A further point is that if, as seems likely, Eq. (10–73) is strongly dependent on Eq. (10–74), the engineer should beware of using Eq. (10–73) in natural rivers, where the slope may not have reached its stable value. This is particularly true in coarse alluvium.

10.7 The Natural River

Introduction

The engineer's interest in the stable channel problem is not confined to the design and operation of artificial canals. He must also deal with natural rivers, which are more irregular and less tractable in their behavior. Rivers present a variety of problems for the attention of the engineer, some of which have been outlined in the Introduction to this chapter, but the basic underlying problem is that of stable channel formation. The engineer must give the river what assistance he can to develop a stable well-defined channel, despite the risks of scour and deposition which are inherent in natural river processes. If a reasonably stable channel cannot be formed, schemes for flood control, navigation improvement, and the like will have little prospect of success.

Traditionally the processes of natural river formation are the interest of the geomorphologist, whose methods have in the past owed little to quantitative concepts such as are the concern of the engineer. However, in more recent years geomorphologists have developed a more quantitative interest in river processes, and engineers have given more attention to the methods and theories of the geomorphologist. Prominent in the development of this fruitful joint interest in the problem has been the important work of the U. S. Geological Survey, some of which is quoted later in this section.

The Dominant Discharge

Perhaps the most important difference between river and canal work is that river discharge is much more variable. The engineer hoping to apply stable-channel theories to river control work must first decide, therefore, what constant discharge is the equivalent, for channel-forming purposes, of the variable river flow. To this figure, whatever it may be, the name *dominant discharge* is given.

In order to develop a satisfactory definition of the dominant discharge, it is necessary to consider the process of channel formation itself. Suppose that the river has formed a stable single channel—a *single-thread* channel, as it is called. If this channel is only a small one, it will be frequently overtopped by floods and the flood plains, or berms, on either side of the channel will frequently have to carry this flood flow at quite low depths and velocities. Over the berms the flow will therefore have much lower transporting power than the deeper faster flow in the channel, and silt will be deposited on the berms, gradually raising their surface level. It is by this process that flood plains are built up; a complete discussion of the process is given by Wolman and Leopold [48].

The above description neglects bank erosion, which could also be an important factor. The conclusion is that either by bank erosion or berm build-up, the channel will be enlarged until a stable condition is reached, in which the discharge just filling the channel has these properties:

1. It can maintain the channel at its present cross section without scour or deposition;

2. It is not exceeded frequently enough for berm build-up to be appreciable.

This discharge can therefore be conveniently adopted as the average, or dominant discharge, and it is clear from this discussion that the notion of frequency will play an important part in defining the dominant discharge;

the discussion also suggests an empirical method of constructing a numerical definition. The method is to study natural rivers which flow in stable single channels with stable berms or flood-plains, and to determine the frequency of the discharge which just fills the channel. This has been done in only two cases: by M. Nixon [49] in England and by Wolman and Leopold [48] in the United States. Nixon found that the "bank-full" discharge was such as to be equaled or exceeded on 0.6 percent of the days of record, i.e., on one day in 170; Wolman and Leopold found that it had a return period (computed from annual floods) of 1.4 years. Both these figures are averages for a number of rivers, there being a substantial amount of scatter about the average value.

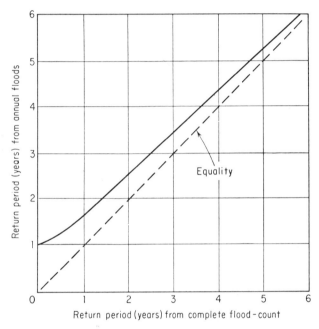

Figure 10-23.   *Relation between Alternative Definitions of Flood Return Period*

Nixon's figure can be said to be equivalent to a return period of six months approximately, since the appropriate discharge is equaled or exceeded once every 170 days, or six months. This figure is based on a count of all floods, whereas Wolman and Leopold's is based on a count of annual maximum floods only, so the two figures cannot be compared directly. However, statistics collected by Langbein and by Ven Te Chow [50], and by the author, indicate a relationship between the two definitions of return period, which is plotted in Fig. 10-23. According to this curve, Wolman and Leopold's "annual flood" return period of 1.4 years is equivalent to a "flood count" return period of about nine months. The difference between this result and

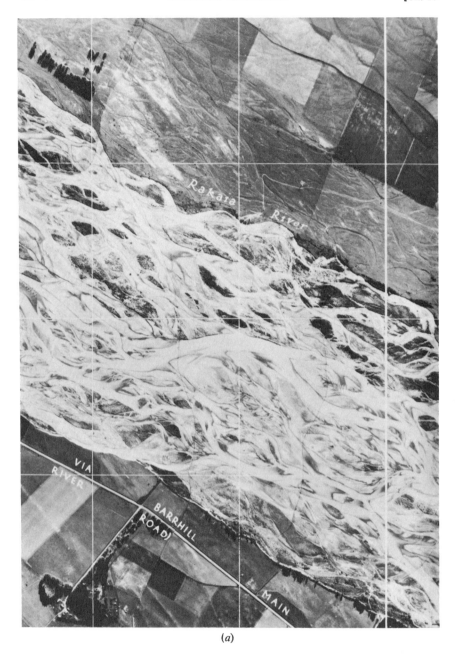

(a)

Figure 10-24. *Typical Natural River Channels, (a) Braided, and
(b) Meandering*

[Courtesy New Zealand Lands and Survey Dept.]

(b)

Figure 10-24. *(Continued)*   [Courtesy George H. Caddie]

Nixon's is not extremely serious, for in many rivers the six-month and nine-month floods differ by less than 20 percent.

These results are by no means conclusive, and more work is needed to resolve the question. The amount of scatter in the field data suggests that some modification of the statistical approach may be necessary, and indeed there is still no proof that it is even possible to "average" the river flow in the way that has been discussed. The problem offers a tempting field for further investigation.

Braided and Meandering Channels

The single-thread channel described above is by no means universal. It is also common for rivers to be "braided," i.e., split into many channels which endlessly divide and rejoin. An example is shown in Fig. 10-24a. Sometimes the biological term anastomosis, which describes the branching and rejoining of blood vessels, is used to describe the process. It most commonly occurs in the steep upper reaches of a river, where the transporting power is high; the braiding process tends to dissipate this power by spreading the flow over a number of channels.

Meandering channels, of which an example is shown in Fig. 10-24b, usually occur on lower, gentler slopes towards the river mouth, and the river flows in one well-defined channel. The secondary currents described in Sec. 7.3 occur at each meander loop, e.g., A-A in Fig. 10-25a, and are shown again in Fig. 10-25b, the single cell contrasting with the multicell pattern of a straight channel, shown in Fig. 10-25c. The outer-bank scour and inner-bank deposition induced by the secondary current in Fig. 10-25b make the meander loops gradually extend outwards, and the more pronounced scour on the downstream faces of the loops makes them gradually migrate downstream. Because of this movement, the geologist regards the meandering form as unstable. The engineer, with a shorter time scale in mind, regards the meandering form as fairly well behaved, for two reasons: first, the gentle river slope and lower transporting power mean that the meander loops can fairly easily be stabilized by protective works; and second, that the secondary flow pattern of Fig. 10-25b tends to keep the channel clear. Straight channels on the other hand have quite a strong tendency to form central bars, and this may be a result of the multicell secondary flow pattern in Fig. 10-25c.

The main objection the engineer has to the meandering form is that if the loops are too pronounced the resistance to flow is too great and the flood levels are too high. In river training work the aim is usually to cut off loops which are too deep, while still preserving, and working in with, the natural meander pattern. Figure 10-25d shows a typical example of this operation.

A section like B-B in Fig. 10-25a is termed a "crossover." Here the water is of uniform depth where the "thalweg," or line of maximum depth, is at its median position while swinging from one side of the channel to the other. At the crossovers the mean depth is less and the velocity is greater than at the loops; in fact the flow at the crossovers may approach the critical condition. This variation in mean depth occurs also in straight channels, in the "pool and riffle" formation shown in Fig. 10-26a. Indeed, the two phenomena seem to be closely akin, for it is found that pools and riffles in straight channels have much the same wavelength as the meander pattern seen from above. An apparently straight channel may also have a concealed meander pattern in which the thalweg follows a sinuous path between straight banks, Fig. 10-26b.

Measurements of the horizontal dimensions of meander patterns show

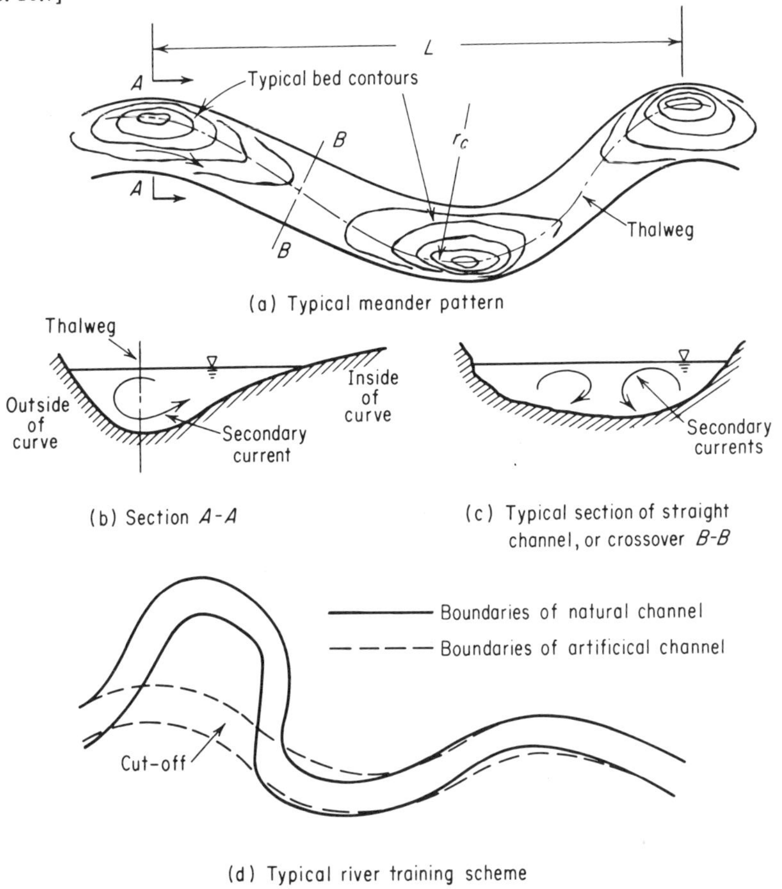
L
A
A
Typical bed contours
B
B
r_c
Thalweg
(a) Typical meander pattern
Thalweg
Outside of curve
Secondary current
Inside of curve
(b) Section A-A
Secondary currents
(c) Typical section of straight channel, or crossover B-B
Boundaries of natural channel
Boundaries of artificical channel
Cut-off
(d) Typical river training scheme

Figure 10-25. *Characteristics of Meandering and Straight Channels*

relations between certain of the parameters which stay remarkably consistent through a large range of stream size, from laboratory streams a foot wide to the Mississippi River nearly a mile wide. Many observers have noted such relations, and their work is summarized by Leopold and Wolman [51]. The wavelength L and loop radius r_c (Fig. 10-25*a*) are related to the width B by these equations

$$\frac{L}{B} = 7 \text{ to } 11 \tag{10–87}$$

$$\frac{r_c}{B} = 2 \text{ to } 3 \tag{10–88}$$

the range shown for the right-hand numbers being not as large as the full range of values found in the field. Values of the ratio r_c/B, for instance, are found to be as small as 1.5 and as large as 10; but the median occurs in the range 2–3, and it is within that range that the river engineer should look when planning to simulate natural meanders in river training works. Similarly for the range of 7 to 11 for the ratio L/B. The width B is the bank-full surface width or its statistical equivalent.

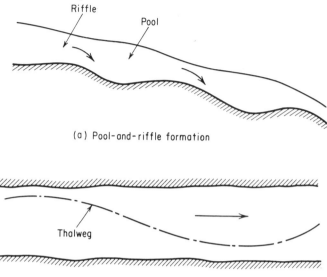

(a) Pool-and-riffle formation

(b) Concealed meander pattern in a straight channel

Figure 10-26.   *Characteristics of Natural Channels*

The origin of meanders has been the subject of much speculation. Although opinions may vary on matters of detail, the general basis of any explanation must apparently lie in the fact that any meander system tends to extend and perpetuate itself by the scouring action of the flow on the outside of each curve. Accordingly, any slight initial departure from the straight-channel form should lead to the eventual development of a meander pattern, and this in fact is observed to be the case. Irrigation engineers find that straight canals have to be maintained with care to prevent meandering, and river engineers too have found that it is better to train a river into a sinuous meander pattern than to attempt a straight alignment. One form of instability in straight channels has already been mentioned: the tendency to form a central bar, round which the flow divides into two streams. Eventually one stream captures the waters of the other and develops into a single meander loop; from this one loop a complete meander system will grow. Alternatively the initial disturbance may take the form of a small hole scoured out of one bank; this too can grow into a complete meander loop through the sequence curved flow → secondary flow → scouring of outer bank → more pronounced curvature. A complete discussion of this mechanism is given by Leliavsky [52].

But sediment transport is by no means the only significant factor. Leopold and Wolman [51] point out, in a remarkably complete and critical discussion of the question, that meanders are also carved in glacier ice by streams of meltwater flowing on the glacier surface, where few if any stones are present to initiate scour in the usual sense. More remarkable still, meanders are observed in ocean currents, including the Gulf Stream. And in both these cases the wavelengths and radii conform to Eqs. (10–87) and (10–88). It appears therefore that some explanation must be sought in the dynamics of the flow itself, as well as in the mechanism of sediment transport. No satisfactory explanation of this kind has yet been forthcoming.

The origin of braiding is also a matter of concern. The obvious explanation is that on steep slopes where the transporting power is high the banks are most vulnerable to attack, and that in fact the river dissipates its surplus power by attack on the banks and the formation of multiple channels. While this simple explanation has been criticized, there is no doubt about the association between braiding and steep slopes, and it is reasonable to assume that high transporting power has some influence on braiding.

Leopold and Wolman [47] gathered data from a large number of American and Indian rivers in coarse alluvial beds (median bed size $\frac{1}{4}$ in. or greater). They found that according to this sample, the criterion

$$S = 0.06Q^{-0.44} \tag{10–89}$$

distinguishes between braided and meandering rivers. For the former S is greater, and for the latter it is less, than the value given by Eq. (10–89), where Q is the bank-full discharge. Straight rivers were found at slopes both above and below this value. Henderson [39] attempted to refine this criterion by taking the size of the bed material d into account. He found that if the ratio $S/0.06Q^{-0.44}$ were plotted against the median bed size d (in feet) than a line could be drawn having the equation

$$S = 0.64d^{1.14}Q^{-0.44} \tag{10–90}$$

such that two-thirds of the points representing straight and meandering channels fell very close to this line. All the braided channels had values of S substantially greater than indicated by Eq. (10–90).

The similarity between Eqs. (10–90) and (10–66) is remarkable; the difference in the size of the coefficient is readily accounted for by differences in the definition of d, which is the median size in Eq. (10–90) and about the D-75 size in Eq. (10–66). The conclusion suggested is that there is a close correspondence between the need for a wider (type A in Fig. 10-21) channel created by slopes steeper than in Eq. (10–66), and between the braiding created by slopes steeper than Eq. (10–90). A river, in seeking by bank attack to turn a type B channel into a type A channel, will in fact create a braided channel.

However, this conclusion must be a tentative one in view of the uncertainties inherent in data collected from natural rivers. Nevertheless, Eqs. (10–66) and (10–90) offer some guidance to the river engineer who wishes to know whether a certain reach of river is likely to respond to control. This question is discussed more fully later. Finally, it should be noted that Eq. (10–90), unlike Eq. (10–89), discriminates not between braided and meandering channels, but between braided and single-thread channels, the latter including both straight and meandering channels. This classification accords well with physical observation, according to which straight channels may have such important features in common with meander channels (e.g., pools and riffles or even concealed meanders) that there is little point in distinguishing between them.

Width-Discharge-Slope Variations

The observations of Leopold and Maddock [53] on a number of United States rivers indicated some support for Eq. (10–73), at least as to the size of the exponent of Q. They found that in moving downstream along any particular river, width and depth varied with discharge of a given frequency in this way:

$$\text{Width} \propto Q^{0.5} \tag{10–91}$$

$$\text{Depth} \propto Q^{0.4} \tag{10–92}$$

Note the use of frequency as a means of standardizing the discharge. The exponents 0.5 and 0.4 were average values, which varied somewhat from river to river; the coefficients of $Q^{0.5}$ and $Q^{0.4}$ varied quite considerably from river to river. The conclusion appears to be that while each river may have adjusted its slope till the S-Q relationship was the correct one for Eq. (10–91) to be true, it does not follow that all rivers will have adjusted their slopes to the *same* S-Q relationship, producing the same coefficient of $Q^{0.5}$ in Eq. (10–91).

The question is what the S-Q relationship should be. If the conditions are live-bed, then Eq. (10–74), due to Lacey, is apposite, i.e., $S \propto Q^{-1/6}$. If the conditions are of the fixed-bed type better described by the tractive-force method, then Eq. (10–66) is the correct one for the narrow type B channel of Fig. 10-21, i.e., $S \propto Q^{-0.46}$. It is readily shown (Prob. 10.16) that an S-Q relationship of closely similar form, i.e., $S \propto Q^{-0.43}$, will be true if the channel is very wide. Of course, the position is further complicated by the variation in bed material, which usually becomes finer as one moves downstream. However, the marked difference between the exponents of Q in these two S-Q relationships suggests that they may be used to distinguish between those river channels which have shaped themselves according to live-bed criteria and those others (probably in coarse material) which have shaped themselves according to the conditions obtaining at the threshold of motion. But it must be conceded that so far no complete theory has been presented explaining the mechanics of the process.

River Training and Control

Most rivers, if left completely to their own devices, will develop qualities that are at best inconvenient to man and his activities, and at worst a threat to life and property. One of the commonest faults is the development of an irregular course which offers high resistance to floods with consequent high flood levels, and which creates navigation difficulties for any boat traffic that the river may carry. Both these factors may be of considerable economic importance; they are coupled together here because from both points of view the remedy is to train the river into a more regular course, usually by confining it to some extent, but without interfering too drastically with the river's own natural inclinations. For this purpose it is useful to have some quantitative description of these inclinations, and for this description it is natural to turn to equations like those developed in this and the preceding section.

Much river-control work consists of measures like bank protection, which depend little on quantitative knowledge of parameters like discharge, width, etc. But there are many problems which call for a quantitative solution. If, for instance, a positive attempt is to be made to confine a river to a regular gently sinuous course as in Fig. 10-25d, the ideal at which the engineer aims is a low-flow channel which just contains the dominant discharge, with berms onto which floods can overflow, the whole floodway being if necessary confined within stopbanks or levees. The question is, how wide should the low-flow channel be? If it is narrow, the rapid berm build-up mentioned previously may reduce the floodway capacity, and the channel itself may scour so vigorously that an unduly large amount of material is subsequently deposited downstream. If it is too wide, aggradation will ensue because the scouring capacity is insufficient.

To some extent the river will set its own width in response to the engineer's control measures, which normally consist of invading the riverbed with transverse barriers, permeable or impermeable, pushed out from the convex banks, together with suitable armoring of the concave banks against excessive scour. In this way the river is confined to a predetermined meander pattern, and it will of itself develop sections at the bends like that of Fig. 10-25b; sections of this sort have the desired features of a deep well-defined low-flow channel and an overflow berm, even though the width of the low-flow channel is not well defined.

Although this width is ill defined, the engineer must clearly have some approximate value in mind before the work is begun. He must also have estimated values of the wave length and loop radii of his proposed meander pattern. These will be obtained from Eqs. (10–87) and (10–88) and modified where necessary as the pattern is fitted by eye to the existing course of the river, so as best to take advantage of its existing features. A width value for the low-flow channel is still required, if only because it is this width which must be inserted in Eqs. (10–87) and (10–88).

The equations developed in this chapter indicate that width is a function of discharge, and for the low-flow channel it is logical to use the dominant discharge, although there is no doubt room for improvement in its statistical definition. If the riverbed is of coarse noncohesive alluvium, the threshold of motion appears to offer the soundest design criterion, and the width is then calculated from Eq. (10–69). Note that the slope is thereby treated as an independent variable, and no assumption is made as to whether slope adjustment has been carried through to any equilibrium condition.

If the riverbed is of sand or silt, a live-bed theory is required and the regime equations appear applicable. In particular, Eq. (10–73) suggests itself as an attractively simple means of determining the width both of the low-flow channel and of the major floodway, using the appropriate discharge value in each case. Certainly the evidence in favor of Eq. (10–73) is strong. It consists of Lacey's original observations on irrigation canals, supplemented since then by many more observations on similar Indian and American canal systems having discharges of up to 10,000 cusecs; and of observations on natural rivers made by Lacey, by Leopold and Maddock [53], and others.

However, there is much scatter in the natural river data. While Leopold and Maddock found that the ratio $P/\sqrt{Q}$ remained constant over the length of many individual rivers, the value of the ratio varied considerably from river to river, by a factor of approximately 6. The most likely explanation appears to lie in the slope-dependence of Eq. (10–73). If this equation is true only when the slope has reached a stable value given by Eq. (10–74), it may not be true in natural rivers where the slope has not become stable. Evidence in support of this inference can be found in the many natural rivers which conform to the first two regime equations, Eqs. (10–70) and (10–72), but not to the third, Eq. (10–73). See, for example, the closure to Ref. [39].

The engineer proposing to apply Eq. (10–73) to a natural river would therefore be wise to verify whether the slope conforms to Eq. (10–74). For this purpose he must have a value for the silt factor f, whose original definition by Eq. (10–71) was somewhat tentative. Later work on the definition of f was referred to briefly at the end of Sec. 10.6; it will now be summarized here. A slight change was found necessary in the coefficient of Eq. (10–70), from 1.17 to 1.15; if the velocity v is now eliminated between Eqs. (10–70) and (10–72) the result is

$$f = 193R^{1/3}S^{2/3}$$

Closer examination of the field data then showed that the f given in this equation differed from that in Eq. (10–70); accordingly there are two silt factors f_1 and f_2, given by

$$v = 1.15\sqrt{f_1 R} \tag{10–93}$$

and

$$f_2 = 193R^{1/3}S^{2/3} \tag{10–94}$$

It is readily shown (Prob. 10.17) that the second of these has a much more important influence than the first one on the slope equation, Eq. (10–74); in fact in using this equation it is accurate enough to set $f = f_2$.

The dependence of f_2 on the size d and the concentration q_s/q has been explored at length. Investigators differ on the precise effect of q_s/q, but agree that the effect is slight (see, for example, Prob. 10.18); they also agree that the effect of d is very close to that given by Eq. (10–71), for values of d up to 1 in. and possibly greater. The end result is that Eq. (10–71) may safely be substituted into Eq. (10–74). Some doubt remains as to whether the discharge substituted into Eq. (10–74) should be the dominant discharge or the flood discharge. However, this uncertainty is not a matter of great concern, for the exponent of Q in Eq. (10–74) is small.

Assuming now that the river slope has been compared with the value given by Eq. (10–74), and found materially different, the question is whether Eq. (10–73) should be modified. The soundest guide appears to be offered by Eq. (10–57b), according to which the ratio B/S^2 should remain constant if Q, d, and q_s/q are all given. Since B approximates closely to P in natural rivers, it follows that Eq. (10–73) becomes

$$P = 2.67\sqrt{Q}\left(\frac{1750Q^{1/6}S}{f^{5/3}}\right)^2$$

$$= \frac{127Q^{5/6}S^2}{d^{5/3}} \tag{10–95}$$

substituting from Eq. (10–71) and expressing d in feet. It may be repeated here that the live-bed theory, on which Eq. (10–95) is based, is best applied to sand or silt beds of grain size less than $\frac{1}{4}$ in. For coarser material, Eq. (10–69) is probably more satisfactory, although it may lead to widths that are too great if a substantial bed load arrives from upstream and has to be passed on downstream. However, the engineer can always keep in mind the possibility of reducing the width if the river behavior seems to call for such a measure.

In fact, any estimates made by equations like Eq. (10–69) and (10–95) must always be qualified by observations made of the river's past behavior. Previous generations of engineers may have excavated cutoff channels whose subsequent behavior has shown them to be too wide or too narrow, according to the criteria mentioned previously; such works are a valuable guide to the present generation.

In steep mountainous rivers the existence of braiding complicates the issue. The question is, can a heavily braided channel be trained into a single-thread meander channel? The answer must be yes, because Nature does it quite often. Examples can be found in the upper reaches of many rivers, where according to Eq. (10–90) the slope is steep enough to qualify the river as braided, as indeed it always is where the valley is wide enough to permit braiding; but where the river is confined in a narrow gorge the river takes up a single-thread meander form. It is therefore quite possible to train a braided channel into a meander form if the banks provided are strong enough; but to provide man-made banks as strong as the rock walls of a mountain gorge would hardly be economic except in a heavily populated region of high property values.

Equation (10–90) can offer some guidance to the river engineer contemplating the control of a braided channel. If the slope is much greater than the value given by Eq. (10–90), the braiding is likely to be vigorous and training works costly and difficult. If the slope is of the same order as indicated by Eq. (10–90), there is a possibility that the braiding will be less vigorous and the river amenable to channel training.

Scour Around Bridge Piers

When bridge piers are set in an erodible bed, the high local values of water velocity around the upstream end of the pier create local scour, which in times of high flood can go very deep. The situation is shown in Fig. 10-27. The plan

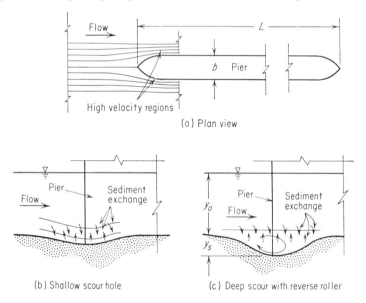

Figure 10-27. *Scour Around Bridge Piers*

view of the pier shows the high-velocity regions near the leading shoulders of the pier, and the two elevations the resulting scour hole. The flow through the hole will be in the same general direction as the river flow when the hole is comparatively long and shallow, as in Fig. 10-27*b*, or a reverse roller may form, as in Fig. 10-27*c*, if the hole is comparatively short and deep. In either case the action is basically the same: the hole must deepen until the velocity and shear stress across the base of the hole are reduced to the point where the net scouring action over the whole plan area of the whole is equivalent to that over the same area of riverbed at some place remote from the piers.

It is helpful to examine this mechanism in more detail, recalling that the essence of sediment transport is a balance between material which is continually being picked up from the bed into the flow, and that which is continually falling back to the bed. The two movements are indicated by two opposing sets of arrows in each of Figs. 10-27b and c; in the equilibrium condition, when the scour hole has reached its maximum depth, the upward sediment movement produced by the scouring action in the hole is equal to the downward movement from the overlying flow. Each of these rates of movement is equal to the corresponding rate produced over the same area of riverbed at some place away from the influence of the pier.

Suppose that such an equilibrium condition exists, and that, as implied in the above description, the upstream riverbed is " live " with an active sediment load containing the same material as the bed does—i.e., the river is " flowing in its own silt." Suppose now that with all depths and velocities remaining the same the mean size of the bed material were suddenly increased, so as to halve the rate of sediment transport upstream. Velocities and shear stresses remain unchanged, but because of the increase in sediment size they are everywhere half as effective in moving the sediment as they were previously. It follows, therefore, that the rate of scour inside the scour hole will also be halved, and that the balance between sediment upflow and downflow shown in Figs. 10-27b and c, will remain unchanged. The scour hole will therefore remain in equilibrium without further scouring or filling.

The conclusion is an interesting one: that the depth of scour is independent of the sediment size. Although it may appear paradoxical, it is well supported by experiment. However, this conclusion is true only in the live-bed case considered above; in the clear-water case, where there is no bed movement upstream, the scour hole will deepen until the shear stress on the walls of the hole drops to the " threshold " value appropriate to the bed material. There is then no movement of sediment into or out of the hole. The finer the sediment the lower is this threshold stress and the deeper the hole must become in order to lower the velocity and shear inside the hole to the correct value. In the clear-water case, therefore, the depth of scour is greater for fine sediments than for coarse. In the terms of this discussion clear-water flow means any flow in which the bed material upstream of the piers is undisturbed, even though the water may contain a heavy suspended load of material finer than the bed material.

Prediction of the amount of scour in any particular situation must depend largely on experimental results, for the problem is too complex in its geometry to be theoretically tractable. The total scour depth y_s will depend on the upstream depth y_0, on the pier width b and the pier nose geometry, on the upstream velocity v, on g, and in the clear-water case on the sediment size d.

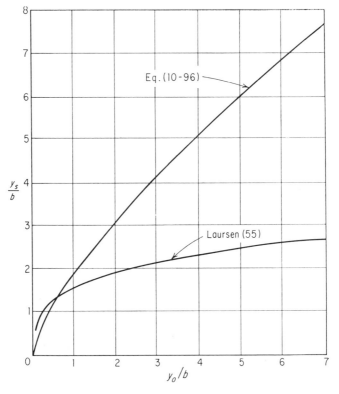

Figure 10-28.   *Scour at Bridge Piers with Live Bed Upstream*

Dimensional analysis therefore yields the result

$$\frac{y_s}{b} = \phi\left(\frac{y_0}{b}, \text{Fr}, \frac{d}{b}\right)$$

where Fr is the upstream Froude number $v/\sqrt{gy_0}$. This last will play a significant role only if there is an appreciable change in the water-surface level as the flow passes over the scour hole; this in turn will be true only if Fr is fairly close to unity, and when Fr is small its effect will also be small.

The experiments of Laursen [55] on model bridge piers set in sand indicated that the Froude number has no material effect, and that in the live-bed case y_s/b is related to y_0/b alone. The design relationship recommended on the basis of these experiments is shown graphically as the lower line in Fig. 10-28. Many similar model studies have been made in India and a number of them are referred to in the discussion of Ref. [55]. For some of them the empirical relation

$$\frac{y_s}{b} = 1.8\left(\frac{y_0}{b}\right)^{3/4} \tag{10–96}$$

has been put forward; this is also plotted on Fig. 10-28 and it is seen to give values of y_s substantially higher than Laursen's design curve. An alternative relation for other Indian experiments [56] is given as

$$\frac{y_s}{b} = 1.7\left(\frac{q^{2/3}}{b}\right)^{0.78}$$ (10–97a)

which can be rewritten as

$$\frac{y_s}{b} = 4.2\left(\frac{y_c}{b}\right)^{0.78}$$ (10–97b)

where y_c is the critical depth and q is the discharge per unit width upstream of the piers. This equation indirectly brings the Froude number into consideration, for it can be rewritten

$$\frac{y_s}{b} = 4.2\left(\frac{y_0}{b}\right)^{0.78} \mathrm{Fr}^{0.52}$$ (10–97c)

Some of the Indian investigators have expressed uncertainty about the relative importance of y_0/b and Fr; this may be due to the confounding of the two effects by the common dependence of the two parameters on the upstream depth y_0. Variation of this depth during a series of experiments tends to vary the two parameters simultaneously in a correlated way, so that their effects are no longer truly independent.

An alternative proposal by the Indian school [56] is that the scour depth y_s should be taken as twice the regime depth according to the Lacey system of equations. From the value of the regime depth obtained in Example 10.3, it follows that

$$y_s = \left(\frac{Q}{f}\right)^{1/3}$$ (10–98)

approximately; this means that the scour depth would be slightly dependent on the sand size d, and independent of the pier width b. This result is not in accord with any of the equations given previously.

If the piers are placed at an angle α to the flow the scour depth will be increased substantially. The effect of angle of attack as measured by Laursen [55] is shown in Fig. 10-29; the scour depth for a pier with zero angle of attack is multiplied by the factor K_α. The effect of pier nose shape is shown in Table 10–4; as in Chap. 7, the "lenticular" nose is the sharp-edged type formed by two circular arcs. The factor K_s is applied to the basic scour depth given by Fig. 10-28.

All of the above data apply to an isolated pier in which the flow pattern is entirely dependent on the pier shape and any neighbouring piers have no influence. If the distance between the piers were steadily reduced, the scour holes would eventually overlap, and at a closer distance still the scour depth would become uniform in the narrow space between the piers. Laursen

TABLE 10-4 Shape Factor K_s for Pier Nose Forms

Nose form	$\dfrac{Nose\ length}{Pier\ width}$ ratio	K_s
Square	0	1.00
Semicircular	0.5	0.90
Elliptic	1.0	0.80
	1.5	0.75
Lenticular	1.0	0.80
	1.5	0.70

analyses this case also, which he describes as the "long contraction," as it forms a convenient basic problem whose solution could be applied to the single-pier case by the application of suitable empirical coefficients. The problem may be solved by the consideration of Eqs. (10–55) and (10–56), preferably replacing the latter by the Manning form, in which $q \propto R^{5/3}$. The aim is to derive an expression for q_s/q comparable to Eq. (10–57), but in this case we are concerned with the difference in depth between two states of flow (upstream of, and within, the contraction formed by the piers) and wish to avoid consideration of the change in slope. Therefore we eliminate the slope S between the two equations for q and q_s, leading to

$$\frac{q_s}{q^6} \propto R^{-7}$$

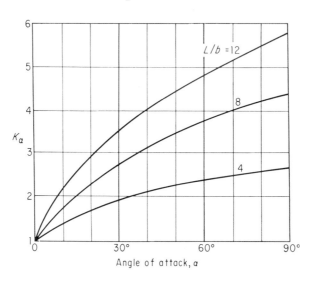

Figure 10-29. *Effect of Angle of Attack on Scour Around an Isolated Bridge Pier, after E. M. Laursen* [55]

or
$$\frac{q_s}{q} \propto q^5 R^{-7} \propto \frac{Q^5}{B^5 R^7} \qquad (10\text{–}99)$$

when the scour is completed, q_s/q will remain the same upstream of, and in, the contraction; from the approximation $R = y$ it follows that

$$\frac{y_1}{y_0} = \frac{y_s}{y_0} + 1 = \left(\frac{B_0}{B_1}\right)^{5/7} \qquad (10\text{–}100)$$

applying the subscripts 0 and 1 to the upstream and contracted flow respectively. Laursen gives a slightly different form based on his own bed-load formula [27].

Other cases treated by Laursen [55] are the problems of scour at a bridge abutment and at "overbank constructions," i.e., structures blocking the berm flow during a flood and forcing it into the main channel.

For the clear-water scour case, where the sediment size d is significant, complete and reliable data are not yet available. A promising beginning has been made by Laursen [57] who presents an analytical approach based first on the long contraction case and then modified to apply to isolated piers, abutments, etc. However, there is little experimental confirmation so far, and complete development of this approach must await further experimental work.

An interesting consequence of scour around bridge piers is that the increase in the flow section due to scour reduces the "backwater" effect discussed in Sec. 7.5. Laursen's experiments [55] with a live bed upstream showed that the amount of the backwater, or change in water-surface level, was usually somewhat less than the upstream velocity head even for large values of the width-contraction ratio. The results of Sec. 7.5, on the other hand, indicate values of backwater equal to many times the upstream velocity head for larger values of the contraction ratio.

References

1. D. B. Simons and E. V. Richardson. "Forms of Bed Roughness in Alluvial Channels," *Proc. Am. Soc. Civil Engrs.*, vol. 87, no. HY3 (May 1961), p. 87.

2. A. J. Reynolds. "Waves on the Erodible Bed of an Open Channel," *J. Fluid Mech.*, vol. 22, part 1 (May 1965), p. 113.

3. F. M. Exner. "Zur Dynamik der Bewegungsformen auf der Erdoberfläche," (On the Dynamics of Various Forms of Movement on the Earth's Surface), *Ergebnisse der Kosmischen Physik*, Vienna, vol. 1 (1931), p. 373.

4. J. F. Kennedy. "Dunes and Antidunes in Erodible-bed Channels," *J. Fluid Mech.*, vol. 16, part 4 (August 1963), p. 521.

5. A. Shields. "Anwendung der Aehnlichkeitsmechanik und der Turbulenzforschung auf die Geschiebebewegung" (Application of Similarity Principles and Turbulence Research to Bed-Load Movement), *Mitteilungen der Preuss.*

Versuchsanst fur Wasserbau und Schiffbau, Berlin, no. 26 (1936). Available also in a translation by W. P. Ott and J. C. van Uchelen, S. C. S. Cooperative Laboratory, California Institute of Technology, Pasadena, Calif.

6. C. M. White. "Equilibrium of Grains on the Bed of a Stream," *Proc. Roy. Soc.* (London), (A), vol. 174 (1940), p. 322.

7. E. W. Lane and E. J. Carlson. "Some Factors Affecting the Stability of Canals Constructed in Coarse Granular Materials," Proceedings of the Minnesota International Hydraulics Convention (September 1953), Joint Meeting of International Association for Hydraulic Research and Hydraulics Division, American Society of Civil Engineers (August 1953), p. 37.

8. Ven Te Chow. *Open-Channel Hydraulics* (New York: McGraw-Hill Book Company, Inc., 1959), Chap. 7.

9. E. T. Smerdon and R. P. Beasley. "Critical Tractive Forces in Cohesive Soils," Winter Meeting, American Society of Agricultural Engineers, Chicago (December 1959).

10. A. C. Carter. "Critical Tractive Forces on Channel Side Slopes," U. S. Bureau of Reclamation, Hydraulic Laboratory Report Hyd-366 (February 1953).

11. A.S.C.E. Task Committee, "Sediment Transportation Mechanics: Suspension of Sediment," *Proc. Am. Soc. Civil Engrs.*, vol. 89, no. HY5 (September 1963), p. 45.

12. V. A. Vanoni and G. N. Nomicos. "Resistance Properties of Sediment-Laden Streams," *Trans. Am. Soc. Civil Engrs.*, vol. 125 (1960), p. 1140.

13. C. Elata and A. T. Ippen. "The Dynamics of Open Channel Flow with Suspensions of Neutrally Buoyant Particles," Technical Report No. 45, Hydrodynamics Laboratory, M.I.T., Cambridge, Mass. (January 1961).

14. Mikio Hino. "Turbulent Flow with Suspended Particles," *Proc. Am. Soc. Civil Engrs.*, vol. 89, no. HY4 (July 1963), p. 161.

15. H. A. Einstein and N. L. Barbarossa. "River Channel Roughness" *Trans. Am. Soc. Civil Engrs.*, vol. 177 (1952), p. 1121, with discussion by T. Blench et al.

16. N. H. Brooks. "Mechanics of Streams with Movable Beds of Fine Sand," *Trans. Am. Soc. Civil Engrs.*, vol. 123 (1958), p. 526, with discussion by H. A. Einstein, Ning Chien et al.

17. V. A. Vanoni and N. H. Brooks. "Laboratory Studies of the Roughness and Suspended Loads of Alluvial Streams," Report No. E-68, Sedimentation Laboratory, California Institute of Technology, Pasadena, Calif. (December 1957).

18. D. B. Simons, E. V. Richardson, and W. L. Haushild. "Depth-Discharge Relations in Alluvial Channels," *Proc. Am. Soc. Civil Engrs.*, vol. 88, no. HY5 (September 1962), p. 57.

19. R. A. Bagnold. "The Movement of a Cohesionless Granular Bed by Fluid Flow over It," *British J. Appl. Phys.*, vol. 2 (February 1951), p. 29.

20. H. W. Smith, New Zealand Ministry of Works, personal communication.

21. I. Pliskin and H. Brenner. "Experiments on the Pressure Drop Created by a Sphere Settling in a Viscous Liquid," *J. Fluid Mech.*, vol. 17, part 1 (September 1963), p. 89.

22. S. M. Thompson. "The Transportation of Gravel by Turbulent Water Flows," M.E. Thesis, University of Canterbury, Christchurch, N. Z., 1963.

23. H. A. Einstein. "Formulas for the Transportation of Bed Load," *Trans. Am. Soc. Civil Engrs.*, vol. 107, (1942) p. 561.

24. C. B. Brown. "Sediment Transportation," in H. Rouse (ed.) *Engineering Hydraulics* (New York: John Wiley & Sons, Inc., 1950), Chap. 12.

25. H. A. Einstein. "The Bed-Load Function for Sediment Transportation in Open Channel Flows," *U. S. Dept. of Agriculture Technical Bulletin No. 1026* (September 1950).

26. A. A. Kalinske. "Movement of Sediment as Bed Load in Rivers" *Trans. Am. Geophys. Union*, vol. 28, no. 4 (1947), p. 615.

27. E. M. Laursen. "The Total Sediment Load of Streams," *Proc. Am. Soc. Civil Engrs.*, vol. 84, no. HY1 (February 1958), p. 1530–1.

28. M. Selim Yalin. "An Expression for Bed-Load Transportation," *Proc. Am. Soc. Civil Engrs.*, vol. 89, no. HY3 (May 1963), p. 221.

29. B. R. Colby and C. H. Hembree. "Computation of Total Sediment Discharge, Niobrara River near Cody, Nebraska," U. S. Geological Survey Water Supply Paper No. 1357 (1955).

30. R. A. Bagnold. "A Gravity-Free Dispersion of Large Solid Spheres in a Newtonian Fluid under Shear," *Proc. Roy. Soc.* (London), A, vol. 225 (August 1954), p. 49.

31. R. A. Bagnold. "The Flow of Cohesionless Grains in Fluids," *Phil. Trans. Roy. Soc.* (London), A, vol. 249 (1957), p. 235.

32. V. A. Vanoni, N. H. Brooks, and J. F. Kennedy. "Lecture Notes on Sediment Transportation and Channel Stability," California Inst. of Technology Report No. KH-R-1 (January 1961).

33. B. R. Colby. "Practical Computations of Bed-Material Discharge," *Proc. Am. Soc. Civil Engrs.*, vol. 90, no. HY2 (March 1964), p. 217.

34. J. R. Hardin et al. "Old River Diversion Control—A Symposium," *Trans. Am. Soc. Civil Engrs.*, vol. 123 (1958), p. 1129.

35. E. M. Laursen. "Sediment-Transport Mechanics in Stable-Channel Design," *Trans. Am. Soc. Civil Engrs.*, vol. 123 (1958), p. 195.

36. R. E. Glover and Q. L. Florey. "Stable Channel Profiles," U. S. Bureau of Reclamation Hyd. Lab. Report Hyd-325 (September 1951).

37. E. W. Lane. "Progress Report on Results of Studies on Design of Stable Channels," U. S. Bureau of Reclamation Hyd. Lab. Report Hyd-352 (June 1952).

38. E. W. Lane. "Design of Stable Channels," *Trans. Am. Soc. Civil Engrs.*, vol. 120 (1955), p. 1234.

39. F. M. Henderson. "Stability of Alluvial Channels," *Trans. Am. Soc. Civil Engrs.*, vol. 128, part I (1963), p. 657.

40. G. Lacey. "Stable Channels in Alluvium," *Proc. I.C.E.* (London), vol. 229 (1929–30), p. 259.

41. G. Lacey. "Uniform Flow in Alluvial Rivers and Channels," *Proc. I.C.E.* (London), vol. 237 (1933–34), p. 421.

42. G. Lacey. "A General Theory of Flow in Alluvium," *J. I.C.E.* (London), vol. 27 (1946), p. 16.

43. G. Lacey. "Flow in Alluvial Channels with Sandy Mobile Beds," *Proc. I.C.E.* (London), vol. 9 (1958), p. 145.

44. T. Blench. "Regime Behaviour of Canals and Rivers," (London: Butterworths Scientific Publications, 1957).

45. D. B. Simons and M. L. Albertson. "Uniform Water Conveyance Channels in Alluvial Material," *Proc. Am. Soc. Civil Engrs.*, vol. 86, no. HY5 (May 1960), p. 33.

46. L. B. Leopold and J. P. Miller. "Ephemeral Streams—Hydraulic Factors and Their Relation to the Drainage Net," U. S. Geological Survey, Professional Paper No. 282-A (1957).

47. L. B. Leopold and M. G. Wolman. "River Channel Patterns: Braided, Meandering and Straight," U. S. Geological Survey, Professional Paper No. 282-B (1957).

48. M. G. Wolman and L. B. Leopold. "River Flood Plains: Some Observations on their Formation," U. S. Geological Survey, Professional Paper No. 282-C (1957).

49. M. Nixon. "A Study of the Bank-Full Discharges of Rivers in England and Wales," *Proc. I.C.E.* (London), vol. 12 (February 1959), p. 157.

50. W. B. Langbein. "Annual Floods and the Partial-Duration Series," *Trans. Am. Geophys. Union*, vol. 30 (1949), p. 879, with discussion by Ven Te Chow et al.

51. L. B. Leopold and M. G. Wolman. "River Meanders," *Bull. Geol. Soc. Am.*, vol. 71 (June 1960), p. 769.

52. S. Leliavsky. *Introduction to Fluvial Hydraulics* (London: Constable & Co., Ltd., 1959), Chap. 8.

53. L. B. Leopold and T. Maddock, Jr. "The Hydraulic Geometry of Stream Channels and some Physiographic Implications," U. S. Geological Survey, Professional Paper No. 252 (1953).

54. Ning Chien. "A Concept of the Regime Theory," *Trans. Am. Soc. Civil Engrs.*, vol. 122 (1957), p. 785.

55. E. M. Laursen. "Scour at Bridge Crossings," *Trans. Am. Soc. Civil Engrs.*, vol. 127, part I (1962), p. 166.

56. Sir Claude Inglis. "The Behaviour and Control of Rivers and Canals," Research Publ. No. 13, Central Water Power Irrigation and Navigation Report, Poona Research Station, India, 1949.

57. E. M. Laursen. "An Analysis of Relief Bridge Scour," *Proc. Am. Soc. Civil Engrs.*, vol. 89, no. HY3 (May 1963), p. 93.

Problems

10.1. Prove Eq. (10–8) for flow in a circular pipe.

10.2. Assuming that $s_s = 2.65$ and $\nu = 1.2 \times 10^{-5}$ ft²/sec (as for water) show that at the point $F_s = 0.056$, Re* $= 400$, on Fig. 10-3, the grain size $d = \frac{1}{4}$ in. approximately, and exceeds $\frac{1}{4}$ in. for higher values of Re*.

10.3. At each of the three points Re* $= 10, 40, 100$, on the Shields curve of Fig. 10-3, calculate d and RS, assuming that $s_s = 2.65$ and $\nu = 1.2 \times 10^{-5}$ ft²/sec. Verify that these calculated values correspond to points on the curve of Fig. 10-5.

10.4. An irrigation canal is to be excavated on a slope of 0.0001 through country consisting of coarse silt having a D-75 size of $\frac{1}{32}$ in., and $s_s = 2.65$. For this material Manning's $n = 0.014$ provided the bed remains flat. The discharge is to be 50 cusecs; if no sediment transport is to be allowed, determine a suitable width for the channel, assuming that it is wide rectangular $(R = y)$ and the banks are grassed and protected.

10.5. The situation is similar to that of the previous problem, except that the bed material D-75 is $\frac{1}{2}$ in., the bed slope 0.002, and the discharge 100 cusecs. Determine a suitable width.

10.6. A channel which will carry a discharge of 1,000 cusecs is to be cut on a slope of 0.0005 through coarse, well-rounded alluvium having a D-75 size of 1 in., and $s_s = 2.65$. The base width of the channel is not to be less than 10 ft; determine whether a trapezoidal channel with unlined side slopes will be suitable, and if so choose suitable values for base width and side slopes, using the same safety factor as in Example 10.2.

10.7. The situation is the same as in Prob. 10.6, but the bed slope is to be 0.005. Assuming that an unlined trapezoidal section is to be used, determine suitable values for the base width and side slopes, and hence comment on the desirability of using such a section. (*Note*: If the base width required for an unlined trapezoidal section is very much greater than that required for a grassed-bank rectangular section, then the latter is clearly preferable.)

10.8. An irrigation channel which will carry a discharge of 300 cusecs is to be cut on a slope of 0.01 through coarse alluvium having a D-75 size of $1\frac{1}{2}$ in., and $s_s = 2.65$. The banks are to be grassed, the bed unlined, and it can be assumed that the channel will be wide enough to make $R = y$. (a) Determine what width the channel will have to be if no drop structures are to be used. (b) If the channel width is made equal to half the value calculated in (a), determine what proportion of the total fall has to be accomplished by means of drop structures.

10.9. Determine what the coefficient of Eq. (10–14) would be for $s_s = 2.1$, and recalculate Probs. 10.4 through 10.8, assuming that the bed material has this specific gravity.

10.10. A river with a sandy bed having a D-35 size of 0.3 mm and a D-65 size of 1 mm is 300 ft wide and approximately rectangular in section; the bed slope is 3 ft per mile. Using Fig. 10-11, prepare a rating curve showing how the discharge varies with depth up to a depth of 15 ft. To obtain a value of Manning's n for the grain resistance, apply Strickler's formula with a coefficient of 0.034 to the D-65 size of the bed material. Assume that $R = y$ (i.e., neglect the bank resistance).

10.11. A fluid contains a suspension of high-penetration spherical grains which are continually being projected upwards from a horizontal surface and falling back to the surface again. Their average upward velocity is v_u and their average downward velocity is v_d; the grain diameter is d and at any instant the volumetric ratio of grains : grains plus fluid is c. Show that the vertical pressure gradient in the fluid is equal to

$$\gamma \left\{ 1 - c \left[1 + \frac{3C_D v_u v_d (v_u - v_d)}{4gd(v_u + v_d)} \right] \right\}$$

where C_D is the drag coefficient on each grain, assumed constant throughout the motion. The fluid has no motion other than that induced by the grains.

10.12. The complete grading curve for the bed material in Prob. 10.10 is shown on the accompanying graph. Using the rating curve obtained in that problem, compute and plot curves of q_s against q for this river, using each of the following bed-load formulas: Du Boys', Shields', Einstein's (Fig. 10-15), Kalinske's, Laursen's, and Yalin's. Apply the last-named formula in two ways: fraction-by-fraction computation of q_s as in Laursen's method, as well as by direct substitution of the gravimetric mean size given by Eq. (10–53).

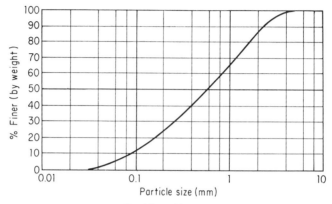

Problem 10-12.

10.13. Prove Eq. (10–58), and from this equation and Eq. (10–15) derive a y-θ relationship for the bed profile of Fig. 10-20. Making the substitution $dy/dx = \tan \theta$, integrate this equation and derive Eq. (10–59). Hence prove Eqs. (10–60) through (10–62), given that

$$E = \int_0^{\pi/2} \sqrt{1 - \sin^2 \phi \, \sin^2 \alpha} \; d\alpha$$

Taking $\phi = 35°$, verify the numerical results of Eq. (10–63), given that $E = 1.432$.

10.14. For canals 1 and 2 in Example 10.3, determine the channel cross section and the slope by each of the Lacey, Blench, and Simons-Albertson methods. Compare the results so obtained with the actual observed values given in Example 10.3.

10.15. In the situation of Probs. 10.6 through 10.8, determine the critical value of bed slope as given by Eq. (10–66) and compare it with the three different values of slope in the above three problems. Determine whether the type B section of Fig. 10-21 would have been a suitable solution to any of these problems.

10.16. For a wide fixed-bed channel to which Eq. (10–69) is applicable, show that if $P \propto Q^{1/2}$, then $S \propto Q^{-0.43}$.

10.17. By eliminating P and R from Eqs. (10–73), (10–93), and (10–94), prove that the introduction of two silt factors f_1 and f_2 has the effect of replacing $f^{5/3}$ in Eq. (10–74) by $f_1^{1/6} f_2^{3/2}$.

10.18. Assuming a resistance equation of the form

$$v \propto R^{3/4} S^{1/2}$$

and a wide rectangular channel, prove from Eq. (10–55) that Eq. (10–75) is exactly true for a given grain size d and concentration q_s/q; also prove that f_2 in Eq. (10–94) varies approximately as the fourth root of q_s/q.

10.19. At very low values of $\text{Re} = vd/v$, the drag force on a sphere of diameter d moving with velocity v through a fluid of viscosity μ is equal to $3\pi\mu vd$ (Stokes' Law). Show that when this law holds, Bagnold's penetration distance x_p is given by

$$\frac{x_p}{s_s d} = \frac{\text{Re}_0}{36}$$

where Re_0 is the Reynolds number at the initial instant of the motion. From the standard C_D–Re curve for a sphere, (to be found in any elementary text on fluid mechanics), read off values of C_D for a number of points in the range $1 < \text{Re} < 1000$, and by numerical methods calculate $x_p/s_s d$ for each case. Hence plot a graph of $x_p/s_s d$ against Re_0 and determine whether $x_p/s_s d$ ever exceeds the value given by Eq. (10–37).

Chapter 11

Similitude and Models

11.1 Introduction

Perhaps more than any other branch of fluid mechanics, the field of open channel flow presents problems which, being insoluble by theory or by reference to standard empirical data, call for the use of hydraulic models. The essential difficulty of the problem may lie in the complexity of the boundary conditions—e.g., flow over a spillway where lateral convergence of the approaching flow makes the problem a three-dimensional one, or it may lie simply in our ignorance of the basic nature of the flow phenomenon, as in the case of sediment transport. In the latter case, as we shall see, present knowledge is so uncertain as to introduce doubts about the proper development and application of model relationships.

The open channel flow problems which call for model treatment are usually of substantial economic importance. The discharge of a flood over a spillway is an event of potentially great danger to a very costly structure, to say nothing of the danger to human life. The correct design of the spillway is therefore a matter of critical importance. Improvement schemes for river channels and harbor works, made in order to reduce flood danger and improve navigation, are also of great significance in the protection of human life and property.

11.2 Basic Principles

Reference was made in Chap. 1 to dimensional analysis, together with a brief mention of hydraulic similitude and model operation. It will be assumed in this chapter that the reader is familiar with the basic principles involved, but they will be outlined briefly here for the sake of completeness.

In Sec. 1.7 it was noted that the dimensional analysis of a fluid-flow situation always leads to at least one dimensionless number, namely $\Delta p/\rho v^2$, which in this form is essentially a drag coefficient. In the inverse form, $v/\sqrt{(\Delta p/\rho)}$, it is a velocity coefficient such as might be applicable to a nozzle. If the flow has a free surface the gravitational constant will be involved, giving rise to the Froude number

$$\text{Fr} = \frac{v^2}{gL}$$

and if the viscosity μ is effective, the Reynolds number

$$\text{Re} = \frac{vL\rho}{\mu}$$

will be introduced. If both gravity and viscosity are effective, the drag or velocity coefficient first introduced will be a function of both Fr and Re:

$$\frac{\Delta p}{\rho v^2} = \phi(\text{Fr, Re}) \qquad \qquad \textbf{(11–1)}$$

The influence of surface tension σ and compressibility E_{co} will introduce the Weber and Cauchy numbers

$$\text{We} = \frac{v^2\rho L}{\sigma} \quad \text{and} \quad \text{Ca} = \frac{v^2\rho}{E_{co}}$$

Equation (11–1), supplemented by We and Ca if applicable, provides the complete theoretical basis for model studies. Since $\Delta p/\rho v^2$ is an unknown function of Fr, Re, ..., a model study can be interpreted only if Fr, Re ..., are severally given the same value in the model as in the large-scale original of the model, called the " prototype." This will mean that $\Delta p/\rho v^2$ has the same value, i.e., drag and velocity coefficients will be the same, in the model as in the prototype. Herein lies the key to the interpretation of model tests.

An alternative parallel approach to the question is through the concept of *dynamical similarity*. In a sense it can be said that a draftsman is making a model study every time he lays out on his drawing board a trial general arrangement of various components. His confidence that if the parts fit on the drawing, or "model," they will also fit on the prototype, stems from an implied confidence in the principles of *geometrical* similarity. Where the motion of fluids is involved, hydraulic models can be built and operated if principles of *dynamical* similarity can be developed in which equal confidence can be felt.

If the model is to be similar to the prototype in this complete sense of the term, it is not enough that the solid boundaries be geometrically similar in the two systems; it is also necessary that the two flow patterns be similar, and it is then said that *kinematic* similarity holds between the two systems. An example is shown in Fig. 11-1. In this case all velocities and accelerations must have the same prototype : model ratio—i.e., in the case shown we have, using the subscripts m and p to indicate model and prototype respectively:

$$\frac{(v_a)_p}{(v_a)_m} = \frac{(v_b)_p}{(v_b)_m} \qquad \qquad \textbf{(11–2)}$$

and of course $(v_a)_p$ and $(v_a)_m$ must have the same direction.

The state of kinematic similarity can be maintained if, and only if, the corresponding force ratios remain constant. That is, if F_a and F_b are the net forces exerted on the fluid elements at A and B,

$$\frac{(F_a)_p}{(F_a)_m} = \frac{(F_b)_p}{(F_b)_m} \qquad (11\text{--}3)$$

Each of these net forces may be thought of as an inertia force, mass × acceleration. It is made up of a number of different forces (those due to gravity,

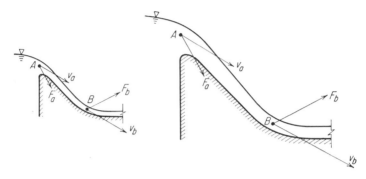

Figure 11-1.   *Basic Model-Prototype Relationships*

viscosity, etc.), all of which vary in different ways with v, L, ρ, etc. If each of the ratios in Eq. (11–3) is to be kept constant at all points in the field of flow, then the various components of force must bear a constant ratio to one another. Now it can be shown (Prob. 11.1) that the Froude number is the ratio of inertia force to gravitational force, expressed in general dimensional form; similarly that the Reynolds, Weber, and Cauchy numbers are the ratios of the inertia force to viscous, capillary, and compression forces respectively. The final conclusion is the same as that already drawn from Eq. (11–1): that if a certain type of force is effective in a certain flow situation, the appropriate dimensionless number must be given the same value in the model as in the prototype.

Secondary Scale Ratios

The detailed interpretation of model measurements requires that scale ratios be available for translating model values of various quantities, e.g., velocity, discharge, etc., into the corresponding prototype values. Scales can be deduced for all physical quantities if scales are known for mass, length, time, and the physical properties of prototype and model fluids. It is convenient to introduce here the subscript r to indicate the ratio of prototype : model quantity, e.g., if model lengths are one-tenth those of the protoype, then $L_p/L_m = L_r = 10$, the subscripts p and m indicating prototype and model as before. Now it is always true that the mass ratio $M_r = \rho_r L_r{}^3$, so we have scale ratios for mass and length. The time scale T_r is deduced indirectly from the relationship between velocity scale and length scale dictated by the fact that the appropriate dimensionless number, e.g., the Froude number, must be kept constant.

In open channel flow the presence of a free surface means that the Froude number Fr is always significant, indeed dominant. The secondary scale ratios based on the constancy of Fr and its corollary

$$v_r = L_r^{1/2} \tag{11-4}$$

will therefore be applicable, although they may be modified in some case by the action of influences other than gravity. A complete list of scale ratios is therefore as follows

$$\left.\begin{array}{llll}
\text{Mass} & M_r & & = \rho_r L_r^3 \\[4pt]
\text{Length} & L_r & & = L_r \\[4pt]
\text{Velocity} & v_r & & = L_r^{1/2} \\[4pt]
\text{Time} & T_r = L_r v_r^{-1} & & = L_r^{1/2} \\[4pt]
\text{Discharge} & Q_r = v_r L_r^2 & & = L_r^{2\frac{1}{2}} \\[4pt]
\text{Force} & F_r = M_r L_r T_r^{-2} & & = \rho_r L_r^3 \\[4pt]
\text{Pressure} & p_r = F_r L_r^{-2} & & = \rho_r L_r
\end{array}\right\} \tag{11-5}$$

as the reader can verify (Prob. 11.2).

The Influence of Forces Other Than Gravity

Compressibility effects are never significant in open channel flow models. Surface tension effects are appreciable only when radii of curvature of the liquid surface, and the distances from solid boundaries, are very small. They will therefore be negligible in all real prototype situations, and care must be taken to keep them negligible in model situations. This is accomplished by keeping model water depths no less than an inch or two, and similarly for channel widths. Beyond the taking of this precaution, capillary effects do not warrant any further attention.

Viscosity is much more important, and exerts its influence in many different situations. The term *scale effect* can be introduced here; it is the name given to the slight distortions introduced by forces—for example, viscosity—other than the dominant one, such as gravity. Such effects are often slight without being altogether negligible. For example, the flow over a spillway will encounter some slight viscous resistance down the face of the spillway, although resistance will be negligible at the crest itself, where the discharge-head relation is determined.

The only perfect way of dealing with the effect of viscosity is to keep both

Fr and Re the same in model and prototype, and this is a practical impossibility (Prob. 11.4). In this problem the use of different fluids in model and prototype is suggested, but this expedient is found to be ineffective as well as inconvenient. Accordingly the same fluid, water, is normally used in model and prototype, the only modification of this practice being the occasional substitution of fresh water for salt water. If the fluids are the same, it is a trivial deduction from Prob. 11.4 that $(Re)_r = L_r^{3/2}$, so that Re is much smaller in the model than in the prototype. However, satisfactory approximate methods are available for dealing with the difficulties raised by viscous effects. Underlying all these methods must be an appreciation of the way that Re exerts its influence, as shown by typical drag coefficient−Re curves, shown schematically in Fig. 11-2. The ideal, as suggested by Eq. (11–1), is to make the

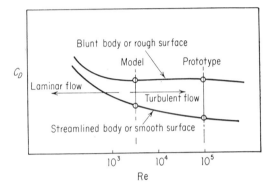

Figure 11-2. *Typical C_D− Re Curves and Their Effect on Model Operation*

drag coefficient the same in model and prototype. Since Re is smaller in the model than in the prototype, the model drag coefficient must be higher if surface drag is appreciable. If only form drag is appreciable, the drag coefficient may be the same in the model as in the prototype. In either case the inference to be drawn from Fig. 11-2 leads to what is an important rule in model studies: that the model Reynolds number should be kept as high as possible, high enough at least to make the model flow fully turbulent, even if not high enough to escape the effects of surface drag.

A simple case of the effect of form drag alone is provided by a spillway model in which the effect of the downstream flow on the riverbed is to be investigated. If the channel is lined with large stones, or "rip-rap," the drag coefficient on each stone will be constant provided that Re is high enough—which means 1,000 or more. Therefore if the model Re (based on the diameter of the stone) is made equal to at least 1,000, the drag coefficient on the stones will be the same in the model as in the prototype. It remains to verify (Prob. 11.6) that the movement of the stones will be faithfully reproduced by the model; this will be true if the prototype : model force ratio for the stones is the same as the force ratio for a purely gravitational model.

Another simple example is provided by bridge piers. It has been seen in Sec. 7.5 that the resistance to flow of normal bridge pier shapes is such that the drag coefficient is well over unity; this implies that form drag is a substantial part of the total drag, so that the Reynolds number will be relatively unimportant.

When surface drag is appreciable, as in the case of long river channels, other means must be found to resolve the difficulty. These will be discussed in the next section.

11.3   Fixed-Bed River and Structural Models

Under this heading will be discussed those river models in which it is not required to reproduce bed movement, and "structural" models, i.e., models of weirs, spillways, etc.

The Surface Drag Problem

We can best explore this problem by considering the Manning resistance formula. If it is expressed in terms of prototype : model scale ratios, the constant 1.49 drops out, and we have:

$$v_r = \frac{R_r^{2/3} S_r^{1/2}}{n_r} \tag{11-6}$$

If the model and prototype are geometrically similar, then $S_r = 1$ and $R_r = L_r$. For the Froude number to remain constant, Eq. (11–4) must be true; from this and the above results it follows that

$$n_r = L_r^{1/6} \tag{11-7}$$

which is the condition that must be met if dynamical similarity is to be maintained. In fact it might have been deduced from Eq. (4–22), according to which the Manning n is proportional to the sixth root of the size of the roughness projections; seen in this light, Eq. (11–7) simply indicates that the notion of complete geometrical similarity is applied to the texture of the surface as well as to the shape of its general outline.

However, the adjustment of the model n indicated by Eq. (11–7) will not solve the problem, for this will merely make the ratio k_s/L constant, where k_s is the height of roughness projection, as in Chap. 4, and L is a lateral dimension of the flow section. The lower value of model Re will still produce higher values of drag coefficient, even at the same value of k_s/L. This can be verified indirectly by considering some typical figures for n_r and L_r. On an actual spillway structure, n will be 0.013 or less; the lowest obtainable value of n

even for very smooth surfaces like polished wood or glass is about 0.009. Hence if a model of this material were used to represent a concrete spillway the scale ratio would have to be

$$L_r = \left(\frac{0.013}{0.009}\right)^6 = 9$$

which is too small for convenience or economy. (It is unusual for a structural model to be built to a scale smaller than 40 : 1.)

The conclusion is the same as that suggested by the previous argument and by inspection of Fig. 11-2, namely that the model has proportionately more resistance than the prototype. In structural models, where resistance losses are small anyway, this question is not usually given much attention. In the case of river and harbor models, resistance is all-important, but it turns out that this problem is overshadowed by another one whose solution disposes also of the resistance problem.

Scale Distortion

Consider the choice of length scale for a model which is to be made of a large river. A convenient example is the Mississippi River, of which a model has in fact been built by the U. S. Corps of Engineers. The Mississippi basin is approximately 2,000 miles in each direction, and a scale of 2,000 : 1 was chosen for horizontal dimensions. The model, which is built out of doors, therefore covers an area approximately one mile square, and even this is none too large, for the model river width will still only be two or three feet in places where the prototype river width is one mile.

Consider now the vertical dimensions of prototype and model. If the same scale of 2,000 : 1 is used, then a prototype depth of 20 ft will mean a model depth of $\frac{1}{8}$ in., which is far too low for satisfactory operation; quite apart from surface tension effects, which may be appreciable, the flow will probably be laminar, and thus completely different in character from the turbulent flow in the river.

The remedy, which appears a radical one at first glance, is to use different scales for vertical and horizontal dimensions. The Mississippi model has a vertical scale of 100 : 1, which produces depths of at least 1 in., and satisfactorily turbulent flow. This scale distortion, or vertical exaggeration, does not seriously distort the flow pattern, and in fixed-bed models anyway it gives perfectly satisfactory results.

It remains to see how the resistance problem appears in the light of this new development in the argument. The model relationships must now be stated in terms of a vertical scale Y_r and a horizontal scale X_r. The Froude relationship must take the form

$$v_r = Y_r^{1/2} \tag{11-8}$$

since it is vertical rather than horizontal distances which measure the effect of gravity on velocity. Nevertheless the resultant velocity is essentially horizontal in direction, so that the time scale becomes:

$$T_r = X_r/v_r$$

$$= X_r/Y_r^{1/2} \tag{11-9}$$

while for the discharge scale we have

$$Q_r = v_r X_r Y_r$$

$$= X_r Y_r^{1\frac{1}{2}} \tag{11-10}$$

Considering now the generalization of Eq. (11-6), the slope ratio S_r will normally be equal to Y_r/X_r, because the vertical scale Y_r will apply to *all* vertical dimensions, bed height z as well as depth y. If $S_r \neq Y_r/X_r$, then $z_r \neq y_r$, and difficulties of two kinds will be introduced. First, the total energy H, which contains both z and y, cannot be correctly represented except at one particular discharge. The second is illustrated by the cross section in Fig. 11-3.

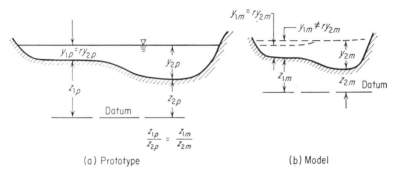

(a) Prototype   (b) Model

Figure 11-3. *The Effect of making* $z_r \neq y_r$ *on the Relationship between Berm and Channel Depths*

It is a matter of elementary geometry to show that if the relative magnitudes of berm and channel depths y_1 and y_2 are correctly represented, then the transverse water surface will not be horizontal in the model. If it is horizontal, the ratio y_1/y_2 will be distorted. Despite these difficulties, S_r is sometimes made greater than Y_r/X_r in movable-bed models, in order to assist the sediment movement; this practice is known as "tilting" the model. However it is seldom if ever employed in fixed-bed models, such as are now being discussed.

Equation (11-6) therefore becomes

$$v_r = \frac{R_r^{2/3} Y_r^{1/2}}{n_r X_r^{1/2}} \tag{11-11}$$

which, combined with Eq. (11–8), leads to

$$\frac{R_r^{2/3}}{n_r X_r^{1/2}} = 1$$

or
$$n_r = \frac{R_r^{2/3}}{X_r^{1/2}} \qquad\qquad (11\text{–}12)$$

In the case of a very wide channel $R_r = Y_r$, but this cannot usually be assumed to be the case in the model. If not, R has to be calculated independently in model and prototype (Prob. 11.7).

The general effect of Eq. (11–12) is that n_r will be distinctly less than unity, because $X_r > Y_r$ and R_r depends more heavily on Y_r than on X_r. It follows that the model has to be rougher than the prototype, so there is no longer any difficulty such as arose out of Eq. (11–7). In fact some effort usually has to be expended to make the model rough enough, by setting pebbles in the model riverbed (which is usually of a cement mortar), by placing folded strips of wire mesh in the path of the flow, or by various other means. While Eq. (11–12) gives a useful guide to the degree of roughening required, the final adjustment must be by a trial and error process. Past floods are simulated in the model by using the recorded flood hydrographs, and the roughness adjusted until the recorded flood levels are reproduced in the model. When berm flow accounts for a material proportion of the flood flow, the surface roughness of the berms must be adjusted with as much care as that of the main channel, and at least two flood discharges must be verified in order to obtain the correct ratio of channel and berm roughness (Prob. 11.8). Problems 11.9 through 11.12 illustrate other aspects of the general problems involved in the choice of length scales and the determination of time, discharge, and roughness scales. It is of paramount importance that the model value of Re shall be high enough to ensure turbulent flow; Allen [4] recommends that $(vy/v)_m$ should be not less than 1,400, where y is the depth.

Short structural models like those of spillways are normally built without scale distortion and no attempt is made to adjust the roughness beyond making the model surface as smooth as possible. This means that water arriving at the base of the model spillway will be moving a little more slowly than it should for true dynamical similarity. This slight error is not of great concern, for it is overshadowed by another scale effect, namely the effect of air entrainment, discussed in Sec. 6.3. This phenomenon is not reproduced on a model because it depends on the existence of high velocities. The combined effect of flow resistance and air entrainment on the action of energy dissipators does not appear to be extreme, for comparisons between model and prototype performance show good correspondence between the two. Typical of these comparisons is the one made by the observations of Peterka [1].

It is appropriate at this point to describe the method of construction normally used for a fixed-bed river model. Templates are cut from plywood or sheet metal to the cross-sectional shape of the river. As shown in Fig. 11-4, the template is female in form—i.e., it lies below ground level and the river section is cut out of it. The space between the templates is then filled with some inexpensive material (e.g., gravel) and the surface is then finished with a plaster of cement mortar, a clearance of about half an inch having been allowed for this purpose when the templates were first cut. For accurate finishing of the surface, steel pins may be fixed to the templates as in Fig. 11-4, the tops of the pins being flush with the finished surface.

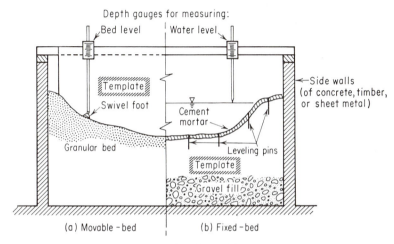

Figure 11-4.   *Typical Details of Model Construction*

Movable-bed models, in which the bed is made of loose sand, are usually molded by means of a male template suspended from rails as in Fig. 11-4. Rails of this sort are used in both fixed-bed and movable-bed models to provide a reference level from which water and bed levels may be measured.

11.4   Movable-Bed Models

Introduction

The design and operation of a movable-bed model are much more difficult, and involve greater uncertainties, than the design and operation of a fixed-bed model, in which the only real complication is created by the need for a trial adjustment of the model roughness. There are two major difficulties in the operation of a movable-bed model: first, the model must faithfully reproduce the sediment movement as well as the water movement of the prototype; second, the bed roughness cannot be arbitrarily controlled by the model operator, but will be determined by the state of the bed and of the sediment motion.

Despite these special difficulties, one of the very first successful hydraulic models was of the movable-bed type. It was constructed by Osborne Reynolds [2] in 1885 to represent the estuary of the River Mersey in England, and appears to have been first built only with the intention of showing the pattern of the tidal currents in the estuary. However that may be, the sand bed of the model did in fact shape itself, after some thousands of tidal cycles, into the same pattern of sandbanks and channels that characterized the prototype estuary. The model was a very small one, with $X_r = 31,800$ and $Y_r = 960$, a vertical exaggeration of 33 : 1. A larger model, having $X_r = 10,560$ and $Y_r = 396$, was even more successful. A noteworthy feature of the operation of each model was Professor Reynolds' realization that the time scale for the tidal cycle must be given by Eq. (11–9); this led to model tidal periods of approximately 40 and 80 sec for the first and second models respectively. Only with these tidal periods did the flow of water and sediment reproduce that in the real estuary.

A similar study was made by Vernon-Harcourt [3] in 1888 of the River Seine estuary. He found that the first sand he tried in the model did not successfully reproduce the prototype conditions, and that some trial and error was needed before a sand was found that reproduced faithfully the features of the prototype estuary. Neither Vernon-Harcourt nor Reynolds had any systematic basis for the choice of the right grade of sand, and it must be assumed that Reynolds' immediate success had an element of good fortune in it which may well be the envy of later generations of workers in this field.

Another point of interest that emerges from these early projects concerns the time scale. Vernon-Harcourt remarked in his paper, when discussing the rate at which changes occurred in the estuary sandbanks and channels:

It would be impossible to determine by experiment the time any changes in an estuary would occupy. The figures, in fact, giving the number of tides during which each experiment was worked, are not even intended as an indication of the rate of change in the model, and much less as any measure of the period required for such changes in an estuary, but merely as a record of the comparative duration of each experiment.

What this means is that the time scale governing the fluid flow (which can be predicted from the length scales) is not necessarily the same as the time scale governing sediment motion; in fact the latter will have to be deduced from the model observations themselves. This distinction between "hydraulic time" and "sedimentation time" is still a well-recognized feature of movable-bed model operation.

Verification by Trial

As the above examples have shown, the operation of movable-bed models preceded by many decades the development of any systematic approach to the general problem of sediment transport; indeed, it was the lack of any such approach that prompted the first ventures into model operation. However, the intelligent use of models appears to require some conceptual basis, sufficient at least to develop the form of the dimensionless numbers which should govern the interpretation of the model tests. But this basis was lacking during the first model studies and the more pragmatic approach of verification by trial has grown up since that time. The essence of this method is that if the model can be made to reproduce events which are known to have occurred in the past, it may be expected to predict future events with reasonable accuracy.

Although the general principle is made clear in this last statement, the operational details can involve difficulties. In many cases, particularly in nontidal river models, it is difficult for the low water velocities in the model to scour the bed material, and it is necessary to use bed material which is substantially less dense than the natural sand (specific gravity 2.65 approximately) which occurs in most natural rivers. Bituminous compounds, coal dust (specific gravity 1.3 approximately) and plastics (specific gravity 1.2 or less) of various kinds are commonly chosen.

Scale distortion presents another problem. On fixed-bed models the ratio $X_r : Y_r$ can be of considerable size without ill effect, but in a movable-bed model distortion may increase the slopes of model sandbanks beyond their angle of repose so that they will no longer stand. One remedy is to make the sides of such banks out of rigid material, but this is allowable only if the prototype banks are known to be stable. Another effect of scale distortion is to increase the longitudinal slope of the river, making increased roughness necessary. But as was mentioned previously the bed roughness is determined by the bed movement itself and is not subject to arbitrary adjustment by the operator. For these reasons the distortion ratio used in river models is normally low, ranging between two and seven and going up to ten in special cases. In harbor and estuary models a greater distortion may be allowable because the prototype sandbank slopes are usually very small, as are the longitudinal slopes of the tidal channels. Indeed, Reynolds' first two models of the Mersey estuary had distortion ratios of 33 and 26.6 respectively and he also remarked in his discussion of them: "From my present experience, in constructing another model I should adopt a somewhat greater exaggeration of the vertical scale."

The exact choice of scales and of bed material has become to some extent a matter of individual judgment based on local experience, subject always to the overriding condition that the model flow must be strong enough to move the model bed material. The following procedure for the initial design of river models is used by one establishment and is typical of the local rules and practices developed by individual model laboratories from long experience:

Coal dust is used for the model-bed material. The model water velocity required to move this material has been found to be in the range 0.3–1.0 ft/sec with an average value of 0.5 ft/sec. The vertical scale Y_r must be chosen to give a model velocity of this order of magnitude. The model bed slope required to fit this velocity to the Manning equation is then determined from the fact that the coal dust has a Manning n of 0.018. The horizontal scale is then determined from the relation $S_r = Y_r/X_r$.

Once this basic design process is completed, and the model built, the trial verification proceeds by putting past flood hydrographs through the model (usually over a prototype period of a year or two) and adjusting the sedimentation time scale or the slope until these flood hydrographs produce the known behavior of the bed. Adjustment of the sedimentation time scale means, of course, adjusting the time base of the flood hydrographs. It will often be found that different time scales must be used for high and low discharges; this is a consequence of the $q_s \propto q^2$ relation deduced in Eq. (10–57). Slope adjustment, or "tilting" of the model, is not difficult to carry out, being done by changing the slope of the guide rails in Fig. 11-4 and reshaping the bed, but as mentioned previously it brings interpretive difficulties in its wake. Finally, attention may be drawn here to the fact that in the above design process the model operator has exercised only one degree of freedom—the choice of bed material. Given this material, both vertical and horizontal scales are determined. In practice this system must be modified to allow for adjustments to the scales as dictated by the area and water supply available.

While details may vary from one establishment to another, the above process may be described as typical. This is particularly true of the verification part of the process, which normally proceeds in the way described by adjustment of the time scale and slope. The hydraulic time scale may also need adjustment. Because local practice and experience play such an important part in the matter, the belief has grown up that movable-bed model design and operation is more of a personal art than a communicable science. Like all such generalities, this one contains an element of truth, but it seems to have had the unfortunate effect of inhibiting both the publication of details of local practice, and the development of more general methods of approach to the problem. A notable exception is the work of Allen [4] whose book is a comprehensive and detailed guide based on experience at the University of Manchester which has developed continuously since the pioneering work of Osborne Reynolds.

Another interesting disclosure of detailed model technique is that of Jonte [5] who describes a model of the River Isère upstream from Grenoble, France, built and operated by the Neyrpic organization in Grenoble. The complete account is too lengthy to be reproduced here, but one feature of the study—the choice of bed material and vertical scale—will be summarized.

A horizontal scale of 1 : 250 was chosen because of space considerations. The remaining elements were determined by what was termed the equilibrium section method. A calibration channel 5 meters long representing part of the course of the Isère was filled with a load of trial bed material, and a constant discharge passed through until erosion ceased and the bed was stable. The slope and the mean depth were then measured, and the results from a number of similar runs compared with the " critical " values of prototype discharge, slope, and depth at which the bed is on the point of motion. These last were determined from field observation and will be denoted by Q_{cp}, S_{cp}, and y_{cp}.

The detailed form of the comparison is shown in Fig. 11-5. The parameters

$$\left(\frac{Q_m X_r}{Q_{cp}}\right)^{2/3} \quad \text{and} \quad \frac{y_m}{y_{cp}}$$

determined from a series of the model runs described above, are each plotted

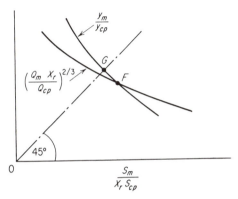

Figure 11-5. *Determination of Vertical Scale by Equilibrium Section Method, after M. Jonte* [5]

against the ratio $S_m/X_r S_{cp}$, the subscript m indicating model quantities as usual. The two curves intersect at the point F; at this point

$$\frac{Q_m X_r}{Q_{cp}} = \left(\frac{y_m}{y_{cp}}\right)^{3/2} = \frac{1}{Y_r^{3/2}}$$

and Eq. (11–10) is satisfied. However, the condition $S_r = Y_r/X_r$ will not be satisfied unless the line OF makes an angle of $45°$ with each axis; this condition is required in order to make

$$\frac{1}{Y_r} = \frac{S_m}{X_r S_{cp}}$$

i.e.,

$$S_r = \frac{Y_r}{X_r} \tag{11–13}$$

The aim therefore is to find a sand which will make the line OF coincide with the bisectrix OG. If these two lines do not coincide, then choice of the point G for the operating condition gives a correct slope scale but a distorted discharge scale; choice of the point F gives a correct discharge scale but requires the model to be tilted.

This technique led finally to a satisfactory choice of vertical scale and bed material. The final choice had to be made with due regard to the limits imposed by the need for a sufficiently high Reynolds number, and by other conditions. For details of other aspects of the work, the reader should consult the original reference [5]; a final comment to be made here is that the system described above, like that discussed previously, allows the operator one degree of freedom. In this case it has been exercised by the choice of the horizontal scale X_r.

Explicit Design Methods

Although the trial methods discussed above are now well-established practice, suggestions have been made in more recent years that explicit design methods can be developed, using as a basis the general studies that have been made in sediment transport since Shields' work in 1936. In particular, the dimensionless parameters that are now recognized as significant determinants of the sediment transport process can presumably be used, just like the Froude and Reynolds numbers, as a basis for model studies.

Consider the plane shown in Fig. 10-3, on which Shields' results are displayed as a plot of the entrainment function

$$F_s = \frac{1}{\psi} = \frac{\tau_0}{\gamma(s_s - 1)d} \tag{11-14}$$

against the particle Reynolds number

$$\text{Re}^* = \frac{v^* d}{v} = \frac{\sqrt{\tau_0/\rho}\, d}{v} \tag{11-15}$$

It was suggested in Sec. 10.3 that the state of the bed, both in shape and in activity, was completely determined by position on the $F_s - \text{Re}^*$ plane. While this contention is not yet established with certainty, the evidence for it continues to accumulate and it can at least be regarded as a useful guiding assumption in a design process.

The aim therefore is to make F_s and Re^* the same in the model as in the prototype. Since the state of the bed will be the same in each case, it can also be assumed that the equivalent roughness k_s will bear the same ratio to the particle size on the model as in the prototype, i.e., that the ratio of grain resistance to bar resistance remains the same. This has two consequences: that $n_r = d_r^{1/6}$, and that $(R'/R'')_r = 1$, where R' and R'' are defined as in Sec. 10.5. The first of these means that Eq. (11-12) can be rewritten

$$d_r^{1/6} = \frac{R_r^{2/3}}{X_r^{1/2}} \tag{11-16}$$

and the second means that $\tau_{0r} = \gamma_r R_r S_r = \gamma_r R_r Y_r / X_r$. It follows from Eqs. (11–14) and (11–15) that

$$\frac{R_r Y_r}{\alpha_r X_r d_r} = 1 \tag{11–17}$$

$$\frac{R_r Y_r d_r^{\,2}}{X_r v_r^{\,2}} = 1 \tag{11–18}$$

setting $\alpha = s_s - 1$ for convenience. R_r is a function of Y_r, X_r, and the channel geometry, so that if $v_r = 1$ the three equations (11–16) through (11–18) involve four independent variables, d_r, Y_r, X_r, and α_r. The designer is therefore free to choose only one of them, the other three being determined from the three equations. This result confirms the validity of the trial processes described previously, in which the designer had one degree of freedom only. This is in contrast with the fixed bed case, in which two of three ratios X_r, Y_r, and n_r could be chosen arbitrarily, the third being determined from Eq. (11–12). In the movable bed case two more equations are added, i.e., Eqs. (11–17) and (11–18), but only one more unknown, α_r. It is noteworthy that according to the above system of equations it is essential that the scales be distorted, i.e., $X_r \neq Y_r$, in order to give the model designer any freedom of choice at all, or indeed to make a reduced-scale model possible at all. If $X_r = Y_r$, then the above equations can only be satisfied by making all scale ratios equal to unity—i.e., by making model and prototype identical.

If the designer fixes the ratio α_r in advance, i.e., if he chooses the material of the bed grains but not their size, it follows from Eqs. (11–17) and (11–18) that (Prob. 11.13)

$$d_r = \alpha_r^{-1/3} \tag{11–19}$$

i.e., if the model bed material is lighter than the prototype's, the model grain size should be larger. In view of this rather surprising result, it is interesting to note that Laurent [6] achieved very close verification of prototype conditions with a river model whose bed material was of Bakelite ($s_s = 1.2$) with grain sizes twice those of the corresponding prototype grains. This makes $\alpha_r = 8$ and $d_r = \frac{1}{2}$, results which fit Eq. (11–19) exactly.

If R_r is not assumed equal to Y_r, then X_r and Y_r have to be found by trial, using the geometry of the river cross section (Prob. 11.14). But if the river channel is very wide and $R_r = Y_r$, then it follows from Eqs. (11–16) through (11–18) that (Prob. 11.13)

$$\left.\begin{array}{l} X_r = \alpha_r^{5/3} \\ Y_r = \alpha_r^{7/6} \end{array}\right\} \tag{11–20}$$

giving a distortion ratio X_r / Y_r equal to $\alpha_r^{1/2}$.

Equations (11–16) through (11–18) form the core of a theory which was presented in fully-developed form by Einstein and Ning Chien [7]. The simple form of the argument given above is somewhat inflexible; for instance it assumes Eq. (11–13) to be true and does not allow for the possibility of tilting the model, which would give the designer somewhat more freedom of choice. Other distortions may be carried out in order to increase the designer's freedom, and in this respect the Froude law itself is not sacrosanct. If the Froude number is low gravitational effects are not pronounced, for changes in the channel section do not create material alterations in water level. It follows that slight changes may be made in the Froude number without serious effect. Thus if Fr = 0.3 in the prototype, a value of Fr = 0.4 could be tolerated in the model if it were necessary in order to allow the designer more flexibility in his choice of scales. A complete algebraic development allowing for all these possibilities is given by Einstein and Ning Chien [7]. Another matter of detail is the sedimentation time scale; this the reader can work out for himself (Probs. 11.15 and 11.16). The curious result is that the sediment concentration q_s/q is higher in the model than in the prototype. The reader can also verify (Prob. 11.17) that a slight adjustment to the Froude number, of the order of magnitude mentioned above, can give the designer a great deal more freedom in his choice of length scales.

Bogardi [8] goes a step further by proposing to relax somewhat the requirement that F_s and Re* should have exactly the same value on the model as on the prototype. He points out first that if $(Re^*)_m \geq 100$ the flow around the grains is fully turbulent (see Fig. 10-3) and it is unnecessary to make $(Re^*)_m = (Re^*)_p$. This conclusion is exactly analogous to the one applying to the ordinary Re in fixed-bed or pipe-flow models. Second, he cites recent experimental evidence in support of the statement that the bed formation is a function only of a parameter β, defined by the equation

$$\frac{1}{\beta} = d^{0.88}\left(\frac{v^{*2}}{gd}\right) \qquad (11-21)$$

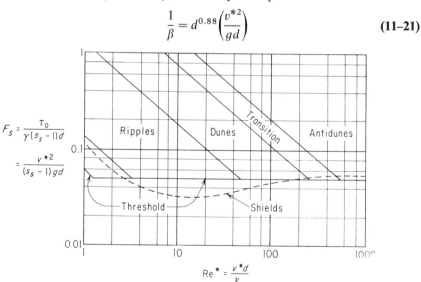

Figure 11-6.   *Relation of Bed Formation to Position on the $F_s - Re^*$ Plane, after J. Bogardi [8]*

where the unit of length used is the centimeter. Because all the evidence quoted is from grain-water systems having the same values of s_s and v, it is possible to recast Eq. (11–21) in completely dimensionless form by inserting these values, which were $s_s = 2.65$ and $v = 0.011$ cm^2/sec. The equation becomes

$$F_s = \frac{v^{*2}}{gd(s_s - 1)} = \frac{456}{\beta^{1.41}}\left(\frac{v^*d}{v}\right)^{-5/6} \tag{11–22a}$$

or
$$F_s = G(\text{Re*})^{-5/6} \tag{11–22b}$$

Replacing $456/\beta^{1.41}$ by a single coefficient G. The values given by Bogardi appropriate to the various formations are as follows:

TABLE 11–1

Condition	β	$G=456/\beta^{1.41}$
Threshold of motion	550	0.060
Ripples form	322	0.130
Dunes develop	66	1.22
Transition occurs	24	5.13
Antidunes develop	9.7	18.5

While β is expressed in centimeter units, G is dimensionless. These values of G define a series of bands on the $F_s - \text{Re*}$ plane, as shown in Fig. 11-6. According to Bogardi, the lines bounding these bands, which by Eq. (11–22) have a slope of $-\frac{5}{6}$, all terminate on the horizontal line $F_s = 0.05$ approximately, which indicates the threshold of motion. While further experiments are no doubt needed to improve the definition of the zones in Fig. 11-6, the general conclusion of the above argument is a reasonable one: that to achieve similarity the points representing model and prototype do not have to coincide, but merely have to lie in the same zone.

There is however a certain conflict between the two criteria suggested by Bogardi. The first one is that the value of Re* is immaterial provided that $(\text{Re*})_m \geq 100$; but the second criterion qualifies the first by laying diagonal bands across the region $\text{Re*} \geq 100$ in Fig. 11-6. It is possible therefore that both model and prototype could have different bed formations even though Re* exceeds 100 in both cases.

Komura [9] by adopting a similar line of argument, develops a series of equations from which all relevant scale ratios can be obtained. A particular feature of interest is the use of empirical relations obtained by other Japanese investigators relating the ratio k_s/d to the entrainment function F_s. Chauvin [13] presents a development essentially the same as is outlined above in Eqs. (10–16) through (10–20), and argues further that it is unnecessary to make $(\text{Re*})_r = 1$, provided only that $(\text{Re*})_m > 60$. This represents a further relaxation of the criterion proposed by Bogardi [8], and doubts may be felt about whether it is justified. Chauvin also gives particular attention to the possibility, implied in the work of Einstein and Ning Chien [7], of achieving a further degree of freedom by adjusting the sediment-size distribution. This is made possible by recognizing two distinct characteristic sand sizes: the size which characterizes the bed roughness, as in Eq. (11–16), and the one which determines its transport properties, as in Eqs. (11–17) and (11–18). In Einstein's method, described in Sec. 10.5, these are taken as the D-65 and D-35 sizes respectively, and this convention is retained by Einstein and Ning Chien [7]; however, Chauvin prefers to use the D-90 size and the mean size given by Eq. (10–53). He describes the application of the method to two river models constructed at the Laboratoire National d'Hydraulique, Paris.

The methods described above are promising but do not yet offer a complete solution. Their practical effect is usually to propose a model bed material larger and lighter than the prototype material, but many successful models—notably the original one of Osborne Reynolds—have operated under bed conditions quite different from these. Reynolds used a natural sand ($s_s = 2.65$) of a size about three-quarters that of the prototype sand, and Allen [4] confirmed by systematic trial that this sand-size relationship does give uniformly good results in tidal models. Allen's general approach to the fixing of scales in a tidal model was to determine the horizontal scale from laboratory space considerations, and to choose the vertical scale so as to give model velocities large enough to make the flow turbulent (i.e., Re $\geq$ 1,400) and to move the model sand for an appreciable part of the tidal cycle. For this latter purpose Allen used an empirical formula developed for the purpose; it could now be replaced by the Shields criterion.

The success of this approach is undoubtedly due to the use of a very high degree of scale distortion, up to 40 : 1 or more. While this is acceptable in tidal models containing mainly very flat sandbanks, Allen points out that much lower degrees of distortion would be required where local scour effects around groines, dikes, etc., create slopes steep enough to make the angle of repose a determining factor.

One general criticism must be made of the methods based on the use of the F_s–Re* plane. While bed formations in model and prototype can be made of the same general type, they are probably not geometrically similar. Ripples, for instance, form over a certain range of wavelength and height, but at no greater size. If a model displays bed ripples, the prototype will not contain a scaled-up version of them. For this reason, much of the argument based on the F_s–Re* plane is somewhat oversimplified and needs further refinement.

The Regime Theory in Model Studies

In Sec. 10.6 the results of the Lacey regime theory were set out. They indicate that in a survey of a self-formed "regime" channel, in which the discharge varies along the channel length, it will be found that the cross section and slope vary systematically with the discharge. In particular, it appears that

$$
\left.
\begin{aligned}
&\text{Width} \propto Q^{1/2} \\
&\text{Depth} \propto Q^{1/3} \\
&\text{Slope} \propto Q^{-1/6}
\end{aligned}
\right\}
\tag{11--23}
$$

These results can be related to model theory, for it is logical to regard the smaller, low discharge channel sections as models of the larger sections. The first observation that springs to mind is that the model will be a distorted one, for according to Eq. (11–23) the smaller sections are proportionately narrower and deeper than the larger sections. This observation was originally made by Reynolds in connection with his early model studies, and was enlarged on by Professor Gibson, Reynolds' successor at Manchester, in the following remarks [10]:

It is perhaps worth noticing in passing that what is in effect a distortion of scale is usual in nature, since small streams flowing through alluvial ground have much steeper side slopes and gradients than large rivers of similar regime in similar ground. In a very large river such as the Mississippi, the Ganges, or the Irrawaddy, the maximum depth will rarely exceed 1 : 50 of the maximum width, while in a small stream in similar ground this ratio will seldom be less than 1 : 5.

A further observation of interest is that the three members of Eq. (11–23) are related in a consistent way which does not require the hypothetical model under discussion to be tilted. For the first two members of the equation indicate, in effect, that

$$
X_r = Q_r^{1/2}
$$
$$
Y_r = Q_r^{1/3}
$$

whence
$$
S_r = Y_r/X_r = Q_r^{-1/6}
\tag{11--24}
$$

a result consistent with Eq. (11–13) and with the third member of Eq. (11–23). A further consequence is that

$$
Y_r = X_r^{2/3}
\tag{11--25}
$$

which appears to offer an immediate means of fixing the vertical scale from the horizontal scale. It follows also from Eqs. (11–23) and (11–25) that

$$
Q_r = X_r^2 = Y_r^3 = X_r Y_r^{1\frac{1}{2}}
\tag{11--26}
$$

just as the Froude law requires. The Lacey equations are therefore remarkably consistent with the normal requirements of similitude. It would appear also that some support for Eq. (11–25) comes from Eq. (11–20), according to which $Y_r = X_r^{0.7}$. However, this comparison is not entirely sound, for in the Lacey equations the "model" has the same bed material as the "prototype." Equation (11–20), on the other hand, is based on using differing bed materials; if the bed material were to be the same, Eq. (11–20) would lead to the result $X_r = Y_r = 1$, according to which the model *is* the prototype.

A general account of this application of regime theory is given by Blench
[11]. Its main feature is of course the interesting possibility of fixing Y_r by
means of Eq. (11–25). However, Allen points out [4] that in a large number of
successful models the choice of scales did not conform to Eq. (11–25).

11.5   Unsteady-Flow and Wave Models

The movement of surges and flood waves in rivers hardly needs any
explicit consideration, for it is covered by implication in the treatment of
Secs. 11.3 and 11.4. Nothing in these sections precludes the representation of
unsteady flow in river models, for the movement of surges and flood waves
is governed essentially by the Froude number and by the resistance; no further
dimensionless number is introduced by the unsteadiness of the flow. The only
possible complication arises from the difference between hydraulic and sedimentation time scales. The corresponding need to distort the flood hydrographs may possibly change their subsidence properties, but this effect is
rarely found to be serious because subsidence is appreciable only over very long
river reaches. Movable-bed models on the other hand usually cover only
short reaches a few miles in length.

Oscillatory waves are an important consideration in the operation of
coastal models relating to beach and harbor works. Their interpretation is a
straightforward application of the Froude law, the operator's intervention
being limited to the setting of the wave period, calculated from Eq. (11–5), in
the appropriate wave-generating mechanism. Once this is done, the waves
assume the correct velocity and wavelength.

The question arises whether scale distortion is possible in wave models. To
settle this question, we must consider the various modes of wave action in
more detail. They are:

1. The interaction of river and tidal currents with wave-generated currents,
e.g. in scour problems and attack on breakwaters, etc. The velocities of the former
will depend on the vertical scale, as given by Eq. (11–8); those of the latter will be
related to the wave velocity c given by

$$c^2 = \frac{gL}{2\pi} \tanh \frac{2\pi y}{L} \tag{8-71}$$

where L is the wavelength. From Eqs. (11–8) and (8–71) it follows that if the two
types of currents are to remain in the same proportion it is necessary that the wavelength scale must be the same as the vertical scale, i.e.,

$$L_r = Y_r \tag{11-27}$$

However, we should note that where angle of repose is of direct importance, as in
wave attack on breakwaters, scale distortion would not be permissible.

2. Wave refraction patterns. These are governed by the refraction Eq. (8–90),
according to which the ratio c/c_0 should everywhere be the same in model and

prototype, where c and c_0 are the local and deep-water wave velocities respectively. It follows from the condition

$$\frac{c^2}{c_0{}^2} = \frac{L}{L_0} \tanh \frac{2\pi y}{L_0} \tag{8–87}$$

that the ratio y/L must be the same in model and prototype, i.e., that

$$L_r = Y_r \tag{11–27}$$

as in case 1.

3. Basin oscillation. It is clear from the treatment in Sec. 8.7 that the character of basin oscillation depends on the ratio of wavelength to the horizontal dimensions of the basin. It follows that for proper reproduction of this effect in the model it is necessary that

$$L_r = X_r \tag{11–28}$$

The only way to satisfy both Eq. (11–27) and Eq. (11–28) is to make $X_r = Y_r$ and avoid scale distortion altogether. Since practically all wave models are of situations in which both wave refraction and basin oscillation are significant, or in which angles of repose are important, scale distortion is not generally used. However, it does seem possible that a distorted wave model might be used in those situations which call for a fixed-bed model and in which *either* refraction *or* basin oscillation alone is important. An example of the former would be an open coastline on which it was required to study the effect of offshore bed contours on the concentration of wave action on certain shore-line features. No complication would arise in the interpretation of such a model, except for the distinction between the time scales for wave period and tidal period (Prob. 11.20).

11.6   General Notes

It would be beyond the scope of this text to provide detailed guidance on model mechanisms and their operation. There are many sources of information on these matters. Allen's comprehensive text [4] is an important one, and the A.S.C.E. manual on the subject [12] contains much useful information. Apart from these standard texts, regular reports on particular model studies are published by working agencies, particularly in the United States. Prominent in this field are the U. S. Army Engineers Waterways Experiment Station, Vicksburg, Miss., the Beach Erosion Board, Washington, D. C., and the U. S. Bureau of Reclamation, Denver, Colo.

Reports of this kind, as well as those of a more general nature [14], provide much useful general background information, particularly on instrumentation and wave- and tide-generating mechanisms. These latter have reached a degree of complexity and sophistication that is a long way in advance of the simple hinged tray with which Reynolds produced the tides on his original model of the Mersey estuary.

References

1. A. J. Peterka. "Performance Tests on Prototype and Model," Symposium on Morning-Glory Shaft Spillways, *Trans. Am. Soc. Civil Engrs.*, vol. 121 (1956), p. 385.
2. Osborne Reynolds. "On Certain Laws Relating to the Régime of Rivers and Estuaries, and on the Possibility of Experiments on a Small Scale," British Association, *Reports*, 1887, p. 555, or *Papers on Mechanical and Physical Subjects*, vol. 2 (London: Cambridge University Press, 1901), p. 326.
3. L. F. Vernon-Harcourt. "The Principles of Training Rivers through Tidal Estuaries, as Illustrated by Investigations into the Methods of Improving the Navigation Channels of the Estuary of the Seine," *Proc. Roy. Soc.* (London) vol. 45 (1888–89), p. 504.
4. J. Allen. *Scale Models in Hydraulic Engineering* (London: Longmans, Green & Co., 1947).
5. M. Jonte. "Note sur l'étude des coupures des boucles de l'Isère en amont de Grenoble," (Proposed cutoffs of the Meanders in the Isère River Upstream of Grenoble), *La Houille Blanche*, vol. 4, spec. no. A, (1949), p. 376.
6. J. Laurent. "Sur la représentation des mouvements de matériaux solides en modèle reduit. Similitude de l'evolution des fonds, et des actions de triage granulometrique" (On the Representation of Bed Movement in a Scale Model. Similitude of Bed Formation and Grain-Size Sorting), *Compt. Rend. de l'Acad. Française*, vol. 236 (January 12, 1953), p. 180.
7. H. A. Einstein and Ning Chien. "Similarity of Distorted River Models with Movable Beds," *Trans. Am. Soc. Civil Engrs.*, vol. 121 (1956), p. 440.
8. J. Bogardi. "Hydraulic Similarity of River Models with Movable Bed," *Acta Tech. Acad. Sci. Hung.*, vol. 24 (1959), p. 417.
9. S. Komura. "Similarity and Design Methods of River Models with Movable Bed," *Trans. Japan. Soc. Civil Engs.*, no. 80 (April 1962), p. 31.
10. A. H. Gibson. "The Use of Models in Hydraulic Engineering," *Trans. Inst. Water Engrs.*, (London) vol. 39 (1934), p. 172.
11. T. Blench, "Scale Relations among Sand-Bed Rivers including Models," *Proc. Am. Soc. Civil Engrs.*, vol. 81, sep. no. 667 (April 1955).
12. "Hydraulic Models," *A.S.C.E. Manual of Engineering Practice No. 25* (July 1942).
13. J.-L. Chauvin. "Similitude in Movable-bed River Models" (in French), *Bull. Centre Rech. Essais*, Chatou, no. 1 (1962), p. 64.
14. "Hydraulic Laboratory Practice," *Engineering Monograph No. 18*, U. S. Bureau of Reclamation, 1953.

Problems

11.1.   Deduce the form of the Froude, Reynolds, Weber, and Cauchy numbers by treating each one as the ratio of inertia force $= \rho L^3 v^2/L$ to gravitational, viscous, capillary and compressibility forces respectively, the term v^2/L representing acceleration.

11.2.   Prove that in a gravitational model the secondary scale ratios will be those given in Eq. (11–5).

11.3.   A movie film is to be made of the flow in a spillway model, and it is desired that when the film is projected the events it shows will take the same time as corresponding events on the prototype—e.g., a particle of water will take the same time as it does on the prototype to fall from the crest to the toe of the spillway. If $L_r = 64$, what is the required ratio between the speeds of taking the film and of projecting it?

11.4.   Show that if both the Froude number and the Reynolds number are to have the same value in model and prototype, it is necessary that

$$\nu_r = L_r^{3/2}$$

Verify by reference to tables of fluid properties that if the prototype fluid is water, the maximum attainable value of L_r is approximately 5, and that this would require the use of mercury as the model fluid.

11.5.   A 1 : 48 scale model of a spillway is to be built and operated. If the prototype discharge is 50,000 cusecs, find the model discharge. Sluice blocks are mounted in the stilling basin to assist hydraulic-jump formation; in the model the combined force on the blocks is measured and found to be 1.6 lb. Calculate the corresponding force on the prototype blocks.

11.6.   In the situation of Prob. 11.5, the downstream channel is lined with rip-rap approximately 2 ft in size. The model velocity near the rip-rap is found to be 1 ft/sec; verify that the model Reynolds number is high enough for the drag coefficient on the model stone to be the same as in the prototype. Show that the force ratio will then be equal to $\rho_r L_r^3$, as given by Eq. (11–5).

11.7.   A fixed-bed model is to be made of a river reach that is approximately rectangular in section and 200 ft wide. The scale ratios are to be $X_r = 200$, and $Y_r = 20$. Manning's n on the prototype is 0.030; find the required value of n for the model if the prototype depth of water is (a) 5 ft, (b) 3 ft.

11.8.   Show that for a fixed-bed model of a river in which berm flow is appreciable, it is necessary that

$$(\Sigma K)_r = X_r^{1\frac{1}{2}} Y_r$$

where K is the conveyance, $= 1.49 \, AR^{2/3}/n$, of each subsection of the flow, and the summation is carried out over the whole section.

   A fixed-bed model is to be made of the river reach described in Prob. 5.20. The scale ratios are $X_r = 200$, $Y_r = 20$. At river mile 14.00, determine suitable values for n_m in the main channel and on the berm by considering flow at water-surface levels of 63.0 and 66.0 ft above datum. With these values of n_m, will the above equation be satisfied when the water-surface level is 64.5 ft? If not, determine the percentage difference between the two sides of the equation.

11.9.   A fixed-bed model is to be made of a river reach that is approximately parabolic in section, the surface width being 200 ft when the center depth is 5 ft. The scale ratios are to be $X_r = 200$ and $Y_r = 20$. Manning's n on the prototype is 0.030; find the required value of n_m if the prototype center depth is (a) 5 ft, (b) 3 ft. Compare the results with those of Prob. 11.7, and note the influence of channel-section shape on the sensitivity of n_r to variation in river stage.

11.10. A fixed-bed model is to be built of a river estuary and part of the river itself. Scales are $X_r = 250$ and $Y_r = 25$. The prototype river has a Manning n of 0.035, and is so wide that $R = y$ on both prototype and model. The prototype tidal period is $12\frac{1}{2}$ hr, and the maximum river discharge is 95,000 cusecs. Determine the required model values of tidal period, Manning n, and maximum river discharge.

11.11. A fixed-bed model is to be built of a certain river reach, and laboratory space considerations have fixed the horizontal scale ratio at 250 : 1. The maximum prototype discharge is 75,000 cusecs, and the maximum flow available in the laboratory is 1.25 cusecs. The Manning n of the prototype is 0.035, and of the cement plaster used in the model is 0.014; this plaster can easily be roughened so as to increase the value of n. Find the limits within which the vertical-scale ratio Y_r must lie; select a suitable value to use in practice and calculate the corresponding model values of n and maximum discharge. It can be assumed that $R = y$ on both model and prototype.

11.12. A fixed-bed model is to be made of a certain river reach which has a slope of 2 ft per mile, $n = 0.035$, a width of 2,000 ft, and a maximum flood discharge of 150,000 cusecs. The cross section is approximately rectangular. The depth of water in the model at maximum discharge is to be at least 3 in., and the model discharge is not to exceed 1 cusec. Choose suitable horizontal and vertical scales, and calculate the value of n required in the model.

11.13. Prove Eqs. (11–19) and (11–20) from Eqs. (11–16) through (11–18).

11.14. A movable-bed model is to be made of a river reach that is approximately parabolic in section, the surface width being 200 ft when the center depth is 5 ft. The specific gravity of the prototype sand is 2.65, and it is decided to use for the model-bed material a plastic having a specific gravity of 1.18. Using equations (11–16) through (11–18), determine d_r and (by trial) X_r and Y_r, assuming that the center depth in the prototype is to be (a) 5 ft, (b) 3 ft.

11.15. Using the Einstein-Brown formula, Eq. (10–47), and assuming that the factor G in Eqs. (10–45) and (10–46) is the same in model and prototype, prove that the scale ratio for the rate of sediment transport per unit width is equal to:

$$q_{sr} = \frac{R_r{}^3 Y_r{}^3}{\alpha_r{}^{2\frac{1}{2}} d_r{}^{1\frac{1}{2}} X_r{}^3}$$

where $\alpha = s_s - 1$ as in Eq. (11–17). Show that according to the theory of Eqs. (11–16) through (11–18), $q_{sr} = 1$ for the wide channel where $R_r = Y_r$. Hence show that the ratio of hydraulic time to sedimentation time is $Y_r{}^{3/2}$ times greater in the model than in the prototype.

11.16. For the situation of Prob. 11.14, determine the scale ratios for hydraulic time and sedimentation time in both cases, (a) and (b).

11.17. Prove that if the ratio of prototype to model Froude number is a factor k, then Eqs. (11–16) through (11–18) are modified to give the results

$$X_r = \alpha_r{}^{5/3} k^{-6}$$
$$Y_r = \alpha_r{}^{7/6} k^{-3}$$

corresponding to Eq. (11–20), but that Eq. (11–19) remains unchanged.

A movable-bed model is to be made of a river reach that is very wide, so that $R = y$ in both model and prototype. The specific gravity of the prototype sand is 2.65, and of the corresponding model material 1.18. The prototype Froude number is 0.3, and it is decided that any value of model Froude number between 0.2 and 0.4 will be acceptable. Determine values of X_r and Y_r if the model Froude number is to be (a) 0.2, (b) 0.3, (c) 0.4.

11.18. Combine the results of Probs. 11.15 and 11.17 to determine ratios of hydraulic time to sedimentation time in the three cases (a), (b), and (c,) of Prob. 11.17.

11.19. The prototype river reach of Prob. 11.17 has a slope of 1.5 ft per mile, a water depth of 8 ft and a water velocity of 4.8 ft/sec. Assuming that $\nu = 1.2 \times 10^{-5}$ ft²/sec locate the corresponding point on the $F_s - \text{Re}^*$ plane (Fig. 11-6) and, if F_s is to be the same in model and prototype, determine how much variation in Re* is allowable in the model according to Bogardi's method. Hence determine the allowable range in the scales X_r and Y_r if the Froude number is to be the same in model and prototype, and compare this range with that obtained in Prob. 11.17.

11.20. A fixed-bed model is to be built to study wave refraction effects on a certain stretch of open coastline. The model is to be distorted, with $X_r = 250$, $Y_r = 25$. In the prototype the tidal period is $12\frac{1}{2}$ hr, and the waves of most practical interest have a period of 10 sec. Determine the corresponding model periods, and the deep-water wavelength in both model and prototype. Will an aerial photograph of the model show proportionately more or fewer wave crests than a similar photograph of the prototype, or the same number?

Index